GW00472002

THE EVOLUTION OF LIFE HISTORIES

THE EVOLUTION OF LIFE HISTORIES
Theory and Analysis

DEREK A. ROFF

Chapman & Hall
New York   London

First published in 1992 by
Chapman & Hall
One Penn Plaza
New York, NY 10119

Published in Great Britain by
Chapman & Hall
2-6 Boundary Row
London SEI 8HN

© 1992 Routledge, Chapman & Hall, Inc.

Printed in the United States of America on acid-free paper.

All rights reserved. No part of this book may be reprinted or reproduced or
utilized in any form or by any electronic, mechanical or other means, now
known or hereafter invented, including photocopying and recording, or by an
information storage or retrieval system, without permission in writing from the
publishers.

Library of Congress Cataloging-in-Publication Data

Roff, Derek A., 1949–
    The evolution of life histories ; theory and analysis / by Derek A. Roff.
        p.   cm.
    Includes bibliographical references (p.   ) and index.
    ISBN 0-412-02381-4 (cloth)—ISBN 0-412-02391-1 (paper)
        1. Variation (Biology)—Mathematical models.   2. Natural selection—
    Mathematical models.   3. Biological diversity—Mathematical
    models.   I. Title.
    QH323.5.R62   1992
    575.01′62—dc20   92-13507
                                                                              CIP
        10 9 8 7 6 5 4 3 2

British Library Cataloguing-in-Publication Data also available

Please send your order for this or any **Chapman & Hall book to Chapman & Hall,
29 West 35th Street, New York, NY 10001, Attn: Customer Service Department.**
You may also call our Order Department at 1-212-244-3336 or fax your purchase order
to 1-800-248-4724.

For a complete listing of Chapman & Hall's titles, send your requests to **Chapman &
Hall, Dept. BC, One Penn Plaza, New York, NY 10119.**

For Daphne, Graham, and Robin

He that would seriously set upon the search for truth ought, in the first place, to prepare his mind with a love of it. For he that loves it not, will not take much pains to get it; nor be much concerned when he misses it. There is nobody in the commonwealth of learning who does not profess himself a lover of truth; and there is not a rational creature that would not take it amiss to be thought otherwise of. And yet, for all this, one may truly say, there are very few lovers of truth for truth's sake, even among those who persuade themselves that they are so. How a man may know whether he be so in earnest is worth enquiry: and I think that there is one unerring mark of it, viz, the not entertaining any proposition with greater assurance than the proofs it is built upon will warrant. Whoever goes beyond this measure of assent, it is plain, receives not truth in the love of it; loves not truth for truth's sake, but for some other by-end.

Locke
Essay Concerning Human Understanding (1691)

Contents

ix

Preface

There are many different types of organisms in the world: they differ in size, physiology, appearance, and life history. The challenge for evolutionary biology is to explain how such diversity arises. This book is concerned with the diversity of life histories and in particular with the ways in which variation can be analyzed and predicted. The central thesis is that natural selection is the principle underlying force molding life history variation. This is not to say that organisms are perfectly adapted to their environments—only that this assumption will generate predictions that are acceptably close in most instances. This book is both an exposition of ways in which to view the world and an account of what we have learned about the evolution of life histories.

Chapter 1 examines two different approaches to analysis, the genetic approach and the optimization approach. Though genetic aspects are discussed throughout the book the principal method of analysis is optimization. The second half of chapter 1 discusses the broad scope of constraints within which life histories can vary. Though much can be achieved without explicit recourse to genetic argument, a proper understanding of the evolution of life histories necessitates an appreciation of the implicit genetic assumptions. These are discussed in chapter 2. Chapter 3 presents the general framework of life history theory, primarily the concept of fitness and trade-offs. The mathematical tools by which predictions can be made and tested are described in chapter 4. Fundamental to life history theory is the characteristic equation that describes how the age schedules of birth and death determine the rate of increase. Equally important is the hypothesis that reproduction carries a cost in terms of survival and/or future reproduction. Factors affecting the age schedule of birth and death are described in chapter 5 and costs of reproduction in chapter 6. Chapters 7 through 10 each cover a particular aspect of the life history: age and size at maturity, reproductive effort, clutch size, offspring size. These categories do not exhaust the possibilities but do represent the major components of a life history.

Three topics not covered in depth are senescence, comparative methods, and parental care. Organisms in the wild do not generally live long enough to grow old, and thus, though the topic is of great interest with respect to the evolution of trade-offs, it is not particularly relevant to the analysis of variation in nature. (It is indeed possible that evolution has acted to postpone senescence beyond the age normally reached in the wild state, but this makes it of historical rather than topical interest in the analysis of variation in the wild.) An exhaustive coverage of the topic is given by Rose (1991). Variation within a taxon may represent independent evolutionary events or be the result of a single event in the common ancestor. The statistical problems introduced by this possibility are noted and addressed in particular instances in this book but the reader is referred to the recent review by Harvey and Pagel (1991) for a detailed account of the comparative method. The evolution of parental care is examined in several sections where particular aspects are relevant but a full account is beyond the scope of the present book. A thorough survey of this topic is given by Clutton-Brock (1991).

In selecting analyses I have concentrated on those theoretical developments that have been tested experimentally. Mathematical analysis is an important component of life history theory and analysis, and as a consequence this book contains numerous equations. I have attempted to ease the burden of equations by providing, in addition, both verbal and pictorial arguments. Covering a large number of papers I have not been able to follow the symbolism of each individual paper. Rather than use numerous greek symbols for constants I have used c subscripted appropriately to distinguish it from other constants within an equation; in some cases specific symbols have been used for particular values where I wish to emphasize their importance or where they have a conventional symbol.

This book would not have been possible without the continuing support and encouragement of my wife and colleague, Dr. Daphne Fairbairn. I should also like to thank Dr. Greg Payne for suggesting that I attempt this enterprise. Sharon David provided invaluable logistical support, and the following colleagues patiently read and commented on various chapters: Drs. D. Berrigan, M. Bradford, E. Charnov, J. Gittleman, R. Huey, I. McLaren, L. Mueller, J. Myers, D. Reznick, B. Sinervo, J. Smith, and S. Stearns.

1

Life History Variation: A First Look

The natural world is composed of a vast array of organisms displaying an enormous diversity of life histories. Plants and animals show profound variation in all aspects of their life histories: age at maturity, age-specific fecundity, survival rate, size at birth, etc. This variation is evident at both the inter- and intraspecific levels. For example, at the interspecific level, species of flatfish range in size from 2-cm long tropical species that reproduce within their first year of life to behemoths such as the Pacific halibut (*Hippoglossus hippoglossus*) which exceed 200 cm and take over 10 years to mature. Though the range in variation within a species is not as dramatic as between species, it is still impressive, as illustrated by variation in the flatfish, *Hippoglossoides platessoides*. In this species maturation occurs at age 3 years at a length of 20 cm in populations off the coast of Scotland while the same species requires 15 years to reach maturity at a length of 40 cm on the Grand Banks of Newfoundland. Longevity and maximum size are equally different in the two populations, Scottish fish reaching a maximum length of 25 cm and an age of 6 years, compared to 60 cm and 20+ years on the Grand Banks (Bagenal 1955a,b; Pitt 1966; Roff 1982). Similar observations on variation in life history characteristics could be made in most taxa. But though the diversity of life histories is readily apparent, attempts to understand its origin and maintenance are still in their infancy. The evolution of diversity in life histories is the subject of this book.

The basic hypothesis underlying most analyses of the evolution of life history traits is that variation is constrained in large measure by trade-offs between traits. These trade-offs can be defined and the evolution of the traits can be predicted either by a genetic model or by one which assumes that selection maximizes some measurable metric that defines fitness. The latter approach, frequently referred to as the optimality approach, has been much used in the last three decades and has, or so I shall attempt to demonstrate, been highly successful in advancing our understanding of the observed patterns of life history variation. Evolution cannot proceed without genetic variation, and thus an important topic of study is the genetic basis of life history traits. The incorporation of genetic models into life history analysis is still in its infancy and present approaches are not entirely satisfactory. The elements of genetic modeling are described in chapter 2, and the theoretical bases of the two approaches in chapter 3. Thereafter, I concentrate upon the optimality approach, discussing, where appropriate or data permit, implications of genetic considerations. Chapter 4 provides a review of the mathematical tools appropriate for the analysis of life history variation, and the remainder of the book uses these tools and the general framework elucidated in chapter 3 to analyze the evolution of life history traits.

The primary goal of any organism is to reproduce. A central aim of life history analysis is thus the understanding of how the age schedule of reproduction evolves. An important component in this evolution is the age schedule of mortality since this both shapes and is shaped by the age-specific expenditure of reproductive effort. Chapter 5 looks at how these two age-specific functions are described and the general factors that should be taken into consideration in an analysis of how they vary. Chapter 6 expands on the themes introduced in chapter 5, dealing in depth with the cost of reproduction with respect to both survival and fecundity.

In the life of any organism the first "decision" that it must make with respect to reproduction is when to start reproducing. The age at first reproduction is therefore a convenient starting place for analysis (chapter 7). Maturity may involve little commitment by the individual, but in others the decision may profoundly influence future alternatives. Pterygote insects, for example, essentially cease to grow upon eclosion into the adult form (there are no more molts and the only growth that can occur is a relatively small change in weight); therefore, the commitment to become mature may limit future possible gains in fecundity or male attractiveness that accrue by virtue of increased size. On the other hand, indeterminate growers such as most fish species can mature and divert just a small fraction of energy into reproduction and hence not sacrifice significant future growth. This brings us to the second aspect of reproduction, age-specific reproductive effort, a topic dealt with in chapter 8. For females this will typically mean the amount of energy invested directly into offspring in terms of biomass and perhaps parental care, while for males it is the amount of energy invested into securing a mate plus, if relevant, the amount invested in parental care. Reproductive effort is clearly quantitative: an organism can expend very little or so much as to cause death after reproduction. Demographically, reproductive effort can be divided into two functional relationships: that between reproductive effort and survival and that between reproductive effort and number of offspring. Reproductive effort can be treated as a fairly abstract quantity, but operationally it must be translated into the number of offspring produced at each age. These offspring may be produced in one or several clutches. Increases in clutch size represent increased reproductive effort which may or may not have significant effects on survival and future reproduction. As a consequence of these impacts on demography clutch size will evolve: this is the subject of chapter 9. The same reproductive effort can be divided in a variety of ways, one of the most important of which is the division between the number and size of propagules. Small propagules permit an increased number but their mortality and time to maturity may be increased. Further, the optimal combination may vary with both the environment and the conditions of the mother. Chapter 10 completes the life cycle by examining the evolution of propagule size.

In this chapter I present a brief overview of the mathematical analysis of life history variation (a discussion expanded upon in chapters 2–4) and then a brief review of constraints, specific examples of which are discussed in greater detail elsewhere in the book when describing the analysis of particular life histories.

1.1. Mathematical Analysis of Life History Variation

Mathematical analysis is a primary tool in the study of the pattern and evolution of life histories. This is to most population biologists now a self-evident fact; but the use of mathematics in ecological investigations has had a much rockier road than its use in genetic analysis, and its general acceptance as an important tool dates only from the 1960s. Kingsland (1985) provides an excellent historical survey of the rise of mathematical approaches in ecology from the work of Lotka in the 1920s to the studies of MacArthur up to 1970. The importance of the mathematical approach to the understanding of genetic variation is amply illustrated by Provine's review of the history of population genetics (Provine 1971) and by the biographies of three of the most influential geneticists of this century: Fisher (Box 1978), Haldane (Clark 1984), and Wright (Provine 1986).

Even by the latter half of the 1940s mathematical thinking had still not made a significant impact on ecological theory; Allee et al. (1949, p. 271) observed that "theoretical population ecology has not advanced to a great degree in terms of its impact on ecological thinking." An early antipathy to the use of mathematical analysis may account in part for the delay in the merging of the ecological and evolutionary perspectives in what is now commonly known as "life history analysis." An influential factor encouraging the use of mathematical investigation into life history variation was

Lamont Cole's 1954 paper, "The Population Consequences of Life History Phenomena," which set out the basic mathematical framework by which the consequences of variation in life history traits can be analyzed. Cole's paper ushered in an era of research predicated on the integration of mathematics and biology in the study of the evolution of life history patterns. The success of this approach can be gauged from the enormous increase in publications on life history evolution: citations for life history studies for the years 1960 to 1980 indicate a doubling time of 4.7 years, a rate that is two to three times the rate for science as a whole (Stearns 1980).

In his review, Cole analyzed how changes in demographic attributes, such as the age at first reproduction, influenced the rate of increase of a population. Except for citations of its historical importance, Cole's paper seems to have gained widespread notice because of an apparent paradox with respect to the value of semelparity versus iteroparity: "*For an annual species, the absolute gain in intrinsic population growth which could be achieved by changing to the perennial reproductive habit would be exactly equivalent to adding one individual to the average litter size*" (Cole, 1954, p. 118, Cole's italics). The resolution to this paradox is very simple (see chapter 8, section 8.1), but its importance lay in drawing attention to the value of mathematical analysis of life history phenomena. Cole's paper enunciated two important principles that are the basis of life history analysis:

> The birth rate, the death rate, and the age composition of the population, as well as its ability to grow, are consequences of the life-history features of the individual organisms. These population phenomena may be related in numerous ways to the ability of the species to survive in a changing physical environment or in competition with other species. Hence it is to be expected that natural selection will be influential in shaping life-history patterns to correspond to efficient populations.

Thus natural selection is seen as maximizing some quantity, here termed "efficient populations," but elsewhere in the paper identified as the rate of population growth. This is not to be taken as indicating that Cole favored the idea of group selection: the tenor of his paper makes it clear that his use of population can be understood in modern terms to be equivalent to genotype. Thus Cole is making the point that selection favors those genotypes that have the highest rates of increase. The second important principle put forward by Cole is that natural selection favors those patterns of birth, death, and reproduction that maximize the rate of increase. This observation was certainly not unique to Cole and can be traced back to Fisher (1930) and in verbal form to Darwin and Wallace. Andrewartha and Birch (1954) emphasized the importance of the potential for increase, devoting a whole chapter to the concept in their book, *The Distribution and Abundance of Animals.* Birch later stressed the relationship between the genotype and its rate of increase, *r*:

> Natural selection will tend to maximize *r* for the environment in which the species lives, for any mutation or gene combination which increases the chance of genotypes possessing them contributing more individuals to the next generation (that is, of increasing *r*) will be selected over genotypes contributing fewer of their kind to successive generations. (Birch 1960, p. 10).

Mathematical modeling has been, and continues to be, an important component of the analysis of life history variation (Stearns 1976, 1977; Parker and Maynard Smith 1990). No model is constructed to capture all the intricacies of the real world, for if it did so it would be as difficult to understand as the real world itself and little would be gained. The purpose of model construction is to address a particular aspect of the real world, ranging from a very detailed analysis of a very specific circumstance to an assessment of a general proposition. All models necessarily are simplifications of reality, and to ensure that the results are robust relative to the assumptions, Levins (1966, p. 423) recommended the use of several different models incorporating different assumptions: "Then, if these models, despite their different assumptions, lead to similar results we have what we can call a robust theorem which is relatively free of the details of the model. Hence our truth is the intersection of independent lies."

There are two basic approaches to the analysis of life history variation, which I shall call "the genetic approach" and "the optimization approach." These methods are neither entirely distinct, nor do they address *exactly* the same questions. The genetic approach is, in large measure, concerned with local events in the sense of describing and predicting variation at small scales of time. The optimization approach addresses the issue of what combination of traits is most favorable in the long run. But there is overlap: quantitative genetic methods have been employed to predict allometric relationships between morphological traits (Lande 1979), and optimization has been used to predict changes in the age schedules of reproduction and growth in response to environmental fluctuations that occur within the life of an organism (Stearns and Koella 1986).

1.1.1. Genetic Approach

Consider a trait whose expression is governed by a single locus with two alleles. A selection coefficient representing their relative contribution to the next generation can be assigned for each of the three genotypes. From this it is a trivial matter to predict the frequencies of the three genotypes at equilibrium (e.g., Hartl 1980). This example encapsulates the genetic approach: define the genetic mechanism determining the phenotypic trait and then simply crank the model through the appropriate mathematical machinery to obtain the equilibrium frequencies. Given a proper knowledge of the genetic architecture of a trait this is obviously a more satisfactory procedure than assuming that some quantity is being maximized as is required for the optimality approach. But the critical problem is the correct definition of the genetic architecture.

Most traits of ecological interest—fecundity, age at maturity, clutch size, egg size, etc.—are continuous in character. Even traits that appear dichotomous, such as liability to disease (Cavalli-Sforza and Bodmer 1971; Curnow and Smith 1975), wing dimorphism (Roff 1986a), diapause (Mousseau and Roff 1987), and sex ratio (Bull et al. 1982; Trehan et al. 1983), are best understood as being the result of some underlying, continuously varying factor exceeding or not attaining a threshold for the expression of the trait. The expression of traits that show continuous variation is not, in general, the result of a single gene, nor two genes, but of a large number of genes that, acting additively, produce a continuous spectrum of phenotypes. The analysis of such traits is the domain of quantitative genetics. This is largely a statistical approach to genetic variation and is founded upon a mathematical analysis of variation rather than an understanding of how groups of genes interact to determine a particular trait. The delineation of the parameters describing quantitative genetic variation—additive, dominance, and epistatic effects—requires breeding data which for many organisms are not available, and may represent formidable technical difficulties.

A second difficulty is that the expression of a trait changes with the environment in which the organism is raised. Such variation is termed "phenotypic plasticity." If an organism produces a phenotype that varies as a continuous function of the variation in the environment then the phenotypic plasticity is termed a "norm of reaction" (Stearns 1989). However, such a distinction is not really tenable since the definition of continuous variation is largely a matter of opinion, and because threshold traits produce discrete morphs but the underlying factors may vary in a continuous manner with the environment (i.e., show norms of reaction). Therefore, I shall refer to any change in phenotype across a gradient a norm of reaction even if the changes are abrupt. There are very few traits that do not vary with their environment, and hence phenotypic plasticity is an integral part of life history variation (Birch 1960; Levins 1963; Bradshaw 1965; West-Eberhard 1989). An additional complication is that reaction norms are themselves functions of the environment, and frequently the reaction norms of different genotypes cross, a phenomenon known as genotype-by-environment interaction ($G \times E$). Incorporating the concept of reaction norms and genotype-by-environment interactions into quantitative genetic theory is not in principle difficult but considerably complicates the theory (for G × E see James 1961; Yamada 1962; Freeman 1973; Hill 1975; Zuberi and Gale 1976; Via and Lande 1985; Westcott 1986; Jong 1990a; and for phenotypic plasticity see Scheiner and Lyman 1989; Jong 1990a,b), and the technical difficulties of experimentally measuring the relevant parameters in a realistic setting are not to be taken lightly (Schlichting 1986; Noordwijk 1989).

Thus far, the discussion has centered upon single traits; but fitness is a composite trait resulting from the integration of age at maturity, reproductive effort, egg size, etc. Therefore, to understand how life histories evolve we must understand how the components of the life history interact. In a genetic setting we must measure the genetic and phenotypic covariance between all traits of interest. Providing we assume that this variance-covariance matrix remains stable over time we can use quantitative genetic theory to predict the outcome of selection acting on a composite trait such as fitness.

The use of quantitative genetic methods to analyze life history variation is not a simple task. If we can ignore, at least as a first approximation, the genetic architecture of the traits under study and approach the study from the perspectives of maximization of fitness within a set of constraints and trade-offs, a solution can be obtained relatively easily. The important issue is whether the two approaches lead to the same result, simply by different avenues of analysis. In the next section I present an overview of the second approach—optimization theory. In chapter 2 some necessary background on quantitative genetic theory is given and in chapter 3 I present a more detailed discussion of optimization theory and its relationship to quantitative genetic theory, showing that current research does indeed indicate that in general the solutions obtained by the two methods will be the same.

1.1.2. Optimization Approach

The concept of trade-offs is central to present theories of how life history traits evolve, for it is such trade-offs that limit the scope of variation. Within the set of possible combinations there will be at least one combination that exceeds all others in fitness. Optimality analysis assumes that natural selection will drive the organism to that particular set. This represents an adaptationist program carried out within a holistic framework. ("Holism" is used here in the sense of multifactorial—see Wilson [1988] for alternate ways in which the term has been used in biology.) Gould and Lewontin (1979, p. 581) more narrowly defined the adaptationist program as one that "proceeds by breaking an organism into unitary 'traits' and proposing an adaptive story for each considered separately." Mayr (1983) criticized this definition as being a caricature of the method as properly applied, noting (p. 327) that "selection does not produce perfect genotypes, but it favors the best which the numerous constraints upon it allow. That such constraints exist was ignored by those evolutionists who interpreted every trait of an organism as an ad hoc adaptation." Obviously it is not necessary to consider every aspect of an organism to understand how each particular trait evolves, but it is necessary to bring together all those factors that directly impinge upon the particular trait under study (Roff 1981a).

Considerable attention has been given to the deficiencies and merits of the optimality approach (Levins 1970; Cody 1974; Maynard Smith 1978; Brady 1979; Oster and Rocklin 1979; Stearns and Schmid-Hempel 1987; Mitchell and Valone 1990; Parker and Maynard Smith 1990). It is not my intention to defend or refute this particular approach: its usefulness depends upon the particular question asked and the data available. Nevertheless, I shall argue that it has been a fruitful approach that has guided much useful research, and it will frequently be used in this book.

To initiate an analysis using the principle of optimality we must designate some parameter to be optimized. In the present case we assume that there is some measure of fitness that is maximized by natural selection. The second step is to construct a set of rules that defines the life history pattern of the organism, hypothetical or real, under study. Within these rules there will exist a variety of possible life histories; the optimal life history is that which maximizes fitness. (The mechanics for arriving at this decision are described more fully in chapter 4.) What do we do if the predicted life history does not correspond to that which is observed? The first point to note is that the principle of optimality is not under test. Failure to get a correct prediction is not taken as evidence that fitness is not being maximized; but it is taken to imply that the model is deficient. Having found that the initial model does not work we enquire into the assumptions of the model—namely, the rules that define the range and scope of life history variation. These rules are changed either arbitrarily or based on further observation until agreement is gained between prediction and observation. Having found congruence we are not able to say that therefore the component relationships are correct. The underlying components must be independently verified; a model is only the logical outcome of a set of interactions and the onus is upon the person specifying those rules to demonstrate that they are indeed valid. Note again that the assumption that fitness is being maximized is not under test, except to the extent that the particular measure chosen may be inappropriate. This does not mean that we assume that all traits and trait combinations are the result of adaptive evolution. But we do choose those that we have a priori reason to suppose are under selection: the life history traits considered in this book are such traits. (A discussion of nonadaptive variation in life history traits among species is discussed in the last paragraph of section 7.1.2.)

A primary purpose of optimality modeling is to organize a program of experimentation and data collection. An adequate fit of a model to data gives us reassurance that a sufficient number of factors have been taken into account. Nevertheless, the validity of a model is continually under question and is challenged by the addition of more information. The more tests the model survives the greater the assurance that it is realistically capturing the important elements that determine the set of traits being studied.

1.2. Constraints

No creature can do all things: it is bound by constraints of its genetic architecture, its history, biophysical and biomechanical factors, and its lifestyle. Within these constraints there are further trade-offs that dictate the set of possible life history traits. This section examines the broad scope of rules within which organisms must exist.

1.2.1. Genetic Constraints

Arguments on the course of evolution, particularly those based on optimality criteria, frequently concern themselves solely with phenotypic variation and assume implicitly that there is sufficient genetic variation to permit the organism to evolve to the most fit combination of traits dictated by underlying constraints and trade-offs. There are actually two problems that must be addressed here: first, can the genetic architecture limit the set of possible traits, and second, having attained the optimal combination, why is genetic variation not completely eroded? These are far from trivial questions (Lewontin 1986), but in large measure they fall outside the scope of this book, which is primarily concerned with phenotypic variation. It is sufficient here to ask if genetic variation is likely to be limiting. A review of mechanisms that preserve genetic variation is presented in chapter 2.

Traits controlled by a single locus may show very limited patterns of variation. In Europe, its native home, there are two forms of the weevil, *Sitona humeralis*—one with wings and one lacking both wings and wing muscles (Jackson 1928)—but in New Zealand only the winged form has been found (Sue et al. 1980). Wing morph within three other coleopteran species, two weevil species and a carabid species, is determined by a single locus with two alleles, brachyptery being dominant (Jackson 1928; Lindroth 1946; Roff 1986a). A reasonable hypothesis is that such a mechanism controls wing form in *S. humeralis*. The stock accidently introduced into New Zealand may have comprised only the winged form (recessive homozygous?). Given that both forms are found in very similar habitats in Europe and New Zealand it seems unlikely that the lack of a flightless form is a consequence of ecological factors. Thus, the evolution of wing dimorphism of *S. humeralis* within New Zealand may well be constrained by the lack of genetic variation for the trait.

As noted previously, most traits of ecological significance, such as development time and fecundity, are not determined by such simple genetic mechanisms: the vast majority of studies indicate that these traits have a polygenic basis. For such characters a quantitative genetic perspective is required, the appropriate measure of genetic variability being heritability in the narrow sense, hereafter referred to simply as "heritability." A full discussion of this parameter is given in the second chapter. Heritability is a measure of the resemblance between parents and offspring. More technically it is the proportion of the total phenotypic variation that is attributable to the effects of additive genetic variance. One method of estimating heritability is to plot the mean offspring value on the midparent value, the slope of the line being heritability of the trait. Note that traits that show no phenotypic variation may be heritable in the sense that parents and offspring share the same characters (e.g., five digits, two wings, etc.) but for such traits heritability in the narrow sense is undefined since there is neither additive nor phenotypic variance. It should also be noted that heritability can only be estimated from parent-offspring regression if there is no environmental correlation between parents and offspring (Feldman and Lewontin 1975; Lewontin 1976): for example, a correlation between the size of parents and offspring may be a consequence of variation in environmental quality shared by both generations. Heritability varies between zero and one: When zero, there is no covariance between parents and offspring and all variation is due to environment or nonadditive sources of genetic variation. When the heritability is one, the average offspring value is equal to the midparent value, and all the phenotypic variance is attributable to additive genetic variance (Falconer 1981). Therefore, a high degree of additive genetic variation is indicated by a relatively high heritability.

Roff and Mousseau (1987) and Mousseau and Roff (1987) examined the distribution of heritabilities among four different categories of traits: morphological, behavioral, physiological, and life historical. Although all of these traits may fall within the purview of life history analysis, the category "life history" has been restricted to those traits such as survival, fecundity, and development rate that are directly and invariably connected to fitness. A simple view of evolution predicts that natural selection will erode genetic variation in traits closely connected to fitness (Fisher 1930; Lerner 1954; Robertson, A. 1955; Falconer 1981). The argument for this is as follows: Suppose that some trait such as fecundity is determined in part by additive genetic variance. Let the number of genes involved be n and at each locus there be two alleles with effects 0 and 1 on fecundity. Those genotypes that are homozygous at all loci for alleles with effect 1 will have the highest fecundity ($2n$) and hence will leave the most progeny. Thus, over time all additive variance will be lost, the population comprising genotypes homozygous at all "fecundity" loci: hence, the heritability of fecundity will be zero. The same process will act on any trait related to fitness but the rate of erosion of variance will depend upon how closely connected the trait is to fitness and hence on how strongly selection acts upon the trait. If selection is weak, then mutation or counterbalancing selection due to correlations between traits may maintain genetic variance. Roff and Mousseau hypothesized that life history traits will, in general, be more closely connected to fitness than morphological, behavioral, or physiological traits. Thus the "classical" theory predicts that life history traits should have the lowest heritabilities. Behavioral and physiological traits are probably under stronger selection than morphological traits. The primary hypothesis under test is that life history traits have lower heritabilities than other types of traits.

The first study (Roff and Mousseau 1987) considered just a single genus, *Drosophila*, for which there is an abundance of data, though the data on physiological traits were too few to derive meaningful conclusions. The average heritability of life history traits ($0.12 \pm \mathrm{SE} = 0.02$) is significantly lower than that for morphological traits (0.32 ± 0.01) and is marginally different from the heritability of behavioral traits (0.18 ± 0.03). These results support the hypothesis of erosion of genetic variance by natural selection. The later study of heritabilities among animals in general (Mousseau and Roff 1987) indicated higher heritabilities—0.51 (morphology), 0.37 (behavior), 0.31 (physiology), 0.27 (life history)—though still the same ranking. A heritability in excess of 0.2 is moderate and more than sufficient to permit fairly rapid response to selection: 50% of the estimates of heritabilities of life history traits exceed 0.2, and approximately 10% exceed 0.5 (Fig. 1.1). Thus, although natural selection appears to be eroding genetic variation of life history traits more rapidly than other traits, there is no reason to suppose that in most instances the existing variation will be insufficient to allow an evolutionary response to changing conditions. Therefore, lack of additive genetic variation is not likely to be a major constraint on life history evolution, and analyses may proceed in the temporary absence of genetic data.

Strictly speaking, optimality arguments assume that the system is at equilibrium and that there is no genetic variability producing suboptimal phenotypes. But we do assume that there is sufficient genetic variation to move the organism to whichever combination is most fit now and in the future. Use of the optimality approach can be justified on the grounds that while mechanisms that maintain and generate genetic variance may also prevent the population from ever reaching the most fit combination they

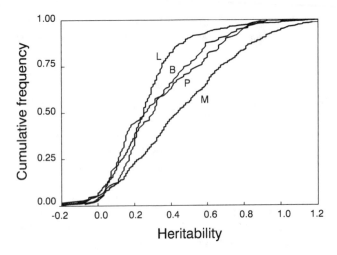

Figure 1.1. The cumulative frequency distributions of the heritabilities of traits divided into four categories: life history (L, $n = 341$), behavior (B, $n = 105$), physiology (P, $n = 104$), and morphology (M, $n = 570$). Data points are joined by straight lines. Although in theory heritability ranges between 0 and 1, it is possible for estimates to extend beyond these ranges. Redrawn from Mousseau and Roff (1987).

do not prevent the population from moving close to this combination. This argument is equivalent to that used to justify the inclusion of some biological features but not others in any analysis. All models are simplifications of reality, and we assume that components excluded do not significantly affect the question under study. If there is a significant discrepancy between prediction and observation then an important life history component may have been omitted: this could indeed be genetic architecture in a particular instance.

1.2.2. Phylogenetic Constraints

History can play a major role in shaping the future course of evolution by limiting possible directions or making some directions much more likely. For example, the evolution of flightlessness in penguins and rheas has clearly led to the loss of genetic variation in wing form to the point that it is highly unlikely that future evolutionary branches of these avian families will become capable of flight. This is not to imply that this is impossible, just that selection is much more likely to favor forms based on the flightless style of life rather than those which would eventually lead back to a way of making a living based on flight. An interesting "exception" to this effect of history may be the Greenland halibut, *Reinhardtius hippoglossoides*. Flatfishes are highly modified for life on the ocean bottom, but the Greenland halibut has become a midwater piscivore. Though still recognizably a flatfish it swims on its "dorsal edge" in the manner of a typical fish (Smidt 1969). This trait, along with a reduced morphological asymmetry, led de Groot (1968, unpublished manuscript) to contemplate the question, "The Greenland halibut, a round flatfish or a flat roundfish?" Such reversals should make us wary of assuming that phylogeny necessarily rules out certain evolutionary pathways.

The basic problem with failing to recognize the effect of phylogeny is that it can lead to an inflation of the number of degrees of freedom in statistical tests. For example, the insect family Siphonaptera (fleas) consists solely of species which are flightless and highly modified as ectoparasites of vertebrates. Radiation of the Siphonaptera into the approximately 2259 species presently known (Arnett 1985) probably occurred after the evolution to a "flea-like" form. Thus the 2259 species cannot be used as separate statistical entities in a test of the correlation between form and function. However, when we find that flightlessness and ectoparasitism are correlated phenomena in a wide range of taxa (Dermaptera, Diptera, Mallophaga and Anoplura), then the hypothesis is supported and merits further study (Roff 1990).

Considerable attention has recently been given to the statistical problems of the analysis of life history variation at various taxonomic levels (Ridley 1983; Pagel and Harvey 1988; Huey 1987; Bell 1989; Gittleman and Kot 1990; Losos 1990; given the rate at which new methods are being proposed, one will shortly be able to carry out a phylogenetic analysis of methods of phylogenetic analysis). A recent book by Harvey and Pagel (1991) provides an extensive discussion of the different methods. Analyses of morphological and life history traits have demonstrated significant phylogenetic effects in lagomorphs (Swihart 1984), squamate reptiles (Stearns 1984; Dunham and Miles 1985; Vitt and Siegel 1985), primates (Harvey and Clutton-Brock 1985), carnivorous mammals (Gittleman 1985), mammals in general (Stearns 1983), North American game birds (Arnold 1988; Duncan 1988), angiosperms (Mazer 1989), and host-parasite relationships (Harvey and Keymer 1991). The demonstration that phylogeny is a significant covariate does not prove that evolution has been constrained by history. It does mean that effects due to history cannot be separated from effects due to other correlated factors. Having demonstrated phylogenetic effects in no way answers questions concerning what factors favor the evolution of particular suites of traits.

1.2.3. Physiological Constraints

Physiological constraints result from processes that act internally within the organism but are above the level of the gene or genes. Development may be constrained by physiology to a prescribed range of options, with one pathway being more likely to be followed than others (Maynard Smith et al. 1985). The demonstration of a physiological constraint is exceedingly difficult. The oft-cited relationship between metabolic rate and body size could represent such a constraint (Case 1978a,b; McNab 1980, 1984). Metabolic rate in endotherms increases dramatically with decreasing body size, leading Pearson (1948) to suggest that 2.5 g may be the lower limit for a free ranging mammal. Pygmy mice, *Baiomys taylori*, are born at a weight of 1.2 g and are unable to thermoregulate below a weight of about 2.5 g, giving support to this hypothesis (Hudson 1974). Kendeigh (1972) examined the lower limit for passerines by considering the relationships of existence metabolism and standard metabolism as functions of body size. Above 2.3 g existence metabolism exceeds standard metabolism but below this value the reverse is true, and hence in principle a passerine smaller than 2.3 g could not survive: in practice the lower limit would be somewhat higher than 2.3 g. The smallest passerines appear to be about 5 g (Kendeigh 1972). Some nonpasserines such as hummingbirds may be as small as 2.5 g, which Kendeigh attributed to a lower rate of standard metabolism in comparison to passerines and the evolution of energy conservation through heterothermy. Schmidt-Nielson (1984, p. 207) argued a contrary view, suggesting that constraints on the rate at which oxygen can be supplied to the tissues may set the lower limit to body size in birds and mammals. In support of this view he noted that shrews and hummingbirds have hearts that are two to three times as large as expected on the basis of allometric scaling. Such an increase in size may be required to compensate for a limitation on the rate at which the heart can contract and hence pump blood.

Physiological constraints have been invoked as a constraint on both growth and activity in very small animals such as larval cyprinids (Weiser et al. 1988) and to account for the evolution of size and shape in mammals such as mustelids (Brown and Lasiewski 1972; Sandell 1989), for differences in the rates of growth of the brains of birds and mammals (Dunbrack and Ramsay 1989a), and for the evolution of small neonate size in bears. (See chapter 10, section 10.2.1.) Most laboratory examinations of the cost of reproduction are based on physiological constraints—e.g., differences in longevity between virgin and mated females (reviewed in Reznick 1985; Bell and Koufopanou 1985). Under the benign conditions of the laboratory environment the only cost that will be evident in these experiments is that which occurs as a consequence of some physiological response to reproduction. But the effects of physiological constraints may only become apparent under stressful conditions. For example, in an appropriate thermal environment a mammal smaller than 2.5 g might easily survive, but in the "real world" where temperatures fluctuate, these exceedingly small animals may quickly succumb in times of thermal stress (Tracy 1977). Similarly, the physiological stress of reproduction may not be apparent under favorable conditions, but placed in an environment in which the organism's physiological state is challenged on several fronts, a cost of reproduction may become apparent. In this regard it is significant that reproductive costs have most frequently been demonstrated under conditions of stress such as starvation (Bell and Koufopanou 1985; chapter 6).

1.2.4. Mechanical Constraints

Gravity is a major factor in the design of animal form, setting limits on the design of all types of animals (McMahon 1973; Garland 1983; Mayo 1983; Schmidt-Nielsen 1984; Gaillard et al. 1989). Within this broad constraint there may be finer aspects that pertain to particular life-styles. For example, the battering of waves may significantly influence the size and design of organisms living along the seashore (Denny et al. 1985); shell size can limit the growth of hermit crabs (Bertness 1981a); mechanical limitations of pelvic aperture may account for the increase of egg size with age and size in turtles (see chapter 10, section 10.2.1); the production of spines by the cladoceran *Bosmina longirostris*, while protecting them from predators, also reduces the brood chamber and hence potential fecundity (Kerfoot 1977); and finally, many animals carry their mates for extended periods of time—an activity that may limit the relative sizes of the two sexes (Adams et al. 1985; Marden 1989; Fairbairn 1990).

1.2.5. Ecological Constraints

The first four types of constraints have been termed "internal constraints," and ecological constraints have been called "external constraints" (Barbault 1988; Gans 1989). This division into two groups is not exclusive: for example, the constraint on growth in hermit crabs due to small shell size is a mechanical constraint that is contingent on the distribution of shell size—an ecological factor.

Ecological constraints are those that are a function of the particular environment in which the animal makes a living. *Daphnia* that contain eggs suffer a higher rate of predation than males or females not carrying eggs (Hairston et al. 1983; Koufopanou and Bell 1984). This cost to reproduction is dependent upon two conditions: first, there must be predators in the environment, and second, females must be more visible when containing eggs (Hairston et al. 1983). The increased susceptibility is eliminated when the *Daphnia* are placed against a dark background (Koufo-panou and Bell 1984). Predation may also create a trade-off between growth and survival by limiting foraging time or habitat (Weissberg 1986; Lima and Dill 1990). Cannibalism by adults limits the spatial distribution of nymphal riffle bugs (Wilson et al. 1978), backswimmers (Murdoch and Sih 1978; Sih 1980, 1982; Streams and Shubeck 1982), and waterstriders (Sih et al. 1990), while increased risks of predation restrict small bluegills to weedbeds (Werner et al. 1983) and limit the foraging activities of guppies (Godin and Smith 1988). In ectotherms the thermal environment may also circumscribe the time and places of foraging (Congdon 1989).

1.3. Summary

Life history variation is enormous, but only recently has the power of mathematical analysis been brought to bear on the subject, though the foundations for such an approach can be seen within the genetics literature. Two approaches to the analysis of variation in life histories are genetic analysis and optimization. The former is in principle better because given a particular genetic mechanism the outcome of selection can be evaluated without recourse to assumptions about what selection maximizes. Unfortunately, the technical difficulties of measuring the requisite parameters that define the genetic architecture of a suite of traits are formidable. Furthermore, such parameters are unstable over time and hence predictions of long-term evolutionary changes are highly suspect. An understanding of genetic variation is crucial to an understanding of short-term variation in traits, but questions of what combinations of traits will be favored in the long run are presently tackled using the optimization approach. The two assumptions of this method are that some quantity such as r is maximized by selection, and that trade-offs between traits limit the possible set of combinations.

There are a number of broad-scale constraints within which the evolution of traits takes place. First, evolution requires genetic variation. Though the heritabilities of life history traits are lower on average than those of other traits (morphological, behavioral, physiological), there is sufficient variation to allow evolutionary response. Lack of genetic variation is not likely to be a major concern in most analyses. Phylogenetic constraints, or constraints resulting from evolutionary history, certainly occur, but their significance rests more in the testing of predictions from life history analyses rather than in influencing the method of theoretical development. Physiological constraints are very difficult to measure and may become apparent only under certain conditions such as stress. This means that experiments examining such questions as the cost of reproduction cannot draw extensive conclusions from single experiments, particularly those carried out in benign laboratory settings. Mechanical constraints are more evident than physiological constraints and should certainly be considered in particular models (e.g., the adverse influence of protective spines on life history characteristics, or the limitations imposed by mate carrying). Finally, the survival and reproduction of an organism is a function of the particular ecological setting, as indeed are the effects of physiological and mechanical constraints. Ecological constraints thus form the backdrop for all life history analyses.

2

Quantitative Genetic Background

Life history theory and quantitative genetics are not mutually incompatible approaches and the future of evolutionary theory will undoubtedly see a melding of them. Attempts at unification have begun. A number of authors have begun to examine the underlying genetic assumptions of life history theory and the extent to which both approaches lead to common predictions. To understand these analyses it is first necessary to introduce some basic concepts of quantitative genetics: this is the subject of the first part of the present chapter. Life history theory assumes that, in general, genetic variance is not limiting. The question of what maintains genetic variation, as opposed to phenotypic variation, is not a central issue of this book, but it is clearly relevant, and in the second part of this chapter I briefly review the evidence for mechanisms that have been hypothesized to restrain the erosion of genetic variance.

2.1. Quantitative Genetic Framework

There are many general texts that deal with quantitative genetic theory (particularly useful ones are Hedrick 1983; Hartl 1980; Bulmer 1985a; Falconer 1989), and here I present only an overview of the theory.

In the last 15 years there has been an increasing interest in the use of quantitative genetic theory in the analysis of life history variation. Much of this interest results from the work of Lande (1976), who brought the framework of quantitative genetic analysis into the domain of evolutionary theory. Though the theoretical foundation of quantitative genetics has a very long history, starting with the work of Yule (1902, 1906) and Fisher (1918), it was largely applied to questions of animal or plant improvement. The extension of the theory to ecological and evolutionary questions has spurred attempts to take quantitative genetics into the field, at this stage primarily by quantifying the impact of natural selection in the wild (Endler 1986). Optimality theory has been criticized for sweeping genetic architecture under the rug (Rose et al. 1987). However, quantitative genetic theory must necessarily make simplifying assumptions, and it is by no means clear just how much of the architecture is actually dragged out from under the rug (Lee and Parsons 1968; Nordskog 1977; Robertson 1977; Kempthorne 1977, 1983; Barton and Turelli 1987). The two most important parameters in the present context are heritability and genetic correlation.

2.1.1. Heritability

Most life history traits, such as age at first reproduction, fecundity, and survival are not determined by simple Mendelian mechanisms such as single-locus, two-allele systems. More generally, they show continuous variation, or if the trait is dichotomous it can be best interpreted under a threshold model in which there is an underlying, continuously varying trait (Falconer 1989). Nevertheless, the operation of a polygenic mechanism is not essentially different from the simpler case of a single locus with two alleles. In the one-locus case the phenotype can be determined by the additive action of the alleles, or one allele may show dominance (Fig. 2.1). When one allele is dominant the effect on the phenotype can be decomposed into two components, an additive effect, designated by the dotted line in Fig. 2.1, and a dominance component representing the deviation from this line. With two or more loci controlling a trait there may be interaction between

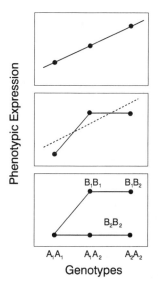

Figure 2.1. Hypothetical models of the interaction within and between loci in the expression of a metric trait. Upper panel: The phenotype is determined by a single locus with the alleles acting additively. Middle panel: Dominance of one allele. The action of the alleles can be divided into two components—an additive component (dashed line) and a dominance component. Bottom panel: The phenotype is determined by two loci, with epistatic interaction. After Hedrick (1983).

loci. These interactions, termed epistasis, can produce a wide range of responses, one possible one being illustrated in Fig. 2.1.

For the two-locus case, the genotypic value, GV, can be written (Bulmer 1985a)

$$GV = m + X_1 + X_3 + X_2 + X_4$$
$$+ X_{13} + X_{24} \tag{2.1}$$
$$+ X_{12} + X_{14} + X_{23} + X_{34} + X_{123} + X_{134} + X_{124} + X_{234} X_{1234}$$

where the terms have been stacked according to their effects: additive (row 1), dominance (row 2), and epistatic (row 3). The term m is the mean value, X_1 and X_3 are the main effects of the two alleles at locus 1, and X_2 and X_4 are the main effects of the two alleles at locus 2. $X_1 + X_3$ is the additive effect of locus 1, and $X_2 + X_4$ is additive effect of locus 2. X_{13} is the interaction between the main effects at locus 1 and represents the dominance deviation; similarly, X_{24} is the dominance deviation at the second locus. All other terms denote epistatic interactions. Typically, epistatic interactions are ignored. (Here we have the first assumption that may not be correct.) The remainder of this overview makes this assumption. Ignoring epistatis, equation 2.1 can be written in the shorthand form

$$GV = m + A + D \tag{2.2}$$

where A denotes the deviation due to additive components and D the dominance deviation. The above relationship can be extended to any number of loci, n. A crucial assumption of quantitative genetic theory is that

> the pairs of random variables (A_i, D_i) at different loci will be statistically independent (whether or not the loci are linked), so that by the bivariate form of the central limit theorem A and D will become bivariate normal when n is large. Since they are uncorrelated they must become independent normal variates as n increases by a well-known property of the bivariate normal distribution. (Bulmer 1985a, p. 123)

This assumption is pivotal to the theoretical basis of quantitative genetics as presently used: "by *assuming* that the effects of all the loci follow a multivariate Gaussian distribution, the evolution of a polygenic trait can be described completely in terms of the mean vector and covariance matrix of this distribution. This greatly simplifies genetic analyses of many important evolutionary problems" (Barton and Turelli 1987, p. 158). But under selection this property of independence is lost since a correlation between loci will be built up (Turelli, personal communication); the significance of this result remains to be fully explored. A second criticism of this model is that there are unlikely to be enough alleles at all loci for the Gaussian approximation at each locus to be valid (Turelli 1984, 1985, 1988).

Assuming no epistasis and statistical independence of loci, the variance of the genetic value is

$$V_G = V_A + V_D \qquad (2.3)$$

where the subscripts refer to the relevant components of equation 2.2. In practice few traits are determined entirely by the underlying genetic mechanism; more generally there will be an environmental component, a term that refers to any influence that cannot be subsumed under the above model. (Hence the term "environment" is used by geneticists in a slightly different sense than used by ecologists.) Thus the phenotypic variance can be decomposed into three components (still assuming no epistasis),

$$V_P = V_A + V_D + V_E \qquad (2.4)$$

where V_P is the phenotypic variance and V_E is the environmental variance. Note that in the above formulation genotype and environment are assumed to be uncorrelated.

From Fig. 2.1 it is apparent that the average similarity between parent and offspring can be measured by the linear regression of offspring on parent: the additive component of the genetic value. By the assumption of multivariate normality, this result can be immediately extended to any number of loci. Let Y be the mean phenotypic value of the offspring and X the midparent phenotypic value. Now $Y = GV + E$ and the regression of the mean offspring value on the midparent value is (Bulmer 1985a, p. 125)

$$Y = (1 - h^2)m + h^2 X \qquad (2.5)$$

where the slope of the regression, h^2, is an estimate of heritability in the narrow sense. More generally, heritability is defined as

$$h^2 = \frac{V_A}{V_P} = \frac{\text{Additive genetic variance}}{\text{Total phenotypic variance}} \qquad (2.6)$$

Heritability in the narrow sense, or generally simply heritability, must not be confused with heritability in the broad sense, V_G/V_P, which is a measure of the overall variance in the trait attributable to all genetic influences. Obviously h^2 varies between zero and one: at zero there is no resemblance between parent and offspring due to the additive effects of genes, while at one the mean offspring value equals the midparent value. The importance of heritability for evolutionary theory is obvious: The higher the value of h^2 the faster genetic changes in the population can occur. Most importantly from the point of view of life history theory, if heritability is zero for a particular trait or traits under consideration then the optimal combination cannot be attained because the effects of selection on parents are not manifested in the offspring.

For nonzero heritabilities the rate at which the appropriate combinations can be realized depends in part upon the value of the heritability and the intensity of selection, measured as the *selection differential, S*—the difference between the mean of the population and the mean of the selected parents.

The actual change in a trait due to direct selection on that trait can be obtained directly from equation 2.5. Suppose we choose as parents those which are on average S units (where S may be positive or negative) from the mean, m: from equation 2.5,

$$Y = (1 - h^2)m + h^2 (m + S)$$
$$= m + h^2 S \tag{2.7}$$

Writing $Y - m$ as the response to selection, R, we have

$$R = h^2 S \tag{2.8}$$

This result depends "only on the assumption of the linearity of the biparental regression and . . . not . . . on any genetic assumptions" (Bulmer 1985a, p. 145). The importance of assumptions concerning the genetic basis of the trait comes into play when we wish to quantitatively predict the result of long-term selection on a trait and the responses of traits that are genetically correlated. In the present context the critical problem is the assumption of a constant heritability: since selection changes the gene frequencies, heritability must change at each generation of selection. This problem is avoided in the application of quantitative genetic theory to natural systems by assuming that selection is weak and population size large, thereby permitting mutation to replace variation eroded by selection (Lande 1976, 1980, 1982).

2.1.2. Genetic Correlation

Consider two traits that are phenotypically correlated (e.g., size and number of eggs in a clutch); by standard statistical theory, the phenotypic correlation, r_P, between these two traits is equal to

$$r_P = \frac{P_{XY}}{\sqrt{P_{XX}P_{YY}}} \tag{2.9}$$

where P_{XY} is the phenotypic covariance between the two traits X and Y, and P_{XX}, P_{YY} are the phenotypic variances of the respective traits. The reason for the double subscripts XX and YY is that the variance is by definition equivalent to the covariance between X and itself; the present notation is used for consistency with later developments. (Note also the change in notation from V for variance: again this is for simplicity in future equations.) The phenotypic covariance, P_{XY}, is the sum of the genetic, G_{XY}, and environmental covariance, E_{XY},

$$P_{XY} = G_{XY} + E_{XY} \qquad (2.10)$$

Rearranging equation 2.9 and noting that the covariance between X and Y is equal to the correlation between X and Y times the square root of their variances, equation 2.10 can be rewritten as,

$$r_P\sqrt{P_{XX}P_{YY}} = r_G\sqrt{G_{XX}G_{YY}} + r_E\sqrt{E_{XX}E_{YY}} \qquad (2.11)$$

where r_G is the genetic correlation between traits X and Y,
 r_E is the environmental correlation between traits X and Y,
 G_{XX}, G_{YY} are the genetic variances,
 E_{XX}, E_{YY} are the environmental variances.
 The genetic correlation expresses the extent to which two characters are determined by the same genes. If r_G is unity the two characters are genetically identical in that they are controlled by a common set of genes; if r_G is zero there are no genetic links between the two characters and any correlation is entirely due to effects of the environment. It is important to note that the sign of the phenotypic and genetic correlations can in principle be opposite in sign, a phenomenon of great significance when analyzing the existence of trade-offs expected by life history theory. Equation 2.11 can be written in terms of heritability by noting that $G_{XX} = P_{XX}h^2$ and $E_{XX} = (1 - h^2)P_{XX}$, with similar terms for Y: substitution leads to

$$r_P = \sqrt{h_X^2 h_Y^2}\, r_G + \sqrt{(1 - h_X^2)(1 - h_Y^2)}\, r_E \qquad (2.12)$$

Because of genetic correlation, traits not directly under selection may nevertheless change by virtue of changes in the frequency of the shared pool of genes. Consider two traits X and Y, where X is under selection and the two traits are genetically correlated with some value r_G. The response of trait X is given by

$$R_X = h_X^2 S_X = \frac{G_{XX}}{P_{XX}} S_X \qquad (2.13)$$

where R_x is the phenotypic response of trait X,

h_X^2 is the heritability of X,
S_X is the selection differential on X,
G_{XX} is the additive genetic variance of X,
P_{XX} is the phenotypic variance in X.

The correlated response of trait Y is (Lande 1979, p. 403; Falconer 1981, p. 286),

$$R_Y = r_G h_Y h_X \sqrt{\frac{P_{YY}}{P_{XX}}} S_X$$

$$= \frac{G_{YX}}{P_{XX}} S_X \tag{2.14}$$

where G_{XY} is the additive genetic covariance between X and Y. The correlated response depends on the heritabilities of the traits, the genetic correlation, and the phenotypic variances. The second notation, involving only variances and covariances, is much simpler and is useful when considering more than two traits and sets of correlated responses, discussed below.

Two traits may change due to direct selection on each trait and a correlated response due to selection acting on one trait having an effect on the other by virtue of the genetic correlation between the two. The expected response in two traits is given by (Lande 1979; Arnold 1983; Lande and Arnold 1983)

$$R_1 = G_{11}\beta_1 + G_{12}\beta_2$$

$$R_2 = G_{21}\beta_1 + G_{22}\beta_2 \tag{2.15}$$

where R_i is the response of the ith trait (frequently denoted by Δz_i),
G_{ij} is the covariance between i and j (when $i = j$, G_{ij} is the variance; also $G_{ij} = G_{ji}$), and
β_i is the selection gradient on character i (explained below). Use of matrix notation simplifies the display of equation 2.15,

$$\begin{pmatrix} R_1 \\ R_2 \end{pmatrix} = \begin{pmatrix} G_{11} & G_{12} \\ G_{21} & G_{22} \end{pmatrix} \begin{pmatrix} \beta_1 \\ \beta_2 \end{pmatrix} \tag{2.16}$$

which by use of bold type can be written in shorthand as

$$\mathbf{R} = \mathbf{G}\boldsymbol{\beta} \tag{2.17}$$

The selection gradient vector can be decomposed into two parts:

$$\boldsymbol{\beta} = \mathbf{P}^{-1}\mathbf{S} \tag{2.18}$$

The first term is the inverse of the matrix of phenotypic variances and covariances; i.e., in the case of two traits it is the inverse of

$$\begin{pmatrix} P_{11} & P_{12} \\ P_{21} & P_{22} \end{pmatrix} \tag{2.19}$$

where P_{ij} is the phenotypic covariance between traits i and j. The second term in equation 2.18 is the vector of selection differentials, i.e., the difference between the character means between parents and offspring. Inversion of a matrix is not always a simple task, though there are abundant software packages for the purpose. In the present case of a 2×2 matrix there is a simple formula, leading to the formula for response to selection for two characters as

$$\begin{pmatrix} R_1 \\ R_2 \end{pmatrix} = \begin{pmatrix} G_{11} & G_{12} \\ G_{21} & G_{22} \end{pmatrix} \frac{1}{P_{11}P_{22} - P_{12}P_{21}} \begin{pmatrix} P_{22} & -P_{12} \\ -P_{21} & P_{11} \end{pmatrix} \begin{pmatrix} S_1 \\ S_2 \end{pmatrix} \tag{2.20}$$

The response of character 1 is thus

$$R_1 = \frac{G_{11}(P_{22}S_1 - P_{12}S_2) + G_{12}(P_{21}S_1 + P_{11}S_2)}{P_{11}P_{22} - P_{12}P_{21}} \tag{2.21}$$

The above formula illustrates the advantage of matrix notation. Extension of the equations given above to more than two characters is immediate and can be readily represented in matrix notation.

To apply the above equations requires the assumption that over the time span of interest the genetic variance-covariance matrix remains stable. As with heritability, this is not strictly correct since changes in gene frequency will necessarily change the genetic correlation (Bohren et al. 1966; Mitchell-Olds and Rutledge 1986). Based on an analytical treatment and computer simulations Bohren et al. (1966, p. 55) concluded, "it must therefore be expected that the static description of a population in terms of additive genetic variances and covariances will be valid over a much shorter period for correlated response than it will be for direct responses." Under artificial selection, genetic correlations have been found to be stable (Bell and Burris 1973; Cheung and Parker 1974) or to vary substantially and unpredictably during the course of selection (Mather and Harrison 1949; Sheridan and Barker 1974). The assumption of a constant genetic correlation is justified in models of natural selection by invoking weak selection and large population size (Lande 1979; Via and Lande 1985), but Turelli (1988, p. 1346) warns that "Until more data are available, evolutionary inferences based on the constant-covariance assumption should be questioned."

Empirical analysis of the genetic variance-covariance structure has shown that the matrix may be variable across both species (Lofsvold 1986; Kohn and Atchley 1988; Cowley and Atchley 1990) and populations within a species (Soller et al. 1984; Berven 1987; Groeters and Dingle 1987; Dingle et al. 1988); therefore, the use of quantitative genetic theory to explore the long-term evolution of traits must be viewed with some skepticism.

In an appendix to the paper of Sheridan and Barker (1974), James examined the change in genetic covariances under the "partition of resources model." This model is based on the premise that two traits draw from a common resource pool; for example, the production of wing muscles and eggs in insects may compete for a common and finite amount of energy (Roff 1986a). Let the total quantity of resource be R, some fraction f being devoted to trait X and the remainder to trait Y. James assumed that R and f are wholly genetically determined but that X and Y contain uncorrelated environmental effects E_X and E_Y. Thus

$$X = fR + E_X \tag{2.22}$$

$$Y = (1 - f)R + E_Y \tag{2.23}$$

The covariance between X and Y is entirely genetic and arises entirely from the partition of resources: hence $P_{XY} = G_{XY}$. G_{XY} is, by standard statistical theory, equal to the covariance between fR and $(1 - f)R$, and hence,

$$G_{XY} = \hat{f}(1 - \hat{f})V_R - \hat{R}^2 V_f + \hat{R}(1 - 2\hat{f})\text{cov}(f, R) \tag{2.24}$$

where the hats denote means; V, variances; and cov, covariances. Assuming that f and R are uncorrelated (i.e., the fraction allocated does not depend upon the total amount of resources), then since $cov(f, R) = 0$,

$$G_{XY} = \hat{f}(1 - \hat{f})V_R - \hat{R}^2 V_f \tag{2.25}$$

Further,

$$\text{cov}(X, f) = \hat{R}V_f \qquad \text{cov}(Y, f) = -\hat{R}V_f$$

$$\text{cov}(X, R) = \hat{f}V_R \qquad \text{cov}(Y, R) = (1 - \hat{f})V_R \tag{2.26}$$

From the above series of equations the results of selecting on X, Y, or some combination of both can be ascertained (Table 2.1; see Sheridan and Barker for details). In all cases the genetic correlation changes, though the direction of change depends upon the value of $\hat{f}$ and the trait under selection. For further discussion on the dynamics of change in the covariance

Table 2.1. Effects of various types of selection on genetic correlation between two traits in the "partition of resources" model

Selection	Effect on the genetic correlation, r_G	
	$f < 0.5$	$f > 0.5$
For X	May increase	Decreases
For $X + Y$	Decreases	Decreases
For $X - Y$	Increases	Decreases
Against X	Uncertain	Increases
Against $X + Y$	Increases	Increases
Against $X - Y$	Decreases	Increases

matrix see Atchley (1984), Cheverud (1984), Riska (1986, 1989), and Stearns et al. (1991).

2.2. Maintenance of Genetic Variation

As demonstrated in chapter 1, there is an abundance of additive genetic variation for life history traits. Such variation permits natural selection to drive organisms toward increasing fitness and the optimal combination of traits. But having arrived at the pinnacle, genetic variation should be eroded, leaving a single genotype expressing itself in a single phenotype or in several phenotypes if the optimal life history requires phenotypic plasticity. The fact that genetic variation is not lost enables relatively rapid response to changes in the environment, but such foresight is not part of selection at the level of the individual. Analysis of the evolution of life history variation can proceed simply on the basis that there is typically sufficient variation for the organism to evolve within the constraints imposed by the trade-offs between traits. However, in the long run the evolution of variation between populations and the maintenance of variation within a population are topics that cannot be entirely disassociated. Therefore, in this section I review some of the more likely and common mechanisms that could maintain genetic variation within a population. These mechanisms can be conveniently divided into five categories:

1. Mutation-selection balance

2. Heterosis

3. Antagonistic pleiotropy

4. Frequency-dependent selection

5. Environmental heterogeneity

2.2.1. Mutation-Selection Balance

Mutation is the process that generates genetic variation. Because of the theoretical work of Latter (1960) and Bulmer (1972, 1985a), until recently it was believed that mutation rates could not maintain heritabilities in excess of 0.5 in the face of stabilizing selection. In 1975 Lande published a paper in which he came to the conclusion that a "large amount of genetic variation can be maintained by mutation in polygenic characters even when there is strong stabilizing selection" (Lande 1975, p. 221). Lande's analysis was based on the "continuum of alleles" model introduced by Crow and Kimura (1964): in this model each locus is assumed to comprise an infinite number of alleles, with a continuous range of effects. Not surprisingly the conclusions of Lande have spawned a plethora of alternate models and analyses (e.g., Turelli 1984, 1985; Barton 1986; Bürger 1988, 1989; Keightley and Hill 1988, 1989; Spencer and Marks 1988; Bulmer 1989). Turelli (1984), using a different set of approximations, came to the same conclusion as Lande, but in a subsequent paper suggested that pleiotropic effects cast doubt on their validity and concluded that (Turelli 1985, p. 190) "Lande's (1975) mutation-selection hypothesis for extant heritable variation must remain an appealing but unsubstantiated conjecture." This cautionary view has generally been reiterated in other analyses (Keightley and Hill 1988; Bulmer 1989), and the mutation-selection hypothesis has been contested by recent analysis of Bürger et al. (1988) and Houle (1989).

2.2.2. Heterosis

Heterotic advantage was one of the first mechanisms suggested for the maintenance of genetic variation in a population. Consider the case of a single locus with two alleles. If the fitness of the heterozygote exceeds that of either homozygote, then, in an infinite population, a stable polymorphism will ensue, though overdominance can actually increase the probability of fixation in a finite population (Ewens and Thompson 1970). Lewontin et al. (1978) showed that heterosis is unlikely to maintain more than two alleles at any given locus. The main focus of investigation, however, has been on the possibility that heterosis across loci can maintain genetic variation (Lewontin 1964; Bulmer 1971a, 1973; Gillespie 1984; Ziehe and Gregorius 1985). As with mutation-selection balance the conclusions drawn differ, Bulmer (1973, p. 12) finding that "a considerable amount of heterozygosity can nevertheless be maintained by quite a small heterozygous advantage," while Ziehe and Gregorius (1985, p. 500) concluded that "we are forced to accept the fact that heterozygote advantage is a concept of limited significance for explaining the maintenance of genetic polymorphisms." These variant conclusions may be due to differences in perspective since Ziehe and Gregorius find that a polymorphism can result from heterosis over approximately 40% of parameter space (estimated from their Fig. 2), which I would not regard as "limited."

Arguments for the selective advantage of heterosis are threefold (Berger 1976; Clarke 1969): (1) the masking of deleterious genes, (2) unique properties of heterozygotes that make them superior to both homozygotes, and (3) a range of gene products that conveys adaptive superiority of the heterozygote in heterogeneous environments. Though heterozygotic advantage has been demonstrated in a variety of species, some examples being given in Table 2.2, the functional basis of the superiority remains obscure.

Table 2.2. Examples in which heterozygous advantage has been found

Species	Heterozygote superior with respect to	Reference
Drosophila pseudobscura	Viability	Dobzhansky and Levene 1955
Drosophila melanogaster	Development time	Bonnier et al. 1959
Drosophila melanogaster	Population size	Carson 1961
Drosophila melanogaster	Fecundity	Watanabe 1969
Drosophila melanogaster	Viability	Tachida and Mukai 1985; Takano et al. 1987
Drosophila melanogaster	Female fecundity, male virility	Bijlsma-Meeles and Bijlmsa 1988
Drosophila melanogaster	Morphological variability[a]	Reeve and Robertson 1953
Drosophila melanogaster	Competitive ability	Miranda and Eggleston 1989
Plectritis congesta (plant)	Morphology[b]	Carey and Ganders 1980
Arabidopsis thaliana (plant)	Growth rate	Langridge 1962
Plantago lanceolata (plant)	Number of scapes	Wolff and Haeck 1990
Fundulus heteroclitus (killifish)	Morphological variability[a]	Mitton 1978
Tribolium casteneum	Survival, development time, body weight	Sokal and Karten 1964
Crassostrea virginica (oyster)	Growth rate	Singh and Zouros 1978; Zouros et al. 1980
Bufo boreas (toad)	Survival	Samollow 1980; Samollow and Soulé 1983

[a]Heterozygotes morphologically less variable than homozygotes. There are numerous studies, reviewed in Palmer and Strobeck (1986), indicating that asymmetry is reduced in heterozygotes. Soulé (1979, 1982) has suggested that this relative lack of asymmetry in heterozygotes reflects better physiological buffering capacity and higher fitness of the heterozygotes.

[b]Heterozygotes morphologically larger than homozygotes.

Nevo, in a monumental series of studies (Nevo et al. 1984; Nevo 1988a; Nevo and Beiles 1988, 1989), demonstrated persistent patterns of heterozygosity with environmental variables such as variability in rainfall, suggesting that heterozygosity is in some manner connected with fitness. Support for the important influence of environmental heterogeneity is provided by the study of Powell (1971), in which it was found that heterozygosity in *Drosophila willistoni* was higher in populations maintained in heterogeneous environments than in constant environments. The possible importance of heterozygosity in maintaining genetic variation cannot be dismissed and requires more investigation.

2.2.3. Antagonistic Pleiotropy

Antagonistic pleiotropy refers to the phenomenon in which a gene has a positive effect on one component of fitness but a negative effect on another: thus, for example, alleles at a particular locus might increase fecundity but decrease longevity. Prout (1980, p. 57) argued that newly arisen and spreading mutations will generally be of this type: "Most new variants are deleterious in *all* manifestations; so it would seem to follow from this, that the next most abundant class would be those where one of the pleiotropic effects results in a net advantage to the genotype, but the remaining pleiotropic effects are deleterious." (See also Simmons et al. 1980.) Antagonistic pleiotropy is important in life history theory because it is the genetic basis of trade-offs. It has been speculated that antagonistic pleiotropy might maintain genetic variation in the population. The argument goes thus: if mutation cannot replace alleles as fast as they are lost under selection, genetic variance will be eroded, pleiotropic or linked combinations of genes that have primarily positive effects on fitness being fixed, leaving variable those genes having opposing effects on fitness (Dickerson 1955; Robertson 1955; Falconer 1985). As attractive as this hypothesis seems to be, the conditions for the maintenance of variation may be quite restrictive, requiring directional dominance (Rose 1982, 1985). This effect can be understood by considering a single locus with two alleles, A and a: directional dominance occurs when the effect on the first fitness component is greatest for one homozygote, say AA, and the effect on the second fitness component is greatest for the second homozygote, aa. (Contrast this with overdominance, where the greatest effect is found in the heterozygote, Aa.) How frequently such effects are found in natural populations is unknown. Negative genetic correlations are putative evidence of antagonistic pleiotropy but mutation accumulation could account for apparent trade-offs in some circumstances (Charlesworth 1984).

2.2.4. Frequency-Dependent Selection

Frequency-dependent selection occurs when the fitness of genotypes vary as a function of the genotypic composition of the population (Ayala and

Campbell 1974; Gromko 1977; DeBenedictis 1978). The most-often-studied form of frequency-dependent selection is the "rare-male advantage" in which rare males are preferred by females (Knoppien 1985; Ayala 1986). Frequency dependence does not necessarily lead to a polymorphism (Lewontin 1958; DeBenedictis 1978), but in theory stable polymorphisms can readily be obtained with this type of selection (Crow and Kimura 1970; Cockerham et al. 1972; Bulmer 1985a; Cressman 1988a,b). Due to a lack of empirical evidence its importance in maintaining genetic variation in natural populations is largely unknown, though the work of Antonovics and Ellstrand (1984) on the grass *Anthoxanthum odoratum* indicates that genotypes can possess frequency-dependent fitness under wild conditions. But frequency-dependent selection may be very important in maintaining phenotypic variation, such as alternate reproductive behaviors, whether or not the variants are different genotypes or different expressions of the same genotype. (See sections 7.1.3 and 8.5.)

2.2.5. *Environmental Heterogeneity*

In 1953 Levene published a paper in which he demonstrated that spatial variation is capable of maintaining genetic variation. Since then, research on the possibility that spatial and/or temporal variation can maintain variation has been both rapid and numerous—sufficiently so that Hedrick was able to review the subject twice within the span of a decade (Hedrick et al. 1976; Hedrick 1986).

Conditions favoring a stable polymorphism in a temporally variable environment were first analyzed by Dempster (1955) and independently studied by Haldane and Jayakar (1963). The essential finding of these studies is that selection favors the genotype having the highest geometric mean fitness. Thus, for a stable polymorphism to occur the heterozygote must have a higher geometric mean fitness than either homozygote. The same result has been found for a variety of models involving temporal variation (Gillespie 1972, 1973a,b; Hartl and Cook 1976; Bryant 1976; Hedrick et al. 1976; Hoekstra et al. 1985) and a combination of both spatial and temporal fluctuations (Gillespie 1975, 1981).

In his analysis of spatial variation, Levene (1953) assumed a simple Mendelian model comprising a single locus with two alleles. The habitat was divided into different niches, individuals from each niche coming together after each cycle of selection and mating randomly. In this model, "a protected polymorphism will exist if the recessive phenotypic fitnesses among the niches have an arithmetic mean greater than unity but a harmonic mean less than unity" (Prout 1968, p. 498). Numerous variants of Levene's model have been analyzed, the two most intensively studied being models allowing for variation in migration rates and habitat choice (e.g.,

Deakin 1966; Maynard Smith 1966, 1970; Christiensen 1974; Gillespie 1975; Taylor 1976; Namkoong and Gregorius 1985; Hoekstra et al. 1985) and models with different rules for the number of parents from each patch (e.g., hard and soft selection; see Karlin and Campbell 1981; Arnold and Anderson 1983; Walsh 1984; Christiansen 1985). The results of these investigations have verified the qualitative conditions for a polymorphism to be maintained in a spatial environment. These conditions are formally equivalent to those required in frequency-dependent selection (Bryant 1976). Three important models, fundamentally different from that of Levene, are those of Bulmer (1971b), Gillespie (1976, 1977, 1978), and Gillespie and Turelli (1989).

Bulmer (1971b) examined the conditions for the maintenance of genetic variation in a metric trait. He assumed a habitat comprised of two patches, from each patch a proportion of the population migrating each generation to the other patch. Under this scenario a stable equilibrium is possible for reasonable parameter values. This model does not appear to have been developed further. However, the finding of genetic by environmental interactions (G × E) in many, if not most, studies in which it has been examined (Table 2.3) argues in favor of environmental heterogeneity playing a role in the maintenance of additive genetic variation. Temporal variation should not, in theory, lead to the maintenance of additive genetic variance since at each cycle there will be truncation at one extreme and hence over time variance should decline. But if the environment is both spatially and temporally variable, migration between habitats could continually restore variation. Even if the combination of environmental variation and migration does not guarantee an equilibrium genetic variance, it may reduce the rate of loss to such a degree that genetic variance will be continually restored by mutation. Experiments by Mackay (1980, 1981) using *Drosophila melanogaster* demonstrated that heritabilities of morphological traits do indeed remain higher in varying environments, though neither the focus of selection nor the mechanism favoring genetic variation could be determined.

Gillespie (1976, 1978) introduced a model in which enzyme activity is an additive function of alleles, and fitness is a concave function of enzyme activity. As a consequence, the heterozygote is intermediate in fitness between the two homozygotes. The fitnesses of the homozygotes are assumed to vary across patches: in the two-niche case, one homozygote, say A_1A_1, is the most fit in the first habitat while the second homozygote, A_2A_2, is dominant in the second habitat. This reversal in dominance is a critical element of the model (reminiscent of antagonistic pleiotropy). Because the function relating fitness to enzyme activity is concave the heterozygote is fitter than the arithmetic mean of the two homozygotes. The

Table 2.3. Some examples of genotype-by-environment (G × E) interactions

Organism	Trait	Environment	Reference
Arabidopsis thaliana	Growth rate	Temperature	1
Sorghum bicolor	Light capture	Temperature	2
Sorghum bicolor	Growth rate	Temperature	3
Hordeum vulgare	Morphology, yield	Density	4
Triticum aestivum	Yield, test weight	Site	5
Papaver dubium	Metrical traits	Soil	6
Salvia lyrata	Number of leaves	Density	7
Drosophila melanogaster	Viability	Temperature	8
Drosophila melanogaster	Power output in flight	Temperature	9
Drosophila melanogaster	ADH activity	Temperature	10
Drosophila melanogaster	Larval survival, development rate, weight	Density	11
Drosophila pseudoobscura	Bristle number, development time	Temperature	12
Tribolium confusum, T. casteneum	Life history components	Food type	13
Liriomyza sativa	Development time	Host plant	14
Uroleucon rudbeckiae	Fecundity	Host plant	15
Various insects (review)	Resource use	Sites	16
Oncopeltus fasciatus	Age at reproduction, Fecundity	Photoperiod	17
Rana sylvatica	Age and size at reproduction	Site	18

References: 1. Langridge 1962; 2. Hammer and Vanderlip 1989; 3. Hammer et al. 1989; 4. Baker and Briggs 1982; 5. Jalaluddin and Harrison 1989; 6. Zuberi and Gale 1976; 7. Shaw 1986; 8. Band 1963; 9. Barnes and Laurie-Ahlberg 1986; 10. Van Delden 1982; 11. Marks 1982; 12. Gupta and Lewontin 1982; 13. Bengi and Gall 1978; Riddle et al. 1986; 14. Via 1984; 15. Service 1984; 16. Futuyma and Peterson 1985; 17. Groeters and Dingle 1987; 18. Berven 1982.

parameter space over which this model produces a stable polymorphism far exceeds that of Levene's model (Maynard Smith and Hoekstra 1980).

Using a standard quantitative genetic model Via and Lande (1985) showed that for the two-patch model, unless the genetic correlation across environments is exactly -1, the population can attain the optimal value for each habitat and hence no stable polymorphism will be maintained by environmental variation. The reasons this is permissible are explained fully in chapter 3: suffice it to note here that the model assumes that there are no trade-offs that restrict the range of additive values. In contrast, Gillespie and Turelli (1989) developed a model that is based on the premise that no single genotype can produce an average phenotype that is optimal in all environments. The essential element of this model is that the mean fitness of a genotype is an increasing function of the number of heterozygous loci; as a consequence stable polymorphisms are easy to maintain.

The ubiquity of spatial and temporal variability suggests that these processes are likely to be involved in the maintenance of genetic variation in populations. Even if the exact conditions for stable polymorphisms are not fulfilled, spatiotemporal variation, coupled with migration between populations, will almost certainly greatly retard the rate of loss of genetic variance. It would be surprising if a single factor could account for the range of variation observed within populations: for example, polymorphism within *Cepea* appears to result from the action of a number of factors (reviewed in Jones et al. 1977). The question, still unanswered is how frequently each of the mechanisms described above operates in the natural world. While we have an abundance of theory we still have a dearth of relevant data with which to test them (Ennos 1983; Nevo 1988b; Pamilo 1988).

2.3. Summary

The two important parameters in quantitative genetic theory are heritability (h^2) and genetic correlation. The former measures the resemblance between parents and offspring, the latter the extent to which different traits are genetically coupled. Direct response to selection is a simple function of heritability and the selection differential. A trait that is genetically correlated to the trait under selection will also change, the amount depending on the genetic covariance between the two traits, the phenotypic variance of the trait under direct selection, and the selection differential. Typically, heritability and the genetic variance-covariance matrix are assumed to be constant. These two assumptions are in theory incorrect, but may not be unreasonable approximations for short-term selection.

There are significant amounts of genetic variance in life history traits. The factors maintaining this variation are poorly understood. Five factors that may play a role are mutation-selection balance, heterosis, antagonistic pleiotropy, frequency-dependent selection, and environmental heterogeneity.

3

Life History Theory: A Framework

Population geneticists do not explicitly define a general measure of fitness: an organism is categorized according to its genotype and for each life history component a fitness measure, such as larval viability or female fecundity, is assigned. (See, for example, Prout 1971.) For simple Mendelian models a well-defined mathematics can then be employed to follow the changes in gene frequency over time. The focus of population genetics is gene frequency, while the focus of the evolutionary ecologist is the trait. In most cases the trait of interest will be controlled by many genes and genetic approaches become either unreliable and/or technically too difficult to utilize. (See chapter 2.) The approach of optimality modeling is to ignore the genetic architecture. Instead it is assumed that there exists some measure of fitness which evolution maximizes. This assumption is by no means new, being given explicit definition by Fisher (1930). Recent developments in population and quantitative genetic theory, described later in this chapter, have given credence to this proposition. The fact that the assumption has led to verified predictions—the subject of chapters 7–10—is itself evidence that the assumption must be reasonably robust relative to the details of the underlying genetic architecture.

This chapter outlines the theoretical rationale by which the evolution of life history traits may be analyzed. The primary framework for such studies is that of optimality modeling, which comprises three components:

1. The assumption that some measure of fitness is maximized
2. That there exist both constraints and trade-offs between traits that limit the set of possible combinations
3. That there exists sufficient genetic variation to permit the attainment of that combination which maximizes fitness, subject to the aforementioned constraints and trade-offs

The third requirement is generally assumed implicitly in analyses, and has been discussed in the first two chapters. The present chapter is divided into three parts: in the first, measures of fitness are discussed; in the second, I address the issue of the relationship between phenotypic trade-offs, which form much of the observational data of life history analysis, and genetic correlations, which are required if the trade-offs are to have evolutionary significance; in the third section problems of measuring trade-offs are examined.

3.1. Measures of Fitness

The starting point of many analyses of evolutionary change, and most particularly that of optimality modeling, is the assumption that there exists some variable maximized by selection. The issue of what is being maximized has been the subject of much discussion, partly because the appropriate measure of fitness changes with circumstance and method of analysis. Broadly, measures of fitness can be separated into two groups—global measures and local measures. A global measure of fitness is one that involves the interaction of all life history components, the best example being Fisher's Malthusian parameter r. Local measures assume that maximization of a fitness component will also maximize the overall fitness of the organism: for example, a common local measure used in foraging theory is the net rate of energy intake (MacArthur and Pianka 1966; Schoener 1971; Westoby 1974; Fritz and Morse 1985). This is an appropriate measure if it can be shown that maximizing this rate does not detrimentally affect other components of fitness, in which case maximizing net rate of energy intake will also increase global fitness. While this may be generally true there are circumstances in which it is not a valid assumption (Houston and Mc-Namara 1986).

Local measures are generally tailored for the particular analysis under consideration, but there now exists a general consensus, and more importantly a sound theoretical rationale, concerning what global measures are likely maximized by natural selection. The following sections describe these measures.

3.1.1. *Density-Independent Measures of Fitness*

3.1.1.1. *Deterministic Environments*

A population growing in an unlimited, homogeneous, and constant environment follows the simple exponential growth function

$$\frac{dN(t)}{dt} = rN(t)$$

$$N(t) = N(0)e^{rt} \tag{3.1}$$

where $N(t)$ is population size at time t, and r is the intrinsic rate of increase, comprising the difference between instantaneous rates of birth and death. Equation 3.1 can also be written as

$$N(t) = N(0)(e^r)^t$$
$$= N(0)\lambda^t \qquad \qquad (3.2)$$

The symbol λ is called the finite rate of increase and is sometimes used instead of r. (Both will be used in the present book.)

Suppose there are two clones with population sizes $N_1(t)$ and $N_2(t)$, respectively, the first with an intrinsic rate of increase of r_1 and the second with r_2, with $r_1 > r_2$. The ratio of population sizes after some time t, given that both clones start with the same population size, is

$$\frac{N_1(t)}{N_2(t)} = e^{(r_1 - r_2)t} \qquad \qquad (3.3)$$

It is clear that as time progresses the above ratio will increase, clone 1 becoming numerically more and more dominant in the combined population. This conclusion does not depend upon the two clones beginning with the same population size: differences in starting condition simply accelerate or retard the rate at which clone 1 increases in frequency relative to clone 2.

For the two clones described above an appropriate measure of fitness is r, since the frequency of the clone with the highest value of r will increase toward unity. Thus any mutation in a set of clones that increases r by changing rates of birth or death will increase in frequency in the population. There are no difficulties in assigning r as a measure of fitness in the above circumstances. Difficulties arise, however, when sexual reproduction and age structure are introduced (Pollack and Kempthorne 1970, 1971). Suppose we have a random mating population in which a mutation arises that increases birth rate or decreases death rate. Since the mutant will initially be rare in the population its fate can be ascertained by considering the birth and death rates of the heterozygote alone (Charlesworth 1974; Charlesworth and Williamson 1975; Christiansen and Fenchel 1977; Reed and Stenseth 1984). If the heterozygote's rate of increase is enhanced, the mutation will increase in frequency in the population, but its ultimate fate depends upon the relative birth and death rates of the homozygotes and heterozygotes bearing the mutant allele. If the homozygote carrying both mutant alleles has a higher birth rate and/or a lower death rate than the heterozygote, the mutant allele will eventually be fixed in the population; otherwise the population will reach a stable polymorphism.

The general assumption, stemming from the work of Fisher (1930), has been that r can be associated with genotypes that follow particular life histories and that selection will favor that genotype with the highest value of r. In an age-structured population the rate of increase is obtained by solving the characteristic equation,

$$\int_0^\infty e^{-rx} l(x) m(x) \, dx = 1 \tag{3.4}$$

where $l(x)$ is the probability of surviving to age x and $m(x)$ is the number of female births at age x. The discrete time equivalent of this is (Goodman 1982)

$$\sum_{x=1}^\infty e^{-rx} l(x) m(x) = 1 \tag{3.5}$$

Note that in the discrete version the initial age is subscripted as one, not zero. The important issue to be considered is the fate of a mutant that increases r. Charlesworth (1973, p. 306) demonstrated that to a rough approximation "the rate of progress of a rare gene eventually becomes directly proportional to its heterozygous effect on r." Charlesworth and Williamson (1975) proved that for a near-stationary population the probability of survival of a mutant allele, p, in an age-structured population is approximately

$$p \approx \frac{2 \sum l(x) m(x)}{V} r \tag{3.6}$$

where V is the variance of the overall offspring distribution and r is the intrinsic rate of increase of the mutant heterozygote. Thus, in the above case the probability of survival of a mutant gene of small effect in a near-stationary population is largely determined by its effect on r. Following a more detailed analysis, Charlesworth (1980, p. 196) concluded "that for the case of weak selection and random mating with respect to age, the intrinsic rate of increase of a genotype or, more generally, the mean of the male and female intrinsic rates, provides an adequate measure of fitness in a density-independent and constant environment."

Lande (1982) tackled the problem of applying quantitative genetic theory to the evolution of r in a population. Assuming weak selection, large population size, and a constant genetic variance-covariance matrix, Lande derived the rate of change of r with time t, dr/dt, to be

$$\frac{dr}{dt} = \begin{pmatrix} \dfrac{\partial r}{\partial \bar{z}_1} & \cdots & \dfrac{\partial r}{\partial \bar{z}_n} \end{pmatrix} \begin{pmatrix} G_{11} & G_{12} & \cdots & G_{1n} \\ \cdot & \cdot & \cdots & \\ \cdot & \cdot & \cdots & \\ \cdot & \cdot & \cdots & \\ G_{n1} & G_{n2} & \cdots & G_{nn} \end{pmatrix} \begin{pmatrix} \dfrac{\partial r}{\partial \bar{z}_1} \\ \cdot \\ \cdot \\ \dfrac{\partial r}{\partial \bar{z}_n} \end{pmatrix} \tag{3.7}$$

The z_is ($1 \le i \le n$) are the n quantitative characters determining the life history characteristics of the organism. The vectors describe the rate of change in r with respect to the mean life history traits, called the selection gradient, and the matrix is the additive genetic variance-covariance matrix. The important result of the above equation is that the rate of change of r is always positive and hence "in a constant selective regime, life history evolution continually increases the intrinsic rate of increase of a population, until an equilibrium is reached" (Lande 1982, p. 611). The intuitive appeal of r as a suitable measure of fitness thus receives qualified support from both population genetic and quantitative genetic theory.

If the population is stationary r is zero, the characteristic equation reduces to

$$R_0 = \int_0^{\infty} l(x)m(x) \, dx = 1 \tag{3.8}$$

R_0 is termed the *net reproductive rate* and is the expected number of female offspring produced by a female over her life-span. The use of R_0 makes analysis easier and can be justified if r is very close to zero or if the density-dependent or other factors that maintain the population at some relatively stable value do not impinge upon the traits under consideration. For example, population size might be controlled by density-dependent mortality in the larval stage, while the object of study is the evolution of female age at maturity. In this case we can examine the relationship between the age at maturity and fitness under the working assumption that genotypes do not differ in the characteristics of their larvae. Since population size is stable, the expected lifetime fecundity, R_0, is then the appropriate measure of fitness.

In some analyses, fitness is determined from the reproductive value at age x, $V(x)$

$$V(x) = \frac{e^{rx}}{l(x)} \int_x^{\infty} e^{-ry}l(y)m(y) \, dy \tag{3.9}$$

The reproductive value of an individual of age x is a measure of the extent to which it contributes to the ancestry of future generations (Taylor et al. 1974). Its discrete version is

$$V(x) = \frac{e^{r(x-1)}}{l(x)} \sum_{i=x}^{\infty} e^{-ri} l(i) m(i) \qquad (3.10)$$

In most analyses the -1 is omitted from the first exponent, leading to a reproductive value that is off by a factor of e^r at every age (Goodman 1982). Williams (1966) postulated that natural selection maximizes r by maximizing reproductive value at every age. This conjecture is supported by the analyses of Goodman (1974) and Schaffer (1974a, 1979a), but a semantic misunderstanding led Caswell (1980) to question it, leading to a flurry of rebuttals (Schaffer 1981; Yodzis 1981). A correct statement of the principle is: "reproductive value at each age is maximized relative to reproductive effort at that age, although not necessarily with respect to effort at other ages (therein lies the misinterpretation of Schaffer's work in Caswell [1980])" (Caswell 1982a, p. 1220).

3.1.1.2. *Variable Environments*

Environments are typically variable in both time and space. While the assumption of a constant environment may be a reasonable first approximation in many cases, there will be many others in which variation cannot be ignored. Thoday (1953) suggested that because of environmental variation the appropriate measure of fitness is the probability of persistence. More recently, Cohen (1966) proposed that in a stochastically, temporally fluctuating environment the correct measure of fitness is the geometric mean of the finite rate of increase. The rationale behind this suggestion is as follows: the size of a population after t time intervals is given by

$$N(t) = N(0)\lambda_1\lambda_2 \ldots \lambda_t$$

$$= N(0) \prod_{i=1}^{i=t} \lambda_i \qquad (3.11)$$

The arithmetic and geometric means are

$$\text{Arithmetic mean} = \left(\sum_{i=1}^{i=t} \lambda_i \right) \Big/ t \qquad (3.12)$$

$$\text{Geometric mean} = \left(\prod_{i=1}^{i=t} \lambda_i \right)^{1/t} \qquad (3.13)$$

It is obvious that the geometric mean describes the long-term population size and hence is the parameter that reflects the fitness of different strategies. If r is used in place of the finite rate of increase, the appropriate measure of fitness is the arithmetic mean of r. This can be demonstrated as follows (Roff 1978): dividing time intervals into small units we can write

$$N(t) = N(0)e^{r_1}e^{r_2} \ldots e^{r_t}$$
$$= N(0)e^{\sum_{i=1}^{t} r_i} \tag{3.14}$$

from which it is readily apparent that the correct measure of fitness is the arithmetic average of r.

Consider two genotypes living in an environment that comprises two types of year, "good" and "bad," each occurring with equal frequency. In "good" years genotype A has a finite rate of increase of 2 and in a "bad" year a finite rate of increase of 0.5, while genotype B has finite rates of increase of 1 and 1.1, respectively. The arithmetic averages of A and B are 2.5 and 2.1, respectively, but the geometric averages are 1 and 1.1. Thus genotype B has the highest long-term fitness although it has a smaller arithmetic finite rate of increase. Genotype A increases more than genotype B in "good" years but suffers a greater reduction in "bad" years. The relatively high fitness of genotype B resides in the fact that although it has a smaller arithmetic average it also has a smaller variance in its finite rate of increase. The importance of this property can be seen from the relationship

$$\text{Geometric mean} \simeq \text{Arithmetic mean} - \frac{1}{2}\frac{\text{Variance}}{\text{Arithmetic mean}} \tag{3.15}$$

Therefore, selection can increase fitness by either increasing the arithmetic average or decreasing the variance (Gillespie 1977; Lacey et al. 1983). Slatkin (1974) coined the term "bet hedging strategies" for life history patterns that reduce variance. Den Boer (1968) advanced the same conceptual framework for fitness in a variable environment and termed it "spreading the risk." Considerable attention has been given to bet-hedging strategies (e.g., Levins 1970; Real 1980; Lacey et al. 1983; Roff 1983a; Seger and Brockman 1987; Phillipi and Seger 1989), but it should not be assumed that bet hedging is unusual. The name itself is somewhat unfortunate since it tends to put the phenomenon into a special category. In a temporally variable environment selection favors the maximization of the geometric average of the finite rates of increase; in a constant environment maximization of the geometric average is redundant but nevertheless gives exactly the same answer as maximizing a constant finite rate of increase.

Therefore, the maximization of the geometric average is the general case, with the model of a constant environment being a special case!

Hastings and Caswell (1979) criticized the use of the geometric mean and advocated the use of the arithmetic mean. They argued thus: suppose that environmental variability is independently and identically distributed from year to year: then the expected long-term growth rate, $E(\lambda)$, is

$$E\left(\prod_{i=1}^{t}\lambda_i\right) = \prod_{i=1}^{n}E(\lambda_i)$$

$$= [E(\lambda)]^t$$

$$= (\text{Arithmetic average of } \lambda)^t \qquad (3.16)$$

The above analysis suggests that in the long term a population will grow at the rate dictated by the arithmetic average of the finite rate of increase. The fallacy in this approach is that after some time t the distribution of population sizes, $N(t)$, becomes highly skewed, and although the mean population size may, if the arithmetic mean λ exceeds 1, increase without bound, the probability of extinction may approach 1 (Lewontin and Cohen 1969; Levins 1969; May 1971; 1973; Turelli 1977; for an age-structured model see Lande and Orzack 1988)—i.e.,

$$\text{as } t \to \infty \; E[N(t)] \to \infty \text{ but } P[N(t) > 0] \to 0$$

This can be illustrated with a simple example: Suppose that λ can take two values, 0 or 3, with equal frequency. The expected value of λ is $(0 + 3)/2 = 1.5$, and hence the expected population size increases without bound as t increases. For example, starting from a single female, after 10 generations

$$E[N(10)] = 1.5^{10} = 57.7 \text{ but either } N(10) = 59049 \text{ or } N(10) = 0$$

and the probability that the population persists for the 10 generations is $(0.5)^{10} = 0.00098$, a very small probability indeed. The arithmetric mean is a very poor guide to fitness. There is no doubt that if the arithmetic mean of a genotype is greater than that of a second genotype, though the geometric means are reversed, there will exist occasions in which there will be a sufficiently long string of good years that the former genotype will come to dominate the population. The extreme skew in persistence times and population size will in most instances, however, favor the genotype that maximizes its geometric mean.

The second type of variability to consider is that of spatial variation. Let the number of patches be n and the rate of increase in patch i be r_i. If patches are selected at random, the number of offspring per female per time unit is (Roff 1978)

$$e^r = \sum_{i=1}^{n} e^{r_i} \tag{3.17}$$

Hence

$$r = \log_e\left(\sum_{i=1}^{n} e^{r_i}\right)$$

$$= \log_e\left(\sum_{i=1}^{n} \lambda_i\right) \tag{3.18}$$

Maximizing fitness in a spatially heterogeneous environment is equivalent to maximizing the arithmetic mean of the finite rates of increase. Because fitness is a function of the arithmetic mean value of the finite rates of increase per patch, spatial variation will have considerably less influence on life history variation than temporal variation.

3.1.2. *Density-Dependent Measures of Fitness*

MacArthur (1962) suggested that in density-dependent models the appropriate measure of fitness is the carrying capacity that can be attributed to a genotype. This suggestion is intuitively appealing and a rough justification for its use is as follows. Suppose there are two clones that in isolation equilibrate at population sizes K_1 and K_2, where $K_1 < K_2$, and are otherwise identical. The two clones are mixed together in equal proportion to make a population initially of size K_1. Since the total population size is equal to the equilibrium population size for clone 1, this clone will show no response, but clone 2 is below its equilibrium and hence will increase in numbers. But now the total population size is above the equilibrium size for clone 1 and hence clone 1 must decline in numbers until the total population size is restored to K_1. Since clone 2 can always increase at this level (i.e., population size $= K_1$), clone 1 will continue to decline, eventually being eliminated, and the population will increase in size to K_2.

Though it is possible to find parameter values that do not lead to the maximization of population size in MacArthur's model (Green 1980), more general analyses of populations with nonoverlapping generations find that population size is maximized (Anderson 1971; Hastings 1978; Iwasa and Teramoto 1980; Turelli and Petri 1980; Desharnais and Constantino 1983). Difficulties arise when there is age structure in the population, and in this case population size may not be maximized (Jong 1984). The important variable is the stage at which the density-dependence occurs, called the *critical age group* by Charlesworth (1972). In "a density-dependent population the progress of selection is controlled by the genotypic carrying capacities . . . selection tends to maximize the number of individuals in the critical age group, subject to the assumption of weak selection" (Charlesworth 1980, p. 168). The tendency for density to increase under these conditions has been termed by León and Charlesworth (1978) the "ecological version of Fisher's fundamental theorem of natural selection."

3.1.3. r- and K-Selection

It should be clear from the first two sections that selection in density-independent and density-dependent conditions may require different measures of fitness. (If, in a density-regulated population, the traits under study are not dependent on density then, as discussed in 3.1.1.1, the appropriate measure of fitness is the expected lifetime fecundity, R_0.) The difference is sometimes referred to as r- and K-selection. The terms are still frequently used though considerable confusion has surrounded both the concept and its application.

A paper by Dobzhansky (1950) has been credited with originating the basic idea (Pianka 1970). Dobzhansky, in a comparison of the tropics and temperate regions, drew the conclusion that populations in the temperate zones are primarily limited by abiotic factors and rarely achieve densities at which intraspecific competition prevails. In contrast, populations in the supposedly constant conditions of the tropics typically attain densities where density dependence becomes a principal factor in population regulation. MacArthur and Wilson (1967) coined the terms r- and K-selection to describe these two conditions. The terms are derived from the standard model of population growth in ecology, the logistic equation

$$\frac{dN(t)}{dt} = rN(t) - \frac{rN(t)^2}{K}$$

$$N(t) = \frac{K}{1 + e^{c-rt}} \tag{3.19}$$

where r is the intrinsic rate of increase,
$\quad K$ is the carrying capacity of the environment, and
$\quad c$ is a constant of integration dependent on the initial population size. I have written the differential in expanded form to indicate that the logistic equation is simply the exponential growth equation with a second-order term added. MacArthur and Wilson, drawing on an earlier study by MacArthur (1962), proposed that alleles could be assigned values of r and K. In an unsaturated environment those genotypes with the highest values

of r would prevail, while in a saturated environment those which could maintain the largest population size would have the advantage, i.e., those with the highest K value. Thus far there is no real problem with the definition since these conform reasonably well to the definitions given in sections 3.1 and 3.2 of this chapter.

Since the analysis by MacArthur and Wilson there have been a great number of further analyses (e.g., Anderson 1971; Clarke 1972; King and Anderson 1971; Roughgarden 1971; Charlesworth 1971; Charlesworth and Giesel 1972a,b; Armstrong and Gilpin 1977; Turelli and Petri 1980; Asmussen 1983; Desharnais and Costantio 1983), most based on the hypothesis, suggested by Gadgil and Bossert (1970), that there is a trade-off between r and K. This hypothesis may not be correct: it receives some support in experiments on *Drosophila melanogaster* (Mueller and Ayala 1981; Mueller et al. 1991) but not in those on the rotifer, *Asplanchna brightwelli* (Snell 1977), the cladoceran, *Bosmina longirostris* (Kerfoot 1977), or the bacteria, *Escherichia coli* (Luckinbill 1984). The foregoing theoretical analyses all demonstrated that K will be maximized and that if the heterozygote has the highest K, a stable polymorphism can be generated. Several analyses have indicated that this may not be universally true (Green 1980; Jong 1982), but the overall conclusion is that in a density-regulated population, selection maximizes population size (or the critical age cohort). To this extent r- and K-selection has been a useful theoretical concept. The problem arises when the concept is applied without refinement to natural populations, most particularly when these definitions are used to define the life history traits of organisms.

Pianka (1970) was the first to propose a set of characteristics expected to be associated with the two types of selection: r-selected organisms were supposed to have (1) rapid development, (2) high rate of increase, (3) early reproduction, (4) small body size, and (5) semelparity (single reproductive episode). K-selective organisms have the opposite characteristics, traits 1 and 2 being traded off for high population size and competitive ability. A rationale for these categories was not given. The categorization ignores the fact that in ectotherms, and to a lesser extent endotherms, development time, size at maturity, and fecundity are intercorrelated. (See chapters 5 and 7.) The rate of increase may therefore be maximized by late reproduction: the actual age will depend upon the various correlations and trade-offs between traits. (See chapter 7.) In support of the above categorization Pianka presented data on the body lengths of vertebrates and insects, claiming that these supported his thesis. But to compare vertebrates and insects is to compare apples and oranges. Differences between taxa as widely separated as vertebrates and insects are undoubtedly due to a multiplicity of causes. A correct analysis would be between different genotypes or at least different populations of the same species.

Mertz (1975, p. 3) objected to the notion of an r-K continuum because "r and K are usually thought of in terms of the logistic equation, and classical logistic theory is plainly incompatible with complex life history considerations"—a view echoed by others (Gill 1972; Wilbur et al. 1974; Whittaker and Goodman 1979; Caswell 1982b; Hall 1988; see also the experimental tests of Pianka's categories by Barclay and Gregory, 1981, 1982). Unfortunately the perceived inadequacies of the r-K concept led others to invent yet more terms: b- and d-selection (Hairston et al. 1970), α-selection (Gill 1974), and h- and T-selection (Demetrius 1977).

The real problem that Pianka's 1970 paper caused was that workers began classifying species on the basis of their life history characteristics rather than by a demonstration of the type of selection operating (reviewed in Parry 1981). Thus the terms r and K have come to be applied either to density-dependent selection or to presumed characteristics associated with r- and K-selection.

I concur wholeheartedly with the view of Mueller (1988a, p. 787):

> Given the substantial problems with the verbal theory of r- and K-selection, it is reasonable to ask if the formal theories that assume logistic fitness functions can be used to predict the evolution of phenotypes other than density-dependent rates of population growth. The answer is probably no. To develop a theory that accounts for the evolution of body size or competitive ability, models with the relevant ecological phenomena must be developed.

The theoretical analysis of Mueller (1988a) is a fine example of this approach, grounded firmly in the biology of a particular organism, *Drosophila melanogaster*, and designed to address the processes observed during density-dependent selection in this species (Mueller and Sweet 1986; Joshi and Mueller 1988; Mueller 1988b, 1990, 1991).

To summarize, the concept of r- and K-selection has been useful in helping to formalize the definition of fitness in density-regulated populations, but attempts to transfer the concept to actual populations without regard to the realities of the complexities in life history have probably been detrimental rather than helpful. The terms r- and K-selection should be interpreted strictly in terms of models of density dependence (Boyce 1984; Elgar and Catterall 1989), and given the confusion that now surrounds the issue, it may be preferable to avoid use of the terms altogether.

3.1.4. Frequency-Dependent Selection

When interactions between clones or genotypes are frequency-dependent, several types will be maintained in the population and fitness cannot be defined by any of the manners described above. The appropriate method of analysis is that of game theory (Maynard Smith 1982: Riechert and Hammerstein 1983), which comprises two essential elements:

1. Particular patterns of behavior will persist in a population provided no mutant adopting an alternate behavior can invade (Maynard Smith and Price 1973). Such stable combinations are termed "evolutionarily stable strategies" (ESS). The concept of the ESS is not unique to game theory: the maximization of r, R_0 and the numbers in the critical age group are all ESSs within the context in which they are appropriate.

2. For each type there must be an assigned gain or loss in fitness when this type interacts with another individual. From this payoff matrix we compute the expected payoff for each behavior. For two behaviors to be evolutionarily stable their fitnesses must be equal.

This approach can be illustrated with the following example. Many animals, such as frogs and crickets, attract a mate by calling from a territory. While this attracts females it also attracts predators. An alternate behavior is to become a satellite of a calling male and attempt to intercept females attracted to the caller. Satellites have a reduced risk of being preyed upon, but are also less successful at obtaining copulations. The behavior is frequency-dependent because without a caller no female will approach a noncaller. The payoff matrix is

	Caller	Satellite
Caller	$Wf/2$	W
Satellite	$1 - W$	0

where the payoffs are to the individuals listed on the left when confronted with individuals listed above. I have for simplicity ignored the situation of caller by itself. (This simply requires that all territories be occupied.) The payoff obtained by a caller next to a satellite is W; that of a satellite is $1 - W$. A caller next to a caller obtains one-half of the matings but also suffers a reduction, f, because the combined calling of two males is more likely to attract a predator. To see if both behaviors will be found in the population we assume that there are p callers in the population (this may represent separate types in the population or individuals adopting the calling behavior with probability p) and test for equality of the expected payoffs— i.e., that

$$p \frac{Wf}{2} + (1 - p)W = p(1 - W) + (1 - p)0 \qquad (3.20)$$

which upon rearrangement gives

$$p = \frac{W}{1 - \frac{W}{2}f} \tag{3.21}$$

Now p must range between 0 and 1, and thus

$$0 < \frac{W}{1 - \frac{W}{2}f} < 1 \tag{3.22}$$

which upon rearrangement gives

$$W < \frac{2}{2 + f} \tag{3.23}$$

Thus, when the above inequality is satisfied, both satellites and callers are predicted to occur in the population. For example, if $f = 0.5$, the W must be less than 0.8 for maintenance of both behaviors. For $W = 0.7$ the predicted frequency of callers in the population is 0.848 (equation 3.21).

Since its conception, game theory, or as it is more popularly (and incorrectly) known, ESS theory, has undergone considerable mathematical refinement (e.g., Riley 1979; Gadgil et al. 1980; Hines, 1980; Cressman and Dash 1987; Vickers and Cannings 1987; see also reviews by Rapoport 1985; and Vincent and Brown 1988) but these do not appear to have greatly influenced the application of the approach to real case studies. For a detailed and lucid description of the application of game theory to real data see Maynard Smith (1982). The most important theoretical question is, to what extent is the method valid in a sexual population. With simple Mendelian models, such as single-locus, two-allele models, an ESS may not be possible (Auslander et al. 1978; Maynard Smith 1981), but more complex genetic models do permit populations at least to approach, if not attain, the predicted equilibrium (Eshel 1982; Bomze et al. 1983; Hines 1987).

3.1.5. *Fitness in a Social Setting*

When survival and/or reproduction depends upon the interactions between individuals that have genes in common it is necessary to take into account the increment in fitness accruing to the individual by virtue of such interactions. The additional component of fitness is termed *inclusive fitness* (Hamilton 1964). There has been considerable confusion over the correct definition of inclusive fitness (Grafen 1982), in part due to the relative obscurity of the text in Hamilton. (Grafen lists 10 major texts that incorrectly define inclusive fitness.) Hamilton (1964, p. 8) wrote,

As to the nature of inclusive fitness it may perhaps help to clarify the notion if we give a slightly different verbal presentation. Inclusive fitness can be imagined as the personal fitness which an individual actually expresses in its production of adult offspring as it becomes after it has been stripped and then augmented in a certain way. It is stripped of all components which can be considered as due to the individual's social environment, leaving the fitness which he [sic] would express if not exposed to any of the harms or benefits of that environment. This quantity is then augmented by certain fractions of the quantities of harm and benefit which the individual himself [sic] causes to the fitnesses of his [sic] neighbours. The fractions in question are simply the coefficients of relationship appropriate to the neighbours whom he [sic] affects: unity for clonal individuals, one-half for sibs, one-quarter for half sibs, one-eighth for cousins . . . and finally zero for all neighbours whose relationship can be considered negligibly small.

Grafen (1982, 1984) describes in some detail the ways in which inclusive fitness has been misunderstood. In practice inclusive fitness can be replaced with Hamilton's rule, which can be considered its operational definition. This rule states that animals are selected to perform actions for which $r^*b - c > 0$, where r^* is relatedness, and b and c refer to the effects of an allele on offspring production: bearers of this allele behave in such a manner that each has c fewer offspring, and the bearer's sib has b more offspring (Grafen 1984). Inclusive fitness is not used in any of the models described in this book (it is presented here for completeness), and the reader is referred to Grafen (1984) and Creel (1990) for further details of its correct use.

3.2. Relationship Between Phenotypic Trade-Offs and Genetic Correlations

A fundamental issue is the extent to which *phenotypic correlations* between traits can be used to predict the optimal combination of traits. For evolution to proceed, trade-offs between traits must translate into genetic differences between individuals. To examine the relationship between the trade-off assumptions underlying the optimality approach and the quantitative genetic approach it is instructive to begin with a model in which only a single trade-off is postulated. First we consider the joint evolution of two traits by using standard quantitative genetic theory, without explicitly invoking trade-offs. The response in the two characters is given by (See section 2.1.2),

$$\begin{pmatrix} R_1 \\ R_2 \end{pmatrix} = \begin{pmatrix} G_{11} & G_{12} \\ G_{21} & G_{22} \end{pmatrix} \begin{pmatrix} \beta_1 \\ \beta_2 \end{pmatrix} \tag{3.24}$$

Selection will eventually move both traits to their respective optimum unless the genetic variance-covariance matrix is singular. (Lande 1970; Maynard Smith et al. 1985; Via and Lande 1985). In the present case singularity occurs when

$$G_{11}G_{22} = G_{12}G_{21} \qquad (3.25)$$

Recall that the genetic correlation, r_G, between the two traits is

$$r_G = \frac{G_{12}}{\sqrt{G_{11}\,G_{22}}} \qquad (3.26)$$

Substituting for G_{12} (note that $G_{12} = G_{21}$), gives the condition for singularity to be $r_G = \pm 1$. Thus, if the genetic correlation is ± 1, evolution in one direction is precluded. This can be easily visualized by considering the regression of the additive genetic values (Fig. 3.1), which for convenience I shall call A_1 and A_2. The regression between these traits can be denoted by

$$A_1 = c_1 + c_2 A_2 + e \qquad (3.27)$$

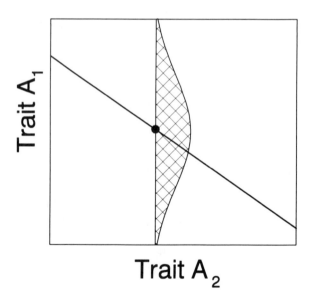

Trait A_1

Trait A_2

Figure 3.1. Hypothetical relationship between the additive genetic values A_1 and A_2. Under the standard quantitative genetic model, values of A_1 are normally distributed about the regression line (e.g., the circle), and hence all values of A_1 are feasible.

where c_1, c_2 are constants and e is a normally distributed variable with mean zero. Unless all points lie exactly on the line it is possible to obtain any value of A_1. The same argument can be applied to A_2. When all points do lie on the line the correlation between A_1 and A_2 is ± 1, and selection can only move the combination of traits along the regression line.

Now consider the cause of a trade-off between the number and size of offspring. Suppose that the organism has a fixed space within its body to house the clutch, or a fixed quantity of resources to convert into propagules. There will be a precise and predictable relationship between the number and size of propagules that can be produced in a clutch. Variation in size of offspring between mothers will necessarily entail variation in number of offspring. A female could produce fewer offspring and hence the set of permissible parameter space is that set of combinations enclosed by the trade-off boundary (Fig. 3.2). In the absence of any other constraints, selection will favor organisms that lie on this boundary and hence both the phenotypic and genetic correlations will be unity. Charnov (1989a) demonstrated how a solution obtained from such an optimality approach will be identical to that obtained from a quantitative genetic analysis. However, to do this he assumed implicitly that the functional constraint observed in the phenotype also applies to the underlying additive genetic values—i.e., that the genetic correlation is -1.

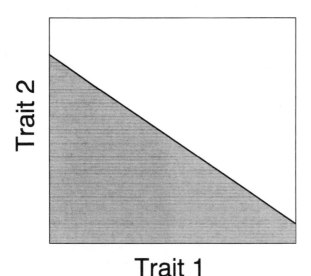

Figure 3.2. Trade-off between two traits such as egg size and clutch size. Combinations lying above the trade-off boundary (solid line) are not feasible. Combinations below the boundary (stippled area) are feasible but confer a reduced fitness.

With two traits, at equilibrium, genetic variation is constrained to lie along the line defining the trade-off between the traits, i.e., a genetic correlation of -1; but as the number of traits involved in the trade-off is increased, genetic correlations can be greater than -1. This is illustrated in Fig. 3.3, where it is assumed that the trade-off involves three traits related such that fitness is equal to a linear combination of the three. If we assume that this functional relationship also applies to the underlying additive genetic values, then these values are free to move across the trade-off surface, and hence variation at equilibrium is not confined to a single line: consequently the genetic correlation between two variables may exceed -1 and indeed, might even be positive. The actual relationship between the genetic covariances depends upon the number of traits and the functional form of the trade-off. Charlesworth (1990) examined this problem for an arbitrary number of traits, assuming that the additive genetic values underlying the phenotypic variation obey the same functional constraints as apply to the phenotypic trade-offs. As with other quantitative genetic analyses, weak selection was assumed. Charlesworth's analysis showed, first, that the genetic analysis predicts the same combination at equilibrium as the optimality model, and second, verified the conjecture of Pease and Bull (1988) that at equilibrium some of the genetic covariances among traits may be positive. However, in the illustrative model that Charlesworth examined, genetic correlations of -1 were obtained between those pairs of traits for which a functional trade-off was explicitly assigned.

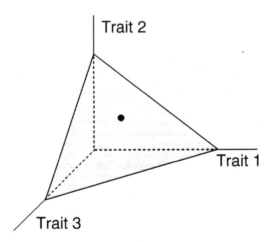

Figure 3.3.   Trade-off involving three traits. The fitness boundary is now a surface and the optimal combination will not in general lead to genetic correlations of -1.

Positive correlations appeared to be generated only where the traits were not functionally constrained. (Chapter 11 discusses this aspect in greater detail.) Thus the set of genetic correlations may give some insight into the possible functional basis of trade-offs.

The existence of positive genetic correlations means that all interactions between traits that may be relevant to the analysis at hand must be thoroughly explored: failure to do so could lead to a misinterpretation of the genetic equilibrium. To regard the organism as a "black box" from which genetic variances and covariances can be extracted simply by the appropriate breeding program is an approach fraught with difficulties of interpretation. To correctly apply quantitative genetic methods to the analysis of life history evolution it is advisable to understand the functional relationship between genes and the traits they determine (Riska 1986, 1989; Clark 1987; Atchley and Newman 1989; Roff 1990b; Stearns et al. 1991). With such relationships the answers from optimality modeling and quantitative genetic analysis should be congruent. The advantage of the optimality approach is that it is experimentally much easier than genetic analysis since phenotypic correlations are statistically less troublesome to obtain than genetic parameters. The *important point to remember is that the trade-offs entering into an optimality model must be grounded in a detailed assessment of the constraints placed upon an organism. Further, these constraints must be demonstrated to have a genetic basis.*

3.3. Measurement of Trade-Offs

The measurement of trade-offs can be divided into four categories (Reznick 1985):

1. Phenotypic correlations: the correlation between two traits measured at the level of the phenotype and involving no manipulation of the organism

2. Experimental manipulations: the direct manipulation of a single factor while keeping all other factors constant, or at least randomly assigned

3. Genetic correlations from sib analysis: estimation of the genetic correlation between two traits by using covariation between individuals within and between families, or covariation between clones or inbred lines

4. Genetic correlations from selection experiments: estimation of the genetic correlation between two traits by using correlated changes in one trait in response to selection on another

The first two categories measure only the phenotypic association between traits while the second two address the issue of whether the trade-off can produce evolutionary change. Reznick (1985) argued that categories 1 and 2 are flawed because of the problem of inferring causation from correlation and because they do not demonstrate that the trade-offs are under genetic control. While I concur that such data must be viewed with caution, the insight that they provide into the functional relationship between traits can be very valuable and they are therefore fruitful avenues of investigation. (See also Pease and Bull 1988.)

3.3.1. Phenotypic Correlations

There are many examples of correlations based on data from unmanipulated situations, but because of the problem of inferring causation from correlation the interpretation of such data is difficult (Partridge and Harvey 1985; Reznick 1985; Noordwijk and de Jong 1986; Pease and Bull 1988). The problem can be illustrated with the example used in the previous section in which there is a trade-off between egg number and size. Suppose this correlation arises because of geometric constraints, e.g., a limited space inside the mother: the relationship will take the form

$$\text{Number of eggs} \propto \frac{\text{Size of mother}}{\text{Size of egg}}$$

Taking logarithms we have

$$Y = B - X \tag{3.28}$$

where Y is log(egg number), B is log(body size), and X is log(egg size); for simplicity the data are assumed to be scaled such that the constant of proportionality equals 1. Provided that the body size of the mother is constant, there will be an observed trade-off between the number of eggs and their size. But suppose body size varies and this fact is not taken into consideration: what will be the observed relationship between egg number and size? If variation in body size is small there will still be an overall negative correlation between Y and X (top panel, Fig. 3.4), but this correlation will decline as variance in body size increases, and if variation is moderate no trade-off may be discernable (middle panel, Fig. 3.4). Furthermore, if there is a correlation between egg size and body size, as has been observed in a wide range of species (Table 10.3), the correlation between the number of eggs and egg size may be reversed (bottom panel, Fig 3.4).

Trade-offs may be obscured because of variation in the rates of allocation and acquirement of resources. Noordwick and Jong (1986) outline the

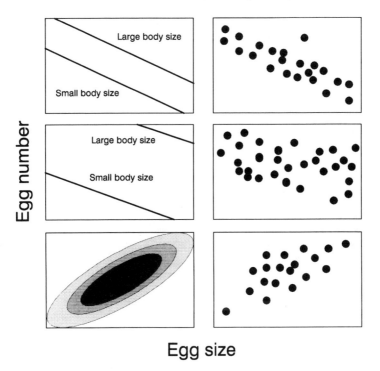

Egg size

Figure 3.4. Possible correlations between egg number and size when effects of body size are not taken into account. The panels on the left show the actual relationship, and the panels on the right what might be obtained by sampling. In all cases egg number decreases with egg size for a fixed body size (solid lines). For a fixed egg size, egg number increases with body size. Top panels: Small range in body size. The observed phenotypic relationship indicates a trade-off. Middle panels: Large range in body size. The negative relationship between egg size and egg number is swamped. Bottom panels: Positive correlation between egg size and body size. (Shading indicates probability of a particular body size/egg size combination.) Small females have small eggs and low fecundities, whereas large females have large eggs and large fecundities. As a consequence a positive correlation between egg size and number is obtained.

mathematical details of how such correlations can arise. They assume there is a fixed quantity of some resource, R that is divided between two traits, X and Y,

$$R = X + Y \qquad (3.29)$$

The quantities measured by an experimenter are X and $R - X (= Y)$, and the covariance of interest, $\text{cov}(X, Y)$ is equal to

$$\text{cov}(X, Y) = \text{cov}(X, R - X) = \text{cov}(X, R) - \text{cov}(X, X)$$
$$= \text{cov}(X, R) - \text{var}(X) \tag{3.30}$$

Individual i is assumed to sequester R_i amount of resource and allocate some fraction F_i to trait X,

$$X_i = F_i R_i \tag{3.31}$$

The total amount sequested by the ith individual can be divided into two components: an amount R that is the mean value for the population, and an amount r_i that represents the deviation in resources obtained by the ith individual relative to the overall average (R). Similarly, the fraction allocated to X can be divided into F, the overall average and f_i the deviation of the ith individual from this mean. Thus,

$$X_i = F_i R_i$$
$$R_i = R + r_i$$
$$F_i = F + f_i \tag{3.32}$$

The quantities r_i and f_i are assumed to be independently distributed with zero means and variances V_r and V_f, respectively. The covariance between X and Y is thus

$$\text{cov}(X, Y) = FV_r - (R^2 V_f + F^2 V_r + V_f V_r)$$
$$= F(1 - F)V_r - R^2 V_f - V_f V_r \tag{3.33}$$

If the variance in acquisition is zero ($V_r = 0$) the covariance is negative and a trade-off is observed, but if variance in allocation is zero ($V_f = 0$) a positive correlation is observed, and the experimenter might incorrectly draw the conclusion that there is no underlying trade-off. From equation 3.33 it can be seen that the sign of the correlation depends upon the relative values of the component parameters and thus correct interpretation of data on the variability between X and Y will depend critically upon a correct interpretation of the functional relationship between them. Bull and Pease (1988) present a method for statistically disentangling effects of resource allocation from resource acquisition in the case of parental investment in sons versus daughters, but at present no general statistical tools are available. Experimental manipulation, if possible, can circumvent many of these statistical problems.

3.3.2. Experimental Manipulations

It is clearly preferable to examine a putative trade-off by manipulating the traits in question, at the same time keeping all other factors constant or randomly assigned to each group. For example, the costs of increasing brood size may be obscured if "better quality" females have larger broods. This problem can be overcome by randomly assigning broods, a relatively easy task with organisms such as birds. Similarly, the effect of reproduction on, say longevity, can be examined by allowing some individuals to mate, while others are kept virgin. Great care must be taken to ensure that the manipulation only involves a single change: increased longevity of virgins, for example, might be a consequence of a lack of interaction with the male rather than the physiological stress associated with reproduction. In an elegant study, Partridge and Farquah (1981) examined the cost of sexual activity in male fruit flies, *Drosophila melanogaster*, by maintaining two types of controls: one in which the males were kept isolated from females, and a second in which males were kept with newly mated females which are sexually unreceptive. There was no significant difference in longevity between the two categories of control, but a difference between virgin and mated males could be statistically detected when the correlation between body size and longevity was factored out. This study illustrates both the power of experimental manipulation and the care needed to isolate possible confounding factors.

The value of manipulative experiments over simple phenotypic correlations is demonstrated by the review of Reznick (1985) on the costs of reproduction: out of 33 phenotypic correlations only 22 (67%) suggested that reproduction bore a cost, while 17 of 20 (85%) manipulative experiments showed a statistically significant negative effect of reproduction.

3.3.3. Genetic Correlations From Sib Analysis

A direct measurement of the genetic basis of a trade-off is the estimation of the genetic correlation between the traits. First it is necessary to be assured that all the relevant interactions involved in the trade-off are measured. Failure to include relevant traits may give the appearance of a lack of a trade-off because some of those traits measured may have positive genetic correlations as a result of correlations with traits not measured. Genetic correlations can be estimated by a variety of breeding designs such as full-sib, half-sib, and parent-offspring regression (Becker 1985). The major problem is that enormous sample sizes are required to estimate the genetic correlation (Tallis 1959; VanVleck and Henderson 1961; Klein et al. 1973). A typical heritability for life history traits is 0.2 (see chapter 1): given two traits with heritabilities of 0.2, a family size of four and 100 families, the standard error of the genetic correlation in a full-sib experiment varies from 0.33 for $r_G = 0.2$ to 0.12 for $r_G = 0.8$. In the former case, over 800 families are required to produce a significant correlation. To obtain estimates with standard errors sufficiently small to be useful may be a daunting experimental task.

The phenotypic correlation is composed of the genetic and environmental correlation. (See equation 2.11.) Hegmann and DeFries (1970) suggested that the genetic and environmental correlations will themselves be correlated. These two observations raise the possibility that the phenotypic correlation will be correlated with the genetic correlation and might be used as an estimate of the latter. Analysis of three data sets reveals a significant positive correlation between the phenotypic and genetic correlations (Table 3.1), but the variability is high and in one set (*D. melanogaster*) the slope of the regression low. Cheverud (1988) argued that much of the discrepancy between genetic and phenotypic correlations stems from statistical inaccuracy due to small sample sizes: with large sample sizes the two estimates are similar and the phenotypic correlation can be taken as an upper estimate of the genetic correlation. Most of the traits analyzed by Cheverud were morphological, and while genetic and phenotypic correlations of such traits may generally have the same sign, this is certainly not true for correlations involving life history traits (Roff and Mousseau 1987). As a "rule of thumb," Cheverud (1988, p. 965) suggested that, "for a set of characters with a geometric mean heritability of 0.33, at least 120 families seem to be indicated." Cheverud's analysis has recently been criticized on methodological grounds (Willis et al. 1991), making this rule of thumb suspect. Even if correct the sample sizes recommended still represent a large investment of effort.

3.3.4. Genetic Correlations From Selection Experiments

A second method of estimating genetic correlations is to select on one trait and measure the response in a second. (See equation 2.13.) Selection

Table 3.1. Estimates of the relationship between the phenotypic and genetic correlations

Data set	r^2	slope	n
Diverse[a]	0.68	0.91	14
Daphnia[a]	0.35	0.96	41
Drosophila melanogaster[b]	0.12	0.63	53

[a]From Bell and Koufopanou (1985): primary source for "Diverse" is Table 19.1 of Falconer (1981), and for "*Daphnia*," Lynch (1984) and Bell (1984a).
[b]From Roff and Mousseau (1987).

experiments require fewer individuals but, depending upon the heritability of the trait and the selection intensity that can be applied, may require many (e.g., at least 10) generations of selection, making them labor-intensive and frequently long-term projects. A more serious problem with selection experiments is that, in general, artificial selection is practiced on only a single character, while natural selection operates on a suite of characters. Correlated responses may be due to unforeseen causes that are not appropriate in a natural setting. For example, there have been numerous attempts to select for changes in the phototactic behavior of *Drosophila melanogaster*: these experiments have generally been successful but detailed investigation of the mechanism producing changes in phototaxis has shown that, at least in some experiments, the actual focus of selection is the eye pigment—i.e., artificial selection has operated by producing flies that are visually impaired (Kohler 1977; Markow and Clark 1984). Similarly, responses for reduced geotactic and phototactic behavior can be a consequence of selection for flies that are physiologically inferior (Pyle 1976, 1978). Selection for such life history traits as longer development time or small size might also be successful because of selection for flies that have reduced fitness relative to controls (e.g., *Tribolium* experiments of Soliman 1972, 1982). Inferences from these types of selection experiments may have little relevance to a natural situation where selection would likely eliminate responses with such serious consequences on other components of fitness. This is not an argument against the use of selection experiments, but it is an argument against simplistic interpretations of the results. In many cases it may be preferable to practice selection in only one direction—e.g., for larger or faster-developing flies but not for smaller or slower-developing flies. More importantly, selection experiments should be accompanied, if feasible, by experiments designed to understand the functional basis of the response.

3.4. Summary

The three basic assumptions underlying life history theory are:

1. There is some measure of fitness that is maximized.
2. Constraints and trade-offs limit the set of possible life histories.
3. There is sufficient genetic variation to permit the attainment of the optimal combination.

Fitness measures can be divided into local and global measures. Local measures consider a subset of traits and maximize a component of fitness on the premise that this does not adversely affect other components of fitness. Global measures are those measures maximized by the interaction of all components of fitness. The most commonly used measures are the intrinsic rate of increase, r, expected lifetime fecundity, R_0, and reproductive value, $V(x)$.

In an unlimited, homogeneous, and constant environment the appropriate measure of fitness is r. In a stationary environment R_0 is the correct choice. Reproductive value is simply an alternative but equivalent definition of these statistics: it may be easier to work with in some cases. The appropriate measure of fitness in a temporally variable environment is the arithmetic average of r (= the geometric average of the finite rate of increase, λ). For spatially heterogeneous environments the correct measure is the arithmetic mean of λ. When a population is governed by density-dependent regulation, selection typically maximizes the total number in that age group subject to the density dependence.

The terms r- and K-selection were coined to describe selection in density-unregulated and density-regulated populations, respectively. However, the terms have come to be attached both to the type of selection operating and particular suites of characters hypothesized to be associated with these. So much confusion is now attached to the terms that they cloud rather than illuminate analyses. I suggest that these terms not be used, analyses being discussed in terms of density-independent and density-dependent selection.

In some cases, particularly situations involving behavioral interactions, the fitness of a particular trait is frequency-dependent. The optimal combination of traits in such circumstances is located using game theory. The additional fitness accruing to individuals by virtue of interactions with relatives is termed inclusive fitness. There is considerable confusion over the correct definition of inclusive fitness, but in practice fitness can be readily calculated using Hamilton's rule.

A fundamental assumption of life history theory is that phenotypic correlations reflect genetic correlations. Quantitative genetic theory predicts the same suite of traits at evolutionary equilibrium as the optimality model provided that the additive genetic values underlying the phenotypic variables obey the same functional constraints as assumed in the optimality model. At evolutionary equilibrium some of the genetic covariances among traits may be positive. Interpretation of genetic architecture with respect to optimal life histories must, therefore, be done cautiously. The best approach is to ground life history analyses in a functional understanding of the trade-offs that limit the set of possible combinations.

Trade-offs can be measured by phenotypic correlations, experimental manipulations, genetic correlations from sib analysis, and genetic correlations from selection experiments. The first are suspect because factors not considered might lead to correlations opposite in sign to those expected. Experimental manipulations can give insight into the functional basis of a trade-off but do not demonstrate that it has evolutionary significance. Genetic analysis can show that trade-offs are genetically based but satisfactory estimation of parameter values generally requires very large sample sizes. Care must also be exercised in interpreting genetic analyses both because, as with phenotypic correlations, not all relevant interactions may be considered, and selection experiments may not adequately mirror the operation of natural selection.

4

Methods of Analysis

The preceding chapters have laid the conceptual framework for the analysis of life history variation. The application of this framework requires a set of tools. This chapter describes those tools and how one compares the model constructed with the real world.

There is no single method of analysis, but those methods in life history analysis most frequently employed can be grouped into four categories: the calculus, graphical analysis, dynamic programming, and matrix methods. In the first portion of this chapter I shall consider the elements of these four approaches; in the second the statistical analysis of prediction and observation.

4.1. The Calculus

The calculus is probably the most important analytical tool for the exploration of problems in life history theory. Unfortunately, for a variety of reasons, not all problems can be tackled with this approach. The essential element in this type of analysis is differentiation. A brief description of all the rules of differentiation used in this book are given in the appendix.

Suppose we wish to find the optimal value of some life history variable, v, such as the age of first reproduction; i.e., that value of v (e.g., age at first reproduction) at which fitness is maximized. For there to be any effect of v on fitness this variable must influence either (or both) survival or fecundity. We can designate this fact by writing the $l(x)$ and $m(x)$ terms as the two functions $l(x, v)$ and $m(x, v)$. The characteristic equation is thus

$$1 = \int_0^\infty l(x, v)m(x, v)e^{-rx}\, dx$$

$$= F(v, r) \tag{4.1}$$

The first difficulty that may be encountered is that the integral cannot be solved. If this is the case one must turn to numerical methods, methods which are described in detail in Press et al. (1986). Assuming that the integral can be solved (the availability of computer programs for personal computers greatly eases the problem of integration), we obtain an equation in terms of v and r, $F(v, r)$. Now, assuming for simplicity that r can be easily extracted from $F(v, r)$, we rearrange equation 4.1 to give

$$1 = f(v) + r \qquad \textbf{(4.2)}$$

which upon further rearrangement gives

$$r = 1 - f(v) \qquad \textbf{(4.3)}$$

This equation defines the relationship between r and v, which should have a maximum at some intermediate value of v (Fig. 4.1). Several alternate possibilities might be obtained (Fig. 4.1). Fitness (r) might increase or decrease monotonically with v, in which case there is something amiss with the formulation of the problem: for example, if v is the age at first reproduction this result implies that selection should favor an individual that delays reproduction forever, or an individual that breeds at age zero. A fourth possibility is that r is independent of v, which may or may not signal an error in the construction of the relationships between the life history variables. Assuming that a local maximum is obtained, the value of v at which r is maximized is found by differentiating with respect to v and setting the differential equal to zero,

$$\frac{dr}{dv} = -\frac{d(f(v))}{dv} \qquad \textbf{(4.4)}$$

The value of v at which r is maximized is obtained by setting dr/dv equal

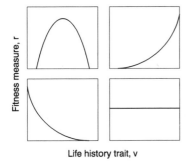

Fitness measure, r

Life history trait, v

Figure 4.1. Four possible relationships between fitness and a life history trait. In the upper left panel an optimal value exists; in the lower right panel fitness is independent of the trait; in the remaining panels fitness is a monotonically increasing or decreasing function of the trait and hence fitness is maximized at extreme values.

to zero and solving for v. If other traits might also vary, dr/dv may be replaced with the partial derivative $\partial r/\partial v$: this does not change the way the equation is differentiated, but simply signifies that the differentiation is made holding all other variables constant. A shorthand way of designating the derivative of some function $f(x)$ is to write it as $f'(x)$, which is taken to mean $df(x)/dx$. The second derivative is likewise written as $f''(x)$. This derivative determines whether the value of x obtained is a local maximum, minimum, or inflection. (See the appendix.) However, it is good policy to plot the relationship between r and v: this both indicates whether a local maximum occurs and also acts as a check on the above analysis, since the approximate value of v at which r is maximized can be read off the graph. In most instances r will not be so easily extracted as a separate factor, and implicit differentiation will be required. The following examples illustrate the methodology outlined above.

4.1.1. Example 1

This and the following example demonstrate that not all analyses will necessarily result in a reasonable biological solution. The value of constructing a model is precisely that it enables one to check the logic of the proposed life history. Suppose the organism in question is semelparous and we wish to predict the optimal age at reproduction. The rate of increase, r, is given by

$$r = \frac{\log_e l(\alpha) m(\alpha)}{\alpha} \tag{4.5}$$

where α is the age at reproduction. Our data indicate that both survival and fecundity are size- and hence age-dependent, with the best fits being

$$l(\alpha) = e^{-c_1\alpha} \tag{4.6}$$

$$m(\alpha) = e^{c_2\alpha} \tag{4.7}$$

where the c_i is a constant. Thus we obtain

$$r = \frac{\log_e(e^{-c_1\alpha}e^{c_2\alpha})}{\alpha}$$

$$= \frac{-c_1\alpha + c_2\alpha}{\alpha}$$

$$= -c_1 + c_2 \tag{4.8}$$

which means that r is independent of α! This is an unlikely situation and suggests that something is wrong.

4.1.2. *Example 2*

Same situation as above but the fecundity function is

$$m(\alpha) = c_3 e^{c_2\alpha} \tag{4.9}$$

Proceeding as before

$$r = \frac{\log_e(e^{-c_1\alpha} c_3 e^{c_2\alpha})}{\alpha}$$

$$= \frac{-c_1\alpha + \log_e c_3 + c_2\alpha}{\alpha}$$

$$= -c_1 + c_2 + \frac{\log_e c_3}{\alpha} \tag{4.10}$$

In this case r declines monotonically with α: clearly this formulation is still unlikely to be correct.

4.1.3. *Example 3*

A more likely relationship between age and fecundity (Fig. 4.2) is

$$m(\alpha) = c_2\alpha^{c_3} \tag{4.11}$$

where the subscripts to the constants are not intended to imply that they take the same values as in the previous example. Now we obtain

$$r = \frac{\log_e e^{-c_1\alpha} c_2\alpha^{c_3}}{\alpha}$$

$$= \frac{-c_1\alpha + \log_e c_2 + c_3 \log_e\alpha}{\alpha}$$

$$= -c_1 + \frac{\log_e c_2}{\alpha} + \frac{c_3 \log_e\alpha}{\alpha} \tag{4.12}$$

Differentiating (see appendix for rules)

$$\frac{dr}{d\alpha} = -0 - \frac{\log_e c_2}{\alpha^2} + \left(\frac{c_3}{\alpha} \times \frac{1}{\alpha}\right) - \frac{c_3 \log_e\alpha}{\alpha^2}$$

$$= \frac{1}{\alpha^2}(-\log_e c_2 + c_3 - c_3 \log_e\alpha) \tag{4.13}$$

There are two values at which $dr/d\alpha$ equals zero

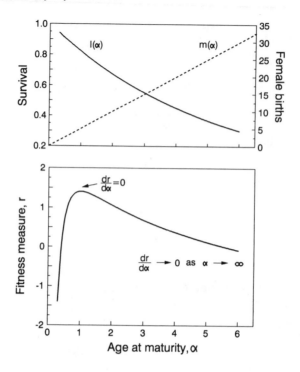

Figure 4.2. A hypothetical life history for a semelparous organism. See text for details. Parameter values are $c_1 = 0.2$, $c_2 = 5$,
$c_3 = 1$.

$$\alpha = \infty$$

$$\alpha = e^{\frac{\log_e c_2 - c_3}{c_3}} \qquad (4.14)$$

The first results from the asymptotic approach of r to $-c_1$ as α tends to
infinity (Fig. 4.2); it is only the second that is relevant.

4.1.4. Example 4

Suppose the population is stationary and we choose to take as our fitness
measure the expected lifetime fecundity, which for a semelparous organism
is

$$R_0 = l(\alpha)m(\alpha) \qquad (4.15)$$

Taking the relationships used in example 2, R_0 is

$$R_0 = e^{-c_1 \alpha} c_3 e^{c_2 \alpha}$$

$$= c_3 e^{\alpha(-c_1 + c_2)} \tag{4.16}$$

It is immediately obvious that there is no local maximum in the above equation: depending on the value of $(-c_1 + c_2)$, R_0 either increases or decreases monotonically (note that unlike example 1, R_0 is not independent of α when c_3 is equal to 1).

4.1.5. Example 5

Taking the relationships used in example 3 and assuming that R_0 is maximized we have

$$R_0 = e^{-c_1 \alpha} c_2 \alpha^{c_3} \tag{4.17}$$

Differentiating gives

$$\frac{dR_0}{d\alpha} = c_1 e^{-c_1 \alpha} c_2 \alpha^{c_3} + e^{-c_1 \alpha} c_2 c_3 \alpha^{c_3 - 1}$$

$$= e^{-c_1 \alpha} c_2 \alpha^{c_3 - 1}(-c_1 \alpha + c_3) \tag{4.18}$$

As before, the derivative is zero at two values of α, but the only relevant value is $\alpha = c_3/c_1$.

4.1.6. Example 6

The following model is taken from Loman (1982). This model attempts to predict the optimal time in the breeding season that a bird should start laying, and the optimal clutch size. Because the female must accumulate resources it is assumed that the total number of eggs a bird can lay is proportional to the delay in the start of laying,

$$f(t) = c_1 t \tag{4.19}$$

where $f(t)$ is the number of eggs that a female can lay if she postpones egg laying until day t of the breeding season. Loman next assumes that incubation does not start until the last egg is laid and that the survival of the offspring is dependent on the amount of time before the end of the breeding season,

$$S(t) = T - c_2(t + c_1 t) \tag{4.20}$$

where $S(t)$ is the survival of young when the female lays its first egg on day t and the term in parentheses is the time at which incubation begins (the time of the first egg, t, plus the time taken to lay all the eggs, c_1t, assuming one egg laid per day). The number of eggs laid and young raised is assumed not to affect the future survival or reproduction of the parent, in which case fitness can be equated to the total number of surviving young in a single breeding season, designated as production, P

$$Production = P = c_1t(T - c_2t - c_1c_2t)$$

$$= c_1Tt - (c_1c_2 + c_1^2c_2)\, t^2 \tag{4.21}$$

Differentiating

$$\frac{dP}{dt} = c_1t - 2t(c_1c_2 + c_1^2c_2) \tag{4.22}$$

Setting dP/dt equal to zero gives the optimal time to start egg laying to be

$$t = \frac{T}{2(c_2 + c_1c_2)} \tag{4.23}$$

4.1.7. Example 7

This example is drawn from Roff (1984a) and is discussed in detail in chapter 7. The problem is to find the optimal age at maturity assuming that selection maximizes r. The model is based upon the life history of fish, though it may have much broader applicability. Mortality comprises two components, an initial mortality during the egg and larval stages which occurs in such a short period as to be considered a "point" event in time, and a later constant source. Thus we have

$$l(x) = pe^{-Mx} \tag{4.24}$$

where p is the proportion surviving the egg and larval stage, and M is the rate of mortality thereafter. Fecundity is proportional to the cube of length, and length increases according to the function $L_\infty(1 - e^{-kx})$ (this describes a concave curve that approaches L_∞ asymptotically; see Fig. 5.6). If the species is semelparous the rate of increase, r, is given by

$$r = \frac{\log_e(l(\alpha)m(\alpha))}{\alpha}$$

$$= \frac{\log_e(e^{-M\alpha}c_1(1 - e^{-k\alpha})^3)}{\alpha} \tag{4.25}$$

where c_1 comprises the product of the proportionality constant in the fecundity/body size function (halved since only female births are considered), the survival fraction p, and the asymptotic length L_∞. For simplicity the terms can be separated:

$$r = \frac{\log_e c_1}{\alpha} + \frac{3 \log_e(1 - e^{-k\alpha})}{\alpha} - M \tag{4.26}$$

Differentiating

$$\frac{dr}{d\alpha} = \frac{-\log_e c_1}{\alpha^2} + \frac{3ke^{-k\alpha}}{(1 - e^{-k\alpha})\alpha} - \frac{3 \log_e(1 - e^{-k\alpha})}{\alpha^2}$$

$$= \frac{1}{\alpha^2}\left(-\log_e c_1 + \frac{\alpha 3\, ke^{-k\alpha}}{(1 - e^{-k\alpha})} - 3 \log_e(1 - e^{-k\alpha}) \right) \tag{4.27}$$

and $dr/d\alpha$ equals zero when the term to the right of $1/\alpha^2$ inside the outer parentheses is equal to zero. It is not possible to separate α from the rest of the terms and its value must be found by numerical methods.

4.1.8. Example 8

In all the examples thus far it has been possible to arrange the equations such that the fitness measure is on one side of the equation and the variable of interest is on the other. This frequently is not the case, and we have to use implicit differentiation. (See the appendix.) The following example illustrates this approach. The model is basically the same as that of example 7; but we assume iteroparity and incorporate a cost of reproduction by assuming that survival and/or growth is decreased by reproduction such that the $l(x)m(x)$ curve can be described by

$$l(x)m(x) = c_1 e^{-Mx}(1 - e^{-k\alpha}) \tag{4.28}$$

One interpretation of this equation is that growth ceases at maturity. But it must be remembered that the equation describes the product $l(x)m(x)$, and it can also be taken to mean that any increment in fecundity accruing from an increase in length is negated by an increase in mortality. (The issue of ascribing unique biological meaning to equations is discussed in greater detail in chapters 6 and 7.) The characteristic equation is

$$\sum_{x=\alpha}^{\infty} c_1 e^{-rx} e^{-Mx}(1 - e^{-k\alpha})^3 = 1 \tag{4.29}$$

Using the standard formula for a geometric series,

$$\sum_{n=1}^{\infty} a^{n-1} = \frac{1}{1 - a} \qquad where \; |a| < 1$$

we obtain

$$\frac{e^{-\alpha(r+M)}(1 - e^{-k\alpha})^3 c_1}{1 - e^{-(r+M)}} = 1 \qquad (4.30)$$

The above can be simplified before differentiating by taking the natural logarithm and, for convenience, multiplying throughout by -1

$$\alpha r + \alpha M + \log_e(1 - e^{-(r+M)}) - \log_e c_1$$

$$- 3 \log_e(1 - e^{-k\alpha}) = 0 \qquad (4.31)$$

It is clear that r cannot be separated in the above equation; therefore we differentiate implicitly (see appendix),

$$r + \alpha\frac{dr}{d\alpha} + M + \frac{e^{-(r+M)}}{1 - e^{-(r+M)}} \cdot \frac{dr}{d\alpha} - 0 - \frac{3e^{-k\alpha}}{1 - e^{-k\alpha}} = 0$$

Rearranging

$$\frac{dr}{d\alpha}\left(\alpha + \frac{e^{-(r+M)}}{1 - e^{-(r+M)}}\right) = \frac{3ke^{-k\alpha}}{1 - e^{-k\alpha}} - r - M \qquad (4.32)$$

Now, provided the term in brackets on the left-hand side of the equation does not equal zero when $dr/d\alpha$ equals zero we can obtain the value of α at which r is maximized by setting the right-hand side equal to zero, giving

$$r = \frac{3ke^{-k\alpha}}{1 - e^{-k\alpha}} - M \qquad (4.33)$$

Note that we still have not arrived at a solution since we now know what value r takes at its maximum, but not yet the value of α. To obtain this we substitute for r in equation 4.31 to finally arrive at

$$\alpha G + \log_e(1 - e^{-G}) - \log_e c_1 - 2 \log_e(1 - e^{-k\alpha})$$

$$where \quad G = \frac{3ke^{-k\alpha}}{1 - e^{-k\alpha}} \qquad (4.34)$$

from which the value of α can be obtained numerically, the values of all other parameters being known.

The predicted optimal age at first reproduction is different for the semelparous and iteroparous cases. If the fitness measure chosen is R_0 the analysis is greatly simplified and, interestingly, the same result is obtained for both semelparous and iteroparous life histories (Roff 1984a).

4.2. Graphical Analysis

The use of a graphical representation can sometimes very clearly indicate the consequences of a set of interactions, but should not be used to obtain quantitative predictions, since these are more accurately obtained using the calculus. As a simple example consider the following problem: two traits v_1 and v_2 (e.g., egg size and number) covary in some manner as shown in Fig. 4.3, the functional relationship between the two being designated as $v_2 = f(v_1)$. Any combination of v_1 and v_2 that lies below the curve is permissible but combinations above the curve are not feasible. This constitutes a "fitness set" (Levins 1962, 1968). Fitness is a function of both traits; for simplicity let this be a linear relationship such as $r = c_1 v_1 - c_2 v_2$. On the plane defined by v_1, v_2 we can construct lines (referred to as isopleths or isoclines) of equal values of r (lower right panel, Fig. 4.3). The combination of v_1 and v_2 that maximizes fitness is the combination at which a fitness isocline is tangent to the curve $f(v_1)$. If the fitness set is concave, intermediate values of v_1 and v_2 will have the highest fitness, but with the convex or linear functions selection will favor extreme values (Fig. 4.3).

The graphical technique very clearly illustrates why an intermediate set of values may be favored. To calculate what the appropriate combination is we proceed as follows: we have

$$v_2 = f(v_1)$$

$$r = c_1 v_1 - c_2 v_2$$

$$= c_1 v_1 - c_2 f(v_1) \tag{4.35}$$

The above can now be differentiated with respect to v_1 (since we are interested in two traits the partial derivative is taken)

$$\frac{\partial r}{\partial v_1} = c_1 - c_2 f'(v_1) \tag{4.36}$$

and the optimum combination is found by setting $\partial r / \partial v_1$ equal to zero. It

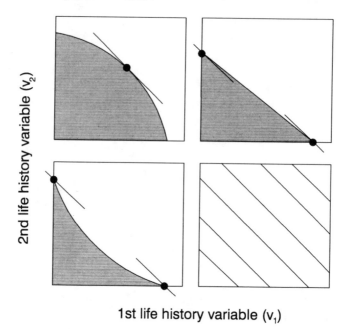

Figure 4.3. Three possible trade-off curves between two life history traits, v_1 and v_2. Stippled regions denote combinations that are permissible. Isoclines of fitness are plotted in lower right panel, with fitness increasing with increasing values of v_1 and v_2. For the trade-off shown in the upper left panel the combination (circle) that maximizes fitness is at the point at which the fitness isocline is tangent to the trade-off boundary. In the other two cases no intermediate optimum occurs, the largest fitness isocline intersecting with the trade-off boundary at the two extremes; both are plotted in the figure but only one will maximize fitness, in the present cases the extreme value of v_1.

is, of course, not necessary to do the graphical analysis first, though this can be instructive.

4.2.1. Example 1

Sibly and Monk (1987) observed considerable intraspecific variation in egg weight of two grasshopper species. They further noted that egg weight decreased with time from hatching to first egg laid (Fig. 4.4) and speculated that differences among sites and species might, because of this trade-off, favor different-sized eggs. To test this idea they developed the following

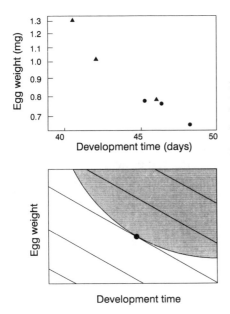

Figure 4.4. Upper panel: Relationship between egg weight and development time in two species of grasshopper, *Chorthippus parallelus* (triangles) and *C. brunneus* (circles), from three different sites. Lower panel: Trade-off between egg weight and development time. The stippled area designates possible combinations. Isoclines of fitness increase with decreases in the two life history traits, the optimal combination being the point (circle) at which the largest isocline is tangent to the trade-off boundary. (Modified from Sibly and Monk 1987.)

model (I have modified slightly the fitness criterion. This does not affect the mathematics.)

Adult and nymphal grasshoppers cannot overwinter, and hence, at the beginning of the season suitable for growth and reproduction, the population consists of a cohort of overwintered eggs. A univoltine life cycle is assumed (i.e., only one generation a year with all eggs entering diapause). Since we are interested only in the mean size of an egg within a population we can consider a population to comprise a range of noninterbreeding genotypes. The fitness of a particular female is then the number of female eggs laid by that individual that enter the next overwintering resting stage (Roff 1980). The life history parameters considered in the model are

1. M_e, mortality rate in egg stage.
2. M_n, mortality rate in nymphal stage.
3. M_a, mortality rate in adult stage.
4. t_e, time taken for eggs to hatch, time being taken from the end of the previous season. (This is arbitrary: since development is halted until diapause is broken by cold weather all eggs start development synchronously.) Since this parameter plays no role in the optimal strategy it could be eliminated and time could be considered to begin at the time of hatching in the spring; it is retained in the following to be consistent with the derivation by Sibly and Monk.

6. t_I, time interval between egg batches.
7. N, the maximum number of egg batches a female can lay.

If the number of eggs per batch is b, then the number of eggs from the first clutch that survive to overwinter is

$$\frac{1}{2} bSe^{-TM_e} \tag{4.37}$$

where T is the time from the first clutch to the end of the breeding season. The probability of survival from this point to first reproduction, S, is

$$S = e^{-(M_e t_e + M_n t_n)} \tag{4.38}$$

The probability that a female survives the interclutch interval (t_I) is

$$e^{-M_a t_I} \tag{4.39}$$

Applying the rules of probability the expected number of female eggs from clutch i, $E(i)$, is

$$\text{clutch 1} \quad E(1) = \frac{1}{2} bS(1)e^{-M_e T}$$

$$\text{clutch 2} \quad E(2) = \frac{1}{2} bS(1)e^{-M_a t_I} e^{-M_e(T - t_I)}$$

$$\text{clutch 3} \quad E(3) = \frac{1}{2} bS(1)e^{-2M_a t_I} e^{-M_e(T - 2t_I)} \tag{4.40}$$

The above form a geometric series, from which it can be ascertained that the expected number of overwintering eggs is

$$\frac{\frac{1}{2} be^{-M_e(T + t_e) - M_n t_n}(1 - e^{N(M_e - M_a)t_I})}{1 - e^{(M_e - M_a)t_I}} \tag{4.41}$$

The next step is to relate egg weight to the number of eggs that can be produced and the time from hatching to first oviposition. Sibly and Monk make the reasonable assumption that if resources are accumulated at a rate k then a total of kT units will be accumulated over the whole season, to be divided among N clutches each containing b eggs of weight w. The number of eggs per batch is thus

$$b = \frac{kT}{mx} = \frac{kt_I}{x} \tag{4.42}$$

This is the heart of the proposed trade-off: larger eggs reduce the time from hatching to first egg and hence increase the likelihood of surviving to lay eggs, but smaller eggs permit a larger number of eggs to be laid. Substituting equation 4.42 into 4.41 and noting that $mt_I = T$ gives

$$\frac{\frac{1}{2} kt_I e^{-M_e(T + t_e) - M_n t_n}(1 - e^{T(M_e - M_a)})}{x(1 - e^{(M_e - M_a)t_I})} \tag{4.43}$$

The time, T, from the time the first clutch is laid to the end of the breeding season is clearly not independent of the development time, t_n. Letting B be the length of the "breeding" season, defined as the time from hatching of the female in the spring to the end of the season (due to frost), we can write $T = B - t_n$. Substituting in equation 4.43 and taking logarithms,

$$R_0 = c + M_e t_n - M_n + \log_e(1 - e^{(M_e - M_a)(B - t_n)})$$
$$- \log_e x - \log_e(1 - e^{(M_e - M_a)t_I}) \tag{4.44}$$

The terms not including x or t_n have been gathered together into a single constant,

$$c = \log_e \frac{k}{2} + \log_e t_I - M_e B - \log_e(1 - e^{(M_e - M_a)t_I}) \tag{4.45}$$

Equation 4.44 defines the isoclines of equal fitness for combinations of x and t_n. This can be seen by rearranging the equation, for convenience letting $z = \log_e t_n$,

$$z = -R_0 + c + t_n(M_e - M_n) + \log_e(1 - e^{(M_e - M_a)(B - t_n)}) \tag{4.46}$$

Except for the last term, z is a linear function of t_n, the elevation depending on the value of R_0. In fact the last term has little influence on the curvature for observed values of the parameters. The trade-off curve of z versus t_n is convex with the permissible values lying above the curve (i.e., a small egg size can give rise to a generation time that has some minimum value). The optimal combination is thus given by the tangent of the fitness isoclines to the trade-off curve (Fig. 4.4).

The optimal combination is thus given by the tangent of the fitness isoclines to the trade-off curve (Fig. 4.4).

To find the combination more exactly we first note that z can be written as a function of t_n, say $z = f(t_n)$. R_0 is differentiated with respect to t_n and $\partial R_0 / \partial t_n$ equated to zero; z can then be found by its relationship with t_n. Differentiating

$$\frac{\partial R_0}{\partial t_n} = \frac{\partial z}{\partial t_n} + M_e - M_n + \frac{M_e - M_a}{1 - e^{(M_e - M_a)(B - t_n)}} \qquad (4.47)$$

Note that the actual functional relationship between z and t_n must be specified. Sibly and Monk actually took a slightly different approach at this point, which is discussed in detail in chapter 10, where the application of the model to the data is examined.

4.2.2. Example 2

This example is drawn from Lloyd (1987), his model being based on that of Smith and Fretwell (1974). Lloyd assumed that a measure of fitness is the number of offspring from a single clutch that survive to reproduce. For a semelparous organism this is, of course, R_0, but for an iteroparous organism the assumption underlying the fitness measure is that the survival or future clutch size of the parent is not influenced by its reproductive allocation. Let the component of fitness be designated w and define the function, $f(q)$, that relates the survival of offspring to investment per offspring, q. If the total amount of resources to be allocated is E then $q = E/N$, where N is the number of offspring. Fitness is thus

$$
\begin{aligned}
w &= \text{Number of offspring} \times \text{Survival of offspring} \\
&= Nf(q) \\
&= \frac{Ef(q)}{q} \qquad (4.48)
\end{aligned}
$$

Rearranging to obtain the relationship between survival and investment per offspring for a constant fitness value,

$$f(q) = q\frac{w}{E} \qquad (4.49)$$

Thus the lines of equal fitness radiate out from the origin with the slope increasing with increasing values of w (Fig. 4.5, upper panel). The relationship between survival and investment is likely to be approximately

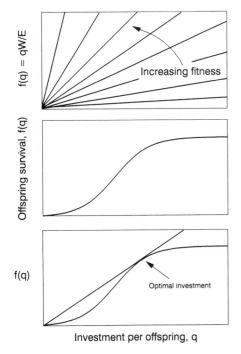

Figure 4.5. Upper panel: Isoclines of fitness in a hypothetical model relating fitness to investment per offspring. Middle panel: Offspring survival increasing as a sigmoidal function of investment in the offspring. Bottom panel: The optimum investment is found by superimposing the above two panels and finding that point at which the fitness isocline is tangent to the offspring survival curve.

sigmoidal (Lloyd 1987), offspring survival being very low if the female places little investment in the offspring (e.g., little yolk) but increasing to a plateau with large investments (Fig. 4.5, middle panel). The optimal combination is given where the fitness isocline is tangent to the trade-off curve (Fig. 4.5, lower panel). The actual value of q is obtained via the calculus

$$\frac{\partial w}{\partial q} = \frac{Ef'(q)}{q} - \frac{Ef(q)}{q^2}$$

$$= \frac{E}{q}\left(f'(q) - \frac{f(q)}{q}\right)$$

$$= 0 \quad \text{when}$$

$$q = \frac{f'(q)}{f(q)} \tag{4.50}$$

4.2.3. Example 3

For an iteroparous organism a critical "decision" is the amount of energy to allocate to reproduction at each breeding cycle. Schaffer (1974a) demonstrated with a simple graphical approach some of the circumstances under which an intermediate investment will be favored. Three assumptions are that reproductive effort and survival rate do not change with age and that fecundity is either constant or grows geometrically (e.g., fecundity might be proportional to size and size increases at some constant rate, generating a geometric series in size and fecundity). Under these assumptions it can be shown (Schaffer 1974a, p. 293; see also chapter 8, section 8.1) that the finite fate of increase, λ, is

$$\lambda = b(E) + s(E)f(E) \tag{4.51}$$

where $b(E)$ is the fecundity of a zero-year-old, $s(E)$ is the survival rate, and $f(E)$ is the fecundity: all are dependent on the amount of reproductive effort, E. Schaffer suggested three possible relationships between $b(E)$ and the product $s(E)f(E)$ (Fig. 4.6). The trade-off between these components can be seen by eliminating the common variable, reproductive effort, to give the graphs shown down the right-hand side of Fig. 4.6. Equation 4.51 can also be rearranged to give $s(E)f(E)$ as a function of $b(E)$ and λ,

$$b(E) = \lambda - s(E)f(E) \tag{4.52}$$

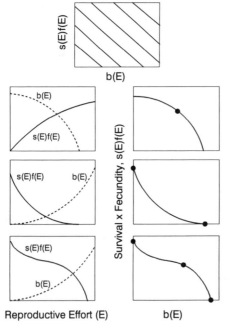

Figure 4.6. Predicting the optimal allocation of reproductive effort. Top panel shows fitness isoclines between the two component functions that determine fitness, $\lambda : \lambda = b(E) + s(E)f(E)$, where E is reproductive effort. Lower three panels on left show some possible relationships between these two components and reproductive effort. Since reproductive effort is common to both components they can be plotted as functions of each other (lower right panels). Superimposition of the fitness isoclines shows that various equilibria are possible. These are indicated by circles, but in each case only a single combination will maximize fitness.

Thus the isoclines of equal fitness are linear, increasing in size as they move away from the origin (Fig. 4.6). The optimal reproductive allocation may be intermediate (Fig. 4.6, upper right panel) or extreme (Fig. 4.6, middle panel) or may depend upon not only the shape of the trade-off but also its position (Fig. 4.6, lower right panel). As before, exact values are obtained by differentiation:

$$\frac{\partial \lambda}{\partial E} = b'(e) + s'(E)f(E) + s(E)f'(E) \qquad (4.53)$$

The results obtained graphically can be obtained by inspection of the above components but they are certainly not as "immediately obvious."

4.2.4. Example 4

The marginal-value theorem (Charnov 1976) has been extensively used to analyze optimal foraging decisions. More recently it has been marshalled to analyze optimal patterns of reproductive effort (Patterson et al. 1980; Townsend 1986) and the evolution of optimal balance between egg and clutch size (discussed in detail in chapter 10). The method can be illustrated using a generalized version of the problem studied by Townsend (1986). Parental care is practiced by males of the Puerto Rican frog, *Eleuthero-dactylus coqui*. These males brood their eggs, keeping them moist and defending them from cannibalistic intruders. But while brooding, the males cannot obtain further mates and hence cannot increase their complement of eggs. There are two components to this problem: first, there is a benefit curve, $f(t)$, represented by the probability of hatching as a function of the time spent brooding (t); and second, a cost, C, represented by the time required to remate, once parental care is abandoned. The benefit curve is most likely to be sigmoidal or concave in shape (Fig. 4.7). Fitness, w, is measured as the proportion of hatching offspring per unit time. This value is found graphically by drawing a line from $-C$ tangent to the benefit curve: the value of t at this point is the optimum brooding time (Fig. 4.7).

The above proposition can be proved as follows: the proportion of hatching offspring per unit time, w, is equal to

$$w = \frac{f(t)}{C + t} \qquad (4.54)$$

To find the optimum brooding time we differentiate the above

$$\frac{dw}{dt} = \frac{f'(t)}{C + t} - \frac{f(t)}{(C + t)^2}$$

$$= \frac{1}{C + t}\left(f'(t) - \frac{f(t)}{C + t}\right) \qquad (4.55)$$

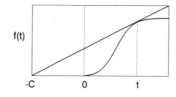

f(t)

-C   0   t

Figure 4.7. The marginal value approach to the analysis of parental care. The benefit curve, $f(t)$, is sigmoidal in this example and represents the proportion of offspring that survive given some parental care t. Fitness is defined as the production of offspring per unit time. This is determined by the proportion of eggs surviving and the time lost between matings. To locate the investment that will maximize offspring production per unit time a line is drawn from the point $-C$, C being the time taken to remate, to a point tangent to the benefit curve: the value of t at this point is the optimal time to invest in parental care.

where $f'(t)$ is simply the shorthand method of writing $df(t)/dt$. The optimum brooding time is when $dw/dt = 0$, i.e., when

$$f'(t) = \frac{f(t)}{C + t} \qquad (4.56)$$

The left-hand side represents the tangent of a line to the curve, $f(t)$, and the right-hand side is that point at which a line drawn from $-C$ is tangent to $f(t)$. Thus the graphical analysis is proven.

4.3. Dynamic Programming

Optimal-control theory has been extensively applied to management problems in ecology but has been relatively little used in the analysis of problems in life history evolution. The two most frequently employed methods are Pontryagin's Maximum Principle, and dynamic programming (Rosen 1967). The former has been most commonly applied to the analysis of how reproductive effort should vary with age (Léon 1976; Goodman 1981, 1982; Schaffer 1983). Dynamic programming has more recently been used quite extensively in behavioral ecology where sequences of choices are involved (Mangel 1987; Mangel and Clark 1988). Both techniques are similar in their underlying mathematical principles. However, application of the Maximum Principle generally involves numerous technical difficulties which can be resolved in general only for very simple cases or by numerical methods (Mangel and Clark 1988). Dynamic programming is both easier to understand and easier to implement and is likely to be broadly applicable to evolutionary problems that involve a sequence of actions such as variation in reproductive effort.

both easier to understand and easier to implement and is likely to be broadly applicable to evolutionary problems that involve a sequence of actions such as variation in reproductive effort.

Under what circumstances is dynamic programming useful? Consider an iteroparous organism that grows according to the function

$$W(x + 1) = W(x) + G(x) - E(x) \qquad (4.57)$$

where $W(x + 1)$ is size (e.g., weight) at age $x + 1$,
   $G(x)$ is the amount of growth possible from x to $x + 1$ in the absence of reproduction, and
   $E(x)$ is the amount of energy, measured in the currency of size, diverted into reproduction at age x.

Further, suppose survival is a function of reproductive effort, the larger the effort the lower the survival rate, and that fecundity is an increasing function of size. For clarity of exposition it is convenient to give particular expressions for these functions. Let survival at age x, resulting from expenditures on reproduction, be a linear decreasing function of reproductive effort, $1 - c_1E(x)$. Mortality is assumed to occur immediately prior to reproduction; thus the number of eggs produced at age x is the product of survival and fecundity. (There will also be a mortality rate that occurs independently of reproductive allocation: this can be ignored in the present example.) Typically fecundity will be both a function of size, which is a measure of past commitments to reproduction, and present reproductive allocation. A simple representation of this is to assume that fecundity at age x is proportional to the product of size and reproductive effort at that age, $c_2W(x)E(x)$.

At any age, an organism has to trade-off future growth and fecundity against high fecundity at the present but low survival. Assuming that fitness can be equated with the total lifetime fecundity, how do we calculate the optimal allocation at each age? The number of different strategies that have to be considered proceeding from birth to death is potentially enormous. For example, suppose that only two growth options are possible: all energy into growth, or all energy into reproduction. If the maximum age is T, there are 2^T possible trajectories. In practice there are an infinite number of possible allocation patterns and hence an infinite number of trajectories to consider if we approach the problem with a numerical "sledgehammer."

There is no way to estimate the present optimal allocation pattern without considering all future consequences. But suppose we approach the problem in reverse. Consider the options of an organism at its final age T and at some designated weight $W(T)$: since it is at the end of its life the final contribution to lifetime reproductive success, denoted as $F(T)$, is

$$F'(T) = c_2 W(T)(1 - 2c_1 E(T))$$

$$= 0 \quad \text{when} \quad E(T) = \frac{1}{2c_1} \tag{4.58}$$

Thus we have the optimal allocation at the final age. Now consider the situation at age $T - 1$,

$$W(T - 1) = W(T) - G(T - 1) + E(T - 1) \tag{4.59}$$

The expected number of eggs produced at age $T - 1$, $F(T - 1)$, is

$$\begin{aligned}
F(T - 1) &= (1 - c_1 E)c_2 W(T - 1)E \\
&= (1 - c_1 E)c_2(W(T) - G + E)E \\
&= c_2 E(W(T) - G + E) - c_1 c_2 E^2(W - G + E) \\
&= c_2(W(T) - G)E + c_2(1 - c_1 W(T) + c_1 G)E^2 - c_1 c_2 E^3 \tag{4.60}
\end{aligned}$$

where for ease of presentation the $T - 1$ has been dropped from $G(T - 1)$ and $E(T - 1)$. The only term in the above equation that is not known is $E(T - 1)$, the parameter that we wish to evaluate. To do this we differentiate, obtaining

$$F'(T - 1) = c_2(W(T) - G)$$

$$+ 2c_2(1 - c_1 W(T) + c_1 c_2)E - 3c_1 c_2 E^2 \tag{4.61}$$

The above is a quadratic equation and hence the value of $E(T - 1)$ for which the expression is zero is easily found. Thus we now have the optimal allocations at ages T and $T - 1$. The process can be repeated in exactly the same manner for $T - 2$, $T - 3$, all the way back to the first year. By back recursion, therefore, we can find the optimal age-specific allocation pattern *given some terminal size* $W(T)$. The procedure is repeated for a range of terminal size from which the relationship between R_0 and terminal size is obtained; that terminal size that produces the highest expected lifetime fecundity is the optimal allocation sequence.

Dynamic programming is a very powerful tool of numerical analysis, greatly reducing the number of possible scenarios that must be considered. Mangel and Clark (1988) provide numerous examples from behavioral ecology on the use of dynamic programming. They provide detailed discussion on the construction of the algorithm and its implementation. The general model can be described as follows: lifetime fitness is described by the function $F(x, t, T)$, where x is the value of the state variable $X(t)$, at time t, and T is the final time. In the example described above the state variable is size and time is equivalent to age. The contribution to fitness at time T is given by the function $F(x, T, T) = f(x)$, ($F(T)$ in the foregoing example). In mathematical notation the backward recursion procedure can be written as

$$\text{Fitness} = \sum_{t=0}^{T-1} R(X(t), A(t), t) + f(X(T)) \tag{4.62}$$

where $A(t)$ is the "action" taken at time t (e.g., the allocation to reproduction), and R is the "reward" for taking action A when the state variable is X at time t (e.g., the expected fecundity at age t). The state variable $X(t)$ is a function of previous actions (here attention is restricted to the immediately preceding state), states, and possibly time,

$$X(t + 1) = G(X(t), A(t), t) \tag{4.63}$$

The general goal is to find the sequence of actions that maximizes fitness. First, we note that at time $T - 1$ the total contribution to lifetime fitness is the reward at time $T + 1$ plus the final contribution $F(x, T, T)$, where the state variable $X(t)$ takes the value x. Now the value of the state variable at time T is given by

$$X(T) = G(x, A, T - 1) \tag{4.64}$$

Hence the total fitness from time $T - 1$ is

$$R(x, A, T - 1) + f(G(x, A, T - 1), T, T) \tag{4.65}$$

from which the optimal value of A can be found, either analytically or numerically. (The above equation corresponds to equation 4.60.) The process can then be repeated for $T - 2$, $T - 3$, etc. The basic computer coding to solve the dynamic programming equation is given in Mangel and Clark (1988, p. 228).

Thus far, fitness has been defined as the expected lifetime production of offspring, R_0. Use of the rate of increase, r, as a measure of fitness presents the difficulty that r must also be estimated. Goodman (1981) provides the following method. First we note that maximizing r is equivalent to maximizing (Schaffer 1974a, 1983)

$$m(x) + p(x) \frac{V(x + 1)}{V(1)} \tag{4.66}$$

where $m(x)$ is the number of female offspring born at age x,
$p(x)$ is the probability of surviving from age x to age $x + 1$,
$V(x)$ is the reproductive value at age x. I have followed Goodman (1982) in designating the first age group as 1. (See chapter 3 for a discussion of the problem of correct notation.) Reproductive value is scaled in units of female births so that $V(1) = 1$. Equation 4.66 can be expanded to

$$m(x) + p(x)V(x + 1)$$

$$= m(x) + p(x)e^{rx} \sum_{i=x+1}^{T} e^{-rx}m(x) \prod_{j=x+1}^{i-1} p(j) \qquad (4.67)$$

For a given value of r the optimal values of $m(x)$ and $p(x)$ (both functions of reproductive effort) can be obtained by proceeding backward using the dynamic programming approach. To deal with the problem of finding the correct value of r we define a function $g(x,s)$

$$g(x, s) = m(x) + p(x) \sum_{i=x+1}^{T} e^{s(i-x)}m(x) \prod_{j=x+1}^{i-1} p(j) \qquad (4.68)$$

This equation can be written in recursive form,

$$g(x, s) = m(x) + p(x)e^{-s}g(x + 1, s) \qquad (4.69)$$

If s is equal to r, then $g(1, s) = e^s$. The method of solution is to find the life table that maximizes the schedule g for different values of s; that one which satisfies the equality $g(1, s) = e^s$ is the optimal life history (Taylor et al. 1974; Goodman 1982).

Ydenberg (1989) used dynamic programming to obtain the optimal fledging age and mass of the common murre, *Uria aalge*. The problem faced by the murre chick is that while survival in the nest is greater than at sea, its growth rate is slower in the nest, and the probability of surviving to breed is an increasing function of the size at the end of the first summer. (Overwinter survival has also been found to be a function of size in fish [Henderson et al. 1988; Post and Evans 1989] and a spider [Gunnarsson 1988].) This problem can be solved using a dynamic programming approach but it is a rather cumbersome method since only one decision is involved and this can be examined analytically or even numerically more easily. Let the size at the end of the summer be denoted by $W(T) = f(x, T - x)$, where x is the age at fledging and $T - x$ is the time spent at sea, T being the length of the growing season. Mortality rates in the nest and at sea are assumed constant (Ydenberg 1989): survival to fledging is thus $\exp(-M_n)$ and at sea, $\exp(-M_o)$, where M_n and M_o are the mortality rates in the nest and at sea, respectively. The survival from the end of the summer to first breeding is an increasing function of size, $S(W)$,

$$
\begin{aligned}
\text{fitness} &= S(W(T))e^{-xM_n-(T-x)M_o} \\
&= S(f(x, T - x))e^{-xM_n-(T-x)M_o}
\end{aligned}
\tag{4.70}
$$

The above can be differentiated with respect to x and the optimal fledging age (and hence mass) found by setting the derivative equal to zero. If the equation involves an integral that cannot be solved (in the present case it does not) a numerical solution can be obtained by forward iteration. There are a maximum of 80 days in the growing season; since there is only a single decision only 80 simulations are required, one for each possible day of fledging. The maximum number of days that would have to be examined is only 80^2, which is small even for a microcomputer. This number can be reduced by using a numerical-search algorithm, available for most micro-computers. The number required using the dynamic programming approach depends upon the increment in terminal masses used, but could certainly be as large. Dynamic programming is a technique that is best reserved for problems involving a sequence of decisions. For other problems either analytical methods or forward enumeration may be better.

4.4. Matrix Methods

The use of matrix algebra to tackle problems in ecology was introduced many years ago by Leslie (1945, 1948). Caswell and his colleagues (Caswell 1978, 1982a,b,c, 1983, 1984; Caswell and Werner 1978; Caswell et al. 1984; Caswell and Real 1987) have recently promoted matrix methods for the analysis of both population dynamics and optimal life history traits. Matrix methods have also been extensively employed in quantitative genetics. (See chapter 2.) The elements of matrix algebra and their general application to population dynamics and evolution are well covered by Caswell's recent book, *Matrix Population Models* (Caswell 1989), and only the salient features of the technique will be presented here.

Changes in the age structure of an iteroparous organism can be represented by the matrix equation

$$
\begin{bmatrix}
N(1,\, t+1) \\
N(2,\, t+1) \\
N(3,\, t+1) \\
. \\
. \\
. \\
N(T,\, t+1)
\end{bmatrix}
=
\begin{bmatrix}
m(1) & m(2) & m(3) & . \; . & . & m(T) \\
p(1) & 0 & 0 & . \; . & . & 0 \\
0 & p(2) & 0 & . \; . & . & 0 \\
0 & 0 & p(3) & . \; . & . & 0 \\
. & . & . & . \; . & . & . \\
. & . & . & . \; . & . & . \\
0 & 0 & 0 & 0 \; 0 \;\; p(T-1) & . & 0
\end{bmatrix}
\begin{bmatrix}
N(1,\, t) \\
N(2,\, t) \\
N(3,\, t) \\
. \\
. \\
. \\
N(T,\, t)
\end{bmatrix}
$$

$$\text{(4.71)}$$

where $N(x, t)$ is the number of females of age x at time t,

$m(x)$ is the number of female births at age x (here assumed to be independent of time),

$p(x)$ is the probability of surviving from age x to age $x + 1$.

A shorthand way of representing this is $\mathbf{N}(t + 1) = \mathbf{A}\mathbf{N}(t)$, where $\mathbf{A}$ is called the Leslie matrix. At a stable age distribution the population increases at a rate of λ (i.e., for a single stage population $N(t + 1) = \lambda N(t)$), which, of course is a measure of fitness. ($\lambda = e^r$, where r is the instantaneous rate of increase.) The value of λ is equal to the maximal real eigenvalue. The stable age distribution is given by the right eigenvector, $\mathbf{w}$, and the distribution of reproductive values by the left eigenvector, $\mathbf{v}$. Thus one can obtain a critical number of key parameters by solving the population matrix. (This will generally have to be done using a computer routine, for which there are abundant commercial packages.)

Certain results may also be most easily obtained using matrix methods. For example, Caswell (1978) derives very easily the result

$$\frac{d\lambda}{da(i,\, j)} = w(i)v(j) \tag{4.72}$$

where $a(i, j)$ is the (i, j)th element in the Leslie matrix. Thus the effect on fitness of changes in survival or fecundity at any stage can be resolved by multiplying the reproductive value of stage i by the representation of stage j in the stable distribution. Previous derivations of this result started by differentiating the characteristic equation and are far more difficult and lengthy. (See for example, Goodman 1971.)

4.4.1. Example 1

The rate of change of fitness (λ) with changes in the age specific reproduction function, $m(x)$, and the age specific survival probabilities, $p(x)$ are given by

$$\frac{\partial \lambda}{\partial m(x)} = v(1)w(x) \tag{4.73}$$

$$\frac{\partial \lambda}{\partial p(x)} = v(x + 1)w(x) \qquad \textbf{(4.74)}$$

(Hamilton 1966; Goodman 1971; Caswell 1978, 1982), where $w(x)$ and $v(x)$ are the xth elements of the eigenvectors $\mathbf{w}$ and $\mathbf{v}$, which have been scaled so that their scalar product equals one.

Now consider the trade-off between survival and fecundity depicted in Fig. 4.8. The tangent to the constraint curve, $dp(x)/dm(x)$, increases in magnitude as survival decreases; thus the value of the tangent, if known, can be used to predict the age-specific mortality rates. The tangent can be

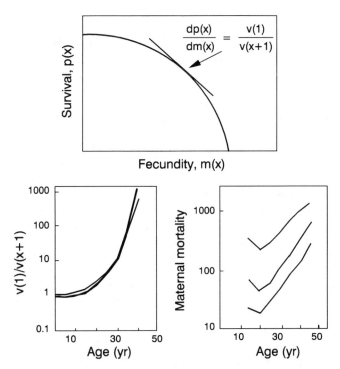

Figure 4.8. Top panel: Hypothesized trade-off between survival and fecundity. Lower left panel: Relationship between $v(1)/v(x + 1)$ and age of U.S. females for the years (top to bottom) 1940, 1950, 1960. Lower right panel: Age specific rates of maternal mortality (per 100,000 births) due to complications of pregnancy for U.S. women (years same as in left panel). After Caswell (1982c).

found from equations 4.73 and 4.74 as follows:

$$\frac{dp(x)}{dm(x)} = \frac{\partial\lambda}{\partial m(x)}\left(\frac{\partial\lambda}{\partial p(x)}\right)^{-1}$$

$$= \frac{v(1)}{v(x+1)} \tag{4.75}$$

Thus if the life history is optimal, survival should vary with age in the same manner as the above ratio of reproductive values. Caswell (1982c) tested this prediction using age-specific mortality rates for human females: the general agreement between prediction and observation is good (Fig. 4.8) though Caswell did not attempt a quantitative assessment. Analysis of the costs of reproduction in red deer, *Cervus elaphus*, by Clutton-Brock et al. (1983) gives qualitative but not quantitative support for the prediction (Caswell 1984).

4.4.2. Example 2

Caswell and Werner (1978) used a matrix approach to examine the optimality of life history patterns in a herbaceous plant, the teasel (*Dipsacus sylvestris*). Teasel is a semelparous plant with no vegetative propagation, and though typically biennial the prereproductive rosette state may last for up to 5 years (Werner and Caswell 1977). One of the questions Caswell and Werner (1978) addressed was, Why should *Dipsacus* be biennial? They begin with a Leslie matrix for a hypothetical perennial teasel and calculate, using equation 4.71, the value of λ for various combinations of annual survival of flowering plants (p) and female birth rate (m, measured as a fraction of biennial seed production), both values assumed to be age-independent. The values of other life history parameters (e.g., prereproductive survival) were estimated from observations on seven populations. Fig. 4.9 plots those combinations of p and m that give the observed rates of increase for each field. Any mutants having p/m combinations above the lines will have higher rates of increase and hence will be favored. The ranges in fecundity and survivorship that are likely to be available to teasel can be determined approximately by comparison with other plants (stippled region in Fig. 4.9). In only one field would the perennial habit be favored over the biennial; as the population was declining in this field ($\lambda = 0.628$) this case is not of great interest. In one field the perennial and biennial habits have about equal fitness (i.e., the critical line passes through the stippled region in Fig. 4.9). For the remaining five fields switching to a perennial habit would decrease fitness.

4.5 Model Testing

4.5.1. Partially and Fully Constrained Models

Models are invariably simplified representations of the real world, and observations are generally made with error. For these two reasons it is

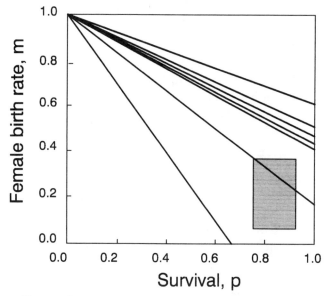

Figure 4.9. Survivorship, *p* (= annual survival of flowering plant), and reproduction, *m* (= fraction of biennial seed production), combinations for a hypothetical perennial *Dipsacus*. Each line gives the critical value of *m*, estimated for each study field, at which the rate of increase of a perennial will equal that of the biennial. Perennials with combinations lying above the lines will have higher rates of increase than the biennial. The stippled area demarcates the set of combinations likely to be available to a perennial *Dipsacus*. Redrawn from Caswell and Werner (1978).

unreasonable to expect predictions and observations to match exactly. It is therefore necessary consider how close must prediction and observation be before the model from which the prediction came is not rejected.

Models can be divided into two categories—partially constrained models and fully constrained models. A partially constrained model is one in which there are a number of parameters that are obtained from the data being predicted. Suppose, for example, that we hypothesize that the rate of change in size, either of a population or an individual, is given by

$$\frac{dx}{dt} = rx\left(1 - \frac{x}{K}\right) \tag{4.76}$$

which is the standard logistic equation,

$$x(t) = \frac{K}{1 + e^{c-rt}} \qquad (4.77)$$

This equation has three parameters, r, K, and c, that are estimated from the sequence of observations on x. Fig. 4.10 shows the logistic equation fitted to data on the growth in height of sunflowers. Clearly the fit is extremely good. But now consider an alternate model,

$$x(t) = \frac{c_1}{2}\left(1 + \frac{2}{\sqrt{\pi}}\int_0^{c_2(t-c_3)} e^{-y^2}\,dy\right) \qquad (4.78)$$

which also has three parameters. This model fits the data at least as well as the logistic (upper panel Fig. 4.10; in fact the sum of squares for this model is less that that of the logistic). As pointed out by Feller (1940, p. 52), from whom this example is drawn, "the recorded agreement be-

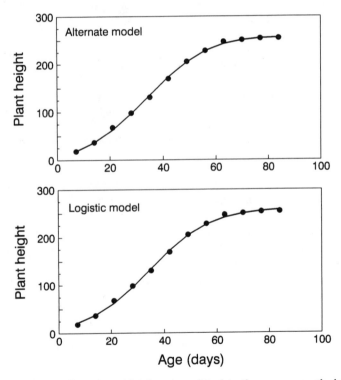

Figure 4.10. Two sigmoidal functions fitted to the same growth data. Data from Feller (1940).

tween the logistic and actually observed phenomena of growth does not produce any significant new evidence in support of the logistic, beyond the great plausibility of its deduction."

The logistic equation is constrained in terms of its general shape (sigmoidal) but the three undetermined parameters mean that it can be fitted to practically any curve that is sigmoidal. The logistic curve can *describe* a curve but it cannot predict the value of some point along the trajectory without recourse to actually using the data to estimate the three parameters. In this sense the logistic model is only a partially constrained model. Acceptance or rejection of the logistic model rests upon the demonstration that the underlying assumptions are correct: if this is so the logistic equation emerges simply as a logical consequence. Failure to appreciate this can lead to some extreme conclusions, as illustrated by the "doomsday" model reported by Von Foerster et al. (1960) in the journal *Science*.

Von Foerster et al. (1960) postulated that, contrary to the generally accepted proposition, increases in density may increase rates of increase within human populations. Such a result, they hypothesized, would come about because of mutual assistance between individuals. They suggested that rates of increase are governed by the equation

$$\frac{dN}{dt} = rN^{\frac{1}{c}} \tag{4.79}$$

where N is population size and c is a constant close to 1. Integration gives

$$N(t) = N(1)\left(\frac{t^* - t_1}{t^* - t}\right)^c \tag{4.80}$$

where $t^* = t_1 + \frac{c}{r} N(1)^{-\frac{1}{c}}$

The parameter t^* is the year in which $N(t)$ becomes infinite (since $t^* - t = 0$). Fitting the above equation, von Foerster et al. predicted "doomsday" to be Friday, 13th November, 2026.

Not unexpectedly this prediction provoked some controversy. (See "Letters to the Editor," *Science* 1961, Vol. 133, pp. 936–943.) In response to their critics Von Foerster et al. (1961) replied

> We believe that support of a hypothesis is gained through compatibility with experimental observation rather than arguments about what should be the case or what should not be the case. This compatibility establishes the relation between theory and reality and serves as a touchstone for accepting or rejecting a hypothesis. If some of our readers express doubt whether or not our simple hypothesis (equation 4.79) has any connection

with reality, we obviously failed to keep them interested in this subject long enough to turn to our Fig. 1, which offers a comparison between theory and observation.

Von Foerster et al. have missed the point: if their equation had not fitted the observed data, the model could be rejected, *but* its ability to *describe* the observed growth in population is not sufficient grounds for accepting its underlying premises. Within the years 0 to 1960 it could be used a predictive tool, but beyond this date it can be used only if the underlying assumptions are demonstrated to be correct and to hold over the future time range. The error of extrapolating beyond observation is well demonstrated by population projections of the logistic equation fitted to U.S. census data from 1790 to 1910: future projections are accurate until 1960, after which point there is considerable divergence. (See Fig. 12.14 in Krebs 1985; and discussion by Hall 1988.) Models that are only partially constrained must be viewed with considerable caution.

The second class of models comprise those in which all parameters are fixed external to the data they seek to predict. For example, Roff (1984a) produced a model predicting the optimal age at maturity in fish

$$\alpha = \frac{1}{k} \log_e\left(\frac{3k + M}{M}\right) \tag{4.81}$$

where α is the optimal age at maturity,
  k is the parameter from the Von Bertalanffy growth equation,
  M is the instantaneous mortality rate.

The theory behind this model is discussed in chapter 7: the important point in the present context is that the two parameters are obtained without recourse to the parameter, age at maturity, being predicted. This does not mean that one does not have to pay considerable attention to the assumptions from which the model is derived (it is these which should be the object of future experimentation and testing), but a fit cannot be ascribed to the action of "free" parameters. To this degree fully constrained models are to be preferred over partially constrained models.

4.5.2. Statistical Testing of Prediction and Observation

There are two broad classes of prediction: first, a model may be used to predict the direction of change in a particular trait; second, a model may be used to predict the actual value of the trait. The likelihood of there being alternate models that produce the same prediction are obviously much greater in the first category. An example from chapter 10 illustrates this point. In many species propagule size has been observed to increase with the size of the female. Two theoretical models make this prediction, but in one mortality is assumed to be negatively density-dependent while in the other it is positively density-dependent. The mortality assumption is critical in both models but the impact is modulated by other aspects of the assumed life history. Thus, one could not conclude on the basis of a correct prediction of direction that a particular model is correct. If the underlying life history structure fits the organism in question one should be motivated to explore the problem further, but one certainly should not stop with such a crude prediction as the direction of change. With a model that predicts the actual value of the trait we can ask whether the correspondence between observation and prediction is statistically adequate.

Suppose we have a model that predicts the value of some trait, say X_p, and we have a set of observed values for the trait, say X_o: for example, a model predicting the optimal age at maturity, and n species for which we have observed ages at maturity. The obvious approach to testing the model is to regress predicted values on observed values, or vice versa. The choice of the dependent and independent variables rests on the use to which the regression is to be put. If the object is to use the regression equation to predict further values of X, then the observed values will be the dependent variable. If the object is simply to test the adequacy of the model the choice of dependent and independent variable is not critical. A significant correlation between prediction and observation indicates that the model is at least partially correct. But a significant correlation is insufficient grounds for concluding that the model is accurate because there may be a systematic bias in the predicted values (top panel, Fig. 4.11).

One approach to resolving this is to compare the number of cases for which the predicted value exceeds the observed. Suppose, for example, 18

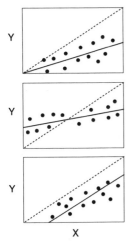

Figure 4.11. Three hypothetical relationships between predicted and observed values. Note that in all three cases there is a significant correlation between Y and X but there is also significant divergence from the 1:1 line (dashed line).

predictions exceed the observed value and 13 are less. This is not significantly different from 50:50 ($\chi^2 = 0.806$, $P > 0.05$). This is a weak test, however, because of the generally large sample size required to detect deviation from a 50:50 ratio and because points may be equally distributed about the 1:1 line while still being biased (middle panel, Fig. 4.11). Another approach is to test if the regression slope is significantly different from 1 and the intercept significantly different from zero. Fig. 4.11 shows the three possible ways in which a significant regression might deviate from the line of equality. The conclusion that one may draw from this analysis may depend upon which variable is chosen as the dependent and which as the independent variable. It is, therefore, good policy to do the analysis both ways. Geometric regression may also be useful in this regard. Of course, all the foregoing discussion is predicated on the assumption that the regression is linear and that variances are homoscedastic. The former assumption is readily tested but the number of data points is unlikely to be high enough to examine the latter.

Failure to obtain a significant regression may mean,

1. The model is wrong.
2. The errors in the prediction, resulting from errors in the measurement of the component parameters, are so large that the data are too poor to exclude the model. As errors in prediction are likely to reduce the correlation, this problem need, perhaps, be considered only if the correlation is low or nonsignificant.
3. The distribution of the estimator is far from normal. This may be the most significant cause for concern. As an example consider equation 4.81, predicting the optimal age at first reproduction. Suppose both k and M are normally distributed. In this case α may be far from normal: this is illustrated in Fig. 4.12 for two hypothetical cases. As can be seen, the distribution of predicted age is skewed to the right, primarily because of the term $1/k$. This problem can be solved by transformation, the appropriate transformation in the present case being to compute the reciprocal of the predicted age, which greatly reduces the skew (upper panels, Fig. 4.12). In the actual data set this transformation increases the variance accounted for by the regression (r^2) from 61% to 71%. (See chapter 7, Fig. 7.4.) Thus for any model which comprises a number of parameters it is useful to examine the possible sampling distribution of the predicted value given the sampling distributions of the component parameters. Data may be available on the sampling distribution of the various parameters, or one might try a variety of distributions. Any model that involves a ratio is likely to be skewed and therefore a transformation will be required.

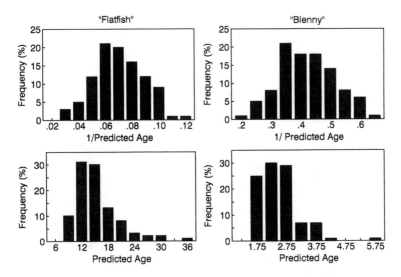

Figure 4.12. Simulated distributions of age at maturity predicted by equation 4.81, for two hypothetical species of fish. Equation 4.81 was assumed to be a correct estimator of the age at maturity, but the parameters k and M estimated with error. Both k and M were assumed to be normally distributed with means applicable to two very different types of fish: a slow growing, late maturing "flatfish" and a fast growing, early maturing "blenny." For the "flatfish": mean value of $k = 0.08$, $SD = 0.03$; mean value of $M = 0.12$, $SD = 0.03$; and for the "blenny": mean value of $k = 0.30$, $SD = 0.10$; mean value of $M = 0.9$, $SD = 0.20$. For each "species" 100 values were drawn at random from the distributions of k and M, and age at maturity estimated from equation 4.81 (lower panels). The data were also transformed using reciprocals (upper panels). For the "flatfish" actual age at maturity and the mean predicted from the untransformed method are 13.7 and 15.3, respectively; and for the "blenny" they are 2.31 and 2.51, respectively. The predicted ages at maturity using the reciprocal transformation are 14.1 and 2.33 for the "flatfish" and "blenny," respectively.

Suppose one has only a single value: for example, Gross (1985) examined the relative fitnesses of precocial ("jack") and late-maturing ("hooknose") coho salmon, *Oncorhynchus kisutch*, in a single stream (Deer Creek Junior, Washington State). For both forms to be maintained in the population the relative lifetime fitnesses must be equal. Lifetime fitness is the product of survivorship to maturity, breeding life-span, and mating success. The relative lifetime fitness is then

$$\frac{w_j}{w_h} = \left(\frac{0.13}{0.06}\right)\left(\frac{8.4}{12.7}\right)\left(\frac{0.66}{1}\right)$$

$$= 0.95 \qquad\qquad (4.82)$$

where w_j is the fitness of the jack, w_h is the fitness of the hooknose, and the figures in the three sets of parentheses refer to the three fitness components. The predicted value of 0.95 is remarkably close to the expected value of 1. This is all the more remarkable given the relative crudity of the data. But the estimate lacks confidence intervals and hence it is difficult to know if the result might not be due to chance. It is, therefore, necessary to establish at least approximate confidence intervals for the prediction (or to do a sensitivity analysis which is discussed in the next section). It may also be noted that the use of a ratio can produce all sorts of nasty distributions, and in the present case it would be better statistically to test the hypothesis that $w_j = w_h$.

If it is not possible to compute exact confidence intervals a computer-intensive method such as the bootstrap or jackknife can be employed. (See Potvin and Roff 1992 for a survey of these methods applied in ecology.) Briefly, the jackknife method is based on the set of all estimates computed after the deletion of a single observation from the data set. (Thus if there are n observations, n estimates can be formed, each estimate comprising $n - 1$ observations.) The bootstrap method consists of computing a large number of estimates by sampling with replacement from the set of observations. Confidence intervals can be constructed from the sets of estimates so obtained.

If the data are too few to apply these methods, an idea of the range in variation likely to occur can be found as follows: first, we assume parameters are distributed in some fashion (e.g., uniform, normal) with known moments (i.e., mean, variance, etc.), these being derived from the available data or "educated guesses." Parameter values are then drawn from these distributions and an estimate is calculated. The process is repeated a large number of times (e.g., 1,000 or more), thereby generating an expected distribution from which confidence intervals can be calculated. (If the distribution is not normal, confidence intervals should be based on the 5% and 95% percentiles, not the standard deviation.)

It may happen that the model parameters are estimated from a variety of sources and the above approaches are not tenable. In this case we resort to sensitivity analysis.

4.5.3. Sensitivity Analysis

Sensitivity analysis is employed for two reasons. First, one may have reasonable estimates of a range for each parameter but no reasonable estimates of the likely frequency distributions. Second, one may wish to examine the sensitivity of the predicted value to variation in each parameter—obviously those parameters that produce the greatest change in the prediction should be those to which further experimentation or estimation should be directed.

Considering the first question, let the predicted trait (e.g., optimal age at maturity) be designated x, the measure of fitness as w, and the set of n parameters (which may traits) within the model that are assumed known be $c_1, c_2, c_3 \ldots c_n$. Thus fitness can be written as function f, of x and c_i, $w = f(x, c_1, c_2, c_3 \ldots c_n)$. For each parameter we are able to designate a most likely value, say c_i^* and at least a maximum range, the range for c_i being denoted by L_i and U_i (for lower and upper value, respectively). The simplest approach is to substitute first L_i and then U_i into the function $f(x, c_1 \ldots c_i \ldots c_n)$ in place of c_i^* and find the two new optimal values for x. The decision as to whether these are significant is not clear-cut; a 10% variation is small while a 100% variation is large, but the point at which percentage variation becomes unacceptably large cannot be defined in the same way that significance levels can be defined. The upper and lower bounds are the extremes of what the modeler believes (and can convince his readers) is likely to occur. The "best" estimate is c_i^* and the likelihood of it being some amount a_i larger or smaller than c_i^* diminishes as a_i increases, but it is unlikely that this likelihood can be assigned. A plot of predicted x on c_i from L_i to U_i is useful to assess how important variation in c_i may be.

The above analysis applies only to variation in a single parameter. There may, however, be nonlinearities in the response and sensitivity may depend upon particular combinations of parameter values. Pairwise variation in two parameters can be displayed by plotting isoclines of x on the two parameters. A maximum range can be obtained by picking that combination of parameter values which gives the smallest value of x (e.g., L_1, $L_2, U_3 \ldots U_n$) and the largest (e.g., $U_1, U_2, L_3 \ldots L_n$). Another possible approach is to make the conservative assumption that each c_i is distributed uniformly between L_i and U_i and draw sets of parameter values on this basis. This technique should indicate whether particular combinations may produce very large deviations from the prediction based on the "most likely" parameter set.

The modeler wishes to know not only if the prediction is sensitive to variation in parameter values but also which parameters are most critical in this regard. This can be done for variation resulting from changes in a single parameter as follows (Tomovic 1963). From the plot of x on c_i the partial derivative, $\partial x / \partial c_1$, is estimated (the simplest method is fit a polynomial through the set of points) for some value of x and c_i, which shall be denoted as $g(x, c_i)$. This measures the sensitivity of x to variation in c_i

but it cannot be directly compared to $g(x, c_j)$ because c_i and c_j may be in different units. What we require is the rate of change for the same proportional change in x. This can be obtained by multiplying $g(x, c_i)$ by c_i/x (and similarly $g(x, c_j)$ by c_j/x). These are called the normalized sensitivities. An example of the application of this method is given by Myers and Doyle (1983) in their analysis of a model that predicts mortality rate in fish based on the assumption that the life history is optimal. (See chapter 5 for a more detailed discussion.) Parameters a and b determine the amount of surplus energy available for growth and/or reproduction (surplus energy $= aWeight^b$), and T is the maximum life-span. The three normalized sensitivities are 1.35, 12.2, and 1.37, respectively. (The data are for St. Lawrence cod.) The predicted mortality rate is most sensitive to variation in b and about equally sensitive to variation in a and T. Thus more effort should be made in determining the bounds for b than for the other two parameters.

4.6. Summary

The most important mathematical tool in life history theory is the calculus. But other methods such as graphical analysis, dynamic programming, and matrix methods may be preferable in some cases.

Models may be divided into those that are partially constrained and those that are fully constrained. The former type contain parameters that are estimated from the same data set as the observations to be predicted, while in the latter all parameter values are estimated without reference to these data. Fully constrained models are to be preferred.

If several predicted values are available, the fit between observation and prediction can be tested with linear regression methods. Where only a single value is predicted it is essential to be able to construct either confidence limits or to undertake an adequate sensitivity analysis.

5

Age Schedules of Birth and Death

The starting point of any comprehensive analysis of life history is the age schedules of birth and death. These two processes are central components of the characteristic equation and hence the calculation of lifetime fitness necessarily entails computing the manner in which birth and death rates change with age, size, etc. In this chapter I present an overview of the birth and death functions, outlining the major factors that play a role in shaping them. In later chapters these components will be drawn together to analyze specific life history traits.

Survival to and reproduction (female births) at age x are specified in the characteristic equation by the two terms $l(x)$ and $m(x)$, respectively. It is therefore convenient to discuss the two components separately. But it should be remembered that in the characteristic equation it is the product function $l(x)m(x)$ that is important, a consequence of which is that several different biological models derived from differing $l(x)$, $m(x)$ functions may give the same $l(x)m(x)$ function. An example of this phenomenon is given in section 5.2.1.

Section 5.1 deals with the age schedule of mortality. Firstly, we examine the ways in which the mortality schedule can be described. In most of this book the age schedule of mortality is used to predict the value of a particular life history trait. But the method can be turned around and life history theory can be used to predict mortality rates: section 5.1.2 gives two examples of this. The $l(x)$ function measures survival as a function of age, but in many, if not most cases, age is simply a surrogate measure of size: section 5.1.3 addresses the question of the relationship between survival and size.

As with the discussion on mortality, the discussion on the age schedule of reproduction opens with a review of general patterns. Section 5.2.2 considers the effect of size on reproduction, demonstrating that among most organisms fecundity increases with size. While death is an all-or-none phenomenon, reproduction can be graded, from diverting no energy into the production of offspring to diverting so much energy that the animal fails to survive to breed again. Methods of operationally measuring reproductive effort and its variation among different taxa are described in section 5.2.3.

5.1. Mortality

5.1.1. Describing Age Schedule of Mortality

The simplest mortality function is that in which the probability of dying during any given interval of time is constant: this can be described by the differential equation

$$\frac{dN(x)}{dx} = -MN(x) \tag{5.1}$$

where $N(x)$ is the number alive at age x, and M is the instantaneous rate of mortality. Integration gives

$$N(x) = N(0)e^{-Mx} \tag{5.2}$$

where $N(0)$ is the number alive at the time of the first census. The proportion alive at time x (e^{-Mx}) is conventionally designated as $l(x)$, and the survivorship curve is the $l(x)$ curve. Taking logarithms we obtain the simple linear equation,

$$\log_e(N(x)) = \log_e(N(0)) - Mx \tag{5.3}$$

If the number in a cohort alive is plotted as a function of age we obtain a straight line whose slope is equal to the instantaneous rate of mortality, M. This fact suggests that survivorship, $l(x)$, curves should be plotted on a log-linear scale rather than an arithmetic (linear-linear scale). While the log-linear plot is a useful method of plotting survivorship, the change in mortality is shown only as the change in the slope of the line, which can be difficult to interpret visually. Therefore, in addition to the survivorship curve it is useful to plot the mortality, or $q(x)$, which is the mortality rate between the age interval, $x, x + 1$, calculated as the ratio of those dying during an age interval to those alive at the beginning of the interval (Caughley 1966). This plot emphasizes the pattern of change in mortality as a function of age: if mortality is constant the $q(x)$ curve will be a straight, horizontal line. Assuming the instantaneous rate of mortality, M, to be at least approximately constant from age x to $x + 1$, M and $q(x)$ can be equated by the function $M = -\log_e(1 - q(x))$.

Various examples of $l(x)$ and $q(x)$ curves are shown in Fig. 5.1. Obtaining such curves for vertebrates is relatively easy because individuals can generally be aged by morphological characteristics such as growth rings on teeth or scales in much the same way as trees can be aged. Such patterns are also found in some invertebrates—e.g., daily cuticular rings are laid down in some insect species, (Neville 1963)—but in many cases survivorship curves are computed on the basis of recognizable life stages (e.g., *Daphnia*) rather than chronological age. The examples shown in Fig 5.1 are somewhat misleading for they do not all contain the survival of the immature stages. On average each female must produce one female offspring that itself survives to reproduce or else the population becomes extinct. Therefore, species characterized by an enormous output of propagules can also be expected to suffer very high mortality rates. Such rates are typically encountered in the immature stage. For example, the softshell clam, *Mya arenaria*, whose survivorship curve is displaced in Fig. 5.1e, releases over 1.4 million eggs in the first year of its life, the number increasing to over 8 million by age 10 years (Brousseau and Baglivo 1988). The probability of an offspring surviving to its first birthday is 1.1×10^{-7}, but thereafter the probability of surviving ranges from 0.5 to 0.97 depending upon its age and location. These extraordinarily high mortality rates are typical of marine invertebrate species with planktonic larvae. Marine invertebrates with nonplanktonic larvae suffer mortality rates several orders of magnitude lower (reviewed by Brousseau and Baglivo 1988). A similar situation is found in fish (Dahlberg 1979; Peterson and Wroblewski 1984; McGurk 1986).

Of particular interest is the deviation of the survivorship curve from a straight line: to what extent does mortality change with age? A simple parameter that indicates the degree of deviation may be derived as follows (Nesse, 1988): beginning at some particular age, such as the age at maturity, and assuming that mortality thereafter is constant, compute the total number of reproductive years expected. (This is equal to the area beneath the curve; Fig. 5.2.) Compare this hypothetical survivorship curve to that observed using the formula (Hypothetical − Observed)/Hypothetical, which will equal 0 when mortality is independent of age and 1 when all individuals die in the first year. An alternative procedure is to fit a linear regression to the log-linear plot of survivorship and compute the mean of the sum of squared residuals. Increasing deviation from constant mortality produces increasingly large residuals (lower panel, Fig. 5.2). The two statistics are strongly correlated ($r = 0.64$, $n = 19$, $P < 0.005$, species listed in Table 1 of Nesse). Analysis of residuals is preferable to the method proposed by Nesse (1988) because it is easier and less subject to errors inherent in attempting to predict rates of mortality from observations on a single or few age classes.

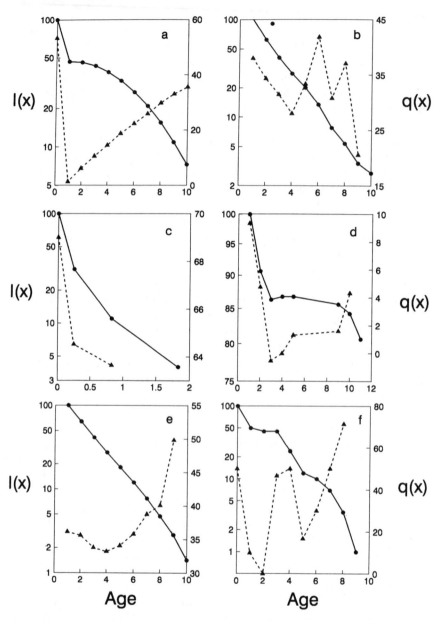

Figure 5.1. Some illustrative schedules of survival, $l(x)$, and mortality, $q(x)$. For convenience the $l(x)$ and $q(x)$ values have been multiplied by 100. All time scales are years except for the one for dragonfly nymphs, which is in months. Organisms displayed are (a) Himalayan thar (Caughley 1966), (b) lapwing (Deevey 1947), (c) lizard (Tinkle and Ballinger 1972), (d) dragonfly nymphs (Wissinger 1988), (e) clam (Brousseau and Baglivo 1988), and (f) grass (Sarukhán and Harper 1973).

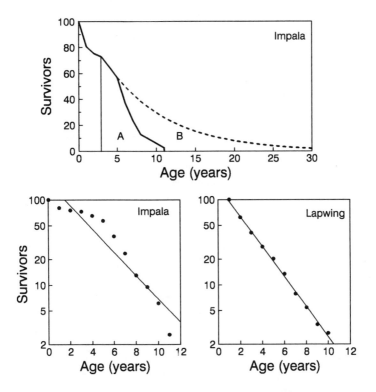

Figure 5.2. Upper panel: Illustration of the method of Nesse (1988). Solid line shows the survival curve of female impala (data from Spinage 1972); the dotted line shows a hypothetical population that has a constant survival rate after maturity, here arbitrarily set at 3 years. Area A is the area enclosed by the observed survival curve, A + B is that enclosed by the hypothetical curve. Lower panels: Illustration of the method of residual analysis. Solid lines show fitted regression; dots show the observed survival fractions. The left panel (female impala) shows a survival curve that deviates strongly from constant mortality; the right panel shows (lapwing, data from Deevey 1947) a survival curve in which mortality rate is very stable throughout life. For the impala the mean squared residual is 0.024 and for the lapwing it is 0.00065.

The assumption of a more-or-less constant mortality rate for birds, first suggested by Nice (1937), supported by the analyses of Deevey (1947) and generally accepted thereafter (e.g, Lack 1943a,b; Davis 1951; Hickey 1952; Gibb 1961; Slobodkin 1966; Bulmer and Perrins 1973), was challenged by Botkin and Miller (1974, p. 108), who noted that with an annual mortality of 3%, the best present estimate, there is one chance in 1,000 that a royal albatross (*Diomedea epomorphora*) breeding in New Zealand today was 25 years old when Captain Cook made his first visit to the island in 1769! If the initial cohort were 10,000 birds, a 302-year-old bird might live in the colony today.

To correct the excessive longevities expected from a constant mortality model they added a constant annual increment to the mortality rate. The required increases range from 0 for the blue tit, having a maximum life-span of 9 years, to 0.01 for the sooty shearwater, which lives a maximum of 30 years (Fig. 5.3). Although for the sooty shearwater the difference between the two survival curves appears quite dramatic, it must be remembered that sample sizes for the older age groups are likely to be very small. There will therefore be considerable variance, and the detection of nonlinearity in the curve will be very difficult. The rate of increase in mortality may also increase with age, reducing the deviation between the two curves in the younger, better-sampled age cohorts. Overall, increases are so small that there is little likelihood of detecting them in natural populations given available methods of censusing birds (Botkin and Miller 1974). From the point of view of life history theory these results are important because they emphasize the potential difficulty of determining from simple census data the increased risk of mortality that may be associated with breeding, a risk that may play a pivotal role in determining the optimal age schedule of reproduction.

Pearl and Miner (1935) classified the survivorship curve into three types (Fig. 5.4):

Type 1: survivorship high until very late in life when survival drops rapidly, possibly as a result of senescence.

Type 2: survivorship constant throughout life, giving rise to the linear function described above.

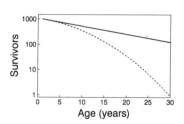

Figure 5.3. Predicted survivorship curves for the sooty shearwater assuming age-independent mortality (annual mortality = 0.07) and age-dependent mortality (annual mortality = annual mortality of previous year + 0.01). The age-dependent factor was chosen by Botkin and Miller (1974) to reduce an initial cohort of 1,000 birds in their first year of adulthood to approximately one bird by age 30 years, the maximum recorded life-span.

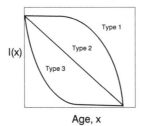

Age, x

Figure 5.4. The three types of survivorship curves proposed by Pearl and Miner (1935).

Type 3: survivorship drops precipitously in the early stages of life until a certain age, size, or life stage is reached, at which point the rate of decline is substantially reduced.

Typical examples of these three types of curves are humans (Pearl and Doering 1923), birds (Deevey 1947), and marine invertebrates (Brousseau and Baglivo 1988). However, the classification into three types does not adequately describe the diversity of survivorship curves. To remove this deficiency Pearl (1940) later added two additional models (low-high-low survivorship, and high-low-high), though there is a general tendency to retain the simple threefold classification. Caughley (1966) argued that, at least for mammals, we know too little about mortality rates to attempt any method of classification, though his analysis and those of Spinage (1972) on African ungulates, Clutton-Brock et al. (1983) on red deer, and Gage and Dyke (1988) on Old World monkeys show that, in general, mammals follow the low-high-low pattern. The early stage of any life history is likely to be a period of relatively high mortality, the young being small, under-developed, and inexperienced; thus, a low-high survivorship component of the $l(x)$ curve can be generally expected.

Siler (1979) combined the three types of age schedules of mortality described by Pearl and Miner (1935) into a single general function relating the instantaneous rate of morality to age,

$$M(x) = a_1 e^{b_1 x} + a_2 + a_3 e^{-b_3 x} \tag{5.4}$$

where $M(x)$ is mortality rate and the three terms individually produce the three types of survivorship curves (Fig. 5.5, upper panel). The first term corresponds to an increasing hazard occurring as a result of senescence (type 1 curve); the second term to a constant hazard, to which the organism cannot adjust (type 2 curve); and the third term to a hazard that decreases as a result of the organism adapting with age to its environment (type 3 curve, Siler 1979). There is no particular biological significance to the functions that comprise this survival function—they are simply mathe-

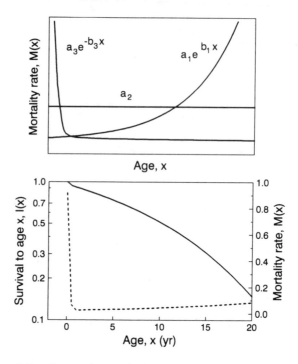

Figure 5.5. Composite survival curve and mortality rate for Old World monkeys. Upper panel: The component functions of the mortality rate function

$$M(x) = a_1 e^{b_1 x} + a_2 + a_3 e^{-b_3 x}$$

where $a_1 = 0.0214$, $a_2 = 0$, $a_3 = 0.895$, $b_1 = 0.0682$, $b_3 = 6.08$. The age scale is the same as shown in the lower panel, but to illustrate the shape of the component functions separate vertical scales (not shown) have been used. Lower panel: the resulting $l(x)$ function (solid line) and the instantaneous mortality rate, $M(x)$ (dashed line).

matically convenient. Siler obtained an excellent fit of the model to data in all species examined (6 mammals, 1 bird, and 1 fish species), though frequently the life-span is too short for the first term (senescence effect) to have an impact. Given the flexibility of the model with respect to variation in shape and five parameters, the satisfactory fit to data is not surprising. It is a very satisfactory descriptor of the survival function but its utility in life history analysis is limited unless the values of the parameters of the model can be made functions of particular life history events such as the age of maturation.

In their analysis of mortality rates in fish, Chen and Watanabe (1989) took a similar approach to Siler (1979) in assuming a mortality rate composed of three components (which they termed, "initial," "stable death," and "death by senescence") but developed the functions on a hypothesized relationship between the rate of mortality, M, and growth. In fish, growth can typically be described by the von Bertalanffy function (Fig. 5.6),

$$L(x) = L_\infty(1 - e^{-k(x-x_0)}) \tag{5.5}$$

where $L(x)$ is length at age x, L_∞ is the asymptotic length, and k and x_0 are constants. Frequently, x_0 is omitted as it is generally very small and has little influence on the growth trajectory. The parameter k measures

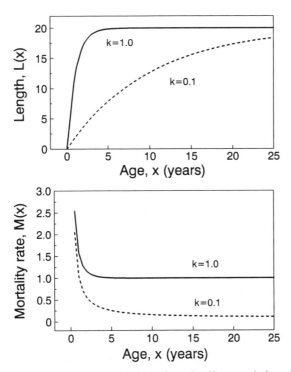

Figure 5.6. Upper panel: Von Bertalanffy growth function for two values of k, where $L_\infty = 20$, $x_0 = 0$. Lower panel: Mortality rate versus age using the model of Chen and Watanabe (1989) and the two growth curves shown in the upper panel.

the rate at which the fish grows, e.g., when $k = 1.0$ the fish attains 90% of its asymptotic length in 2.3 years (assuming $x_0 = 0$), while it takes 23 years to achieve this percentage when $k = 0.1$. The von Bertalanffy function is a general description of growth and, in addition to fish, it has been applied to crustaceans, molluscs (Hughes and Roberts 1980), reptiles (Schoener and Schoener 1978), birds (Ricklefs 1968, 1973a), and plants (Richards 1959).

Chen and Watanabe (1989) predicted that mortality, $M(x)$, could be described by two functions—one relevant to the early and middle phases of a fish's life, and the second acting during the later "senescent" stage, defined as ages greater than

$$-\frac{1}{k} \log_e |1 - e^{kx_0}| + x_0 \tag{5.6}$$

Prior to this age the mortality rate is given by

$$M(x) = \frac{k}{1 - e^{-k(x - x_0)}} \tag{5.7}$$

The biological rationale for these equations is developed in Chen et al. (1988), but as the paper is in Japanese I cannot comment on its validity. I therefore restrict myself to a general discussion of the rationale's feasibility and ability to predict mortality rates. The general shape of the mortality curve is the same as that given by the formulation of Siler, the increasing limb of the curve depending on k and x_0. In the examples shown in Fig. 5.6 x_0 is set at zero and hence the mortality rate declines to an asymptotic value of k, and there is no ascending limb. According to equation 5.7 mortality increases with the rate of growth (k) but decreases with body size (the denominator). A decrease in mortality with size is a common observation (see section 5.1.3), and a positive correlation between k and M has been long recognized (Beverton and Holt 1959; Adams 1980; Gunderson 1980; Kawasaki 1980). Thus the two components in Chen and Watanabe's equation have empirical support; the contribution of this study is to bring them together in this particular relationship.

An important distinction between the models of Siler (1979) and Chen and Watanabe (1989) is that in the latter model, all the parameters are defined without reference to the observed mortality curve. In Siler's model an adequate fit can be simply ascribed to the flexibility of the curve and "free parameters." This is not so in Chen and Watanabe's model, where the parameters are estimated from the growth trajectory and not the survival curve.

The mortality rate in fish is usually considered to be constant after an initial period of high larval mortality (Ricker, 1975), though this is only an approximation (Beverton and Holt 1959; Vetter 1988), sufficient for most analyses of population dynamics (but perhaps not for life history analysis). From Chen and Watanabe's formula the average mortality rate during the period of "stable" mortality is given by

$$\overline{M}(x, x + t) = \frac{1}{t} \log_e\left(\frac{e^{k(x+t)} - e^{kx_0}}{e^{kx} - e^{kx_0}}\right) \tag{5.8}$$

where x and $x + t$ are the two ages considered. Chen and Watanabe (1989) obtained a remarkably close fit between the predicted mortality and observed average mortality, obtained by fitting a linear regression through the observed numbers at age (Fig. 5.7). In fact, given the problems of estimating mortality rates the correspondence between observed and predicted values is greater than one might have expected. To test the model further I used data on 40 species (7 species being represented by several

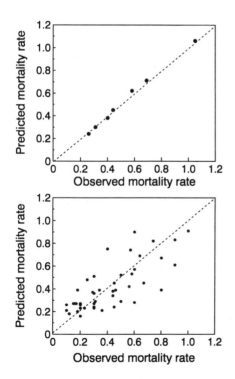

Figure 5.7. A comparison of observed and predicted mortality rates based on the model of Chen and Watanabe (1989). The upper panel shows the data set presented by Chen and Watanabe; the lower panel shows that obtained by using data from Pauly (1978). From the general distribution of age at maturity on maximum length (Fig. 11 of Roff 1988) I set the initial ages, x, for estimation as follows: $x = 1$ when $L_\infty < 25$ cm, $x = 2$ when $L_\infty < 50$ cm, $x = 3$ when $L_\infty < 100$ cm, $x = 4$ when $L_\infty > 100$ cm. Based on the age spans used by Chen and Watanabe in their analyses the time span over which mortality was estimated was determined from the expected age at which the cohort at age x was reduced to 25% of its size at age x. (Varying this rule did not appreciably change the results.) Thus the age span considered was x to $x - (1/M)\log_e 0.25$, where M was the observed mortality rate.

stocks, making a total of 48 values) obtained from Pauly (1978). Data on the age groups used to compute mortality rate were not available and so I estimated the initial and final ages according to the rules described in the caption to Fig. 5.7. I omitted all species in which the mortality rate exceeded 1 since less than 15% of any cohort will survive 2 years (i.e., e^{-2} = 0.14) and no reasonable age spread can be obtained. The fit between observed and predicted mortality is very good (r = 0.79, n = 48, $P <$ 0.001). Although mortality rates at the low end of the range tend to be overestimated, the distribution of points about the 1:1 line (22 above the line and 26 below) shows no bias. This model deserves further investigation.

5.1.2. Estimating Mortality by Using Life History Theory

Myers and Doyle (1983) approached the problem of estimating natural mortality as a function of age in fish by assuming that fitness, in this case expected lifetime fecundity, is maximized. They assumed:

1. Mortality at age, $q(x)$, is determined by the amount of surplus energy (= energy available after maintenance costs have been met) devoted to reproduction. Myers and Doyle selected a mortality function which ensures fitness is maximized by allocation of surplus energy to both growth and reproduction, as actually observed in fish. (The theoretical rationale for this is discussed in chapter 8.) But though the shape of the function (convex) was selected to match biological reality the particular algebraic formulation is arbitrary (Fig. 5.8, upper panel),

$$q(x) = q_1 + q_2 u(x)^{c_1} \qquad (5.9)$$

where q_1 is the mortality in the absence of reproduction,
q_2 is the additional mortality if all surplus energy is channeled into reproduction (i.e., $u(x)$ = 1),
$u(x)$ is the proportion of surplus energy channeled into reproduction at age x ($0 < u(x) < 1$), and
c_1 is a constant that modulates the effect of reproductive expenditure on mortality.

2. Weight at age $x + 1$, $W(x + 1)$, is given by the simple partitioning equation (Fig. 5.8, middle panel),

$$W(x + 1) = W(x) + g(x) - g(x)u(x) \qquad (5.10)$$

where $g(x)$ is surplus energy, estimated according to the method of Ware (1980).

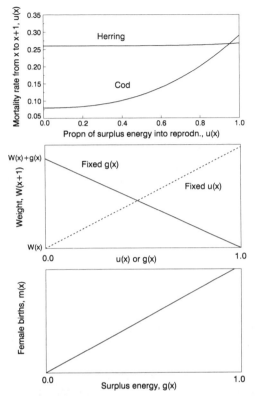

Figure 5.8. Component functions for the model of Myers and Doyle (1983). Top panel: Relationship between mortality rate, $q(x)$, and proportion of surplus energy devoted to reproduction, $u(x)$, at age x, for two species from the Gulf of St. Lawrence. The general model is

$$q(x) = q_1 + q_2 u(x)^{c_1}$$

Parameter values are herring, $q_1 = 0.26$, $q_2 = 0.007$, $c_1 = 5.0$; cod, $q_1 = 0.08$, $q_2 = 0.210$, $c_1 = 2.5$. Middle panel: Solid line shows relationship between weight at age $x + 1$, $W(x + 1)$, and the proportion of surplus energy devoted to reproduction at age x, for a given value of surplus energy, $g(x)$. Dashed line shows the relationship between weight at age $x + 1$ and the amount of surplus energy, $g(x)$, at age x, for a fixed allocation fraction, $u(x)$. The general formulation is

$$W(x + 1) = W(x) + g(x) - g(x)u(x)$$

Lower panel: Relationship between the number of female births, $m(x)$, and surplus energy, $g(x)$, at age x:

$$m(x) = c_2 g(x)$$

3. The number of female offspring produced by a female of age x is proportional to the amount of surplus energy (Fig. 5.8, lower panel),

$$m(x) = c_2 g(x) \tag{5.11}$$

where c_2 is a constant.

There are four unknown parameters in the model; q_1, q_2, c_1, c_2. These may be estimated as follows: first, since $W(x)$, and $g(x)$ are known, $u(x)$ can be estimated for each x using equation 5.10; trial values of the unknown parameters are then selected and the expected lifetime fecundity estimated using equations 5.9 and 5.11 according to the relationship,

Expected lifetime fecundity

$$= \sum_{x=1}^{\omega} l(x)m(x) \text{ where } l(x) = \prod_{y=1}^{x-1} (1 - q(y)) \tag{5.12}$$

and ω is the last age. The set of parameters that maximize expected lifetime fecundity can then be found numerically.

There is considerable variation in the predicted relationship between allocation to reproduction and mortality. Over the range 0% to 100% allocation to reproduction, herring show little variation in predicted mortality, while the mortality rate of cod changes approximately fourfold (Fig. 5.8). In the latter case the actual mortality rates are much less because the predicted allocation to reproduction varies only from 0% at age 3, to approximately 60% at age 19, giving an annual mortality ranging from 8% to 13% per year.

Observed mortality rates are estimates averaged over several age classes. Therefore, comparison of predicted and observed mortality rates requires averaging the predicted rate across ages to take into account the changing allocation to reproduction, and hence changing mortality rate. Unfortunately, Myers and Doyle do not supply sufficient information to do this computation. An approximate test can be made by comparing predicted mortality in the absence of reproduction ($M = -\log_e(1 - q_1)$, see equation 5.9), against observed average mortality (Fig. 5.9). Despite an obvious deviation from a slope of one, the two mortality rates are reasonably close. However, the results of the foregoing analysis by Myers and Doyle cannot be used to conclude that mortality changes with age. The reason is that the model utilizes the product of $l(x)m(x)$ and not just $l(x)$; therefore, effects attributed to changes in mortality might in fact be due to changes in fecundity. Such an assumption forms the core of the second model they analyzed. In this model mortality rate is assumed to be constant and fe-

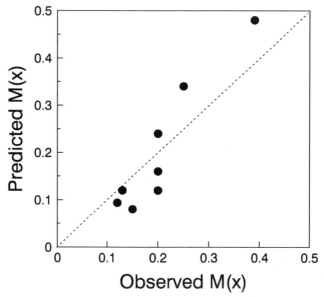

Figure 5.9. Predicted rates of mortality in the absence of reproduction versus the observed instantaneous rates of mortality in
various fish species. The observed rates encompass rates experienced by reproducing fish. Predictions are from the model of
Myers and Doyle (1983); data are from Table 1 of Myers and
Doyle, except for American plaice—correct number is 0.2, not
0.13 as reported by Myers and Doyle (Roff 1984). Also note ty-
pographical error in the case of plaice ($q_1 = 0.09$ not 0.9) in
Table 1 of Myers and Doyle. Dashed line is 1:1 reference line.

cundity to be a function of surplus energy and a concave egg conversion
function, $h(u(x))$ (Fig. 5.10). As with the mortality function of the previous
model (equation 5.9) the only criterion for the choice of $h(u)$ was that its
shape be such as to favor allocation of energy to both growth and reproduction. The age schedule of female births is thus given by

$$m(x) = h(u(x))g(x) = (u(x) - c_3 u(x)^{c_4})g(x) \tag{5.13}$$

where c_3 and c_4 are constants to be estimated. Mortality rate must also be
estimated, the method of analysis being the same as described for the
earlier model. The mortality rate predicted from this model is virtually
identical to that predicted by the first model in the absence of reproduction
(Fig. 5.11).

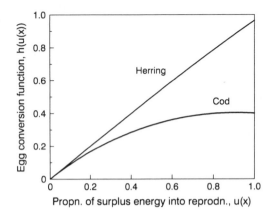

Figure 5.10. Two examples of the egg conversion function used by Myers and Doyle (1983). The number of female births is given by $m(x) = h(u(x))g(x)$, where the egg conversion function, $h(u(x))$, is determined by

$$h(u(x)) = u(x) - c_3u(x)^{c_4}$$

where the parameter values are herring, $c_3 = 0.035$, $c_4 = 4.0$; cod, $c_3 = 0.600$, $c_4 = 1.8$.

The two models of Myers and Doyle (1983) are based on different biological premises but they are mathematically very similar and hence both fit the data. Thus, while one can assume a particular relationship and use the optimality argument to estimate its parameters, this should not be taken as sufficient evidence for the reality of that relationship, unless it is the only relationship in the analysis (which is unlikely).

The concept that selection will produce a mortality schedule that is consistent with the maximization of fitness was also used by Alerstam and Högstedt (1983) to predict the mortality rates in animals with parental care. They assumed

1. Selection maximizes the expected number of offspring per brood. This assumption is appropriate for a semelparous species or for an interoparous species in which breeding episodes are independent—i.e., the reproductive allocation at one reproductive event does not influence survival to, or allocation at, another event.

2. The instantaneous rate of mortality from birth to independence is proportional to brood size. Alerstam and Högstedt argued that the number of begging calls, visits to the nest, etc., will increase

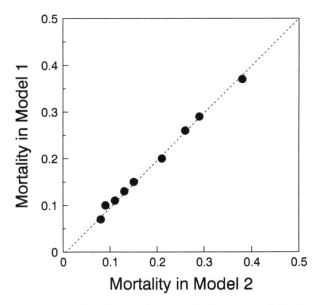

Figure 5.11. Predicted mortality rates in the two models of Myers and Doyle (1983). Model 1 gives the mortality rate in the absence of reproduction, q_1, from the model in which mortality increases with the allocation to reproduction. (See upper panel of Fig. 5.8, and Fig. 5.9, which shows that q_1 is a reasonable estimate of the observed mortality.) Model 2 gives the mortality rate from the model in which mortality rate remains constant with age but fecundity varies in a concave fashion with surplus energy allocated to reproduction. (See Fig. 5.10.) Dashed line is 1:1 reference line.

linearly with brood size and hence mortality rate will show a similar pattern. The expected production per brood is then

$$P_b = Sxe^{-Mx} \qquad (5.14)$$

where S is the proportion of offspring that survive from independence to reproduction,

   M is the rate of mortality from birth to independence for a brood of size 1, and

   x is brood size. To obtain the optimal brood size, P_b is differentiated with respect to x, and the result set equal to zero,

$$\frac{dP}{dx} = Se^{-Mx} - SxMe^{-Mx}$$

$$= 0 \quad \text{when } x = \frac{1}{M} \tag{5.15}$$

At evolutionary equilibrium the survival from birth to independence is thus

$$e^{-M\frac{1}{M}} = e^{-1} \tag{5.16}$$

or the rather extraordinary result that at maximum reproductive success the survival to independence will be $e^{-1} = 0.368$. At brood sizes of around six, survival to independence is indeed close to 0.368, but it deviates for this value with increasingly larger or smaller broods (Fig. 5.12).

Ricklefs (1983) criticized the model of Alerstam and Högstedt on the grounds that mortality rate is unlikely to be proportional to brood size. In an earlier paper (Ricklefs 1977a) Ricklefs had suggested that reproductive output, P_b, in altricial birds could be described by the equation

$$P_b = c_1 x e^{-c_2 x^{c_3}} \tag{5.17}$$

where c_1, c_2, and c_3 are empirically determined constants. The choice of this particular equation was based solely on its flexibility, and not on any

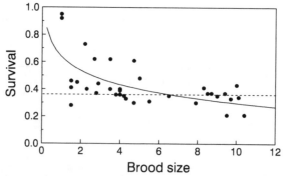

Figure 5.12. Relationship between survival from birth to independence and brood size in birds and mammals. Data from Table 2 of Alerstam and Högstedt (1983), with human survivorship data omitted. The fitted curve (solid line) is
$Y = 0.639 - 0.149 \log x$, $r = 0.62$, $n = 36$, $P < 0.001$. The dashed line is $e^{-1} = 0.37$.

biological considerations. Ricklef's model is equivalent to that of Alerstam and Högstedt when $c_3 = 1$. From empirical fits to 17 species of birds, Ricklefs found that c_3 varied between 1.5 and 12.8 with a mean of about four, from which he concluded that a value of $c_3 = 1$ "is not empirically justified" (Ricklefs 1983, p. 284). Ricklefs, however, did not demonstrate that his model actually produces a statistically better fit than that of Alerstam and Högstedt; therefore, while I concur with Ricklefs that mortality is unlikely to be exactly proportional to brood size I do not accept his conclusion that a distribution of survival rates of e^{-1} is fortuitous. The fit observed in Fig. 5.12 suggests that mortality rates may be approximately proportional to brood size, though the model is clearly too simplistic to capture the observed spread in values. The model, therefore represents a useful starting point for additional analysis.

Another criticism of the Alerstam and Högstedt model came from Stenseth (1984), who argued that their model is appropriate only for a semelparous organism, which as explained above is not the case. A more cogent criticism advanced by Stenseth is that the assumption of no correlation between adult survival and brood size is unlikely to be correct. In rebuttal Alerstam and Högstedt (1984) argued the data on brood size and adult survival suggest that this factor is of minor significance. Most of the lack of significance between these two life history variables can be attributed to poor statistical design in the experiments, particularly low sample sizes. (See chapter 6, section 6.2.2 and Table 6.5.) It would be premature to conclude that a reduction in adult survival with increasing brood size is not of considerable importance.

5.1.3. Size and Survival

There is abundant evidence that mortality rates are age-specific, but it seems unlikely that in most cases age per se is the important factor; it is far more likely that experience, developmental stage, size, and reproductive status are the factors that are functionally related to survival. Experience may play a role in the survival of endothermic vertebrates and perhaps some ectothermic vertebrates, but for most organisms (invertebrates and plants) experience will be of no consequence. Many, if not most, organisms are not born as miniature replicas of their parents. They are generally undeveloped at birth and hence relatively unable to flee from predators or to move to alternate habitats if necessary. This ontogenetic factor undoubtedly is an important component of the high mortality rates experienced by most organisms during the early phase of their lives. Another factor is size: most predators take only prey within a particular size range, and hence the growing organism must pass through a gauntlet of

Table 5.1. Allometric relationship between survival and body size in birds, mammals, and reptiles

Taxon	Trait	a	b	n	r	Reference
Eutherian mammals	Life-span in captivity	2.96	0.20	63	0.77	1
Eutherian mammals	Life expectancy at birth	0.23	0.24	29	0.87	2
Eutherian mammals	Life expectancy at maturity	0.36	0.24	29	0.87	2
Artiodactyls	Life-span	1.33	0.22	14	0.92	3
Primates	Life-span	2.45	0.24	14	0.88	3
Carnivores	Life-span	2.85	0.17	17	0.86	3
Heteromyids	Life-span	5.04	−0.15	7	0.46	4
Birds	Life-span in captivity	7.62	0.19	58	NG	5
Passeriformes	Life-span	3.58	0.26	71	NG	5
Nonpasseriformes	Life-span	4.79	0.18	81	NG	5
Birds	Adult mortality rate	6.60	0.18	14	0.70	6
European birds	Juvenile survival rate	0.27	0.07	56	0.39	7
European birds	Adult survival rate	0.38	0.09	107	0.58	7
Anseriformes	Adult survival rate	0.24	0.14	18	0.60	7
Passeriformes	Adult survival rate	0.34	0.10	32	0.52	7
Charadriiformes	Adult survival rate	0.51	0.07	27	0.45	7
Reptiles	Life-span	2.99	0.23	NG	NG	8

Relationships are given in the format $y = ax^b$, where y is survival measured on a yearly scale and x is body weight in grams.

References: 1. Sacher 1959; 2. Millar and Zammuto 1983; 3. Western 1979; 4. Jones 1985; 5. Lindstedt and Calder 1976, 1981; 6. Calder 1983; 7. Saether 1989; 8. Calder 1976.

NG: not given.

predators. An increase in mortality associated with reproduction may also be important and will be considered in chapter 6.

Age-related changes in survival may be a consequence of increased experience and development, but size itself can be a significant component of mortality rates. Interspecifically, there is a strong correlation between body size and survival in mammals (Ohsumi 1979; Promislow and Harvey 1990; Table 5.1), birds (Table 5.1), and fish (Beverton and Holt 1959; Beverton 1963; Pauly 1980; Adams 1980), though the mechanism(s) generating these relationships is far from clear. Peterson and Wroblewski (1984) developed a model for the mortality of fish within the pelagic ecosystem, which McGurk (1986) later showed empirically to be applicable to invertebrates and marine mammals (whales). Based on the following assumptions—

1. The rate of change in numbers is primarily due to predation.
2. The weight of prey is a constant fraction of the weight of the predator.
3. The system is in steady state.

Peterson and Wroblewski derived the relationship

$$M(W) \propto W^{-c} \qquad\qquad (5.18)$$

where $M(W)$ is the instantaneous rate of mortality and w is weight. The constant of proportionality depends upon a number of metabolic parameters, and the exponent c is equal to the exponent in the allometric relationship between growth and metabolic rate. Empirical estimates of these parameters for pelagic fish lead to the equation $M(W) = 1.92\ W^{-0.25}$, where the time scale is per year and weight is measured in grams. Over 16 orders of magnitude the equation of Peterson and Wroblewski predicts the observed mortality rates with very credible accuracy (Fig. 5.13). Allometric relationships between survival and body size are recorded for mammals, birds, and reptiles, the exponent ranging from -0.15 to 0.26, with a mean of 0.16 (Table 5.1). This variation is not reduced significantly

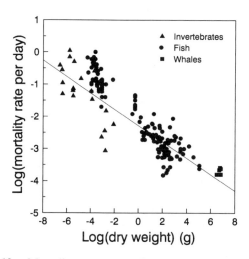

Figure 5.13. Mortality rate versus dry weight for three groups of marine animals: invertebrates, fish, and whales. The solid line shows the values predicted by the model of Peterson and Wroblewski (1984). Data from McGurk (1986).

by considering only data sets with large sample sizes: for example, the exponent for eutherian mammals is 0.2 ($n = 63$), while that for adult European birds is 0.09 ($n = 107$). Differences in the measures of survival (life-span, annual survival rate, instantaneous survival rate) preclude comparison of the constants of proportionality and may also reduce the value of comparisons between exponents. Whether there exists a general allometric relationship between size and survival among terrestrial animals comparable to that found by Peterson and Wroblewski remains to be determined, though the data are not encouraging.

The foregoing analyses all show that interspecific mortality rates decline with size. Studies within species also support the hypothesis that large size can be a protection against predators and environmental stress. Within the subtidal region, large size can be a refuge from predators (Connell 1970; Menge 1973; Vermeij 1974; Miller and Carefoot 1989; Palmer 1990). If sufficiently large, barnacles and mussels can withstand attacks from their principle predators—starfish (*Pisaster*) and gastropods (*Thais*) (Connell 1970, 1972; Dayton 1971; Paine 1976). Species within the bivalve genera, *Venus*, *Cerastoderma*, and *Modiolus*, can also achieve a size refuge (Ansell 1960; Seed and Browne 1978). Because it swallows its prey whole, prey size in the opisthobranch, *Navanax inermis*, is limited, and large size in potential prey can make them invulnerable to this predator (Paine 1965). Similarly, shell-crushing crabs cannot crush the shells of larger hermit crabs (Bertness 1981b,c).

Size-selective predation and a size refuge have been demonstrated in a number of studies of larval amphibia. Invertebrate predation is a major source of mortality in larval amphibia (Travis 1983a; Werner 1986) and may frequently be size-selective (Pritchard 1965; Travis 1983b). Size-selective predation has been demonstrated in dragonfly nymphs (Calef 1973; Caldwell et al. 1980; Brodie and Formanowicz 1983), the backswimmer, *Notonecta undulata* (Licht 1974), and larvae of the predacious diving beetle, *Dytiscus verticalis* (Brodie and Formanowicz 1983). Vertebrate predators may also be important, and in these limitations of gape can restrict the size of prey. A size refuge from salamander predation has been demonstrated for the larvae of *Rana aurora* (Calef, 1973), *Hyla gratiosa* (Caldwell et al. 1980), *Rana clamitans*, *Rana catesbiana* (Brodie and Formanowicz 1983), and *Ambystoma maculatum* (Stenhouse et al. 1983). The importance of size in the survival of amphibia can extend beyond the larval period, postmetamorphic survival in the wood frog, *Rana sylvatica*, being dependent on size but not age (Berven and Gill 1983).

The spines of sticklebacks are generally acknowledged to be morphological structures that have a defensive role, increasing manipulation time and hence the probability of escape (Werner 1974; Zaret 1980; Hoyle and Keast 1987). An analysis of scars resulting from unsuccessful attacks on sticklebacks showed that injuries were absent among juvenile fish (<50 mm), rare among subadult fish (50–70 mm), and common among adults, the frequency of scars increasing with body size (Reimchen 1988). From the pattern of scarring Reimchen concluded that at least one-third of the unsuccessful attacks were from avian piscivores. Further, since sticklebacks of all sizes are preyed upon, the pattern of scarring reflects an increasing likelihood of surviving an attack with body size (Reimchen 1988).

Perhaps one of the most well-known (certainly the most reanalyzed) cases of size selective mortality is that of Bumpus's sparrows. Bumpus (1899) collected the bodies of house sparrows, *Passer domesticus*, killed after a severe winter storm and compared these with the size of the birds that survived. Reanalysis of his data and the analysis of further samples showed that males were subject to directional selection, the larger individuals surviving, while the surviving females were both smaller and less variable (Johnston et al. 1972; Fleischer and Johnston 1984). The actual selective factors have not yet been isolated but large size is thought to be favored because of increased fasting ability (Murphy 1985). This cannot explain the decreases in size and variability of the females, which remain enigmatic.

A severe drought on the Galapagos island of Daphne Major greatly reduced the population of Darwin's medium ground finch, *Geospiza fortis*: The size of birds surviving this population crash was larger than the size of birds prior to the drought (Boag and Grant 1981). Concomitant with the drought and population decline, the seeds upon which the finch feeds decreased in abundance and increased in size. Boag and Grant (1981) hypothesized that the large birds survived best because they were able to crack the large and hard seeds that predominated in the drought. Although large adult size may be favored, further studies suggest that selection may favor small juveniles, though the mechanism generating size-specific mortality is unknown (Price and Grant 1984).

The importance of predation as a factor in size-selective mortality has been demonstrated in the freshwater zooplankters, in which it is the *larger* forms that are differentially removed by vertebrate predators (Galbraith 1967; Brooks 1968; Hall et al. 1976; Lynch 1977, 1980a; Mittelbach 1981). Lakes with fish have smaller-sized zooplankters than those without (Hrbáček and Hrbáčková-Esslová 1960; Brooks and Dodson 1965), while the introduction of fish or increase in fish abundance also leads to a shift to smaller zooplankters (Brooks and Dodson 1965; Wells 1970; Warshaw 1972). The effect of fish predation occurs not only across species but also within a population: for example, the mean size of *Daphnia lumholtzii* in the stomachs of the fish *Alestes baremose* was 1.28 mm (SE = 0.026), compared to 1.06 mm (SE = 0.024) in the plankton (Green, 1967). Increased visibility as a consequence of increased size may be important

(O'Brien et al. 1976, 1979), but in at least one species, *Bosmina longirostris*, it is an increased eye size rather than total size that appears to be critical (Zaret and Kerfoot 1975; Hessen 1985).

5.2. Reproduction

5.2.1. Describing Age Schedule of Reproduction

The age schedule of reproduction is conventionally denoted by the function $m(x)$, where only females are considered. Two general patterns of reproduction can be distinguished: semelparity and iteroparity. The former is colloquially termed "big bang" reproduction because it is applied to the situation in which an organism breeds once and then dies. Examples of this type of reproduction are mayflies, Pacific salmon, the European and American eels, octopods, bamboo, agaves, and leeks. Iteroparous organisms breed more than once. Iteroparity is probably more common than semelparity, though the distinction can become blurred. For example, only 15–20% of capelin and blackhead minnows survive to spawn more than once (Markus 1934; Winters and Campbell 1974); should we consider these iteroparous or semelparous species? The term "semelparity" could be reserved only for those organisms which die after reproduction whether or not they are maintained in an environment favorable for continued survival (e.g., Pacific salmon). Death in these cases appears to be a direct consequence of reproduction. Alternatively, we might include all organisms for which the inclusion of bouts of reproduction subsequent to the first have no impact on the optimal life history because survival is too low—i.e.,

$$\sum_{x=\alpha}^{\infty} e^{-rx} l(x) m(x) \approx \frac{log_e(l(\alpha)m(\alpha))}{\alpha} \tag{5.19}$$

where α is the age at first reproduction. It is preferable to retain the term "semelparity" only for the former types of organisms (criterion 1) and note that particular organisms might be considered semelparous for the purpose of analysis (criterion 2).

The shape of the age schedule of reproduction can be divided into three classes: uniform, asymptotic, and triangular (Fig. 5.14). Organisms that produce a constant number of offspring per breeding episode generate a uniform $m(x)$ function (top panel, Fig. 5.14): examples of this type of pattern are large mammals and birds. In some birds the broods of the first-time breeders may be smaller than older animals, generating a function that may be more like the second category, "asymptotic," or even in some instances a "triangular" function due to a decrease in brood size in older birds (Klomp 1970).

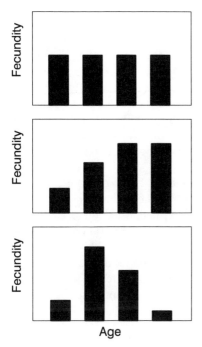

Figure 5.14. Schematic illustration of the three types of age schedules of fecundity. Each rectangle represents a breeding episode. Upper panel: uniform; middle panel: asymptotic; bottom panel: triangular.

Many ectotherms and plants continue to grow throughout life and increase their fecundity with size; as a consequence fecundity increases with age, but since growth generally slows down with age and approaches an asymptote, so does age-specific fecundity (middle panel, Fig. 5.14). Examples of species with an asymptotic fecundity curve are some populations of the grass *Poa annua* (Law et al. 1977), lizards (e.g., Tinkle and Ballinger 1972; Tinkle 1973; Vinegar 1975), the Iceland scallop, *Chlamys islandica* (Vahl 1981), and most species of fish.

Some organisms, most particularly insects, cease to grow at maturity and their age schedule of fecundity is triangular in shape, e.g., the beetle, *Pterostichus coerulescens* (van Dijk 1979), milkweed beetles, *Oncopeltus* (Landahl and Root 1969), the cabbage butterfly, *Pieris rapae* (Jones et al. 1982), and crickets (Roff 1984b). Other taxa in which a triangular $m(x)$ function is found are nematodes (Woombes and Laybourn-Parry 1984), *Artemia* (Browne 1982; Browne et al. 1984), rotifers (Jennings and Lynch 1928; Bell 1984b), small mammals (Krohne 1981), and many species of plants (Harper and White 1974). A useful descriptor of this curve is (McMillen et al. 1970a,b)

$$m(x) = c_1(1 - e^{-c_2 x})e^{-c_3 x} \tag{5.20}$$

where c_1, c_2, c_3 are constants, and x is age measured from the production of the first offspring. Notice that the first part of the function is an asymptotic function, mathematically the same as the von Bertalanffy function. (See Fig. 5.6.) This component, by itself, describes the age schedule of fecundity when size increases according to the von Bertalanffy equation and fecundity increases in proportion to size. It may also be used to approximate a uniform age schedule of reproduction, particularly if the first brood is smaller than subsequent broods.

Although the three shapes of the $m(x)$ function appear rather different these differences may become less distinct when the product $l(x)m(x)$, the expected number of female offspring at age x, is considered. Recall that the mortality function, $l(x)$, can be decomposed into several components, one of which specifies a constant risk, so that the probability of surviving to age x is $l(x) = e^{-Mx}e^{-F(x)}$, where M is the instantaneous (constant) rate of mortality, and $F(x)$ is some function of age (or size, etc.). The expected number of offspring for a triangular age schedule of reproduction is thus given by

$$l(x)m(x) = c_1(1 - e^{-c_2 x})e^{-c_3 x}e^{-Mx} e^{-F(x)}$$
$$= c_1(1 - e^{-c_2 x})e^{-x(c_3 + M)} e^{-F(x)} \tag{5.21}$$

The above function is equivalent to a fecundity function specified only by the first component (i.e., a uniform or asymptotic function) with a constant mortality rate not of M but $M + c_3$. Mathematically, therefore, the $l(x)m(x)$ curves may be very similar in form even though the biological assumptions about the shape of the $m(x)$ function may be quite different. This is illustrated in Fig. 5.15 in which are plotted the $m(x)$, $l(x)$, and $l(x)m(x)$ functions for two bird species—one, the European Tree sparrow, having a uniform fecundity schedule and the other, the yellow-eyed penguin, having an asymptotic age schedule of reproduction. In both cases the $l(x)m(x)$ functions are triangular; indeed it would be a peculiar life history that did not show this property! Nevertheless, this observation is extremely important in the analysis of optimal life histories because it means that an analysis may be far more general than implied by the specific biological assumptions initially made to generate a model.

Some organisms, particularly amphibia, reptiles, and fish, do not breed every year (Bull and Shine 1979). Females more frequently than males skip reproduction. In most cases, reproduction involves an accessory activity, such as migration, egg brooding (throughout incubation), and live-bearing. To what extent such activities are found in species that do not

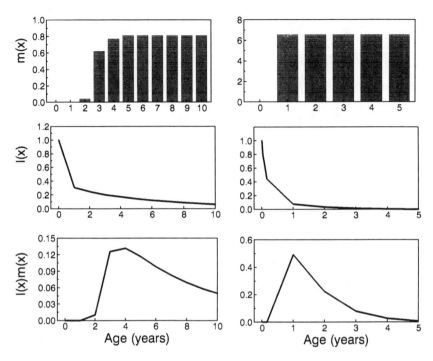

Figure 5.15.   Life tables for two species of bird: left panels, yellow-eyed penguin, *Megadyptes antipodes*; right panels, European tree sparrow, *Passer montanus*. Data from Tables 4 and 5 of Ricklefs (1973b).

skip episodes of reproduction is not clear. Among the urodele amphibians that exhibit a low frequency of reproduction there is a significantly higher frequency of maternal egg-brooders (67%) than among urodeles in general (37%, $P < 0.005$), and in snakes, live-bearing is significantly more frequent in species that skip years than snakes in general (80% versus 31%, $P < 0.005$). Bull and Shine (1979) speculated that a low frequency of reproduction is a consequence of energetic constraints on breeding, reproduction, and associated accessory activities depleting energy supplies to a level that cannot be replenished within a single year. In snakes of the temperate regions, the probability of reproduction is dependent upon the amount of lipid stored before winter; reproduction can itself prevent sufficient storage for the next year, resulting in biennial reproduction (Derickson 1976). Nonreproductive Arctic char, *Salvelinus alpinus*, of Nauyuk Lake in northern Canada lost 30% of their fall reserves of energy by the following spring, while postspawners, upon migration back to the sea, contained as much as 40% less energy than the nonreproductive fish (Dutil 1986). Postspawners required more than one summer to replenish their supplies and hence exhibited a 2-year cycle of reproduction—a feature common to northern fish species (reviewed in Dutil 1986).

5.2.2. Size and Reproduction

We have seen that the probability of survival may depend upon size rather than age, and so it is with reproduction: both the age at maturity and reproductive output may be more related to size than age. The importance of size in determining maturation is frequently cited but there are comparatively few data to support the claim, though it is a reasonable hypothesis. Metamorphosis to the adult form appears to be size-dependent in a variety of insects (Wigglesworth 1934; Nijhout and Williams 1974a,b; Nijhout 1975, 1979; Blakley and Goodner 1978; Woodring 1983), while maturation may depend to some extent on the attainment of a critical size in fish (Roff 1991a), rats (Childs 1991), reindeer (Skogland 1989), the plant teasel (Werner 1975), and plants in general (Harper 1977, p. 687; Kachi and Hirose 1985; Yokoi 1989).

Among ectotherms and plants a common finding is that fecundity increases with body size. In contrast to the relationship between size and maturity, that between size and fecundity has been extensively studied and I provide only a representative sample: plants (Primack 1979; Watkinson and White 1985); Crustacea (Green 1954; Jensen 1958; Daborn 1975; Bertness 1981c; Corey 1981; Rhodes and Holdich 1982); Mollusca (Glynn 1970; Green and Hobson 1970; Hughes 1971; Spight and Emlen 1976); Annelida (Creaser 1973); Echinodermata (Gonor 1972; Rutherford 1973; Menge 1974); Arachnida (Peterson 1950; Kessler 1971); Insecta (Chiang and Hodson 1950; Tyndale-Biscoe and Hughes 1968; Clifford and Boerger 1974; Waage and Ng 1984; Cooke 1988; Marshall 1990); Tunicata (Wyatt 1973); fish (Bagenal 1966; Wootton 1979); Amphibia (Tilley 1968, 1972; Salthe 1969; Wilbur 1977) and Reptilia (Tinkle 1967; Lemen and Voris 1981; Gibbons et al. 1982; Ford and Seigel 1989).

The relationship between clutch size and body size is frequently plotted on an arithmetic scale: while such a regression may be statistically appropriate if the relationship is linear within the observed range, a better method is to use a log/log transformation. The reason for this is that when the range in body size is large it is generally found that the relationship is allometric—i.e., is best expressed in the form,

$$\text{Fecundity} = a\text{Length}^b \tag{5.22}$$

where a and b are constants. This relationship is linearized by taking logarithms,

$$\log(\text{Fecundity}) = \log(a) + b \log(\text{Length}) \tag{5.23}$$

If fecundity varies allometrically but the range in length (or some other body dimension) is small there may be no statistical difference between the linear/linear and log/log regressions. However, for purposes of comparison the latter regression is to be preferred as a general procedure. Is there any biological reason to expect fecundity to vary allometrically? The simplest explanation is that the amount of space available for eggs increases with size. On such a scenario the value of b should be about three, the amount of space increasing as the cube of length. This prediction depends on size dimensions increasing isometrically—i.e., upon there being no change in shape with size and upon egg size remaining constant. The only extensive analysis of the distribution of the exponent b is that on fish by Wootton (1979). The range in values of b is enormous, from about one to seven, though this reflects both real and sampling variation, and the true range might be much smaller. The modal class is 3.25–3.75, suggesting that fecundity increases slightly faster than expected according to the geometric argument; but given the variability in the data it would be premature to draw any definite conclusions. An exponent of three need not necessarily imply a limited amount of space for eggs. In many fish species energy for egg production is stored in organs such as the liver, and the size of this organ might constrain fecundity, rather than actual space for the maturation of the eggs. The rate at which resources can be gathered might be proportional to weight, which would also produce an exponent of three.

Endothermic vertebrates do not typically show an allometric relationship between size and fecundity. Why is this so? The obvious answer is that the range in body size is small, the relative size of offspring is large, and the clutch size is small. Large mammals, such as deer, goats, and marine mammals, tend to have only one or two young: considering the size at birth it would take a relatively enormous increase in body size to accommodate another offspring. Small mammals are frequently quite fecund, but their range in body size and fecundity doesn't match that of ectothermic vertebrates. Nevertheless, a positive correlation between litter size and the mother's body weight has been demonstrated in various cricetine rodents: *Peromyscus longicaudus* (Hamilton 1962), *P. leucopus* (Svendsen 1964; Lackey 1978), *P. maniculatus* (Svendsen 1964; Myers and Master 1983), *P. yucatanicus* (Lackey 1976), *P. polinotus* (Kaufman and Kaufman 1987), *Neotoma floridana* (McClure 1981), and *Sigmodon hispidus* (McClenaghan and Gaines 1978): and in the microtine rodents, *Microtus ochrogaster* (Fitch 1957; Keller and Krebs 1970), *M. pennsylvanicus* (Keller and Krebs 1970), *M. townsendii* (Anderson and Boonstra 1979), and *Clethrionomys rufocanus* (Kalela 1957). A positive correlation between body size and litter size is also found in mountain hares, *Lepus timidus* (Iason 1990).

In the analysis of life history variation within a species the relationship between female body size and fecundity must clearly be acknowledged.

Equally important is the relationship between size and mating success in males. Though a large size advantage has been observed in mammals and birds (Byrant 1979; Clutton-Brock et al. 1983; Price 1984; Sauer and Slade 1987), much of the detailed investigations have dealt with ecototherms, where numerous studies have shown that large males are able to secure more matings (Table 5.2). A small male advantage has been detected in two species of pyralid moths (Marshall 1988), but this appears to be unusual. However, a common finding is that mating is assortative with respect to body size (reviewed in Ridley 1983; Fairbairn 1988; Crespi 1989). Given the greater fecundity of large females, size-assortative mating should differentially increase the fitness of large males. The advantage of large size in males does not enter into the characteristic equation, and hence this approach to analysis is not appropriate. Male size-selective advantages will be frequency-dependent and hence the optimal body size can be found using the game-theoretic approach (Maynard Smith and Brown 1986).

Table 5.2. A survey of species in which large males have been shown to have a mating advantage

Taxon	Advantage in contests[a]	Mating advantage
Arachnids	3	—
Insects	13	17
Crustacea	3	7
Fish	—	23
Amphibia	2	16

[a]Species in which large males have been shown to be able to defeat smaller males but an advantage in terms of securing more matings has not been demonstrated.

References consulted:

Arachnids: Beebe 1947; Rovner 1968; Buskirk 1975; Riechert 1978.

Insects: Ewing 1961; Mason 1964, 1969; Scheiring 1977; McCauley and Wade 1978; Borgia 1980, 1982; Thornhill 1980a, 1981b; Eberhard 1982; Johnson 1982; Gwynne 1982; McCauley 1982; Partridge and Farqhar 1983; Ridley 1983; Thornhill and Alcock 1983; Dodson and Marshall 1984; Otronen 1984; Crespi 1986; Simmons, L.W. 1986, 1988; Flecker et al. 1988; Santos et al. 1988; Brown 1990a,b; Pitafi et al. 1990.

Crustacea: Hazlett 1968; Ridley and Thompson 1979; Ridley 1983; Ward 1989.

Fish: Constanz 1975; Warner and Downs 1977; Perrone 1978; Schmale 1981; Downhower et al. 1983; Noonan 1983; Warner 1984a,b; Keenleyside et al. 1985; Berglund et al. 1986; Thompson 1986; Hikada and Takahashi 1987; Myers and Hutchings 1987; Bisazza and Marconato 1988; DeMartini 1988; Foote 1988; Hastings 1988; Petersen 1988; Bisazza et al. 1989; Cote and Hunte 1989; Magnhagen and Kuarnemo 1989; Marconata et al. 1989; Petersen 1990.

Amphibia: Wells 1977a; Davies and Halliday 1978, 1979; Howard 1978, 1980, 1984, 1988; Wilberg et al. 1978; Ryan 1980, 1983; Berven 1981; Fairchild 1981; Arak 1983a; Ridley 1983; Robertson 1986a,b; Given 1987, 1988a; Morris 1989; Wagner 1989.

5.2.3. Reproductive Effort

An organism can channel energy into three avenues: maintenance, growth, and reproduction. The age schedule of reproduction and, at least in some cases, survival, is dependent upon the allocation of energy into growth versus reproduction. It is, therefore, necessary to have some measure of this allocation: such measures have been commonly termed "reproductive effort," though a number of definitions have been suggested. One definition of reproductive effort, RE, is "the proportion of total energy, procured over a specified and biologically meaningful time, that an organism devotes to reproduction" (Hirshfield and Tinkle 1975), which in symbols can be written (Swingland 1977; Tuomi et al. 1983),

$$RE = \frac{P_r}{P_r + P_g + R} \tag{5.24}$$

where P_r is the production diverted into reproduction,
P_g is the production diverted into growth, and
R is respiration.

In the analysis of the optimal allocation to growth and reproduction, reproductive effort is frequently defined to include only the surplus energy: the amount of net energy left after all additional costs of living—namely, standard metabolism and costs of locomotion—have been discounted (Ware 1980). In an immature organism, surplus energy is equivalent to total somatic growth whereas after sexual maturity it represents the energy diverted into both growth and reproduction. An organism can divert energy from the maintenance of body components or may break down tissues to pay for present reproduction. Defining reproductive effort only on the basis of surplus energy makes the assumption that no such sacrifice occurs.

Another problem with the above definition of reproductive effort is that it is a proportion and hence does not take into account the costs of acquiring the energy. An organism might devote a lower fraction of its productivity to reproduction than another, but the costs of obtaining that energy might be greater in the former. By the above definition(s) the latter organism has a higher reproductive effort, but a case can certainly be made that in reality the former has the higher effort. There is no simple solution to this problem; one can only remember that the term must be precisely defined in any analysis and that comparative analyses must consider the issue of costs of acquiring energy, either by using large sample sizes (i.e., many different species, etc.) in which the costs can be considered random effects or by using organisms in which it can be reasonably assumed that the costs are similar.

Apart from the philosophical problems of definition, there are serious experimental problems in determining reproductive effort as defined above. Harper and Ogden (1970) suggested the alternate, operational definition,

$$RE = \frac{\text{Total weight of propagules}}{\text{Total biomass at maturity}} \tag{5.25}$$

where reproductive effort is typically calculated on a per clutch basis. The above definition assumes

1. Weight of other reproductive structures is negligible.

2. Weight of propagules is proportional to production diverted to reproduction (P_r).

3. Total productivity is proportional to total weight by the same amount as 2.

4. There is no parental care.

The last assumption does not hold for most birds and mammals, nor for many fish, some amphibians and reptiles, and some invertebrates. The weight of other reproductive structures, particularly in plants, may be significant, though since these may be photosynthetic there are obvious problems in assessing their impact on total reproductive effort. Thompson and Stewart (1981) discuss this problem and suggest that in plants the appropriate currency is mineral nutrients, rather than calories or biomass. These problems notwithstanding, the relative biomass of propagules is readily measurable and does provide a useful, if crude, index of reproductive effort in females. However, since it does not include parental care, which can involve significant costs in endotherms, the use of the term "reproductive effort" may be misleading. In the fisheries literature the relative weight of the eggs plus associated ovarian tissue has been termed the "gonadosomatic index" (GSI). Studies on mammals, reptiles, and frequently plants generally measure only the weight of the clutch in relation to female weight and use the term "relative clutch mass" (RCM).

The gonadosomatic index has been calculated in two ways,

$$GSI_1 = \frac{\text{Gonadal weight}}{\text{Gonadal weight} + \text{Somatic weight}} \tag{5.26}$$

$$GSI_2 = \frac{\text{Gonadal weight}}{\text{Somatic weight}} \tag{5.27}$$

One can be converted to the other by the transformations

$$GSI_1 = \frac{GSI_2}{1 + GSI_2} \qquad (5.28)$$

$$GSI_2 = \frac{GSI_1}{1 - GSI_1} \qquad (5.29)$$

Relative clutch mass has been calculated in an analogous fashion. Unfortunately, it is not always clear which measure (GSI_1 or GSI_2) has been adopted. For small values the two measures are very similar, but for values much larger than 20% the difference increases rather steeply. Except where indicated, I shall use GSI_1 and simply refer to it as the gonadosomatic index (GSI) or, where appropriate, relative clutch mass (RCM).

Reproductive effort, however defined, can be calculated on a per-clutch or per-unit-time basis. The advantage of calculating it on a per-clutch basis is that it gives a measure of the potential stress that may be placed on an organism. For example, the relative clutch mass of the common lizard, *Lacerta vivipara*, is 29% (Avery 1975): the sheer weight of a clutch of this size must impede the movement of the lizard and require increased foraging to supply the necessary energy for the growth of the developing young. It is significant that while this lizard typically uses a flight response to escape predators, the gravid female switches to a dependence on crypsis (Bauwens and Thoen 1981). If an organism produces more than one clutch per year the calculation of reproductive effort on an annual basis gives us some idea of the increase in foraging required, and hence the potential increased risk from predators.

Table 5.3 shows the allocation of resources to maintenance, growth, and reproduction in a flatfish and three lizard species. The proportion of annual productivity devoted to egg production (Rep1 in Table 5.3) corresponds

Table 5.3. Percentage allocation of annual production to maintenance, growth, and reproduction in females of four ectotherms, a flatfish (data from MacKinnon 1972) and three lizard species (data from Tinkle and Hadley 1975)

Species	% of annual production				
	Maint.	Growth	Rep1	Rep2	RE
Hippoglossoides platessoides	66	19	15	44	9–13
Sceloporus jarrovi	71–83	7–18	10–13	35–65	15
Sceloporus graciosus	74–77	0–2	23–24	92–100	26
Uta stansburiana	78–79	2–3	19	86–90	20–22

Reproductive effort (RE) calculated as gonadosomatic index for the flatfish and relative clutch mass for the lizards. Percentage allocated to reproduction estimated based on total production (Rep1) and surplus energy = growth + reproduction (Rep2).

quite closely to reproductive effort as measured by the gonadosomatic index or relative clutch mass. The ranking of the reproductive effort indexes also corresponds roughly with reproductive effort measured as the fraction of surplus energy devoted to reproduction (Rep2 in Table 5.3). But there are clear quantitative differences, and hence these reproductive effort indexes (gonadosomatic index or relative clutch mass) could not be used to estimate the proportion of surplus energy devoted to reproduction, as required in the model of Myers and Doyle (1983, see 5.1.2) or those discussed in chapter 8.

Browne and Russell-Hunter (1978) estimated reproductive effort in molluscs by two methods: first, the percentage of nonrespired assimilation that is diverted into reproduction annually, and second, the ratio between the amount of carbon that is channeled into reproduction and the amount of carbon contained in the average adult female. The second measure approximates the gonadosomatic index, except that it is measured over a number of clutches, and hence can exceed unity. The former method does not include maintenance costs, but the two measures are highly correlated ($r = 0.82$, $n = 8$, $P < 0.02$, data from Tables 1 and 2 of Browne and Russell-Hunter 1978), further supporting the use of the gonadosomatic index as an index of reproductive effort.

How variable is the gonadosomatic index or relative clutch mass—i.e., how variable is reproductive effort, as indicated by GSI or RCM? Among plants, reproductive effort indexes may vary from 0.2 to 66% (Table 5.4), with a similar range among invertebrates, ranging from 9% in starfish to as high as 75% in a crab spider (Table 5.5). Among fish, the gonadosomatic index ranges from 2% to 31% in females and 0.2% to 16% among males (Fig. 5.16). The average gonadosomatic index of male fish is much lower than that of females (3.7% versus 12.2%)—a pattern that is found consistently between sexes within species (Fig. 5.16). The relative clutch mass of lizards extends from 5% to 40%, with an average value of 21%, which is considerably higher than fish (Fig. 5.17). Among snakes, relative clutch mass is even higher, ranging from 10% to 60%, with a mean of 31% (Fig. 5.17). Blueweiss et al. (1978) found that clutch mass increased allometrically with body mass in reptiles: clutch mass = 0.35 Body weight$^{0.88}$. This equation must be viewed with caution since there appear to be differences in relative clutch mass associated with taxon (e.g., the separation between snakes and lizards suggested by Fig. 5.17). An exponent less than one indicates that the relative clutch mass decreases with size.

The frequent practice of regressing reproductive effort indexes against size is to be discouraged since it violates the basic assumption of regression analysis of independence between measurements of the dependent and independent variables (i.e., the dependent variable contains 1/x: for a discussion of this problem see Klinkhamer et al. 1990). However, if one

Table 5.4. Reproductive effort (RE), as measured by weight of reproductive parts, in various plant species

Scientific name	RE[a]	Reference
Semelparous (annuals + monocarpic perennials)		
Chamaesyce hirta	7–26	1
Dipsacus sylvestris	27	2
Lupinus nanus	61 (29 S)	3
Helianthus annuus	14–30 S	4
Carthamus tinctorius	13–25 S	4
Senecio vulgaris	14–27 S	4
Chrysantheum segetum	26–27 S	4
Calendula officinalis	27 S	4
Matricaria matricarioides	35 S	4
Allium porrum	33	5
Polygonum cascadense	38–58	6
Agave palmeri	59	7
8 grass species	41–66 S	8
Iteroparous (perennials)		
Lupinus variicolor	18 (5 S)	3
Lupinus arboreus	20 (6 S)	3
Solidago (5 species)	2–10	9
Grasses (31 species)	0.2–31 S	8
Achillea milefolium	22	10
Artemisia vulgaris	2	10
Cirsium arvense	7	10
Taraxacum officinale	24	10
Tussilago farfara	26	10
Astrocaryum mexicanum	1–40	11

References 1. Snell and Burch 1975; 2. Caswell and Werner 1978; 3. Pitelka 1977; 4. Harper and Ogden 1970; 5. Boscher 1981; 6. Hickman 1975; 7. Howell and Roth 1981; 8. Wilson and Thompson 1989; 9. Werner 1976; 10. Bostock and Benton 1979; 11. Pinero et al. 1982.

[a]Values followed by "S" are for seeds only.

wishes to obtain the relationship between the reproductive effort index and body size a regression of this index against body size is permissible provided that the correlation coefficient is high and that any tests of statistical significance are based on the bias expected by regressing $1/x$ against x. In the aforementioned reptile data the correlation coefficient is very high ($r = 0.98$; unfortunately no sample size is given, but it is probably in excess of 50), and in the absence of the raw data an estimate of the relationship between relative clutch mass and body size can be obtained by dividing throughout by body weight to obtain

$$\text{Relative clutch mass} \approx 0.35 \text{ Body weight}^{-0.12} \qquad \textbf{(5.30)}$$

Table 5.5. Reproductive effort indexes in some invertebrate species

Common name	Scientific name	RE (%)	References
Water flea	*Daphnia magna*	61–70	1
Isopod	*Ligia oceanica*	23	2
Crab spider	*Misumena vatia*	75	3
Starfish	*Leptasterias hexactis*	9	4
Starfish	*Pisaster ochraceus*	10	4
Sea urchin	*Strongylocentrotus drobachiensis*	15–17	5
Winkle[a]	*Littorina* spp. (4 *spp.*)	9–60	6
Beetle	*Callosobruchus maculatus*	56	7

References 1. Tessier et al. 1983; 2. Saudray 1954 cited in Lawlor 1976a; 3. Fritz and Morse 1985; 4. Menge 1974; 5. Vadas 1977; 6. Hughes and Roberts 1980; 7. Mitchell 1975.

[a]Based on annual production, not per brood as in others.

A 10-g reptile is predicted to have a relative clutch mass of about 27% and a 100-kg animal to have an RCM of 9%. That snakes typically seem to have relative clutch masses in excess of 27% (Fig. 5.17) further indicates that this equation may not be fully representative. For fish there is no correlation between adult size and the gonadosomatic index ($r < 0.0001$, $n = 91$, data from numerous literature sources). Within fish species there is a correlation between body weight and gonad weight, but the slope of the allometric regression is greater than one, indicating that the gonadosomatic index increases with size (Roff 1983b), a phenomenon found in other taxa also. (See chapter 8.) This illustrates the general caution that interspecific comparisons should not be used to infer intraspecific relationships.

In mammals there is a very definite association between litter mass and body weight, analysis of 250 species giving the relationship

$$\text{Litter mass} = 0.11 \, \text{Body weight}^{0.77} \quad r = 0.96 \qquad (5.31)$$

where all masses are in kg (Millar 1981). Again, to estimate relative clutch mass we divide throughout by body weight,

$$\text{Relative clutch mass} = 0.11 \, \text{Body weight}^{-0.23} \qquad (5.32)$$

As with the reptile model, the above RCM is calculated relative to the somatic weight alone and does not consider uterine tissues such as the placenta. In the house mouse, *Mus musculus*, the placenta and associated fluids weigh 1.6 g, the total weight of young is 11.2 g, and the female weights 21 g immediately after parturition (Myrcha et al. 1969). The relative

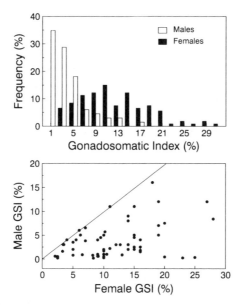

Figure 5.16. Upper panel: Frequency distributions of gonadosomatic indexes for males and females of various fish species (*n* =
106 for females, *n* = 66 for males). Lower panel: Gonadosomatic
index of male versus gonadosomatic index of female for 60 species of fish (*r* = 0.39, *P* < 0.005). Solid line is 1:1 reference line.

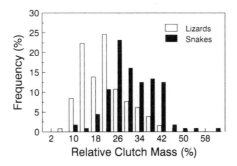

Figure 5.17. Relative clutch mass among lizards and snakes.
Lizard data from Huey and Pianka (1981), Vitt and Price (1982),
and Ananeva and Shammakov (1986). Snake data from Siegel
and Fitch (1984) and Seigel et al. (1986). For lizards *n* = 130, for
snakes *n* = 112.

clutch mass with the placenta is thus 61% compared to 53% if the placenta is excluded (38% versus 33% if total weight is used to compute RCM). In the bat *Tylonycteris pachypus*, litter weight, including all fetuses, fluid, and membranes, measures 1.4 g while litter weight is 1.2 g (Medway 1972). Thus exclusion of the placenta is not likely to greatly change the estimated relative clutch masses.

According to the above equation a small mammal such as a deer mouse (*Peromyscus*) weighing 30 g will have a relative clutch mass of 25%, while a mammal of 100 kg is expected to have an RCM of just 4%, and an elephant reaching a weight of 6,000 kg will achieve an RCM of only 1.5%. These results are fully consistent with the relative clutch masses calculated directly for these weights from the data of Millar (1981). No bats were included in Millar's data set; given the stringent requirements for flight capability over terrestrial modes of locomotion we might expect that bats would have lower relative clutch masses. Such is not the case, the allometric relationship between clutch mass and body size for bats being not significantly different from that of other mammals (Kurta and Kunz 1987).

Birds produce only one egg at a time and hence it would be misleading to compute the gonadosomatic index based on clutch size. Based on 1,244 observations, Rahn et al. (1985) estimated the relationship between egg mass and body mass in passerines as

$$\text{Egg mass (g)} = 0.258 \text{ Body weight}^{0.73} \qquad r = 0.96 \qquad \textbf{(5.33)}$$

and for nonpasserines (n = 557)

$$\text{Egg mass (g)} = 0.399 \text{ Body weight}^{0.72} \qquad r = 0.94 \qquad \textbf{(5.34)}$$

Rearranging these equations (noting again that there are serious statistical problems with algebraic manipulations of regression equation; but in this instance the correlation coefficients are so high that these can be ignored) gives

$$\text{Relative egg mass} = 0.258 \text{ Body weight}^{-0.27} \qquad \textbf{(5.35)}$$

$$\text{Relative egg mass} = 0.399 \text{ Body weight}^{-0.28} \qquad \textbf{(5.36)}$$

for passerines and nonpasserines, respectively. Passerine body weights recorded by Rahn et al. (1985) range from 4 g to 1,200 g, giving a variation in relative egg mass from 18% to 4%, with the highest density of body sizes between 20 and 60 g, which translates into a typical relative egg mass of 11% to 9%. For nonpasserines body weights range from a 3-g hummingbird to the 92,000-g ostrich, from which the range in relative egg mass is 29% to 2%. Most nonpasserines fall in the weight range from 100 g to 1,200 g, for which the range in relative egg mass is 11% to 5%. Thus, for passerines and many nonpasserines, a single egg represents a fairly substantial investment of biomass. The basal metabolic rate of passerines increases allometrically with body weight, with virtually the same exponent (0.71) as the egg mass/body weight relationship (0.73). From this Rahn et al. (1985) estimated that the cost of producing an average passerine egg is equal to 41% of the daily metabolic rate of the adult. The figure is much larger for nonpasserines: in ducks each egg is equivalent to 180% of the basal metabolic rate (Table 5.6). The cost in time of producing an egg can be quite considerable—that for seabirds ranging from 10 to 40 days—and can significantly influence the potential clutch size (Goodman 1974).

The principal energetic cost of reproduction in birds and mammals is in supplying food to the offspring prior to weaning/fledging. Birds appear to increase their rate of activity from approximately 2.6 times their basal metabolic rate (BMR) when nonreproductive to between 3.4 and 3.9 times BMR when raising young (Table 5.7), a rate that is approximately equivalent to that of heavy human labor (Drent and Daan 1980). Drent and Daan (1980) speculated that above 4.0 times basal metabolic rate a bird would experience severe physiological problems. This level is achieved by a 53% increase, a value exceeded by the house martin and the long-eared owl (Table 5.7), by tree swallows, *Tachycineta bicolor* (Williams 1988), and by some seabirds (Roby and Ricklefs 1986). Though immediate effects of this stress have not been recorded, future effects such as reduced survival are quite likely.

Table 5.6. Typical values for relative egg mass (REM—egg weight as % adult body weight) and energy requirements of egg laying in several groups of birds

Group	REM (%)	Typical clutch size	Cost per egg (% above BMR)	% Increase[a]
Passerines	10	4–5	45	17
Galliformes	6	10–12	126	48
Hawks and owls	7	2–4	39	15
Ducks	6	10–12	180	69
Shorebirds	17	4	149	57
Gulls and terns	16	3	170	65

Data from Ricklefs (1974).

[a]Computed assuming that a nonreproductive bird typically operates at 2.6 × BMR per day (Drent and Daan 1980). At least for galliform birds this may be an overestimate, King (1973) citing maintenance cost of 1.7 × BMR per day. Thus the increases may be larger than given above.

Table 5.7. The energetic cost of raising young in some birds

Common name	Scientific name	Cost per day (% BMR)	% increase over[a] nonreproductives
Purple martin	*Progne subis*	180–210	8–19
House martin	*Delichon urbica*	300	54
Chaffinch	*Fringilla coelebs*	250	35
Ring dove	*Streptopelia risoria*	160 (1 young)	0
		250 (2 young)	35
Long-eared owl	*Asio otus*	>300	>54
Glaucous-winged gull	*Larus glaucescens*	280	46

Data from Drent and Daan (1980).

[a]Percentage increase based on a daily rate of 2.6 × BMR for nonreproductive birds (Drent and Daan 1980).

It is somewhat easier to measure the energetic costs in mammals since these supply milk to their offspring. (For recent reviews see Loudon and Racey 1987; Gittleman and Thompson 1988.) Milk yield per day increases allometrically with body size with an exponent that ranges from 0.64 (data from Oftedal 1984; and Oftedal et al. 1987) to 0.77–0.79 (Hanwell and Peaker 1977; Martin 1984), differences reflecting taxonomic variation and the inclusion of domesticated animals. For the present discussion I shall use the equation of Hanwell and Peaker (1977), which excluded domesticated species,

$$\text{Milk yield (kg/day)} = 0.084 \text{ Body weight}^{0.77} \qquad (5.37)$$

where body weight is in kg. Because the exponent is less than one the relative amount of milk produced per day decreases with body size. Thus a 5-g shrew produces a mass of milk equivalent to 28% of its weight, while a 1,000-kg ox produces only 2% of its weight in milk per day. Though it forms a decreasing proportion of body weight the energetic cost of producing the milk remains relatively constant, the energy required to support lactation increasing from 50% to 150% during lactation independently of body size (Table 5.8). An increase in food intake is probably the main mechanism by which small mammals support lactation costs, while large mammals, producing a relatively small fraction of milk per unit weight, can draw upon body reserves to a much greater extent (Hanwell and Peaker 1977; Gittleman and Thompson 1988). Thus a pygmy shrew (*Sorex minutus*) weighing 5 g must more than double its daily intake and eat as much as four times its own weight each day (Hanwell and Peaker 1977).

The increased demands of lactation almost certainly place severe physiological demands on a mammal, particularly a small mammal, and though

Table 5.8. Percentage increase in daily energy requirements of pregnant or lactating mammals above those of nonreproductives: $100(x - y)/y$, where y is the maintenance costs and x the cost during pregnancy or lactation

Common name	Scientific name	Pregnancy	Lactating
Cow	*Bos taurus*	29	—
Pig	*Sus scrofa*	—	100
Human	*Homo sapiens*	10–23	50
Sheep	*Ovis aries*	81	146
Rabbit	*Oryctolagus cuniculatus*	27	—
Norway rat	*Rattus norvegicus*	46	152
Cotton rat	*Sigmodon hispidus* (5 young)	25	66
	(7 young)	19	111
Wood rat	*Neotoma floridana*	22	65[1]
Squirrel	*Spermophilus saturatus*	—	55 (193)[2]
Common vole	*Microtus arvalis*	32	133 (300)[3]
Bank vole	*Clethrionomys glareolus*	24	92 (137)[4]
Hamster	*Phodopus sungorus* (<5 young)	33	45[5]
	(>5 young)	33	90[5]
House mouse	*Mus musculus*	34	111
Deer mouse	*Peromyscus leucopus*	18	140[6]
Shrew	*Crocidura olivieri*	—	34 (113)[7]
Shrew	*Crocidura viaria*	—	49 (106)
Shrew	*Crocidura russula*	27	49 (128)
Shrew	*Sorex coronatus*	—	139 (285)
Shrew	*Sorex minutus*	—	130 (302)

Animals are listed in approximate descending order of weight. Except where indicated, data taken from Randolph et al. (1977).

[1]McClure 1987; [2]Kenagy 1987; Kenagy et al. 1990, peak requirements shown in parentheses; [3]Migula 1969; [4]Kaczmarski 1966; [5]Weiner 1987; [6]Millar 1975, 1978; [7]all shrew data from Genould and Vogel 1990.

these demands might be satisfied under normal conditions, any environmental challenge increases the likelihood of the female succumbing to hypothermia, disease, predation, etc. The same argument can be applied to the cost of producing young in general, but this is difficult to measure from an index of reproductive effort alone. Physiological stress on any mature organism may arise because it cannot supply all the demands of the developing gonads from surplus energy and must degrade its own body tissues. This fact may not be evident from the reproductive effort index since it depends not upon whether the index is large or small but upon its magnitude in relationship to the surplus energy. Furthermore, reproductive effort, measured by the gonadosomatic index, relative clutch mass, or proportion of calories diverted into reproduction, ignores the possible importance of particular compounds required by the developing young. Thus indexes of reproductive effort must be interpreted with caution.

The above discussion has centered on the reproductive effort of females: the reproductive effort of males is much more difficult to measure as it generally involves behavioral components that are hard to assess from an energetic perspective. Testis weight is generally lower than clutch weight. In fish there is a significant, though weak, correlation between the female gonadosomatic index and that of the male (Fig. 5.16), with the average male GSI being 3.7%. Although the male gonadosomatic index is consistently below that of the female it can achieve the relatively large value of 16%. Litter mass and testis weight are both highly correlated with body mass in primates: for females

$$\text{Litter mass} = 0.084 \text{ Body weight}^{0.69} \qquad (5.38)$$

($r = 0.91$, $n = 37$, $P < 0.001$, data from Millar 1981, all weights in kg); and for males

$$\text{Testis weight} = 0.0038 \text{ Body weight}^{0.67} \qquad (5.39)$$

($r = 0.77$, $n = 33$, $P < 0.001$, data from Harcourt et al. 1981, all weights in kg). Primates range in size from the diminutive *Nycticebus murinus*, weighing a mere 0.085 kg, to *Gorilla gorilla*, in which the females weigh 60 kg and the males 169 kg. Over this range, relative litter masses decrease from 18% to 2% and relative testis weight from 0.9% to 0.07%. There are only 12 species common to both data sets and hence an analysis of variation between the sexes is not very conclusive: the correlation between the relative clutch mass and testis weights is positive but fails to reach the 5% level of significance ($r = 0.42$, $0.1 > P > 0.05$).

Analysis of testis weight in 133 species of mammals covering a wide range of orders gives a allometric relationship similar to that found for primates alone, the coefficient and exponent being 0.0051 and 0.72, respectively (Kenagy and Trombulak 1986). Furthermore, no order differs significantly from the overall regression, though some individual species may deviate markedly from the regression, the testes of some rodents being as much as 8% of the body mass. In primates alone, all mammals considered together, and mammals excluding primates, there is an association between relative testis weight and mating system: males in multimale breeding systems have relatively large testes in comparison to males in monogamous or single-male systems (Harcourt et al. 1981; Kenagy and Trombulak 1986).

In some species of Orthoptera (crickets, grasshoppers, and their kin) and Anura (frogs and toads), females are attracted to the males by the male calling song. The metabolic cost of calling may be as great as 20-fold greater than resting costs (Table 5.9), which represents at least a short-term drain on energy reserves. This drain is evidenced by a considerable weight loss in male anurans during the breeding season (Table 5.10). Although this drain apparently does not retard future growth in *Rana temporaria* (Ryser 1989), small male carpenter frogs, *Rana virgatipes*, do appear to suffer a loss in growth related to the amount of reproductive effort (Given 1988b).

Male sagebrush crickets, *Cyphoderris strepitans*, provide their mates with two forms of nutrient investment, a spermatophore, and while copulating, hemolymph that oozes from a wound made by the female. Following mating, calling rate is significantly reduced, illustrating both the large nutrient investment (spermatorphore plus body fluids) given to the female and the

Table 5.9. Fractional increase in energy metabolism (calling/resting) during calling in some insects and two frog species

Species	Scientific name	VO_2 calling/resting	Reference
Katydid	*Neocephalus robustus*	18	1
Katydid	*Euconocephalus nasutus*	13	1
Bladder cicada	*Cystosoma saundersii*	20	2
Cricket	*Anurogryllus arboreus*	13	3
Cricket	*Oecanthus celerinctus*	8	3
Cricket	*Oecanthus quadripunctatus*	8	3
Cricket	*Gryllotalpa australis*	13	4
Cricket	*Teleogryllus commodus*	4	4
Frog	*Physalaemus pustulosus*	6	5
Frog	*Hyla versicolor*	21	6

References: 1. Stevens and Josephson 1977; 2. MacNally and Young 1981; 3. Prestwich and Walker 1981; 4. Kavanagh 1987; 5. Bucher et al. 1982; 6. Taigen and Wells 1985.

Table 5.10. Weight loss during the breeding season in some male anurans

Species	Percentage loss	Reference
Bufo bufo	21	Arak 1983b
Bufo calamita	13	Arak 1983b
Rana clamitans	0–30[a]	Wells 1978
Rana temporaria	25	Ryser 1989
Ranidella signifera	33	MacNally 1981
Ranidella parisignifera	36	MacNally 1981

[a]Weight loss significantly correlated with size, larger males losing the highest percentage.

high energetic cost of calling (Sakaluk and Snedden 1990). Nevertheless, in the frog, *Physalaemus pustulosus*, Ryan et al. (1983) estimated that the female expended 13 times more energy during reproduction than the male. In the lizard, *Uta stansburiana*, the female expends twice as much energy as the male on reproduction (Nagy 1983).

Male sage grouse, *Centrocerus urophasianus*, display to females at a lek, the display comprising both strutting and vocalization. Loss of weight by the males over the breeding season suggests that displaying may be energetically expensive (Gibson and Bradbury 1985). Using the doubly labeled water technique of measuring CO_2 production in free-ranging animals, Vehrencamp et al. (1989) estimated that, over a 24-hour time period, the energetic expenditure of a vigorous male sage grouse may equal four times basal metabolic rate. Assuming a normal expenditure of $2.6 \times$ BMR (Drent and Daan 1980), the cost of display represents a 54% increase, a level approximately equivalent to the energetic cost of raising young. (See Table 5.7.) This figure agrees well with the estimate of 40–50% obtained by Vehrencamp et al. (1989)

5.3. Summary

The life table can be conveniently divided into two components—the age schedule of survival and the age schedule of female births. The former is traditionally designated $l(x)$, the latter $m(x)$, where x is age. Selection, however, operates on the product of these and hence different biological assumptions on the shape of the $l(x)$ and $m(x)$ functions may nevertheless give the same product function $l(x)m(x)$.

Two methods of displaying the change in mortality with age are the $l(x)$ and $q(x)$ curves. The former describes the probability of surviving to age x, the latter the mortality rate between ages x and $x + 1$. Plotted on a log-linear scale the $l(x)$ curve is linear when mortality rate is constant with age, the slope of the line being M, the mortality rate. Given a constant M, a similar plot of the $q(x)$ curve produces a horizontal line. Survival curves are typically classified into three types (Fig. 5.4), though these are rather simplistic and should be viewed only as very crude descriptors. Several authors (Siler 1979; Chen and Watanabe 1989) have produced mathematical descriptions of the $l(x)$ curve that are sufficiently flexible to fit most observed patterns. The model of Chen and Watanabe is of particular interest because its parameters are derived from the growth curve, and hence mortality rate is predicted rather than simply fitted as in the model of Siler.

Myers and Doyle (1983) estimated natural mortality rate in fish under the assumption that natural selection maximized lifetime fecundity R_0.

Their model was based on the premise that mortality increases with the amount of surplus energy devoted to reproduction. Though the predicted mortality rates correspond well with observed rates, this cannot be taken as indicative that the model components are correct, for an alternative model based on the premise that mortality is independent of allocation to reproduction but that fecundity is a concave function of allocation produces identical predictions. The reason for the congruence in predictions is that the relevant parameter is the product of the $l(x)$ and $m(x)$ functions: though the two models are based on different biological premises they produce the same $l(x)m(x)$ function. Alerstam and Högstedt (1983) used the same general principal to estimate mortality in animals with parental care. They arrived at the extraordinary result that survival to independence will evolve to 0.368. Reasons why their model may be inadequate are discussed.

Age brings morphological development and experience—two factors that probably increase the chances of survival. Size alone may play a significant role in survival probabilities as mortality rates typically decrease with size both inter- and intraspecifically.

Two general patterns of reproduction can be distinguished: semelparity and iteroparity. The former denotes that pattern of reproduction in which the organism breeds once and dies, the latter the situation in which there are repeated episodes of reproduction. The $m(x)$ function can be divided into three classes: uniform, asymptotic, and triangular (Fig. 5.14). However, the product $l(x)m(x)$ will typically be triangular in shape, and hence analyses that assume a particular $m(x)$ function may nevertheless have greater applicability than implied by the $m(x)$ function chosen (as, for example, in the case of the model of Myers and Doyle).

As with survival, so with reproduction: larger size typically increases reproductive success. In ectotherms fecundity generally increases with size. Such increases have also been reported for small mammals. Large males of both ectotherms and endotherms also enjoy increased reproductive success, largely through increased mating success.

Reproductive effort refers to the allocation of energy into reproduction. It has been defined in a variety of ways, which makes its use potentially confusing unless the particular index chosen is clearly indicated. Ideally it refers to the proportion of energy devoted to reproduction, though this may refer to either total energy income or only surplus energy (i.e., the amount remaining after maintenance costs have been met). Operationally, reproductive effort is frequently measured as the relative biomass of propagules to the mass of the female (gonadosomatic index or relative clutch mass).

The gonadosomatic index or relative clutch mass varies enormously across species, from 0.2% in some plants to 75% in a crab spider. In reptiles and mammals relative clutch mass decreases with size, but no such pattern is evident in fish. Birds do not carry their entire clutch of eggs within them but produce eggs singly: relative egg size decreases with body size. The relative size of male reproductive organs is almost always less than the gonadosomatic index or relative clutch mass of the female.

For some organisms, principally birds and mammals, reproductive effort includes parental care, birds foraging for food for their offspring, and mammals providing milk. The energetic cost of parental care can approach the physiological limit of the parent, and if it does not produce immediate physiological problems it could increase susceptibility to future stress or predators.

6

Costs of Reproduction

If reproduction carries no cost either in terms of future survival or fecundity an organism should begin reproducing at the earliest possible age. That this does not seem to happen suggests that there are costs to reproduction. Indeed, common experience would suggest this to be the case, for it is hard to imagine that in the "real world" (as opposed to the benign and artificial world of the laboratory), reproduction does not deplete the resources of an organism, thereby making it more prone to stress-related sources of mortality such as disease, hypothermia, etc.; or that the act of seeking a mate does not place an animal at greater risk from predators either by virtue of the increased time spent away from safe havens such as burrows, or because the activities associated with mate attraction (calling and displaying) also attract predators; or that pregnant females may be more visible or less ambulatory than nonpregnant females and hence less likely to escape detection or be successful in flight from a predator.

The question of costs of reproduction has been recently reviewed by Reznick (1985), Bell and Koufopanou (1985), and for birds by Lindén and Moller (1989). Reznick attacked the issue by dividing tests of the cost of reproduction into four types:

1. Phenotypic correlations based on field or laboratory observations of unmanipulated situations
2. Experiments in which organisms where manipulated to vary the amount of reproductive effort (e.g., virgin versus mated individuals, or manipulations of clutch sizes in birds)
3. Genetic correlations, obtained by sib analysis, between reproduction and some component of fitness such as survival
4. Genetic correlations demonstrated by a correlated response to selection either on the age schedule of reproduction or the cor-

related response of this schedule to selection on a component of fitness

Since evolution can only proceed if there is genetic variation for the traits in question, Reznick (1985; Reznick et al. 1986) argued, quite correctly, that only the last two measures provide definitive proof of a cost to reproduction that is of evolutionary significance. There is, however, merit in the second approach because it can establish the mechanism generating the cost (Pease and Bull 1988; see also chapters 2 and 11), but the first is suspect because of the covariation of traits that could produce false conclusions if such covariation is not taken into account. (See chapter 2.) For example, reproduction may be determined by condition, those in poor condition not breeding; consequently it would not be surprising to find the survival rate of nonbreeders to be less than breeders.

The results of Reznick's survey are presented in Table 6.1. As predicted from theoretical considerations, phenotypic correlations frequently fail to detect a cost. However, this cannot be used to infer that no cost exists since if costs were invariably absent we should expect to find only one or two of the data sets giving significant results, whereas 22 of 33 cases have found costs to reproduction in unmanipulated situations. It is clear that manipulation removes the problem of multiple interactions, the number of cases demonstrating a cost rising to 17 out of 20 studies. The total number of studies that have examined the genetic basis for a cost to reproduction is low—20, of which 10 are questionable (Reznick 1985) and are not included in Table 6.1. Of the 10 remaining cases all indicate that costs are present, and that they are genetically based and hence of potential evolutionary importance.

In many branches of science a single experiment can refute a hypothesis. This is not the case in evolutionary biology. Demonstrating that at a particular time and place an organism does not exhibit a cost to reproduction is insufficient grounds for rejecting the cost hypothesis. There is no reason to suppose that costs will be manifested in the same way either among

Table 6.1. A summary of results of studies of the cost of reproduction, as reviewed by Reznick (1985)

	Costs present	Costs absent
Phenotypic correlations	22	11
Experimental manipulations	17	3
Genetic correlations by sib analysis	5	0
Correlated responses to selection	5	0
Totals	49	14

different organisms or by the same organism in different circumstances. Costs may be a consequence of physiological or ecological factors. In either case, a cost to reproduction is likely to be apparent only when the organism is subjected to a particular set of conditions.

Reznick (1985) noted that costs may be more evident under conditions of stress, a hypothesis examined in more detail by Bell and Koufopanou (1985), who divided the phenotypic correlations obtained from nongenetic studies into those obtained under laboratory conditions and those from natural populations. In the former category only 5 of 29 cases produce evidence of a cost, while 9 of 14 cases in the latter category indicate a cost to reproduction, supporting the stress hypothesis. Reproduction is likely to be an energetically expensive activity (see section 5.2.3); if enough resources are provided the organism may be able to meet all the demands placed upon it, but if resources are in short supply then some functions must necessarily be relatively neglected. Consider, for example, an organism that is given just sufficient food to meet maintenance costs. For this individual to reproduce it must sacrifice body tissue. Initially this may not cause any harm since storage tissues may be able to meet the demands of maintenance and reproduction, but once the storage tissues are depleted the organism must either cease reproduction or degrade its essential body components. Failure to provide energy for maintenance will not only cause a reduction in weight but a general deterioration in physical condition. Continued reproduction under such conditions will likely increase the individual's susceptibility to disease or prevent it from adjusting its internal milieu when challenged by adverse environmental conditions, thereby increasing the likelihood of death.

It should be apparent that the above scenario is not independent of ecological conditions: A mammal stressed by lactation may be capable of maintaining condition in a thermally benign environment but succumb to hypothermia when the temperature is lowered. A cost to reproduction may, however, be generated by the particular ecological circumstances even if the organism is not physiologically stressed. For example, the presence of eggs may increase the visibility of an animal and hence make it more vulnerable to predators. The cost of reproduction in this case is contingent upon the presence of predators, but it is a cost nevertheless. Considerations of physiology and ecology must be important factors in the design of experiments seeking to examine the costs of reproduction. Equally, they should be given due consideration in the interpretation of results. Reflection upon the energetic costs of reproduction, the behavioral components, and the ecological circumstances under which reproduction occurs makes it inconceivable that a cost is generally nonexistent. The question is not whether there are costs, but under what circumstances and in what ways a cost is exhibited.

No experiment can demonstrate that no cost exists. In a trivial sense this is true because the appropriate measurements may not have been taken; e.g., a survival cost may have been tested while the actual cost is a reduction in future fecundity. More importantly, no experiment can ever conclude that there is no difference between two groups: It can provide only a probability statement about the likelihood of any specified difference being observed. Suppose we wish to test the hypothesis that reproduction decreases life-span: One method is to compare a group that has been allowed to mate and reproduce with another group that has been kept virgin. Assuming that life-spans are normally distributed (or suitably transformed to be so), the probability that the difference between the two groups is no greater than expected by chance can be assessed by using a t-test. Let the number in each group be 20. For such a sample size the 5% level of significance is 2.02. There is nothing sacrosanct about the 5% level of significance; it is entirely arbitrary. Suppose in one experiment we obtained a value of $t = 2.03$ and in another $t = 2.01$: it would be silly to build a theory based on the "significance" of the former experiment while accepting no difference in the latter.

Suppose t is much smaller than the 5% level of significance. In this case we cannot reject the null hypothesis of no difference between groups. However, not rejecting the null hypothesis is a far cry from accepting it. Consider the following situation: the mean life-span of the group permitted to reproduce is 12 time units, that of the virgin group is 8 time units, and the standard deviation of both groups is the same at 10 units. The t value is 1.26, which is well below the critical value: in fact, the probability of obtaining this result by chance alone is approximately 20%. But it is obvious that the difference between 12 and 8 is large, the former being 1.5 times as large as the latter. The problem is that the variances are large and the sample sizes relatively small. The relevant question in the present circumstances is, What difference is biologically meaningful and does the present test reject this difference?

Suppose that in fact the samples do come from different statistical populations (i.e., there is a real effect on life-span). Further, suppose that the observed difference and standard deviations are actually equal to the true values. What difference could be detected? Simple algebraic manipulation of the formula for t gives the difference to be 6.39; i.e., a significant difference will be evident only when the difference in means is in excess of 6.39. If on biological grounds we can state that a difference less than 6.39 is biologically meaningful then our test proves only that the difference is less than 6.39 *and not that there is no biologically meaningful difference between the two groups*. This example illustrates the importance of deciding a priori what differences are biologically important, and then, on the basis of preliminary estimates of the parameters, deciding upon the appropriate sample size. In the present case, suppose that we decide that a difference of one is the minimum difference likely to have biological significance; the minimum sample size (obtained by rearrangement of the formula for t) is 816, which is considerably larger than 20! If this sample size cannot be achieved then one might question whether it is worth undertaking the experiment. Of course, there may be circumstances in which no a priori estimates can be made. The important point is that the statistical test tells use what the minimal difference between two samples is, not that the difference is negligible simply because it is nonsignificant.

In practice we do not know the "truth," and our estimation of required sample size must include the probability of accepting no difference when the truth is that the means are different (type II error) and also include the probability of rejecting the null hypothesis when the truth is that the two means are not different (type I error). In general this cannot be done analytically because the alternate hypothesis is not stated precisely. Solutions can be still obtained and are described in detail by Zar (1984), while Peterman (1990) provides a very useful analysis from an ecological perspective. The foregoing discussion may seem obvious, but a surprisingly large number of studies fail to consider the power of a statistical test, a good example being experiments in which brood size is manipulated to test for effects on future survival and fecundity (discussed in detail in section 6.2.2).

It is important not merely to establish that a correlation, positive or negative, exists, but to determine what mechanism is responsible. By determining the mechanism we are in a position to estimate the extent to which the phenomenon is likely to be general. If the mechanism is a limited number of ova then we can suggest that a negative relationship is likely to exist for all of those species in which ova are few in number; thus we can determine the number of taxa in which this circumstance prevails and conduct experiments to further test the hypothesis. On the other hand, if the reduction in future reproductive success is hypothesized to be a consequence of a decline in physiological condition, we can formulate models based on energetic considerations and delineate those species in which such a drain on reserves is most likely to occur. The object should not be to simply demonstrate that in a particular species a reproductive cost exists, for such a demonstration gives us little insight into the prevalence of the phenomenon. With a knowledge of the cause for the correlation we can specify both the conditions under which a cost is to be found *and the conditions under which it will not be observed*.

Costs of reproduction can be conveniently divided into three categories: effects on growth; effects on survival; and effects on future fecundity. Within each of these divisions discussion is organised under three headings: phenotypic correlations; experimental manipulations; genetic correlations.

6.1. Reproductive Effort and Growth

The effect of reproduction on growth may be important because survival and/or fecundity may be size-dependent. (See chapter 5.) Therefore, although an adverse effect of reproduction on growth is not direct evidence of a reduction in fitness, it is important indirect evidence. If energy is limiting there should be a trade-off between present reproduction and future growth. Such a trade-off is certainly reasonable but it is not inevitable, for there could conceivably be limits to the rate at which energy can be diverted into the somatic tissues. It is also possible that energy is not limiting and that an organism's rate of growth is not limited by input. How do we demonstrate that a trade-off exists? As discussed above, analyses can be divided into those based solely on phenotypic correlations and those that involve genetic variation between experimental units. The former may be difficult to interpret because individual variation might mask a trade-off, and because phenotypic variation does not imply genetic variation, which is required for evolutionary changes. The first objection can be overcome by doing manipulations, thereby randomizing effects other than the treatment of interest; the second can only be addressed by further experiments that examine the genetic basis of the trait.

6.1.1. Phenotypic Correlations

Most of the data on the relationship between growth and reproduction are from observations on variation in an uncontrolled field situation and therefore must be interpreted carefully. Here, I review a selected number of studies to illustrate the variety of approaches taken in the analysis of data that are derived from simple observation and the difficulties of interpretation. Three approaches have been taken: first, comparisons made between species; second, correlations estimated from studies within species; third, analytical models relating growth and reproduction.

From a negative regression of weight of seeds per unit leaf area (reproductive effort) on leaf area for 25 species of *Plantago*, Primack (1979) inferred a trade-off between growth and reproduction. However, since his measure of reproductive effort includes the independent variable the regression analysis is invalid. (The procedure adopted actually regresses y/x on x, and, hence a negative relationship is not unexpected.) Cattails (*Typha*) reproduce both vegetatively and sexually, the former accomplished via rhizomes. McNaughton (1975) tested the hypothesis that across species there should be a trade-off between rhizome production and foliage production: in accordance with the hypothesis a highly significant negative correlation was obtained ($r = -0.73$, $P < 0.02$, $n = 11$ comprising three species and a number of separate populations). As with Primack's analysis these data have to be viewed with some scepticism because species may differ in a variety of facets. Interspecific comparisons are too unreliable to be reasonable evidence.

A second approach to measuring a growth cost is to directly measure the relative amounts of growth and reproduction within a population and look for a correlation between growth and fecundity. The following examples illustrate the problem of interpreting the results of such observations.

A feature of many tree species is the phenomenon known as "masting," in which huge crops of seeds are occasionally produced. Following such a season there is a pronounced reduction in growth (Kozlowski and Keller 1966; Harper and White 1974; Waller 1988). Gross (1972) studied the phenomenon in yellow birch, *Betula alleghaniensis*, and white birch, *B. papyrifera*. That there would be a future lack of growth was apparent at the time of fruit production in a significant lack of buds throughout the crown, and indeed terminal growth was significantly negatively correlated to the ratio of fruit per bud site. In the European beech, *Fagus sylvatica*, flowering and fruiting withdraws about twice as much stored carbohydrate reserves as somatic growth, much of these reserves being stored in the branches (Gäumann 1935): a very large seed crop could quite easily deplete these reserves.

The evidence for a trade-off between growth and reproduction in trees is reasonably good, and the interpretation of cause and effect makes sense, but this does not necessarily make it correct. The potential error of inferring causation from phenotypic correlations is well illustrated by observations on growth and reproduction in the barnacle, *Balanus balanoides*. In this species feeding is substantially reduced during egg production and molting occurs at a frequency comparable to that in a starved animal (Barnes 1962). Body weight declined substantially during this period, from 8.0 mg at the start to 3.4 mg at the end, with the organism releasing reproductive material (semen and eggs) amounting to at least 150% of its initial body weight. However, the decline in growth during reproduction in barnacles is probably not a consequence of the lack of sufficient food in the surrounding water mass to sustain both growth and reproduction, but rather the result of the cessation of feeding, possibly to reduce the likelihood of accidently expelling the developing eggs (Barnes 1962). Thus the trade-off may not be one of allocation of limited resources but rather between growth and egg survival.

To estimate the effect of maturation on growth in natural populations of herring, *Clupea harengus*, Iles (1971) plotted an index of the potential effect of gonad maturation on the growth increment of cohorts in their third year (the "gonad effect") against the fraction of fish within these cohorts that matured in their third year. From his analysis Iles (1971, p. 392) concluded "either that no gonad effect can be demonstrated or, if it does exist, it has only a small effect on somatic growth." There are two problems with this analysis: First, the statistical error in the estimation of the relevant parameters is likely to be so large as to mask even substantial effects (the problem of statistical power). Second, and more seriously, we have no idea why some fish matured and some did not: The two groups do not truly represent treatments differing only in whether or not they mature. This latter problem of course bedevils all attempts to interpret unmanipulated sets of observations.

In a review of the effect of maturity on growth rate in fish Alm (1959, p. 124) begins by stating, "As is well known it is a general opinion, and has in many cases been proved as fact that maturity has a growth-inhibiting effect." The basis for this "general opinion," appears to be a belief that there is a marked cessation of growth at the time of maturity. This is certainly true for determinate growers such as holometabolous insects, but the evidence from phenotypic correlations in fish is not readily evident. A sharp decrease in growth is not likely to be generally apparent in natural populations of indeterminate growers such as fish, due in part to members within a cohort maturing over several years. If it were possible to characterize the growth trajectory in the absence of reproduction it would be possible to estimate the growth cost of reproduction. Even better, if we had a model of the effect of reproductive allocation on growth, this model could be compared to the observed growth curve.

This approach, the third of those outlined at the beginning of this section, was taken by Roff (1983b) in the analysis of growth and reproduction in the American plaice, *Hippoglossoides platessoides*. The model is derived from the simple allocation relationship

$$W(x + 1) = W(x) + E(x) - G(x + 1) \tag{6.1}$$

where $W(x)$ is weight at age x,

$E(x)$ is surplus energy available at age x, and

$G(x)$ is weight of the gonads at age x.

In American plaice (but necessarily generally), prior to maturity, the change in length is linear with age. Since weight is proportional to length cubed, the length of an immature fish can be described by the simple linear relationship

$$L(x + 1) = L(x) + c \tag{6.2}$$

and for mature fish by

$$L(x + 1) = \frac{L(x) + c}{(1 + GSI_2(x + 1))^{1/3}} \tag{6.3}$$

where $L(x)$ is length at age x,

c is the maximum yearly increment in length ($= 2.28$ cm/year), and

GSI_2 is the second definition of the gonadosomatic index, gonadal weight/somatic weight.

The gonadosomatic index in American plaice increases slightly with age (ranging from 9% to 11%), but the effect on the predicted size is insignificant and a constant value can be assumed. This model fits the observed growth trajectory extremely closely (Fig. 6.1), the maximum growth increment being inputed as a known variable and GSI_2 being estimated to minimize the error in sums of squares between observed and predicted length at age. The predicted value of GSI_2 is 0.10, which falls within the observed range of 0.09–0.11 (Roff 1983b). Therefore, the initial assumption that growth and reproduction are competing for a limited pool of resources is supported by the concordance between predicted and observed growth rates.

6.1.2. Experimental Manipulations

Though some analyses of the phenotypic correlation between growth and maturation suggest a trade-off there are too many confounding influences to take such data as definitive. Experimental manipulations can at least show if the trade-off is present at the level of the phenotype, though genetic analyses are required to establish the evolutionary significance of such findings. Results from experimental manipulations have demonstrated that at least in some circumstances there exists a trade-off between growth and reproduction.

By removal of flowers and artificial pollination, Montalvo and Ackerman (1987) were able to modulate the rate of fruit set in the orchid, *Ionopsis utricularioides*. Both growth and the probability of not producing flowers in the following year were negatively related to fruit set. A similar treatment

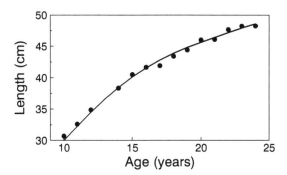

Figure 6.1. Predicted (solid line) and observed (dots) growth curves for St. Mary's Bay American plaice. Modified from Roff (1983b).

by Horvitz and Schemske (1988) using a perennial tropical herb, *Calathea ovandensis*, showed the same effects, but these were statistically nonsignificant. Commendably, they present a power analysis showing that for the given sample sizes the difference in means required to produce significance range from 14% to 31%, while the observed differences range from 4% to 24%. Sample sizes required to obtain significance given the observed differences were two to three times those actually used. Interpretation of the results thus depends upon whether one considers the observed differences to be biologically important: Horvitz and Schemske conclude that they are not; others may conclude otherwise.

Crisp and Patel (1961) examined the growth cost of reproduction in the barnacle, *Elminius modestus*, by growing barnacles in pairs and singletons. Prior to maturity both treatments grew at the same rate, but at maturity the pairs showed a statistically significant decrease in their growth rates relative to the singletons. Crisp and Patel attributed this difference to pairs not being able to sustain growth and reproduction on a reduced food supply. The difference in tissue weight was approximately equal to the calculated loss of tissue in the form of eggs, but the lack of accurate measurements makes this experiment a relatively poor test of the trade-off hypothesis.

In the bluehead wrasse, *Thalassoma bifasciatum*, a common reef fish of the west Atlantic, there are two types of male, termed initial- and terminal-phase males (Warner and Robertson 1978). The former type of male is large and may be the result of sex reversal, but initial-phase males are primary males (i.e., not the result of sex change) which turn into terminal-phase males when they become large. Unlike terminal-phase males, initial-phase males do not hold territories and usually spawn in multimale groups with single females. Terminal-phase males must defend their territories largely against incursions by initial-phase males (Warner and Hoffman 1980a). As reef size increases the proportion of initial-phase males increases and they are able to procure an increasing fraction of the matings (Warner and Hoffman 1980b). Data from tagged and transplanted initial-phase males showed that fish transplanted to reefs of different sizes adjusted their reproductive activity to match that of native initial-phase males. Initial-phase males spent more time on reproductive activities on large reefs and grew less on such reefs. However, females showed no difference in spawning activity with reef size, and as predicted showed no difference in growth rate.

It has been commonly observed that, in fish, a reduction in ration leads to a reduction in fecundity (Hislop et al. 1978; Hirshfield 1980; Wootton 1990), but such effects are difficult to interpret without information on the ration typically taken in nature. In haddock, *Melanogrammus aeglefinus*, both growth and reproduction occur even when the fish are fed on a low ration: thus, although fish grew in length, somatic weight fell and fecundity was reduced relative to fish maintained on high rations (Hislop et al. 1978). This suggests that growth and reproduction are not separate components, and if in the wild food becomes scarce, growth will be reduced by reproduction.

The care that must be taken in interpreting the relationship between growth and reproduction in fish is well illustrated by the experiments of Reznick (1983) using the guppy, *Poecilia reticulata*. Two experiments were performed: in the first, reproducing and nonreproducing females were maintained at three ration levels. Females were mated once a week, the nonreproducing females being paired with a male missing the tip of its anal fin and hence unable to inseminate the female. As expected, the energy content of the reproductive tissues of reproducing females was significantly higher than that of the nonreproducing group. But the lengths of females in the two groups did not differ, "indicating that nonreproducers do not appear to devote any of their "extra" energy to growing longer" (Reznick 1983, p. 867). There was increase in somatic energy but it appeared not in growth but as fat storage, suggesting that guppies may have a "preset" growth trajectory that is independent of reproduction. But this does not mean that variation in the growth trajectory will not affect reproductive effort. To answer this question we need to examine stocks that differ genetically in their growth rate. Reznick compared the growth of four stocks of guppies from four different localities in Trinidad. Only reproducing females were used in this experiment. The total energy (soma + reproductive) of fish was virtually identical for all localities, but there were significant differences in how this energy was allocated to growth and reproduction (Fig. 6.2), supporting the trade-off hypothesis.

6.1.3. Genetic Correlations

The above study by Reznick demonstrated that there is a genetically based trade-off between growth and reproduction. Three other experiments, all with plants, have demonstrated a genetic correlation between growth and reproduction.

Primack and Antonovics (1979) measured the correlation between reproductive effort, measured as weight of seeds produced per unit leaf area, and the number of rosettes produced (vegetative growth) of the plantago, *Plantago lanceolata*. Plants from eight different populations were grown from seed under low and high nutrient conditions. Reproductive effort varied significantly between populations and there were significant negative correlations between reproductive effort and vegetative growth both within the eight individual populations and overall (nested analysis of covariance). This negative phenotypic correlation was matched by a highly significant negative genetic correlation, indicating that the trade-off is genetically determined (Primack and Antonovics 1979).

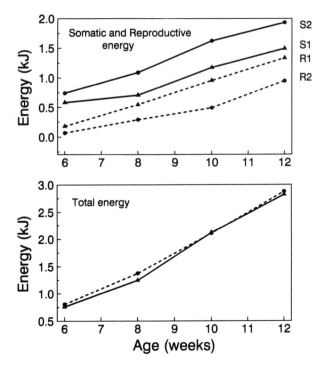

Figure 6.2. Energy allocation by female guppies from two different locations ("*Rivulus* 1" and "*Rivulus* 2") reared under a
common environment. The lower panel shows the total energy
production with age, the solid line indicating *Rivulus* 1 and the
dashed line *Rivulus* 2. The upper panel shows the allocation into
reproductive tissues (dashed lines, R1, R2) and somatic tissues
(solid lines, S1, S2), the two populations being identified on the
right as 1 or 2. Note that there is virtually no difference in total
energy production but the two populations differ in their allocation to reproduction and soma. Modified from Reznick (1983).

Law (1979a) examined the present and future fecundity in the annual
meadow grass, *Poa annua*, basing his analysis on family means. Environmental effects were removed by growing the plants in a completely ran-
domized design (Law et al. 1977), and hence differences in growth rate
between families can be ascribed to genetic differences. Both plant size
and the number of inflorescences per plant in the second season were
negatively correlated with the number of inflorescences per plant in the
first season, thus demonstrating a genetically based trade-off between present reproduction and future growth.

Growth and reproduction in higher plants depend on meristems. Meristems that become reproductive cannot undergo further vegetative growth. Geber (1990) demonstrated that in the annual plant, *Polygonum arenastrum*, early commitment of meristems to reproduction leads to high early fecundity but a reduced growth rate and lowered later fecundity. Commitment of meristems to vegetative growth early in life produces a low initial fecundity but a high growth rate and a high later fecundity. A quantitative genetic analysis showed growth and fecundity to be genetically correlated.

Phenotypic correlations derived from unmanipulated field observations do not provide unequivocal evidence that there is a trade-off between growth and reproduction. With the possible exception of the study by Horvitz and Schemske (1988) manipulation experiments have found a trade-off between fecundity and changes in the soma. The four studies which compared the growth and reproduction of different populations or families maintained under controlled conditions (Primack and Antonovics 1979; Law 1979a; Reznick 1983; Geber 1990) clearly demonstrate a trade-off. These results indicate that there is a genetic basis to the trade-off and hence that it can play a role in the evolution of the optimal life history.

6.2. Reproductive Effort and Survival

A reduction in growth due to reproduction might increase the probability of death by maintaining the organism within a size range that is preferred by predators. But this indirect effect is probably not so important as the likelihood that reproduction itself puts the organism directly at risk from predators, disease, or other environmental hazards to which physiologically inferior organisms may be more likely to succumb.

6.2.1. Phenotypic Correlations

Studies of unmanipulated populations, whether under field conditions or in the laboratory, must be viewed with great caution. Suppose we do observe a negative relationship between reproductive effort and survival; what is cause and effect? Survival might indeed be reduced by the effort of reproduction; on the other hand, it might be that those individuals showing the highest reproductive effort are those that have little chance of breeding again regardless of their present reproductive expenditure, and hence are "going for broke."

In 17 studies of reproduction and survival under field conditions reproduction is negatively correlated with survival in 9 cases, uncorrelated in 4 studies, and positively correlated in 4 studies (Table 6.2). The sex in which the correlation is reported is not evenly distributed among these groupings.

Table 6.2. Phenotypic correlations between reproduction and survival in unmanipulated field populations

Common name	Scientific name	Reference
	Negative correlation between reproductive effort and survival	
Palm	*Astrocaryum mexicanum* (♀)	Piñero et al. 1982
Plant	*Senecio keniodendron* (♀)	Smith and Young 1982
Meadow grass	*Poa annua* (♀)	Law 1979d
Mayapple	*Podophyllum peltatum* (♀)	Sohn and Policansky 1977
Mussel	*Anodonta piscinalis* (herm)	Haukioja and Hakala 1978
House martin	*Delichon urbica* (♀)	Bryant 1979
Great tit	*Parus major* (♀)	Tinbergen et al. 1985
Deer mouse	*Peromyscus maniculatus* (♀)	Fairbairn 1977
Red deer	*Cervus elaphus* (♀)	Clutton-Brock et al. 1982, 1983
	No significant correlation between reproduction and survival	
Bluehead wrasse	*Thalassoma bifasciatum* (♂)	Warner 1984a
Sparrow hawk	*Accipiter nisus* (♀)	Newton 1985
House martin	*Delichon urbica* (♂)	Bryant 1979
Willow tit	*Parus montanus* (♀)	Orell and Koivula 1988
	Positive correlation between reproductive effort and survival	
Milkweed beetle	*Tetraopes* (♂)	McCauley 1983
Magpie	*Pica pica* (♀)	Högstedt 1980, 1981
Song sparrow	*Melospiza melodia* (♀)	Smith 1981
Red deer	*Cervus elaphus* (♂)	Clutton-Brock et al. 1982, 1983

Analysis of the correlation in males has found either no correlation (two studies) or a positive correlation (two studies). It is the male sex in which we might expect the greatest problem of spurious correlation. In competition for mates the most successful male is frequently the largest and strongest (e.g., red deer, Clutton-Brock et al. 1982, 1983; see also Table 5.2) and thus the male that likely also has the highest survival rate. Thus to appropriately measure the survival cost of reproduction in males correction must be made for possible differences in size or general condition.

Reproductive effort depends not simply on the number of offspring produced but also the quality of the mother. Interpretation of the lack of a correlation between reproduction and survival in sparrow hawks is confounded by the fact that brood size is positively correlated with female size (Newton 1985). It is highly likely that in this case reproductive effort does not increase with clutch size. In the magpie, *Pica pica*, the number of breedings is correlated with both clutch size and fledgling production, but these are also correlated with territory quality (Högstedt 1980, 1981). As a consequence the females that have the best territories have the highest survival rates and the largest reproductive success. Unless territory quality can be accounted for, costs of reproduction cannot be properly assessed.

The analysis of Tinbergen et al. (1985) on the great tit is particularly illuminating in demonstrating the importance of ecological conditions in modulating the probability of survival associated with breeding. Analysis of data for the years 1957 to 1978 revealed that the survival cost of reproduction was determined by the size of the beech crop. When the crop is poor a negative correlation between parent survival and number of fledglings produced is found, but in years of high beech-seed yield no correlation is obtained (Fig. 6.3). The importance of the beech crop is that it provides winter food for the birds: a low crop presumably leads to increased starvation, which is exacerbated by the previous breeding effort.

A similar hypothesis has been advanced by Fairbairn (1977) to account for the relatively high mortality of female deer mice that attempt to breed in spring. Fairbairn suggested, based on work by Sadleir (1974), that lack of food in spring causes high mortalities among breeding females because they cannot support both lactation and thermoregulation. This hypothesis is supported by a study of Fordham (1971) in which supplementary food was given to a population of deer mice from February to September. The proportion of early breeding females increased, as did their survival, suggesting that food supply is critical to the breeding and survival of females in spring.

Reproductive effort and reproductive life-span are negatively correlated in the mussel, *Anodonta piscinalis*, for the year 1975 but not for 1976 (Haukioja and Hakala 1978). The average condition index of nonreproducing mussels in 1975 was significantly less than in 1976, suggesting that in 1976 reproducing mussels may not have been as stressed as in 1975. These three cases highlight the importance of time and circumstance in the manifestation of a reproductive cost.

That a survival cost to reproduction should be evident most strongly in conditions of stress is not surprising. But it warns against the naive use of laboratory experiments to examine the cost of reproduction. In the benign conditions of the laboratory where there are no predators or disease and equitable climatic conditions we might not expect to find reproduction to decrease life expectancy. Thus it should come as no surprise to find that attempts to demonstrate a survival cost to reproduction under laboratory conditions have generally been unsuccessful. Although a negative correlation between survival and reproductive effort is reported by 4 studies, no significant correlation has been observed in 11 species and a positive correlation is reported in 4 others (Table 6.3).

Activities specifically associated with reproduction, such as courtship, incubation, and calling or searching for mates may increase the risk of

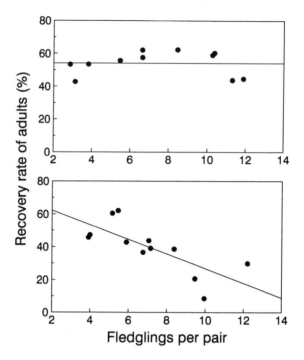

Figure 6.3. Survival rates, measured by annual recovery rates, of adult great tits as a function of the number of fledglings produced per pair in the preceding season. Top panel shows rates when beech-crop index exceeded 30; lower panel, survival rates when the beech-crop index was less than 30. A high index corresponds to a high seed production. Solid lines show fitted regression, nonsignificant in the upper panel, significant in the lower ($r = 0.75$, $n = 12$, $P < 0.05$). Modified from Tinbergen et al. (1985).

predation (Aleksiuk 1977; Burk 1982; Sargeant et al. 1984; Endler 1987; Trail 1987). The demonstration of this increased risk has been based not on a comparison of breeding and nonbreeding individuals but on a showing that the predator locates its prey by the sexual advertisement of the prey, or based on comparing differences in behavior and survival of males and females (Table 6.4).

Reproductive behavior can increase the risk of being preyed upon in three ways. First, parasites or predators may home in on the mating call of the male or female. Taped calls have demonstrated that both parasites and predators can locate males via their acoustical signals. Females may attract males using pheromones; at least one tachinid fly, *Trichopoda pen-*

Table 6.3. Phenotypic correlations between reproduction and survival in unmanipulated laboratory populations

Common name	Scientific name	Reference
Negative correlation between survival and reproductive effort		
Rotifer	*Asplanchna brightwelli*	Snell and King 1977
Bdelloid rotifer	*Philodina* sp.	Bell 1984a
Chicken	*Gallus* sp.	Hall and Marble 1931
Desert wood rat	*Neotoma lepida*	Cameron 1973
No significant correlation between reproduction and survival		
Oligochaete	2 species	Bell 1984a
Rotifer	*Platyias patulus*	Bell 1984b
Copepod	*Daphnia pulex*	Bell 1984a
Copepod	*Mesocyclops*	Feifarek et al. 1983
Ostracod	*Cypridopsis vidua*	Bell 1984a
Ground beetle	2 species	van Dijk 1979
Flour beetle	*Tribolium casteneum*	Mertz 1975
Lace bug	*Gargaphia solani*	Tallamy and Denno 1982
Grasshopper	*Melanoplus sanguinipes*	Dean 1981
Positive correlation between reproductive effort and survival		
Leaf miner	*Agromyza frontella*	Quiring and McNeil 1984
Fruit fly	*Drosophila melanogaster*	Kidwell and Malick 1967
Fruit fly	*Drosophila simulans*	Murphy et al. 1983
Chicken	*Gallus* sp.	Dempster et al. 1952

nipes, uses this cue to locate females on which to lay its eggs (Harris and Todd 1980). Predators might also act as satellites of a calling male and prey upon incoming females, as is found in the gecko, *Hemidactylus tursicus*, preying on females of the decorated cricket, *Gryllodes supplicans* (Sakaluk and Belwood 1984).

Second, instead of homing in on the mating call of the prey a predator may attract males by mimicking the call of the female. The bolas spider, *Mastophora* sp., catches insects by means of a sticky ball suspended on the end of short vertical thread that is attached to a single horizontal line. The spider rests on the horizontal line holding the vertical thread with one front leg and swings the bola at passing insects. The incidence of male noctuid moths captured and the direction of approach of prey (upwind) suggest that the spider may incorporate a chemical that mimics the male-attracting pheromone emitted by female noctuid moths (Eberhard 1977, 1980). More convincing evidence of female call mimicry comes from the study of Lloyd (1965) on female fireflies of the genus *Photuris*. In this genus the females are carnivorous and attract males of the genus *Photinus* by mimicking the

Table 6.4. Reproductive activities that lead to an increased risk of predation

Prey	Predator	Reference
Parasites attracted to call of male or female		
Cricket (♂)	Tachinid fly[a]	Cade 1975, Mangold 1978
Katydid (♂)	Tachinid fly[a]	Burk 1982
Cicada (♂)	Dipteran[a]	Soper et al. 1976
Stink bug (♀)	Tachinid fly[b]	Harris and Todd 1980
Predators attracted to call of male		
Cricket	Various vertebrates[a]	Bell 1979; Walker 1964; Sakaluk and Belwood 1984
Frog	Four-eyed possom[a]	Ryan et al. 1981
Frog	Fringe-lipped bat[a]	Tuttle and Ryan 1981; Ryan et al. 1981, 1982
Males attracted to mimetic call of the female by a predator		
Noctuiid moth	Bolas spider[b]	Eberhard 1977
Firefly (*Photinus*)	Firefly (♀ *Photurus*)[c]	Lloyd 1965
Increased activity increases susceptibility to predator		
Firefly (♀)	Unknown	Wing 1988
Hanging fly (♂)	Web-building spiders	Thornhill 1980b
Cicada (♂)	Web-building spiders	Gwynne 1987
Cricket (♀)	Digger wasps	Gwynne and Dodson 1983

[a]acoustic signal, b: pheromone, c: light flashes.

response flashes of the female of this genus—a phenomenon demonstrated by the use of artificial calls (Lloyd 1965; Lloyd and Wing 1983).

Third, the increased activity of an animal may increase its susceptibility to predation either by making it more visible or simply by increasing the likelihood that predator and prey meet. The former situation occurs in digger wasps, *Palmodes laeviventris*, which provision their nests with more female than male mormon crickets, *Anabrus simplex* (Gwynne and Dodson 1983). The male mormon cricket calls from the protection of a burrow and hence is less likely to be preyed upon by a visually hunting predator. In contrast, the territorial behavior of male digger wasps increases their vulnerability to predators. Significantly more males than females are taken by robber flies (Gwynne and O'Neill 1980). This bias arises because males patrol their territory and approach intruders: if the intruder happens to be the predatory robber fly the wasp may patrol no more.

The situation of increased vulnerability due to increased activity is illustrated by mortality rates in Mecoptera. In the mecopteran *Hylobittacus apicalis*, the female does not hunt but accepts a nuptial gift of prey from the male. To hunt for prey the male must forage among the herbs and this makes it more likely than the female to fly into spider webs, as evidenced by a highly significant male-biased ratio within the webs (Thornhill 1980b). However, nuptial feeding does not occur in the related mecopteran, *Bittacus strigosus*, and females hunt in the same manner as the males; in this species there is no sex-biased mortality from web-building spiders (Thornhill 1980b).

Parental defense of young may also logically place a reproductive individual more at risk than a nonreproductive. Male three-spined sticklebacks, *Gasterosteus aculeatus*, defend their eggs and young against fish that could potentially eat both the young and the parent: this defense could therefore increase the mortality of guarding male sticklebacks (Pressley 1981; Giles 1984). Similarly, mobbing behavior of birds defending their young can be dangerous, as evidenced by numerous accounts of kills occurring during this behavior (reviewed by Curio and Rengelmann 1985).

6.2.2. *Experimental Manipulations*

The effect of present reproduction on future survival has been experimentally assessed by five types of manipulation: comparisons between virgins and mated groups, removal of reproductive organs, alterations in brood size, introduction of predators into cages with both virgins and reproductives, and effects of altered temperature regime. The last method of manipulation is not valid. In the three studies (Hirschfield 1980; Feifarek et al. 1983; Woombs and Laybourn-Parry 1984) noted by Bell and Koufopanou (1985) the subjects of the experiments were ectotherms, which makes the calculation of survival on a chronological (days, etc.) scale inappropriate. For ectotherms, life-span should be compared using a physiological time scale (degree days). In ectotherms there is a critical threshold temperature below which developmental processes effectively cease. (This temperature may vary between processes such as growth rate and egg development rate.) Time is computed as the number of degrees above the threshold temperature per unit time on a chronological scale (generally a day). Development time measured on a chronological scale in ectotherms typically decreases with temperature but remains the same when measured in degree days. Thus the changes in longevity found in the three studies noted above may be artifacts, and while fecundity may change between temperatures, it is possible that longevity remained the same.

The simplest manipulation is to prevent an organism from mating and compare its life-span with that of a mated treatment. In 23 such experiments using nematodes or arthropods (a total of 19 species) the results are consistent: virgins live longer than mated individuals. (See Table 1 of Bell and Koufopanou 1985.) The same experiment applied to two species of fish gives the same result, but two mammalian species produced equivocal results, not only between species but also within a species.

Male pipefish brood their young in a pouch and hence both sexes contribute significant reproductive effort. Under laboratory conditions both virgin females and males of the species *Syngnathus typhle* survived significantly longer than breeding males and females (Svensson 1988). The source of mortality is not known. However, similar results have been obtained for male three-spine sticklebacks, *Gasterosteus aculeatus* (Dufresne et al. 1990), suggesting that the phenomenon might be general among fish species in which the male broods its young.

Virgin male rats do not live as long as mated males (Slonaker 1924; Drori and Folman 1969). But in mice, virgin females lived longer than mated females in two studies (Mühlbock 1959; Suntzeff et al. 1962), while in another, neither virgin females nor males lived as long as mated individuals (Agduhr 1939).

Although virgin females of the grasshopper *Melanoplus sanguinipes* live longer than mated females, there is no difference in egg production between the two types (Dean 1981). This strongly suggests that, in this species, the reduction in survival is a consequence of mating not egg production. The effect of mating on longevity in both male and female fruit fly, *Drosophila melanogaster*, has been elegantly demonstrated in a series of experiments by Partridge and her colleagues.

Partridge and Farquar (1981) examined the cost of sexual activity in male fruit flies *D. melanogaster*, by maintaining two types of controls: one in which males were kept isolated from females, and a second in which males were kept with newly mated females, which are sexually unreceptive. There was no significant difference in longevity between the two types of control but, after correction for effects of body size, there was a significant difference between virgin and mated males, the former living longer. By successively presenting males during their life with females, Partridge and Andrews (1985) were further able to show that changes in survival rate with age are a consequence of mating activity and not senescence.

Fowler and Partridge (1989) tested for an effect of mating on longevity in female fruit flies using two groups: one group of females, the "high-mating" group, was exposed to sexually competent males throughout the experiment, while a second group, the "low-mating" group, was allowed access to these males for 1 day out of 3. On the remaining 2 days the low-mating group was exposed to males that had their external genitalia ablated by microcautery: these males were sexually active with respect to courtship but could not inseminate the female. The two groups did not differ in egg production but the high-mating group mated significantly more frequently than the low-mating group (Fowler and Partridge 1989) and did not live as long. These results do not preclude an additional effect of egg production. Such an effect was observed in experiments by Maynard Smith (1958) and Lamb (1964) on *Drosophila subobscura*.

Another elegant experiment that disentangled the cost of mating from that of oviposition is that of Roitberg (1989) on life-span in female rosehip flies, *Rhagoletis basiola*. Roitberg monitored life-span in three treatments: (1) females provided with males and allowed to oviposit, (2) females provided with males but not permitted to oviposit, and (3) a spinster group that was not provided with oviposition sites. At death all treatments had the same number of eggs in the abdomen but differed in life-span. Females from the three groups lived on average 27.1 days (SE = 1), 43.9 days (2.6), and 47.6 days (2.6). Thus in this case the effect of mating on life-span is insignificant compared to the cost of oviposition.

As emphasized throughout this chapter, experiments that attempt to assess the cost of reproduction are unlikely to measure significant cost if the organisms are maintained under benign conditions. Thus it is not surprising that in the experiments of Calow (1977) on the backswimmer, *Corixa punctata*, and of Lamb (1964) on the fruit fly *Drosophila subobscura*, the difference between virgin and mated individuals disappeared under favorable conditions of culture. This finding is consistent with the hypothesis that the decrease in longevity of the mated group is a consequence of stress induced by attempting to produce eggs under conditions that are not commensurate with both reproduction and maintenance.

Another method of testing for an effect of reproduction on survival is to remove the reproductive organs. Removal of reproductive organs in semelparous organisms such as soybeans (Leopold 1961) and salmon (Calow 1977) is reported to increase longevity. However, death might not in such organisms be a direct result of the energy drain of reproduction. For example, the octopus *Octopus hummelincki* lays eggs, broods them, reduces its food intake, and dies after the young hatch. But if the optic gland is removed after spawning, the female ceases brooding, resumes feeding, and lives for a prolonged period (Wodinsky 1977). Males also have a prolonged life if the optic gland is removed. Furthermore, "a significant proportion of the females with the optic glands removed lived for periods of 2 to 4 months after the cessation of feeding, a period far exceeding the postspawning longevity of normal females" (Wodinsky 1977, p. 951). These results suggest that post spawning death in the octopus is not typically the result of starvation. Rather there would appear to be a deterioration that is "programmed" into the life history. The possible evolutionary reasons for this are not relevant here: what is important is that it cannot be assumed that increased longevity of semelparous organisms following surgical treatment is a consequence of removing the stress of reproduction.

A third type of manipulation is to alter brood size and measure the subsequent survival of the parents. Tallamy and Denno (1982) removed eggs from the lace bug *Gargaphia solani* as they were laid. This bug typically broods its eggs: loss of its brood causes it to lay further eggs. Females that were permitted to brood lived longer than those that continually produced eggs (Tallamy and Denno 1982). Twelve studies have attempted to measure the relationship between reproductive effort and survival by manipulating the brood size of free-ranging birds (Table 6.5). There is a close correspondence between sample size and the result obtained. Studies with sample sizes less than 100 find no significant effect of brood size on survival

Table 6.5. Effect of manipulating brood size on subsequent survival of parent birds

| Species | Brood size | | Sample size | S^a | Reference |
	Normal	Expt			
House sparrow (δ + $\circ$)	1–7	±2	19	NS	1
Swallow tailed gull (δ + $\circ$)	1	±1–2	30	NS	2
Tree swallow ($\circ$)	5–7	7–9	41	NS	3
Tengmalm's owl (δ)	5–6	±1	65	NS	4
Rook (δ + $\circ$)	$?^b$	4	66	NS	5
Kestrel (δ + $\circ$)	5	3–7	72	NS	6
Pied flycatcher (δ)	3–7	9	111	Yes	7
Great tit ($\circ$)	5–13	x2,x0.5^c	147	Yes	8
Blue tit	3–15	3–15^d	216	Yes ($\circ$) NS (δ)	9
Glaucous winged gull (δ + $\circ$)	1–3	1–7	289	Yes	10
Collared flycatcher (δ + $\circ$)		±1–2	320	NS	11
Great tit	5–13	±3–4	483	Yes ($\circ$) NS (δ)	12

References: 1. Hegner and Wingfield 1987; 2. Harris 1970; 3. DeSteven 1980; 4. Korpimäki 1988; 5. Røskaft 1985; 6. Dijkstra et al. 1990; 7. Askenmo 1979; 8. Tinbergen unpublished data cited in Dijkstra et al. 1990; sample size from Tinbergen 1987; 9. Nur 1984a, 1988a; 10. Reid 1987; 11. Gustafsson and Sutherland 1988; 12. Boyce and Perrins 1987; Nur 1988b.

[a]NS = No significant correlation between brood size and survival; Yes = significant negative correlation.

[b]The rook lays 4–5 eggs and normally fledges 2 young. Røskaft (1985, p256) states, "brood size . . . experimentally enlarged to four (almost twice the mean) when the young were aged 12–15 days."

[c]The experimental protocol consisted of grouping clutches initially of equal size into sets of three, dividing one clutch in half, doubling the second, and keeping the third as a control.

[d]The experimental protocol consisted of randomly distributing artificial broods, ranging in size from three to 15.

but among the six remaining studies, in which sample sizes ranged from 111 to 483, five report a significant correlation in females.

To expect to detect differences in survival rate given sample sizes less than 100 for control and enlarged broods verges on fantasy. Suppose that the survival of parents raising normal broods is 50% and those raising enlarged broods is reduced by the effort to 32%. Given a sample size of 50 for each group, the χ^2 value expected, given survival rates as indicated, is 3.35, which is not significant at the 5% level! Survival rates much closer to that experienced by the control group could be biologically highly significant and have a major impact on the optimal life history. It is not sufficient grounds to conclude "no effect" when the significance level is not surpassed. The important statistic is the confidence region about the estimate of relative survival: if this includes a value that is biologically meaningful the results must be interpreted with great care.

The fourth type of manipulation attempts to place the test subject in an ecologically relevant setting: a group of reproductive and nonreproductive individuals are placed in the presence of a potential predator. The effect of being reproductive is then assessed by the relative survival rates of the two groups. Experiments on five invertebrate species have all found that survival of reproductive individuals is less than that of nonreproductives (Table 6.6), though the magnitude of the difference depends upon the type of predator and other ecological variables.

Copepods carry their eggs in sacs at the posterior of their body and may be both more visible and less maneuverable than females without eggs. Winfield and Townsend (1983) presented ovigerous (egg bearing) and non-ovigerous *Cyclops vicinus* individually to two fish species, bream (*Abramis brama*) and roach (*Rutilus rutilus*). With increasing number of trials roach showed an increasing ability to capture ovigerous cyclops but remained relatively incompetent at capturing cyclops without eggs. On the other hand, bream were proficient at capturing both types of cyclops. In both species the survival of ovigerous females was less than that of nonovigerous females, but was statistically significant only in roach (Table 6.6). The reaction distance of both bream and roach is greater for the ovigerous females, reflecting the greater visibility of these. A similar finding is reported by Hairston et al. (1983) for another copepod, *Diaptomus sanguineus* being preyed upon by sunfish, *Lepomis macrochirus*. Increased visibility probably also accounts for the higher susceptibility of egg-bearing females of the copepod, *Eurytemora hirundoides* (Vuorinen et al. 1983).

In the spring and fall, *Daphnia* populations may produce resting eggs that are carried in pigmented envelopes called ephippia. Mellors (1975) found a significantly higher proportion of ephippial female *Daphnia galeata mendotae* in the stomachs of pumpkinseed sunfish, *Lepomis gibbosus* and perch, *Perca flavescens* than in the water column. He tested the hypothesis

Table 6.6. Predation risk associated with reproduction in five invertebrate species

Predator	Condition	RS[a]	Reference
Prey = *Cyclops vicinus*			
Bream	Lab	1.29	1
Roach	Lab	2.73*	1
Prey = *Eurytemora hirundoides*			
Stickleback	10:10[b]	1	2
	3:6	>1*	2
Prey = *Daphnia pulex*			
Sunfish	3 light levels	1.42–1.67[c]	3
Red-spotted newt	Lab	1.6	3
Guppy	Light background	1.27–1.63[d]	4
	Dark background	0.86	4
Stickleback	Light background	1.37*	4
Shiner fry	Light background	1.11	4
Sunfish	Light background	0.67–1.1[e]	4
Backswimmer	Light background	1.28*	4
Brown hydra	Light background	1.06	4
Prey = *Palaemon adspersus*			
Cod[f]	Lab	1.60*	5
Prey = *Gerris odontogaster*			
Backswimmer	Lab	2.8*	6

References: 1. Winfield and Townsend 1983; 2. Vuorinen et al. 1983; 3. Mellors 1975; 4. Koufopanou and Bell 1984; 5. Arnqvist 1989.

[a]RS; the survival of females without eggs, or not in on copula (*Gerris odontogaster*) compared to females with eggs or in copula.

[b]Ratio of ovigerous females to nonovigerous.

[c]Significant in two experiments, NS in one.

[d]Significant in all four experiments.

[e]Not significant in any of three experiments.

[f]In some experiments cod plus another fish species (whiting or sculpin) used.

*Differences are significant at least at the 5% level.

that ephippial female *Daphnia* are more susceptible to predators by exposing equal numbers of ephippial and nonephippial *Daphnia pulex* to either sunfish or a second predator, the red-spotted newt, *Diemictylus viridescens*. In all cases the proportion of nonephippial females in the test group increased as a result of selective predation on ephippial females (Table 6.6), but the difference was significant only in the groups exposed to bluegills at the two highest light levels (0.13 and 0.86 *lx*). The lack of significance at the lowest light level may be due to relatively few *Daphnia* being eaten (9.8 per trial compared to 14.8 and 13.1 at the two higher light levels).

The importance of the type of predator and the environmental conditions is highlighted by the analysis of Koufopanou and Bell (1984), who also used *Daphnia pulex*, but a wider variety of predators than Mellors. Against a light background, guppies, *Poecilia reticulata*, selectively preyed upon egg-bearing females, but against a dark background no selectivity was observed (Table 6.6). The results for other fish species also suggest selective feeding, though statistical significance was achieved in only one case (stickleback). A visually hunting invertebrate predator, the backswimmer, *Notonecta* sp., took more egg-bearing females than expected by chance, but a tactile feeder, *Hydra pseudooligactis*, appeared not to differentiate.

Females carry eggs or young for periods much longer than the duration of copulation. Nevertheless, in some organisms, such as some insects and amphibia, copulation or some type of amplexus may be extended and potentially put the couple at risk because of decreased mobility or increased visibility. This has been experimentally examined by Arnqvist (1989) using the waterstrider *Gerris odontogaster* and a potential predator, the backswimmer *Notonecta lutea*. Survival rates of male and females in pairs were compared to those of female-female pairs. Waterstriders in copula are significantly more likely to be taken by a backswimmer than two separate females (Table 6.6). Furthermore, it is the female, not the male, that is subject to the increased vulnerability (Arnqvist 1989). This makes sense, as the males "rides" on top of the female and the backswimmer attacks from below. Fairbairn (personal communication) has observed a similar phenomenon in the gerrid, *Gerris remigis*: pairs in copula were more likely to be taken by frogs, though in these experiments both males and females were consumed.

Manipulation experiments measuring the cost of reproduction in the presence of predators have been attempted for vertebrates, but flaws in experimental design make interpretation difficult. Shine (1980) examined possible predation costs of reproduction in the lizard *Leiolopisma conventryi* by exposing a gravid female and male of approximately equal body size to a predator, the white-lipped snake *Drysdalia coronoides*. Eight of 16 lizards were taken by the snakes, of which seven were gravid females, a statistically significant difference. While this experiment demonstrates that gravid females may be more vulnerable than males, it does not follow that gravid females are more vulnerable than nonbreeding females. This is a plausible extension but needs verification.

Svensson (1988) examined the risk of predation on two species of male pipefishes. Males within this family (Syngathidae) brood the eggs and are more conspicuous at this time. The experimental protocol consisted of comparing the number eaten by a fish predator within a given time period at two different times of the year, when the males were carrying eggs and when they were out of breeding condition. A problem with this design is that differences due to changes in predator behavior associated with the time of the year cannot be disentangled from differences due to breeding condition of the male pipefish. Thus although in both species brooding males suffered a significantly higher mortality rate, this cannot be unambiguously attributed to their reproductive state.

The above two studies on the potential survival costs of reproduction in vertebrates were poorly designed in comparison to those for invertebrates, but a study by Cushing (1985) very clearly demonstrates that vertebrates may increase their vulnerability to predators when reproductive. Cushing (1985) showed that prairie deer mice, *Peromyscus maniculatus bairdii*, are more likely to be captured by a weasel, *Mustela nivalis*, when in estrus than when out of estrus. These experiments followed a similar protocol to that used in the invertebrate experiments previously described, in that two mice, one in estrus and one in diestrus, were introduced into an arena (1.83 m × 1.83 m) containing a single weasel. Each trial ended when the weasel caught one of the mice. To examine the hypothesis that the weasel was preferentially finding the estrous mouse because of its odor, Cushing (1985) repeated the experiments using diestrous mice that had been painted with urine from an estrous mouse. Under these conditions there was no difference in mortality rate between the estrous and painted diestrous females.

To summarize, among invertebrates there is clear evidence that under laboratory conditions virgins live longer than mated individuals. This may be a function both of mating itself and/or egg production. Data for two species of fish (pipefish and three-spine stickleback) demonstrate a significant reduction in survival with breeding, but the data for two mammalian species (rat and mouse) are equivocal. Manipulation of reproductive effort after eggs have been laid generally supports the cost hypothesis—those studies not finding a correlation being suspect because of very small sample sizes. The analysis of predation on reproductive females demonstrates that, in general, females bearing eggs or in copula suffer higher rates of predation than females not doing so. It is, however, important to note that the cost of reproduction depends upon physiological condition (e.g., starvation increases the cost) and ecological circumstance (e.g., susceptibility to predation depends upon the particular predator and background color). Manipulations should be relevant to the conditions likely to be experienced by the female in her natural habitat.

6.2.3. Genetic Correlations

The primary problem with the interpretation of genetic correlations between survival and reproduction is that results contrary to those which might be expected on the basis of the trade-off hypothesis can be dismissed because the studies used inbred lines or because the stocks were not well adapted to the conditions of culture (Rose 1984a,b; Service and Rose 1985). Since the appropriateness of the culture conditions may be highly subjective it is difficult to divide studies into "good" and "poor" experiments.

The presence of some positive correlations is to be expected but the genetic correlations between the traits for which there is a good mechanistic reason to expect a trade-off should be negative (Charlesworth 1990; see also section 3.2 and chapter 11). This means that attempts to assess costs solely by the estimation of genetic correlations must be considered cautiously: failure to find negative correlations between just two traits may simply not be very informative.

Consistently positive correlations between longevity and fecundity are reported by Giesel and his co-workers for *Drosophila* (Giesel 1979; Giesel and Zettler 1980; Giesel et al. 1982), but the standard errors are very high and the stocks inbred. However, using outbred stocks Tantawy and Rakha (1964), Tantawy and El-Helw (1966), Temin (1966), and Murphy et al. (1983) all obtained similar results, with longevity being positively correlated or uncorrelated with fecundity but never negatively correlated. On the other hand, Rose and Charlesworth (1981a) obtained negative genetic correlations between early fecundity and life-span. Scheiner et al. (1989) found that at 25°C genetic correlations between longevity and peak fecundity, last third fecundity, and total fecundity were all negative (though it is not clear if they were also significantly different from zero) while at 19°C the first two were negative, but total fecundity was positively correlated with longevity. Genetic correlations are affected by temperature conditions (Murphy et al. 1983; Scheiner et al. 1989) and thus the variety of results may reflect differences in rearing conditions. Overall, these data are so confusing as to prevent any reliable conclusion. In all cases the flies were raised under relatively optimal conditions.

A trade-off may, however, only be evident under stressful conditions. Solbrig and Simpson (1974) present data on clonal variation in the dandelion, *Taraxacum officinale*, that illustrate this proposition. From three sites in the Mathei Botanical Garden (University of Michigan), Solbrig and Simpson collected four clones of dandelions, designated A, B, C, and D. Clone A predominated in the two disturbed sites, comprising 73% and 53% of the total sample from each locality, while clone D predominated in the undisturbed site, comprising 64% of the sample. In contrast, clone D was virtually absent from the two disturbed sites (0% and 1%), and clone A made up only 17% of the sample from the undisturbed site. Plants from clones A and B were raised under two conditions: in pure culture and in a mixed culture of 50:50. In pure culture both clones fared about the same, but in mixed culture clone A had a reduced survival and a lower dry weight than clone D, but clone A produced a significantly larger biomass of seeds (Table 6.7). These data suggest a genetically based trade-off between competitive ability, reflected in survival rates, and fecundity.

A second method of estimating the genetic correlation between two traits is to select on one trait and measure the correlated response in a second. (See chapter 2.) Experiments on two insects, *Tribolium* and *Drosophila*, in which individuals were selected for early or late fecundity and changes in survival were monitored, give support to the hypothesis of a genetic trade-off. Both Sokal (1970) and Mertz (1975) selected for early oviposition in *Tribolium* by discarding adults either 3 days (Sokal) or 10 to 20 days (Mertz) after the start of oviposition. Sokal found that selected lines had shorter longevities than the controls but Mertz found no consistent difference. Rose and Charlesworth (1981b) and Rose (1984c) selected for delayed reproduction in *Drosophila melanogaster* in a similar manner as above and also found a correlated increase in longevity.

6.3. Reproductive Effort and Schedule of Reproduction

It has been postulated that a cost to present reproduction is a loss in future reproduction by virtue of a reduced future fecundity. Before surveying the evidence for such a trade-off it is worthwhile to ask how such a cost might arise. First, the number of ova at maturity might be fixed: if this number is low it is possible that the organism could simply run out of eggs. Females that for whatever reason laid a large complement early in life would necessarily lay few eggs at a later age. This problem also applies to the relationship between reproductive effort and survival. Bell's analysis of egg production in the gastrotrich, *Lepidodermella squammata* (Bell 1984a), nicely illustrates the potential confounding influence of this factor. This small metazoan generally produces only four eggs. Let the total number of eggs produced by an individual be E, with the first egg being produced

Table 6.7. Survival and fecundity in two clones (A and D) of dandelion grown under two different conditions

| | Conditions of culture | | | |
| | Pure | | Mixed (50:50) | |
Trait	Clone A	Clone D	Clone A	Clone D
Survival (%)	63.4	61.3	49.5	61.0
Total seed wt. (mg)	10.0	7.1	6.0	0.5

Data from Solbrig and Simpson (1974).

at age α and death occurring at age ω. The rate of egg production (eggs per day) over the adult life-span is thus $E/(\omega - \alpha)$. If we hypothesize that fecundity and survival are negatively correlated we might be tempted to test for a correlation between the rate of egg production measured over the life-span against life-span, the prediction being that females that have a high rate will have a short life-span. This requires that one correlate $E/(\omega - \alpha)$ against ω, which is of course statistically improper since the independent variable also occurs in the dependent. In *L. squammata* the postreproductive life-span is long relative to the prereproductive, and since almost all females lay four eggs, the proposed test amounts to plotting $1/\omega$ against ω; not surprisingly, a very good correlation is observed. Bell suggested that this problem might account for the negative correlation between age-specific fecundity and probability of survival reported by Snell and King (1977) for the rotifer, *Asplanchna brightwelli*. But in this analysis the variables are $m(x)$ and $l(x + 2)/l(x)$, and it is not clear how important statistical autocorrelation might be: this problem should be investigated further using computer simulation.

While care must be exercised in analyzing fecundity and survival relationships, the negative correlation that might arise because of a fixed allotment of ova is still a reproductive cost and must be treated as such. But it is not the only way in which egg production might be constrained. The effort of reproduction could tax an organism and prevent it from being physically capable of gaining sufficient nutrient to furnish the same number of eggs as previously. Studies to date have not examined the reason for correlations between egg production at different stages, and such experiments are needed.

6.3.1. *Phenotypic Correlations*

Bell and Koufopanou (1985) list 11 studies (eight species) in which the phenotypic correlations between early and late fecundity were estimated under laboratory conditions. In only one case was the phenotypic correlation negative. Positive correlations were found in the invertebrates *Aelosoma*, *Cypridopsis*, *Daphnia*, *Gargaphia*, *Philodina*, and *Platyias* and the vertebrate, the domestic chicken. The single negative correlation occurred in the invertebrate *Pristina*. (See Table 3 of Bell and Koufopanou 1985.)

However, phenotypic correlations estimated for 10 field populations are negative in nine cases. The single positive correlation occurred in the pied flycatcher, *Ficedula hypoleuca* (Harvey et al. 1985). But a negative correlation has been found for three other bird species: the great tit, *Parus major* (Kluyver 1963); the Eastern bluebird, *Sialia sialis* (Pinkowski 1977); and the house sparrow, *Passer domesticus* (McGillivray 1983). Negative correlations between early and late fecundity and between fecundity and survival have been found in two plant species—meadow grass, *Poa annua* (Law 1979a; since this study uses family means the correlation obtained is a measure of the genetic correlation between early and late fecundity; it is mentioned here as it is one of those listed by Bell and Koufopanou) and *Senecio keniodendron* (Smith and Young 1982)—and a single mammalian species, the female red deer, *Cervus elaphus* (Clutton-Brock et al. 1983). Three studies of primates report a negative correlation between present and future reproduction. In Japanese macaques, 95% of females that lose infants reproduce the following year while only 8.6% of females not losing their infants reproduce the next year (Tanaka et al. 1970). In rhesus monkeys two studies observed that the date and probability of parturition depended upon the immediate prior reproductive history (Drickamer 1974; Wilson et al. 1978).

Under the relatively benign conditions of laboratory culture the stress of reproduction is probably greatly reduced, and hence the lack of a negative correlation in laboratory studies is not surprising. There is also a distinct phylogenetic difference between the laboratory studies and those of the field, the former comprising 7 invertebrates and 1 vertebrate (domestic chicken), while the latter consist of 2 plant species and 8 vertebrate species. Whether this may also be a reason for the different results is not clear but an analysis of mechanisms would be a profitable course of study in assessing this possibility.

6.3.2. Experimental Manipulations

Manipulation experiments on the consequence of present reproductive output for future output are few. Experimental adjustment on 13 different populations of birds (10 species) have produced negative results (i.e., finding no correlation between present and future clutch size) in six cases and found significant reduction in future brood size in the remaining seven (Table 6.8). Unlike for the survival data, there is no trend for significant effects to be correlated with sample size. This is not surprising because the statistical test is much more powerful. To illustrate, consider the following scenario: the number of eggs in the second brood of the control group is five and that from the group in which the first brood was enlarged is $5 + x$. Both samples have a standard deviation of 0.8. (These figures are based on data from Djikstra et al. 1990.) What is the minimum value of x that can be detected for a particular sample size? Given equal sample sizes of 25 for both control and treatment group, the difference that would be statistically significant (i.e., $t > 1.68$, one-tailed test) is 0.38, a difference of only 7.6%.

Pettifor et al. (1988) reported no significant effect of brood size on future fecundity in the great tit *Parus major*, but a significant effect was obtained

Table 6.8. Effect of manipulating brood size on subsequent brood size in birds

Species	Brood size Normal	Brood size Expt	Sample size	B[a]	Reference
House sparrow	1–7	±2	19	Yes	1
Rook	?[b]	4	66	Yes	2
Tengmalm's owl	5–6	±1	66	NS	3
Kestrel	5	±2	77	NS	4
Canada Goose	1–9	1–9[c]	159	NS[d]	5
Great tit	5–13	±4–5	178	Yes	6
Great tit	5–13	±4–5	179	Yes	7
Great tit	5–13	x2,0.5[e]	213	Yes	8
Blue tit	3–15	3–15[c]	216	Yes	9
Glaucous winged gull	1–3	4–7	302	NS	10
Collared flycatcher	6–7	±1–2	320	Yes	11
House wren	5–8	?[f]	560	NS	12
Great tit	5–13	±3–4	>1000	NS	13

References: 1. Hegner and Wingfield 1987; 2. Røskaft 1985; 3. Korpimäki 1988; 4. Dijkstra et al. 1990; 5. Lessells 1986; 6. Lindén 1988; 7. Smith et al. 1987; 8. Tinbergen 1987; 9. Nur 1988a; 10. Reid 1987; 11. Gustafsson and Sutherland 1988; Gustafsson and Pärt 1990; 12. Finke et al. 1987; 13. Pettifor et al. 1988.

[a]NS = No significant reduction in subsequent broods; Yes = significant reduction in next brood.

[b]The rook lays 4–5 eggs and normally fledges 2 young. Røskaft (1985, p256) states, "brood size . . . experimentally enlarged to four (almost twice the mean (when the young were ages 12–15 days."

[c]The experimental protocol consisted of randomly distributing artificial broods, spanning the usual range in brood size.

[d]Next brood delayed but not significantly smaller.

[e]Each treatment consisted of selecting three broods of equal size. Half of the young from one brood were transferred to one of the other nests, the third being kept as a control.

[f]Not clear from paper.

in three other studies of the same species (Smith et al. 1987; Tinbergen 1987; Lindén 1988). The difference between the four studies is that the first examined variation across years (i.e., did the raising of an enlarged brood in year t decrease the brood raised in year $t + 1$?), while the other three studies tested for effects on the second brood produced within the same breeding season. These findings support the hypothesis that short-term energy expenditures may limit future reproduction but given sufficient time these costs can be recouped.

Minchella and Loverde (1981) were able to manipulate early reproductive effort in the snail, *Biomphalaria glabrata*, by a most unusual treatment.

Snails infected by the trematode parasite, *Schistosoma mansoni*, ceased egg laying 4 weeks after exposure, though unexposed controls continued laying eggs for at least 20 weeks (Fig. 6.4). Exposed snails that did not become infected showed an increase in fecundity for the first 2 weeks, a difference that is marginally significant ($U = 819$, $n = 73$, $P = 0.045$). After week 6 the fecundity of the "exposed but uninfected" group dropped below that of the unexposed controls (Fig. 6.4). These results support the hypothesis that increased early fecundity is associated with decreased later fecundity, though given the marginal significance, further study is warranted.

The reproductive effort of plants is easily manipulated by the removal of inflorescences and artificial pollination. In two experiments of this type, one by Horvitz and Schemske (1988) on the neotropical perennial herb *Calathea ovandensis* and the second by Montalvo and Ackerman (1987) on the orchid *Ionopsis utricularioides*, future fecundity was negatively correlated with present reproductive effort, though the results in the first experiment were not statistically significant. (For a full discussion of these experiments see section 6.1.2.)

Failure to find a cost to reproduction in a single experiment is not sufficient grounds for rejecting the cost hypothesis. As I have stressed throughout this chapter, the real question is not whether a cost exists but under

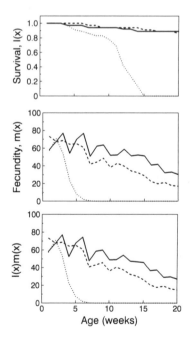

Figure 6.4. Graphical representations of the life tables of the snail *Biomphalaria glabrata*. Dotted line: Snails infected with the trematode, *Schistosoma mansoni* ($n = 35$). Dashed line: Snails exposed but not infected ($n = 36$). Solid line: Unexposed controls ($n = 37$). Data from Minchella and Loverde (1981).

what circumstances a cost is manifested. The experiments reported in this section show that under some circumstances future fecundity is negatively correlated with present fecundity.

6.3.3. Genetic Correlations

Negative genetic correlations between present and future fecundity have been found in meadow grass (Law 1979a), and *Drosophila melanogaster* (Rose and Charlesworth 1981a), but positive correlations are reported on *Oncopletus fasciatus* (Hegmann and Dingle 1982) and *Daphnia pulex* (Bell 1984a; Lynch 1984).

Rose and Charlesworth (1981b) and Luckinbill et al. (1984) have independently verified the negative correlation between early and late fecundity in *D. melanogaster* by selection. Mertz (1975) similarly found a correlated response of decreased later fecundity with selection for early fecundity in *Tribolium casteneum*. Selection experiments on chickens (Lerner 1958; Erasmus 1962; Nordskog and Festing 1962; Morris 1963), geese (Merritt 1962), and mice (Wallinga and Bakker 1978) support the hypothesis that early and late fecundity are negatively correlated.

6.4. Summary

Life history theory assumes that reproduction carries a cost, in terms of either future growth, survival, or fecundity. Tests of this hypothesis can be divided into four categories: phenotypic correlations, manipulative experiments, estimation of genetic correlations by sib analysis, and estimation of genetic correlations by selection experiments. Phenotypic correlations are suspect because there may be other correlated variables that could account for any observed relationship. Manipulative experiments can demonstrate a phenotypic trade-off but do not demonstrate that this trade-off is of evolutionary significance. The last two methods are preferred because they test both the phenotypic and genetic basis of the trade-off.

Both manipulative experiments and genetic analysis support the hypothesis of a trade-off between growth and fecundity. The potential importance of this trade-off is that present reproduction reduces growth and hence reduces future body size, which can diminish future fecundity in plants and ectotherms.

Reproduction increases mortality in three ways: first, the active search for mates increases the vulnerability to predators; second, reproductive females are more likely to be preyed upon because they are either more visible or less mobile; third, reproduction may incur physiological costs that reduce life-span. Evidence for the first source of mortality is provided by extensive field studies (Table 6.4). Manipulative experiments provide support for the second and third mortality factors, though in many experiments the actual cause of death cannot be ascertained. Finally, analyses of individuals from different populations and selection experiments have confirmed that the trade-off between reproduction and survival has a genetic component.

A confounding factor in the above experiments is that costs may be evident only when the organisms are stressed, or in particular environmental conditions. This hypothesis is supported by field studies (Fairbairn 1977; Haukioja and Hakala 1978; Tinbergen et al. 1985) and laboratory experiments (Tables 6.6, 6.7).

Present reproduction may reduce future output. As with previous sources of mortality, this cost may only be expressed under particular conditions. Negative phenotypic correlations between present and future fecundity have not generally been found in organisms maintained in the laboratory, but they predominate in field studies. Experimental manipulations have demonstrated that in some species there is a trade-off between present and future fecundity. Further, sib analysis and selection experiments have consistently obtained negative genetic correlations between present and future fecundity.

7

Age and Size at Maturity

The maximization of fitness requires that an organism's life cycle be optimal with respect to the age schedule of reproduction. This schedule can be conveniently divided into two components: age at first reproduction, and reproductive effort, i.e., the allocation of energy given that an organism is committed to reproduction. The present chapter deals with the first issue; chapter 8 deals with the second. In many organisms reproductive success is closely tied to body size (see chapter 5, section 5.2.2), and, therefore, an analysis of the optimal size at maturity may be more pertinent in some cases than a consideration simply of age. Of course, in simple analyses where growth is deterministic the two variables are entirely congruent, and hence it matters little which we choose to use, though body size is frequently easier to measure. The study of body size itself is a subject of much research (see, for example, the recent books by Peters 1983; Calder 1984; Schmidt-Nielsen 1984), and demographic analyses may in many cases be more appropriately undertaken with a size-based model (Sauer and Slade 1987). This chapter will be restricted to the question of the appropriate size at which to commence reproduction: issues such as how biomechanical factors limit size variation (e.g., Maynard Smith and Savage 1956) or the optimal size at metamorphosis in amphibians (e.g., Wilbur and Collins 1973; Werner 1986) shall not be considered, though the method of analysis is applicable to such questions. Even with the restriction outlined above, this chapter is necessarily selective in its coverage. I have concentrated primarily on studies that provide empirical support for their theoretical arguments, a policy adopted throughout chapters 7–10. This chapter is divided into three major divisions, the first two based on the type of environment: first, analyses predicated on the assumption of a deterministic environment, and second, analyses that consider the consequences of a variable environment. Analyses are formulated under one of three principles: that there exists a single optimum phenotype; that selection is frequency-dependent, thereby favoring the persistence of several phenotypes; or that a given genotype displays a range of phenotypes depending on environmental conditions (phenotypic plasticity/norms of reaction). The question of genetic versus phenotypic variation is generally not explicitly considered; the analyses focus upon the phenotype. Nevertheless, there are important modeling considerations that enter into the analysis of changes predicated on genetic variation versus those that seek to find optimum norms of reaction predicated on the assumption of a single, flexible genotype. The third section of this chapter, Predicting Genotypic Versus Phenotypic Changes, reviews several models discussed earlier in the chapter that highlight the differences in approach.

7.1. Deterministic Environment

7.1.1. Fitness, Development Time, and Fecundity

What changes the rate of increase more, a decrease in the age at first reproduction or an increase in fecundity? This question was addressed by Lewontin (1965) in the context of colonizing ability (an important distinction as we shall see). Lewontin assumed that the $l(x)m(x)$ function can be represented as a triangular function (Fig. 7.1; for a justification of this see chapter 5, section 5.2.1). He then asked "what changes in a single parameter are required to increase r by some specified amount?" Suppose, for example, r is increased from 0.3 to 0.33; for the particular life history parameters chosen by Lewontin the age at first reproduction must be decreased by only 10% whereas the total expected fecundity must be increased by almost 100% (Fig. 7.2). From this and other numerical examples Lewontin concluded that selection should act more strongly on the age at first reproduction than on fecundity. Despite the fact that the limitations of this conclusion have been repeatedly pointed out (MacArthur and Wilson 1967; Meats 1971; Green and Painter 1975; Snell 1978; Caswell and Hastings 1980; Caswell 1982b), authors still make the blanket statement

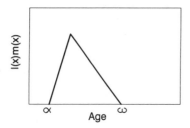

Figure 7.1. Hypothetical $l(x)m(x)$ function. The age at first reproduction is α, and the last age is ω. In this example, drawn to scale, $\alpha = 12$ units and $\omega = 55$ units. Total fecundity, given by the area under the curve $= 780$ offspring.

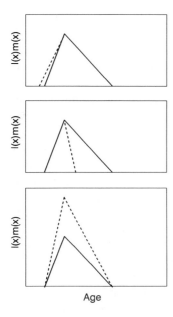

Figure 7.2. Changes in the age at first reproduction, α, the last age, ω, or total fecundity required to increase *r* from 0.30 to 0.33, given initial values of α = 12, ω = 55, and total fecundity = 780. Required changes shown by dashed lines. Top panel: Age at first reproduction, α, decreased to 9.8. Middle panel: Last age, ω, decreased to 34. Bottom panel: Total fecundity increased to 1,350.

that selection will *in general* act to decrease the age at first reproduction more than fecundity.

Lewontin was concerned with the evolution of colonizing ability and therefore considered organisms in which *r* is relatively large. His conclusions are valid in this context. They are not valid when the rate of increase is close to zero. This can be most easily seen by considering the approximation for *r*

$$r = \frac{\log_e(\text{Fecundity} \times \text{Survival})}{\text{Generation time}}$$

$$= \frac{\log_e b}{T} \tag{7.1}$$

where *b* is the "effective fecundity," equivalent to the sum of $l(x)m(x)$, and *T* is generation time. Suppose we double fecundity or halve development time: *r* for the two cases is, respectively,

$$r_b = \frac{\log_e 2b}{T} = \frac{\log_e b}{T} + \frac{\log_e 2}{T} \tag{7.2}$$

$$r_T = \frac{\log_e b}{\dfrac{T}{2}} = \frac{\log_e b}{T} + \frac{\log_e b}{T} \tag{7.3}$$

where the r_b and r_T designate the two changes. The effect of doubling fecundity will be greater than a one-half reduction in the age at first reproduction whenever fecundity, b, is less than two, i.e., $\log_e(b/T) < \log_e(2/T)$. This can be generalized to "if b is multiplied by a factor c and T divided by an equivalent amount $(1/c)$, the effect of a change in fecundity on r will be greater whenever $b < c$." As the rate of increase approaches zero an increase in fecundity becomes increasingly more important in its relative effect on r. Thus for a bird such as the California condor, *Gymnogyps californicus*, which lays only one egg every second year and has an annual rate of increase of only 5%, an increase in fecundity or survival (which would increase the effective fecundity) would be far more significant than an equivalent change in the age at maturity (Mertz 1971).

Mertz (1971), and later in a more rigorous fashion Caswell (1982b), demonstrated that in a declining population the effects of selection will be reversed, favoring a delay in the age at first reproduction. This result can be readily demonstrated by the simple model given above (equation 7.1)— increases in generation time, T, being equivalent to decreases in the age at maturity (assuming no compensatory changes in maximum age). The relationship between r and the generation time is hyperbolic (Fig. 7.3). When r is positive an increase in T decreases r, but when r is less than zero a delay in the age at maturity will increase r (Fig. 7.3). Since continuous decline leads to extinction, selection in a declining population will be of significance only if the decline is temporary. Evolutionary responses to fluctuations in r depend upon the heritabilities of the traits and the periods over which r remains negative or positive.

To obtain the optimal age at first reproduction we must examine the consequences of changes in the age schedules of survival and reproduction. These relationships are biologically different for ectotherms and endotherms, and to a lesser degree between males and females. In ectothermic

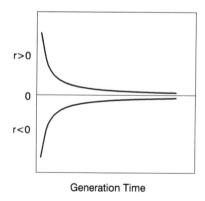

Figure 7.3. The relationship between the rate of increase, r, and generation time.

Generation Time

females (including plants) fecundity typically increases with size (see section 5.2.2.), and since growth is likely to be reduced by reproduction (curtailed in pterygote insects) there is a trade-off between present and future reproduction, regardless of any survival cost to breeding. Though in small mammals there is a correlation between body size and litter size, the effect is far less than in most ectotherms. This can be ascribed primarily to variation in fecundity being very large in ectotherms but small in endotherms. In general, we can assume that among mammals a female gains a negligible increase in litter size by delaying reproduction to grow larger. The effects of reproduction on survival are more likely to be important in shaping the age at maturity in endothermic females. However, the mathematical formulation may subsume both phenomena, the critical factor being not the individual components, $l(x)$ and $m(x)$, but their product, $l(x)m(x)$. In males of both ectotherms and endotherms success at obtaining a mate is frequently associated with large size. (See 5.2.2.) Thus, as with ectothermic females, selection may favor a delay in maturity in males where contests for mates occur, but unlike the former case, selection is likely to be frequency-dependent. Because the forces of selection are different on females and males each will be considered separately.

7.1.2. Age and Size at Maturity in Females

One approach is to ask whether a female's fitness would be increased by maturing 1 year earlier than observed. (See, for example, Tinkle and Ballinger 1972; Wittenberger 1979.) This approach is limited, however, because the question remains on the fitness value of maturing at some still-earlier or -later age. It is better, therefore, to model the effects of reproduction on the age schedules of survival and fecundity and hence predict the optimal age or size at maturity. All of the models within this section assume that fitness is density-independent. Fecundity changes with the age at first reproduction in all models but different mortality regimes are assumed.

The effect of the age schedule of mortality on the optimal age at maturity may depend upon the fitness measure adopted. This can be demonstrated with the following example. The organism in question is semelparous with a fecundity that increases with size and hence age. Let the relationship between fecundity (female births only) and age at maturity, α, be $m(\alpha)$. The age schedule of mortality is divided into two phases: a short period immediately following birth, and the subsequent period. The first period is sufficiently short in relation to the second that mortality over the first period can be considered a point event in time. Mortality rate after this period is constant: the probability of surviving to age α is thus $pe^{-M\alpha}$, where p is the initial proportion surviving, and M is the rate of mortality thereafter. If r is taken to be the appropriate measure of fitness,

$$r = \frac{\log_e m(\alpha)pe^{-M\alpha}}{\alpha}$$

$$= \frac{\log_e m(\alpha)}{\alpha} + \frac{\log_e p}{\alpha} - M \tag{7.4}$$

To find the optimal age at maturity we differentiate the above (see, for example, 4.1.3) to obtain

$$\frac{dr}{d\alpha} = \frac{1}{\alpha}\left(\frac{m'(\alpha)}{m(\alpha)} - \frac{\log_e m(\alpha)}{\alpha} - \frac{\log_e p}{\alpha}\right) \tag{7.5}$$

The optimal age at maturity is found by setting the term in parentheses equal to zero, at which value $dr/d\alpha$ equals zero. Note that the optimal age depends on p but not M.

Now suppose we take for our measure of fitness the expected lifetime fecundity, R_0. We have

$$R_0 = m(\alpha)pe^{-M\alpha} \tag{7.6}$$

Differentiating,

$$\frac{dR_0}{d\alpha} = pe^{-M\alpha}(m'(\alpha) - m(\alpha)M)$$

$$= 0 \quad \text{when}$$

$$\frac{m'(\alpha)}{m(\alpha)} = M \tag{7.7}$$

Note that in this case the optimal age at maturity is independent of p but dependent on M. The important message to draw from this example is that general inferences about how the optimal age of maturity, or any life history trait, should not be made without considerable attention to the assumptions underlying the analysis.

With this cautionary note let us consider how different authors have modeled the evolution of the optimal age at maturity in females and the conclusions to be drawn from these analyses. The first two models assume a semelparous life history, mortality rate being constant in the first and size-dependent in the second.

McLaren (1966) asked if the chaetognath, *Sagitta elegans*, could increase its fitness by maturing at ages 1 or 3 years rather than at the observed age of 2 years. The advantage of delaying reproduction is that a larger size and increased fecundity are achieved, fecundity (= number of female births, the species reproducing parthenogenetically) being allometrically related to body length, L, according to the function $m(L) = 0.115L^{2.46}$. But the increased fecundity is offset by an increased probability of dying before maturity. The average number of eggs produced by a female in her second year is 543. To estimate the probability of an individual surviving to reproduce McLaren assumed the population to be stationary and semelparous. Consequently, the probability of *S. elegans* surviving to its second year is simply 1/543. To estimate the rate of increase of a chaetognath maturing in its first year we need an estimate of fecundity at age 1 year and survival to this age. The first was obtained from the allometric regression of fecundity on body length (integrated over the size distribution of animals in their first year), giving 138 eggs. Survival was estimated by implicitly assuming that mortality can be divided into two components as described in the first example of this section (equation 7.4). The probability of surviving to age 1 (e^{-M}) having survived the first "period" (early mortality of eggs or larvae) was obtained by dividing the number of 2-year-olds in a sample by the number of 1-year-olds, giving a survival probability of 0.384. The first component, p, was obtained by noting that for 2-year-olds, $543p0.3836^2 = 1$—i.e., each individual just replaces itself—and hence $p = 1/(543 \times 0.384^2) = 0.0125$. No direct estimates could be made for individuals maturing in their third year and McLaren assumed mortality rate M remains constant. From these data the rates of increase for the three ages can be calculated (Table 7.1). Individuals maturing at either age 1 or 3 years have both a reduced rate of increase and expected lifetime fecundity, supporting McLaren's hypothesis that *S. elegans* is maturing at an age that maximizes its fitness.

Size-dependent mortality was incorporated by Kachi and Hirose (1985) in a stochastic simulation model of the semelparous plant *Oenothera glazioviana*. Survival comprised three components: seed survival, seedling survival, and annual survival of vegetative rosettes. Emergence rate of

Table 7.1. Estimated life history parameters for the chaetognath, *Sagitta elegans*, maturing at ages 1, 2, or 3 years

Age (years)	Fecundity	p	e^{-My}	R_0	r
1	138	0.0125	0.384	0.66	−0.43
2	543	0.0125	0.147	1.00	0.0
3	1318	0.0125	0.056	0.92	−0.02

seedlings (seed survival) and survival probability of seedlings were assumed constant at the 5-year averages of 0.0205 and 0.48, respectively. The seedling stage was assumed to last one year, after which the plant was classified as a vegetative rosette until bolting occurred (i.e., flowering). Annual survival, S, of the vegetative rosettes increased with rosette diameter, D (cm), according to the equation

$$S = 0.39 \log D + 0.36 \qquad (0.0 \le S \le 0.7) \qquad (7.8)$$

Growth of seedlings and vegetative rosettes was modeled by use of two probability distributions representing the growth rates of the two phases. Growth rate in the seedling phase was independent of size but in the vegetative phase, sigmoidal growth dependent upon rosette diameter best fit the observed growth pattern.

As is typical of fecundity relationships, seed production, N, varied allometrically with rosette diameter,

$$\log N = 2.22 \log D + 0.96 \qquad (7.9)$$

The above model has three stochastic elements—two for growth and one for the survivorship of rosettes. As the critical size at flowering is increased, the probability of surviving to reproduce decreases but seed production increases, the combined effect being to produce a maximum value of r and R_0 at a critical size of 16.6 cm (Table 7.2). This size is close to the size, approx. 14 cm, at which 50% of the plants bolted, and within the size range of reproduction (9–23 cm) in the natural population (Kachi and Hirose 1985). Sensitivity analysis showed that increased seedling sur-

Table 7.2. Summary of the effects of varying rosette size at reproduction on the life history schedule and rate of increase for a hypothetical population of *Oenothera glazioviana*

Size (cm)	Probability of reproduction	Seed production	Generation time (years)	R_0	r
1.0	0.59	17	1.5	10	−1.370
3.2	0.24	154	2.4	37	−0.400
5.0	0.16	357	2.9	57	−0.186
10.0	0.09	1,240	4.0	112	0.014
16.6	0.04	3,140	5.1	126	0.058
32.0	0.01	11,000	6.6	110	0.004

Modified from Kachi and Hirose (1985).

vival favors a smaller size and lower age at maturity. Kachi and Hirose also examined the scenario in which maturity is age-dependent and found that such a pattern is less fit than a size-mediated maturation. This result deserves further investigation.

The two preceding analyses made the simplifying assumption of a semelparous life history and consequently cannot address the relative importance of prematurational versus postmaturational survival rates. The models of Roff (1981a, 1984a) and Kusano (1982) are useful in this regard. These models, though based on different taxa—pterygote insects (Roff 1981a), fish (Roff 1984a), and amphibia (Kusano 1982)—are mathematically very similar.

Because it deals with both semelparous and iterparous life histories the fish model shall be considered first. The model components are:

1. Growth in fish can be described by the von Bertalanffy equation

$$L(x) = L_\infty (1 - e^{-kx}) \tag{7.10}$$

2. Fecundity is typically proportional to the cube of length (section 5.2.2) and hence $m(x)$ can be written as the following function of age:

$$m(x) = c_1(L_\infty(1 - e^{-kx}))^3 \tag{7.11}$$

3. Mortality can be divided into three portions: first, a very high mortality period during the egg and early larval stage, the proportion surviving this period being p. This is followed by a mortality rate during the immature period of M_i that increases at maturity to M_a as a consequence of reproduction. For the semelparous case the latter mortality rate is irrelevant since, by definition, no female survives after reproduction. For iteroparous fish the data are typically not sufficient to distinguish between M_i and M_a and so in the simple iteroparity model I assumed a constant rate M. Effects of reproduction on growth and survival are jointly incorporated into this model by modifying the expected fecundity. (See below.)

From these considerations $l(\alpha)m(\alpha)$ can be written as

$$l(\alpha)m(\alpha) = pe^{-M\alpha}c_1L_\infty^3(1 - e^{-k\alpha})^3 \tag{7.12}$$

To accommodate an iteroparous life history a simple change that still makes the model analytically tractable is

$$l(x)m(x) = pe^{-Mx}c_1L_\infty^3(1 - e^{-k\alpha})^3 \tag{7.13}$$

The only change is to include age-specific survival, $l(x) = pe^{-Mx}$. This revision assumes that the cost of reproduction is such that increases in fecundity due to increases in size with increasing age are offset by increases in mortality. This is likely to be at least approximately true if there is a survival cost associated with reproduction. The optimal age at first reproduction depends upon the fitness measure chosen but not the assumption of iteroparity or semelparity (Roff 1984a). For most fish, populations are probably more or less stationary and hence the most suitable measure of fitness is R_0, from which we obtain the optimal age at first reproduction as

$$\alpha = \frac{1}{k} \log_e \left(\frac{3k}{M} + 1 \right) \tag{7.14}$$

Bell (1980) noted that selection will favor a delay in the age at maturity so long as the product $l(x)m(x)$ increases with age: in this regard it is interesting to note that equation 7.14 is the turning point of the $l(x)m(x)$ curve (Roff 1984a)—i.e., that age at which the product $l(x)m(x)$ begins to decline with age.

If r is taken as the measure of fitness the optimal age at first reproduction is

$$\alpha \left(\frac{3ke^{-k\alpha}}{1 - e^{-k\alpha}} \right) - \log_e(pc_1 L_\infty^3 (1 - e^{-k\alpha})^3) = 0 \tag{7.15}$$

The optimal age at maturity cannot be separated from the other components and hence the above equation must be solved numerically. The optimal age at maturity is independent of adult mortality (M) but dependent on larval survival (p). If we divide M into two components—a mortality rate prior to maturity, M_i, and a rate after maturity, M_a—the optimal age at maturity depends not only on p but also on the difference between the adult and immature rates ($M_a - M_i$). However, postmaturational mortality is also implicitly incorporated by equation 7.13, and therefore the response of age at maturity to postmaturational mortality cannot be judged by M_a alone.

Since the assumption of a stationary population is reasonable for most fish populations and because the predicted age at maturity in this case is based on two parameters (k and M) for which relatively good data are available, equation 7.14 was chosen as the one to be tested. The fit between predicted and observed ages at maturity is encouragingly good (Fig. 7.4). The fish model can be recast in terms of the optimal length at maturity (Roff 1986c), giving

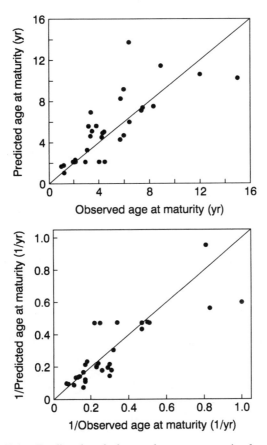

Figure 7.4. Predicted and observed ages at maturity for a variety of fish species. Predicted values obtained from the fish model. For statistical reasons (see chapter 4) the correct comparison is between the reciprocals (lower panel; $r = 0.84$, $n = 31$, $P < 0.001$). Solid line is 1:1 reference line. Data from Roff (1984a).

$$L(\alpha) = L_\infty \left(\frac{3k}{3k + M} \right) \tag{7.16}$$

This model gives an even better fit between observation and prediction than that for age at maturity (Fig. 7.5).

The evolution of the age at maturity depends not only upon which parameters may be changed but also which measure of fitness is being maximized (Table 7.3). An increase in the fecundity coefficient (c_1), early larval survival (p), or asymptotic length (L_∞) decreases the optimal age at ma-

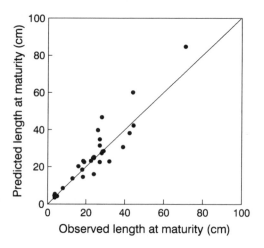

Figure 7.5. Predicted and observed lengths at maturity for the fish species shown in Fig. 7.4 ($r = 0.93$, $n = 31$, $P < 0.001$). From Roff (1986c).

turity if r is maximized but has no effect if R_0 is maximized. Increased immature mortality rate (M_i) or growth rate (k) decreases the optimal age at maturity for both measures of fitness. On the other hand, increased adult mortality favors a delay in maturation when r is maximized but has no effect if R_0 is the appropriate fitness measure.

Pterygote insects differ from fish in three respects: first, growth ceases at maturity (Ephemeroptera undergo a molt as adults but this does not change body mass); second, the age schedule of births is triangular, though fecundity increases with size; third, at least in holometabolous insects, the larval and adult forms may inhabit completely different habitats and hence be subject to independent mortality regimes. The following model was designed to predict the optimal size/age at maturity in a specific insect, *Drosophila melanogaster* (Roff 1981a).

In *D. melanogaster* body size increases with development time and fecundity increases allometrically with body size. The age schedule of female births is triangular in shape (Fig. 7.6), and is described by the equation

$$m(x) = \frac{1}{2} c_1 L^{c_2}(1 - e^{-c_3(x - c_4)})e^{-c_5 x} \qquad (7.17)$$

where the c_is are constants, L is thorax length, and x is age. Thorax length, L, scales the age schedule of reproduction, larger females producing more eggs, but does not change its position. The constant c_1 is the product of

Table 7.3. Effects of changing parameter values on the optimal age at first reproduction for two life history models, one based on fish ("fish" model) and the other on an insect life history ("*Drosophila*" model)

	Symbol in model		Fitness measure	
Parameter	Fish	*Drosophila*	r	R_0
Fecundity coefficient[a]	p, c_1	c_1	−	0
Fecundity exponent	3^b	c_2	+	+
Growth coefficient	L_∞	c_6^c	−	0
Growth exponent[d]	k	c_7	−	−
Immature mortality	M_i	M_l	−	−
Adult mortality	M_a	M_a	+	0

A " + " indicates that the optimal age at maturity is increased by an increase in the indicated parameter, a " − " that it is decreased, and a "0" that it has no effect.

[a]Includes egg viability and very early larval mortality (p in fish model).

[b]This parameter not varied in fish model.

[c]When the fitness measure is r the optimal length decreases, but it is not clear if the optimal age at maturity decreases (probably). When the fitness measure is R_0 the change in the optimal thorax length is matched by the change in c_6 and the optimal age at maturity is not changed.

[d]These two components are somewhat different, increases in k decreasing the time to reach any given size, while increases in c_7 increase development time.

two constants: the coefficient of proportionality within the allometric relationship between length and fecundity, and the proportion of eggs that fail to hatch, equivalent to p in McLaren's analysis and the fish model.

Development time, $d(L)$, scales to body size according to the relationship (Fig. 7.6)

$$d(L) = c_6 L^{c_7} + c_8 \tag{7.18}$$

Thus development time scales allometrically with size except for the constant c_8 which represents time required for the eggs to hatch and the development within the pupa, both of these components being independent of size.

Because information on mortality rates is so poorly known I assumed that instantaneous rates remain constant in the adult and larval phases at M_a and M_l, respectively. Probability of surviving the larval period, $l(L)$, is thus (Fig. 7.6)

$$l(L) = e^{-M_l d(L)} \tag{7.19}$$

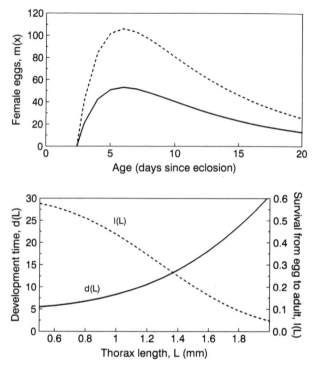

Figure 7.6. Relationship between body size and life history traits in *Drosophila melanogaster*. Upper panel: Age schedule of reproduction for small (solid line) and large (dashed line) females. Lower panel: Development time and survival from egg to adult in relation to adult thorax length.

Drosophila melanogaster is a colonizing species and hence the appropriate measure of fitness is r. Combining the above relationships we obtain the characteristic equation

$$\sum_{x=1}^{\infty} \frac{1}{2} e^{-r[x+d(L)+c_4]-M_l d(L)-M_a(x+c_4)-c_5 c_4} c_1 L^{c_2}$$

$$\times \, (1 - e^{-c_3 x}) e^{-c_5 x} = 1 \qquad (7.20)$$

where, for convenience, age has been rescaled to begin at the first day of egg laying. Though this equation is tediously long its solution presents no great difficulty. Briefly, the method is first to evaluate the series by making use of the fact that it is a geometric progression, and second to differentiate implicitly to obtain the optimal length at maturity. (Full details are given in Roff 1981a.) Unlike the analysis for a semelparous organism (equations 7.4–7.7), the optimum thorax length depends upon adult mortality, larval mortality, and the constant initial "mortality" component (p or c_1). Figure 7.7 shows the relationship between r and thorax length by using values obtained from laboratory stocks and estimates from wild populations. The predicted maximum, 0.95 mm, falls very nicely within the observed range in thorax length of 0.90 to 1.15 mm.

Given the possible error in the estimation of parameter values, the close fit between prediction and observation bears closer scrutiny. A priori, the close fit suggests (1) the right values were fortuitously chosen, or (2) the optimum is relatively insensitive to variation in parameter values. To test the second hypothesis I varied two parameters at a time and computed the change in the optimum thorax length. Wide variation in the three pairs of components that are likely to be poorly estimated produces relatively little variation in the optimum thorax length (Table 7.4). An estimate of the total variation possible can be obtained by selecting the set of extreme values that favor either an increase or decrease in body size. From this the smallest thorax length is 0.60 mm and the largest, 1.38 mm. Although this range is larger than that observed (0.90–1.15) it is still surprisingly small given the range in parameter values. The maximum range obtained covers

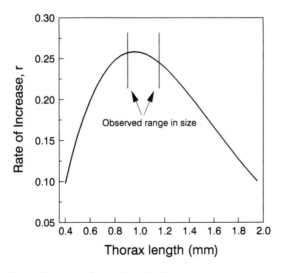

Figure 7.7. Estimated relationship between fitness, as measured by the rate of increase, and thorax length in *Drosophila melanogaster*. (Redrawn from Roff 1981a.)

Table 7.4. Effect of pairwise variation in parameter values on the predicted optimal thorax length of *Drosophila melanogaster*

Parameter 1		Parameter 2		Range in thorax length (mm)
Symbol	Range	Symbol	Range	
c_1	40–140	c_2	2–4	0.85–1.15
c_6	3–6	c_7	3–4	0.80–0.95
M_a	0.0–0.5	M_l	0.0–0.5	0.80–1.10

Data from Roff (1981a).

Explanation of symbols:

c_1: The product of the coefficient in the allometric relationship between thorax length and fecundity (i.e., the value of a in Fecundity $= a(\text{Thorax length})^{c_2}$), and the proportion of eggs that fail to hatch.

c_2: The exponent in the allometric relationship between thorax length and fecundity (see above).

c_6, c_7: The coefficient and exponent in the equation relating development time to thorax length. (See equation 7.18.)

M_a: Instantaneous rate of adult mortality.

M_l: Instantaneous rate of larval mortality.

almost the full range of body size of North American drosophilids, demonstrating that this set of species could easily have evolved from relatively minor changes in parameter values.

Hawaiian drosophilids may be very much larger than the North American species, some being as large as houseflies. The longer development time of these forms is reasonably well predicted by extrapolating the *D. melanogaster* growth curve (Roff 1983c; see lower panel of Fig. 7.16). In the present model large shifts in bodies size can be achieved by reducing the fecundity parameters and the mortality rates considerably below those values in the sensitivity analysis. This is consistent with the considerably reduced fecundities of the Hawaiian species (Robertson et al. 1968; Kambysellis and Heed 1971), which also necessarily implies very low rates of mortality.

The rate of increase predicted in the above analysis is very high; Ricklefs (1982) suggested that the analysis be restricted to cases in which $r = 0$. An alternate approach is to consider first cases in which the rate of increase is low and second an analysis based on the assumption that r is equal to zero. The former approach does not make a significant difference to the analysis (Roff 1983c), but the latter does. Under the assumption that the population is stationary and that selection favors the maximization of expected lifetime fecundity, the optimal thorax length is

$$L = \sqrt[c_7]{\frac{c_2}{M_l c_7 c_6}} \qquad (7.21)$$

and the optimal age at metamorphosis is

$$\alpha = \frac{c_2}{M_l c_7} + c_8 \qquad (7.22)$$

Thus the optimal body size depends on the exponent of the fecundity function (c_2), the parameters describing the allometric component between development time and body size (c_6, c_7), and the larval mortality rate (M_l). The optimal age at metamorphosis does not depend on c_6 but is an additive function of c_8, the time required for egg and pupal development. Adult mortality rate (M_a) and the fecundity/egg viability constant (c_1) no longer influence adult body size or α. This result is consistent with the earlier analysis of the semelparous model (equations 7.4–7.7), the larval mortality rate corresponding to the mortality rate prior to breeding. The optimal thorax length using the same values as in Fig. 7.7 is 1.46, substantially larger than the observed. However, the optimum is very sensitive to the estimate of larval mortality which, as noted above, is one of the parameters for which we have the least amount of data. The value used was 0.1, corresponding to a daily mortality rate of 10%: increasing the rate to 0.31, corresponding to a daily mortality rate of 27%, gives an optimum thorax length of 1 mm. Under natural conditions the higher figure is not unrealistic. (In the sensitivity analysis the range in larval rate considered was 0.0–0.5, Table 7.4.)

Despite their different biological underpinnings, the fish and *Drosophila* models make the same general predictions (Table 7.3). The evolutionary response to changes in the fecundity/egg viability coefficient (c_1) is contingent on whether the population is stationary or increasing. An increase in the rate of egg production per unit body size (c_2) favors a decrease in size at maturity. A reduced age at maturity accompanies an increase in the growth parameters, c_6 and c_7. Increased larval mortality (M_l) favors decreased size at maturity, while the influence of adult mortality (M_a) depends upon the fitness measure.

The optimal age at maturity in the salamander, *Hynobius nebulosus tokyoensis*, was estimated by Kusano (1982) using the same methodology as described for the *Drosophila* model, though the functional relationships are somewhat different. Fecundity varies linearly with length ($L(x)$ = snout-vent length at age x), and hence

$$m(x) = \frac{1}{2}(2.19L(x) - 124) \qquad (7.23)$$

Unlike *D. melanogaster*, salamanders continue to grow throughout their lives. Kusano estimated the growth rate of immature and mature salamanders based on the following observations: (1) growth rate of postmetamorphic juveniles is 8 mm per year; (2) the size at the end of the first year is 27.2 mm, and (3) the growth rate of mature animals is 1.1 mm per year. From these the size at age x is

$$L(x) = 8(x - 1) + 27.21 \qquad 1 \le x \le \alpha$$
$$L(x) = 1.1(x - \alpha) + L(\alpha) \qquad x > \alpha \qquad \text{(7.24)}$$

where α is the age at maturity. The final component needed is survival. This Kusano specified as

$$l(x) = pe^{-M(x-0.5)} \qquad \text{(7.25)}$$

where p ($= 0.062$) is the proportion of salamanders that survive to metamorphosis, occurring at 6 months of age (hence the 0.5 because the time unit is 1 year), and M ($= 0.7$) is the rate thereafter.

The relationship between r and the age at first reproduction is shown in Fig. 7.8; the optimal age at maturity is 5 years, which corresponds to that observed in the natural population. As predicted from the previous analyses, increasing the premetamorphic survival (p) favors a decreased age at maturity. The immature and mature rates of mortality (i.e., postmetamorphic rates) are equivalent to larval and adult rates in the *Drosophila* model but because they are set equal a distinction cannot be made. In-

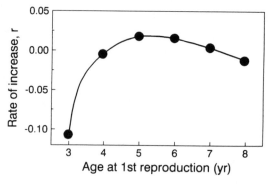

Figure 7.8. Estimated relationship between fitness, as measured by the rate of increase, and age at first reproduction in the salamander *Hynobius nebulosus*. (Redrawn from Kusano 1982.)

creasing postmetamorphic survival (i.e., decreasing M) selects for an increased age at maturity (Kusano 1982)—a result that corresponds to the effect of increasing larval mortality in the *Drosophila* model. This suggests that mortality rate prior to maturation is most significant in molding the life history—a conclusion that is also supported by the relative sensitivity of body size to variation in the two mortality rates (Roff 1981a) and by the fact that the influence of adult mortality disappears if R_0 is taken as the fitness measure (Table 7.3). Such a result in intuitively reasonable, for a female that dies before maturation leaves no offspring, while one that dies after maturation leaves some. Therefore, we might expect, a priori, that if immature and adult survival rates are correlated, selection will favor an alteration that favors a decrease in immature mortality rate more than one which decreases adult mortality rate.

The results for the effect of adult and juvenile mortality rates on the optimal age at maturation (Table 7.3) are noteworthy because they are in contradiction to theoretical analyses which predict that an increase in adult mortality will select for a decrease in the age at reproduction, while an increase in juvenile mortality will favor the opposite response (Gadgil and Bossert 1970; Law 1979b; Michod 1979). This difference is a consequence of how mortality is entered into the models. The *Drosophila* model is constructed on the basis of two different phases in the life cycle—an immature phase and an adult phase. A female incurs the mortality associated with the adult phase only by metamorphosing into the adult form. If adult mortality is high, selection favors females that delay entry into the adult stage, remain as larvae, and grow larger. Likewise, a high larval mortality favors rapid exit from this stage into the adult. The animal thus has the option of varying how long it is subjected to either mortality regime. The same principle applies to the other two models—the mortality rate in mature fish increasing as a consequence of reproduction, while that in the salamander model of Kusano changes with metamorphosis. In contrast, the models of Gadgil and Bossert (1970), Law (1979b), and Michod (1979) are predicated on the premise that mortality rate depends upon age not developmental stage. If mortality is solely age-dependent, a female cannot escape the source of mortality by delaying reproduction; therefore, if mortality in later age groups increases, selection will favor females that reproduce before they are subjected to the increased mortality. If the cost of reproduction in terms of mortality declines with age, selection will act to delay maturation. By appropriate changes in the relationship between age, reproductive effort, and survival a variety of more complex responses can be obtained (Schaffer and Rosenzweig 1977; Law 1979b).

Reznick and Endler (1982) studied the response of guppies, *Poecilia reticulata*, to variation in mortality rates induced by variation in the predator assemblage. Three types of communities were studied, which were named after the dominant predator: (1) Crenicichla (*C. alter*), high predation intensity, predominantly on adult guppies; (2) Rivulus (*R. hartii*), moderate predation, predominantly on juveniles; and (3) Aequidens (*A. pulcher*), low predation on all size classes. Reznick and Endler (1982, p. 162) predicted that "the age at maturity of guppies [would be seen] to rank Crenicichla < Aequidens < Rivulus, and reproductive effort to rank in the reverse order." This prediction is based on the unstated premise that predation cannot be avoided by reducing reproductive effort: in this event a female guppy in the high predator environment should reproduce before she enters the size category at which she is at risk and compress her reproductive effort into this time frame. Results from both field-collected and lab-reared guppies support both predictions (Table 7.5). The data from the laboratory rearing is important in demonstrating that the traits are genetically based, and thus that the observed values for field populations can plausibly be attributed to an evolutionary response.

Edley and Law (1988) experimentally tested the prediction that growth rate will evolve in response to size-selective predation by using two culling regimes on laboratory populations of *Daphnia magna*. In one regime only small individuals were removed, while in the second only large individuals were sieved from the population. After 150 days of culling clonal differences in growth rate were evident. In the clones subjected to predation on small size, growth was rapid during the early life stages, thereby pushing the females rapidly through the vulnerable period. Maturity in these clones was delayed, as expected if there is a trade-off between growth and reproduction. In contrast, populations subjected to predation on large size showed a reduced growth rate and early maturation, a phenology that

Table 7.5. Life history variation in guppies from three sites differing in predation rates

Life history traits	High (*Crenicichla*)		Moderate (*Rivulus*)		Low (*Aequidens*)	
	Field	Lab	Field	Lab	Field	Lab
Minimum size of gravid females (mm)	14.6	14.6	17.4	18.5	18.5	17.4
Size of mature males (mm)	14.9	14.9	16.4	17.8	17.8	16.4
Gonadosomatic index (%)	16.0	16.0	12.5	14.0	14.0	12.5

Data from Reznick and Endler (1982). Data from both field-caught specimens and lab-reared are presented. The data from the field specimens are the grand means from 7 (high), 5 (moderate), and 4 (low) sites. See Reznick and Endler (1982) for statistical tests.

reduces the impact of predation. Unfortunately, the culling regimes also changed densities differentially, and hence it is possible, though Edley and Law think unlikely, that the observed responses were a result of density-dependent selection.

As shown by the models of Myers and Doyle (1983, see section 5.1.2), it is not always possible to distinguish between different biological scenarios from a comparison of predictions and observations alone. This is clearly demonstrated by the analyses of Stearns and Crandall (1981). Stearns and Crandall attempted to predict the optimal age at maturity for a variety of lizard and salamander species by using three models, which they called the salamander model (SAM), the linear fecundity model (LFM), and the quality-of-young model (QYM). The first and last of these models differ from those of Roff (1981a, 1984a) and Kusano (1982), discussed above, in assuming that mortality rate in the immature stage is not constant but varies with the age at maturity.

Assumptions of the salamander model are:

1. Mortality rates: the proportion of juveniles surviving to maturity is a function of the age at maturity, α, and the constant adult mortality rate M_a,

$$l(\alpha) = e^{-M_i\alpha}$$
$$= e^{-(M_a + (c_1/\alpha^{c_2}))\alpha} \qquad (7.26)$$

where M_i is the immature mortality rate and c_1 and c_2 are constants. Equation 7.26 states that the mortality rate of immatures decreases as the age of maturity increases (Fig. 7.9). One possible interpretation of this is that older animals are "better" parents either by virtue of better parental care or because they put more resources into each egg (the quality-of-young model is named after this function). Another way in which such a function could arise is if survival increases with size: delaying maturity will automatically increase the average survival rate per time interval. It is important to remember that a variety of biological phenomena may be described by the same mathematical expression, and hence that model results may not be as constrained as might be gauged from the original biological arguments upon which the model is grounded.

A potential weakness in the above mortality model is that neither the shape nor the particular values are based on empirical data. For the quality-of-young model (described below) no optimum is possible if c_2 is less than one: Stearns and Crandall arbitrarily set c_2 to the next integer, two. The mortality rates of juveniles, M_i, and adults, M_a, were estimated from published data and c_1 was then estimated from

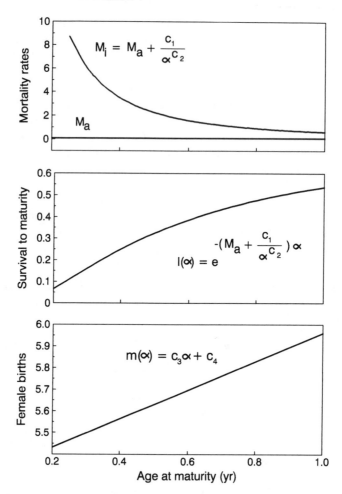

Figure 7.9. Life history components of the salamander model applied to the lizard *Uta stansburiana*. Estimates as given by Stearns and Crandall (1981). Top panel: Adult mortality rate, M_a, a constant at 0.08. Mortality rate of immatures described by the equation shown with $c_1 = 0.54$ and $c_2 = 2$. Middle panel: Survival to maturity increases with the age at maturity. Lower panel: Fecundity a linear function of the age at maturity, where $c_3 = 1.32$ and $c_4 = 10.6$.

$$c_1 = \alpha^2(M_i - M_a) \tag{7.27}$$

2. Fecundity is a linear function of age at maturity, as might occur if fecundity increases linearly with size and size increases linearly with age until the age at maturity, at which point growth ceases (Fig. 7.9). Thus the number of female births, $m(x)$, is

$$m(x) = c_3\alpha + c_4 \tag{7.28}$$

where c_3 and c_4 are constants.

The characteristic equation for the salamander model is thus

$$\int_\alpha^\infty e^{-rx}e^{-(M_a+(c_1/\alpha^2))\alpha}e^{-M_a(x-\alpha)}(c_3\alpha + c_4)\, dx = 1 \tag{7.29}$$

Note that this model is mathematically equivalent to one in which mortality rate is constant throughout life and fecundity increases with the age at maturity according to the relationship

$$m(x) = (c_3\alpha + c_4)e^{-(c_1/\alpha)} \tag{7.30}$$

Such a relationship would occur if growth declined with age and ceased at maturity. This example again emphasizes the point that a variety of biological mechanisms may be subsumed under a single mathematical expression.

The linear fecundity model differs from the salamander model in assuming a constant mortality rate and a fecundity that, after maturity, continues to increase linearly with age:

$$m(x) = c_3\alpha + c_4 + c_5x \tag{7.31}$$

This model allows for continued growth after maturity but at a reduced rate as a consequence of allocation to reproduction. It is similar to the equation derived by Kusano (equation 7.23).

The quality-of-young model differs from the salamander and linear fecundity models in assuming a constant fecundity but is the same as the salamander model with respect to the juvenile survival function (equation 7.26).

These three models can be solved by integrating the characteristic equation and differentiating r with respect to α. All three models predict the optimal age at maturity much better than expected by chance in four lizard species ($n = 7$ because one species is represented by four populations) and a salamander (two populations; Fig. 7.10). The linear fecundity model does not always produce an optimum age at maturity (i.e., the relationship

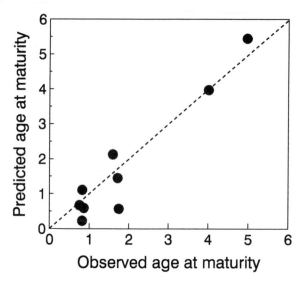

Figure 7.10. Predicted and observed ages at maturity for lizards and salamanders. Dashed line is 1:1 reference line. Predicted ages obtained using the salamander model ($r = 0.96$, $n = 9$, $P < 0.01$; data from Stearns and Crandall 1981).

between r and α is monotonically increasing or decreasing) and the model was successful in only six cases.

Stearns and Crandall (p. 461) suggested that "Both the Linear Fecundity and Quality of Young models can account for much of the variation among these populations and species in age at maturity, but the Salamander Model does so remarkably well." The failure of the linear fecundity model in three cases certainly relegates it to third place but the correlation between prediction and observation is not sufficiently different between the other two so that it can be said that the salamander model is better than the quality-of-young model. (The three correlations are 0.896, 0.929, and 0.956 for the linear fecundity model, the quality-of-young model, and the salamander model, respectively.) Furthermore, the salamander model contains five parameters while the quality-of-young model contains only four: given the lack of statistical difference between the predictions the quality-of-young model may be preferred because of the fewer parameters. But which model we choose should be based upon the realism of the underlying biological assumptions. These assumptions need more rigorous testing. The important message to derive from this analysis is that great caution should be exercised in interpreting a correspondence between observation and prediction as evidence for the validity of the supposed biological mechanisms in the model.

Stearns and Koella (1986) applied a modified salamander model to the analysis of optimal-age-at-maturity in fish. They assumed the same juvenile mortality function (equation 7.26), which for fish is probably best interpreted as survival rate increasing with size. (See, for example, Fig. 5.13.) Fecundity was assumed to be either a linear or a cubic function of length, the $m(x)$ function for the latter thus being

$$m(x) = c_2 L(x)^3 + c_3 \qquad (7.32)$$

Typically the relationship is obtained from a log/log regression, and hence the back-transformed equation does not include the constant c_3. Given the high correlations generally obtained with logarithmic regression, it is likely that c_3 is very close to zero. (Unfortunately the parameter values are not provided in the paper and have been lost [Steve Stearns, pers. comm.].)

Length at age was described by the von Bertalanffy equation (see Fig. 5.6),

$$L(x) = L_\infty(1 - e^{-kx - c_1}) \qquad (7.33)$$

where L_∞ is the asymptotic length and k and c_1 are species- or population-specific constants. The constant c_1 is typically very close to zero and can be omitted without significantly altering the predicted length at age (Roff 1984a). Predicted and observed ages at maturity in 19 stocks of fish are highly correlated (Fig. 7.11), though there does appear to be a bias in the predictions, those below age 2 years lying below the 1:1 line and those above 3 years lying above the line.

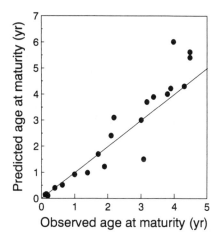

Figure 7.11. Predicted and observed ages at maturity for a variety of fish species. Predicted values obtained from the modified salamander model; 1:1 reference line shown ($r = 0.93$, $n = 19$, $P < 0.001$; data from Stearns and Koella 1986).

Despite their apparent differences in formulation, the modified salamander model (Stearns and Koella 1986) and the fish model (Roff 1984a), are very similar mathematically. This can be shown by consideration of the case of maximizing R_0 for a semelparous organism. The Stearns and Koella model predicts that the optimal age at reproduction is

$$\alpha = \frac{1}{k} \log_e \left(\frac{3k + M_a - \frac{c_1}{\alpha^2}}{M_a - \frac{c_1}{\alpha^2}} \right) \tag{7.34}$$

which differs from equation 7.14 only by the term c_1/α^2, which for field populations is estimated as $M_i - M_a$ (Stearns and Crandall 1981). This difference has relatively little effect on the optimal age at first reproduction (Roff 1984a). The optimal age at maturity in the fish model is the age at which the $l(x)m(x)$ curve has a turning point (i.e., up until this age the product $l(x)m(x)$ is increasing while after this age it is decreasing); the Stearns and Koella model predicts exactly the same turning point, again emphasizing the similarity of the two models.

The *Drosophila* model is a general description of the insect life cycle and may be appropriate to other organisms with determinate growth, while the remaining models are descriptive of organisms with indeterminate growth such as fish (Roff 1984a; Stearns and Koella 1986), amphibians and reptiles (Stearns and Crandall 1981; Kusano 1982), shrimp (Charnov 1989b using the fish model), and plants (Kachi and Hirose 1985). However, as indicated by the identity of their predictions (e.g., Table 7.3), the differences between the two types of models are superficial. The important fact to remember is that the mathematical equation describing the $l(x)m(x)$ curve may have a variety of biological interpretations. Equally important, it must be remembered that a class of models may not include a biological phenomenon that may be significant in some cases; for example, not all the models considered include the type of age-specific mortality regime studied by Reznick and Endler (1982) and Edley and Law (1988). Fortunately, the consequences of this type of variation are intuitively obvious; this may not always be the case, and thus we must be very clear about the assumptions, both mathematical and biological, that underlie our models.

Further refinement of the models is required; future models should include

1. A more accurate representation of the age schedule of mortality prior to maturity. Analyses of mortality rates in the pelagic ecosystem and more generally across taxa (chapter 5) suggest that

mortality rates are allometrically related to size, a fact that needs to be incorporated—and studied empirically with respect to single populations. Size-selective mortality is probably a major factor generating the $l(x)$ curve.

2. A growth function that relates size and the allocation of energy to reproduction. For fish, surplus energy is accumulated in an allometric fashion with size (Ware 1975a, 1978, 1980); whether this is typical is not clear, though it seems likely. Given an expression for the amount of surplus energy available the consequences of allocating a specific portion to growth and the remainder to reproduction can be assessed by using a simple compartment model (Roff 1983b).

3. A function relating growth and survival. An animal that forages more increases its risk of being preyed upon (Weissburg 1986; Godin and Smith 1988; Lima and Dill 1990). In a very general model this risk might be subsumed within the mortality/size relationship.

4. A function that relates fecundity to reproductive effort. An important question in this context is how changes in reproductive effort generate the observed allometric relationships.

5. The advantage of a reduced reproductive effort is a higher future fecundity either because of increased size or increased survival. Thus a function relating survival to reproductive effort may be required. Though there are few data from which to construct such a relationship, theory does predict what the general shape of the function should be. (See chapter 8.)

The approach I advocate is the development of detailed models that explicitly include the costs of reproduction and age/size schedule of mortality. The results of the foregoing analyses suggest that both these factors are critical in the evolution of the age at maturity. Read and Harvey (1989) also concluded from a comparative analysis of life history variation among eutherian mammals that mortality plays a major role in the evolution of life history traits.

Charnov (1991a) and Charnov and Berrigan (1991a) advocate a diametrically different approach to that which I have proposed above: they favor a phenomenological approach. If we assume the population to be stationary, the fitness measure R_0 can be written as

$$R_0 = l(\alpha) \left(\frac{\int_\alpha^\infty l(x)m(x)d(x)}{l(\alpha)} \right) \tag{7.35}$$

The term in parentheses is the reproductive value of a female aged α, denoted by $V(\alpha)$. Therefore, $R_0 = V(\alpha)l(\alpha)$. Thus far, we have simply algebraically manipulated the characteristic equation; Charnov and Berrigan hypothesize that $V(\alpha)$ is an allometric function of the length at maturity $L(\alpha)$,

$$V(\alpha) \propto L(\alpha)^\delta \qquad (7.36)$$

and that the two parameters of the von Bertalanffy growth equation, L_∞ and k, are also allometrically related

$$L_\infty = Ak^{-h} \qquad h < 1 \qquad (7.37)$$

Putting these two equations into the characteristic equation gives

$$R_0 \propto l(\alpha)A^\delta k^{-h\delta}(1 - e^{-k\alpha})^\delta \qquad (7.38)$$

Charnov and Berrigan propose that natural selection acts on this model through variation in α and k. At evolutionary equilibrium h is a function of the relative length at maturity $(L(\alpha)/L_\infty)$. For the Gadidae (cods) h averages 0.57 while theory predicts 0.65, and for the Clupeomorpha (herrings, sardines, and anchovies) h averages 0.40, theory predicting 0.45 (Charnov and Berrigan 1991a). These results are encouraging support for the approach but the question still remains as to why equations 7.36 and 7.37 should be correct.

   The phenomenological approach has been taken one step further by Charnov and Berrigan: they propose that life history variation may be classified by three dimensionless numbers (Charnov and Berrigan 1991b): αM, k/M, and $L(\alpha)/L_\infty$. That life history characteristics such as age at maturity, α; mortality rate, M; length at maturity, $L(\alpha)$; and maximum length, L_∞, are intercorrelated is well known (Peters 1983; Calder 1984). The twist that Charnov and Berrigan give this observation is the suggestion that the three dimensionless numbers may have values characteristic of different taxa (see also Charnov 1991a); for example, the product αM is approximately 2 in fish and pandalid shrimps, 1.5 in temperate snakes, 1.3 in lizards, 0.75 in mammals, and 0.4 in birds (Charnov and Berrigan 1990). Should such a suggestion stand further scrutiny we have a powerful criterion by which to judge life history theories, for such theories must be capable of producing these observed patterns. The same argument can be advanced for the already-established correlations between life history traits; i.e., one aim of life history analysis is to produce models that correctly predict correlations between life history traits, and possibly the values of the dimensionless numbers conjectured by Charnov and Berrigan to characterize the specific taxa.

However, the aim of the mechanistic models presented in this section is not directed at understanding the genesis of broad-scale correlations among life history traits. Systematic variation between species and across taxa is likely due to variation in life-style, and, most importantly, due to allometric constraints. Logarithmic regression shows that many life history, morphological, and physiological variables scale with a power in the region of ± 0.25 or ± 0.75. There is continuing debate over the meaning, and even existence, of such pervasive patterns (Peters 1983; Calder 1984; Schmidt-Nielsen 1984), but they may reflect underlying constraints. Given such constraints, do the observed patterns reflect adaptive evolution? One approach is to assume the operation of the constraints and proceed with optimality arguments, an approach used by Charnov (1991b) in the examination of life history variation among female mammals. From such an analysis we can derive the signs of the correlations between parameters. But we have no null model with which to compare these predictions. An essential criterion for existence is that in the long term $R_0 = 1$; i.e., each female replaces herself. This suggests the appropriate null model: whether or not life histories are optimal, the set of life history parameters must conform to the condition that $R_0 = 1$ (Roff 1981b; Sutherland et al. 1986). Suppose that only two parameters are free to vary: in this case there must be a perfect correlation (possibly after suitable transformation) between the two. As more parameters are free to vary the expected correlation between any two will decline. Analyses of global variation in life history characteristics must consider not only those patterns expected under the assumption that natural selection is maximizing fitness, but also those patterns expected solely on the basis of persistence, taking into account that some parameters may not, for morphological or physiological reasons, be free to vary independently.

7.1.3. Age and Size at Maturity in Males

In many ectotherms and endotherms competition for mates is frequently contest competition in which the victor is generally the larger. (See section 5.2.2.) To grow large an animal must forage—an activity that increases its susceptibility to predators. Therefore, when size is not a critical factor in mating success the optimal pattern of growth is to grow to the minimal size necessary to sustain the investment in gonads and success in the scramble for mates. Consequently, the optimal size and age at maturity of males relative to the female should be smaller in species with scramble mate competition than in species with contest mate competition.

Bell (1980) tested this hypothesis on a large scale, comparing freshwater North American fish with birds and mammals. As predicted, while the vast majority of male fish mature earlier than females, the reverse is true in birds and mammals. Potential problems with differences ascribable to phylogeny can be resolved by considering a single taxon. To this end I collated the relative size at maturity of North American fish species into three groups: those in which the male is territorial, those in which the male exhibits parental care, and those in which the male shows neither of the preceding behaviors. Males in the last group are predicted to be smaller at maturity than the female, and in the first two categories the male will be larger. Both predictions are supported by the data in Table 7.6. Factors other than contest for mates may also shape the relative age and size schedules of maturity; for example, if one sex carries the other for a prolonged period the "carrying sex" may have to be larger than the "carried" for purely mechanical reasons (Fairbairn 1990).

Predicting the optimal age at maturity in males is not generally a simple task and much remains to be done in this regard. One circumstance in which efforts have been successful has been in the prediction of satellite behavior. In many species two distinct male morphs can be distinguished: a large morph that defends a territory (which may comprise simply the immediate area around a receptive female) and a smaller morph that acts as a satellite and attempts to obtain copulations either by intercepting the incoming female or by sneaking between the male and female at the moment the eggs are released. Variations on this basic pattern are numerous: within a population several types of satellite males may occur (Taborsky 1987); differences in morphology between satellites and territorial males may form a continuum rather than distinct classes, and the satellite behavior may consist of simply remaining in the vicinity of the territorial males, taking over the territory after the territory is vacated (Wells 1977b).

Satellite behavior has been found in a wide range of taxa (Table 7.7) and may represent an alternate behavior which has an equal fitness with

Table 7.6. Association between the size of male teleost fishes relative to the size of the female and the type of male reproductive behavior

Size of male relative to female	Territorial behavior	Parental care	Neither
Larger	4	8	0
Equal	1	1	4
Smaller	0	1	26

From Roff (1983b).

Table 7.7. Species in which satellite males have been observed

Class	Species	Reference
Insecta	*Syrbula admirabilis, S. fuscovittatus*	Otte 1972
	Gryllus integer	Cade 1979
	Hetaerina vulnerata, C. maculata	Thornhill and Alcock 1983
	Dryomyza anilis	Otronen 1984
	Hylaeus alcyoneus	Alcock and Houston 1987
Pisces	*Polycentrus schomburgkii*	Barlow 1967
	Poeciliopsis occidentalis	Constantz 1975
	Coryphoterus nicholsi	Cole 1982
	Lepomis macrochirus	Dominey 1980
	Salvelinus malma, Oncorhynchus keta	Maekawa and Onozata 1986
	Oncorhynchus nerka	Foote and Larkin 1988
	Salmo salar	Hutchings and Myers 1988
	Xiphophorus nigrensis	Zimmerer and Kallman 1989
Amphibia	*Rana clamitans, R. catesbiana, R. virgatipes*	Emlen 1968, 1976; Howard 1978, 1984; Wells 1977a; Given 1988a
	Bufo cognatus, B. speciosus, B. compactilis, B. calamita	Axtell 1958; Brown and Pierce 1967; Arak 1988
	Hyla versicolor, H. chysoscelis, H. squirrella, H. cinera, H. regilla, H. crucifer	Pierce and Ralm 1972; Fellers 1975, 1979a,b; Perril et al. 1982; Gerhardt et al. 1987
	Acris crepitans	Wagner 1989
	Uperoleia rugosa	Robertson 1986a
Reptilia	*Sauromalus obesus*	Berry 1974
Aves	*Philomachus pugnax*	Hogan-Warburg 1966; van Rhijn 1973
	Telmatodytes palustris	Verner 1963
	Spiza americana	Zimmerman 1971
	Parus major	Krebs 1971
	Agelaius spp.	Orians 1961
	Indicator xanthonotus	Cronin and Sherman 1977
Mammalia	*Mirounga angustirostris*	Le Boeuf 1974

territorial behavior, or it may represent the "best of a bad lot" situation (Eberhard 1982; Jonsson and Hinder 1982; Dominey 1984). Suppose that for some reason (e.g., a local scarcity of food) a male cannot achieve a size at which it can be successful as a territorial male. In this case the male should adopt the satellite behavior because even though its lifetime fitness will be less than that of a large territorial male it will be greater than that of a small territorial male. For this scenario there is no assumption of equal fitnesses: the behavior is maintained in the population because it is adaptive given unavoidable variation in size at maturity. The behavioral response is an adaptive norm of reaction and implies that there exists or existed genetic variation for the behavioral repertoire.

The second situation is that in which the two behaviors are maintained because they have equal fitnesses. It is possible to construct a circumstance in which the fitnesses would be equal at all frequencies of the two morphs; but such cases must be exceedingly rare because they require a "knife-edge" balance that will be upset with the slightest variation in parameter values. A more usual circumstance will be that in which the relative fitnesses are frequency-dependent, in which case stability is assured. Age has been demonstrated to be an important component in the fitness of alternate male behaviors in several species of teleost. Species in which smaller, younger males mature at an early age and become satellites, while other males mature at a later age, are bigger, and become territorial comprise various species within the subfamily Salmoninae (Scott and Crossman 1973; Gross 1985; Maekawa and Onozato 1986; Hutchings and Myers 1987, 1988) and the swordtails, *Xiphophorus spp.* (Zimmerer and Kallman 1989). Satellite and territorial behavior has been observed in the bluegill sunfish *Lepomis macrochirus* but it is not known if this represents two entirely different patterns of maturation or an ontogenetic shift from satellite to territorial, though various lines of evidence favor the former possibility (Dominey 1980; Gross 1982). In other indeterminate growers such as amphibians the shift between tactics might represent an ontogenetic shift, though there appears to be more flexibility in these cases, with individuals shifting between behavioral modes opportunistically. (See chapter 8.)

The phenomenon of precocial maturation or "jacking" is well known within various salmon species. Although some progress toward understanding the factors favoring this phenomenon is possible using a frequency-independent approach (e.g., Caswell et al. 1984), a game-theoretic framework as outlined above will likely prove the most successful (Myers 1986). Precocial maturation is clearly related to environmental conditions (Lundqvist and Fridberg 1982), but there is not unambiguous evidence that the two forms are genetically distinct. Analysis of the relative fitnesses of jacks and "normal" males supports the hypothesis of equal fitnesses, but analysis is preliminary and flawed by lack of confidence limits (Gross 1985; see chapter 4, section 4.5.2, for further details). The proportion of precocial males varies widely in the Atlantic salmon, *Salmo salar* (Myers et al. 1986), but though Myers (1986) has produced a game-theoretic model the quality of the data is too poor for a meaningful test.

An analysis of the behavioral decision in bluegill sunfish was made by Gross and Charnov (1980) on the assumption that the two morphs are distinct. Let the $l(x)$, $m(x)$ functions for the two morphs be designated by the subscripts s ($=$ satellite) and t ($=$ territorial). Satellites mature at two years and territorial males at 8 years. Assuming a stationary population and the fitnesses of the two morphs to be equal we have

$$\int_2^\infty l_s(x)m_s(x)\, dx = \int_8^\infty l_t(x)m_t(x)\, dx \qquad (7.39)$$

If the proportion of males that becomes satellites is q the ratio of eggs fertilized by satellites to territorials is simply

$$\frac{q \int_2^\infty l_s(x)m_s(x)\, dx}{(1-q) \int_8^\infty l_t(x)m_t(x)\, dx} = \frac{q}{1-q} \qquad (7.40)$$

This proportion is the same as the average proportion of eggs fertilized by a satellite; letting this be h we have the simple expression,

$$\frac{q}{1-q} = \frac{h}{1-h}$$
$$q = h \qquad (7.41)$$

That is, the proportion of males becoming satellites should be equal to the fraction of eggs they manage to fertilize. For the fish of Lake Opinicon 21% ($=q$) of all males become satellites and fertilize 14% ($=h$) of the eggs. Binomial confidence limits (95%) for the former statistic are 11–31%. The hypothesis of equal fitnesses cannot, therefore, be rejected.

Maekawa and Hino (1987) extended the model of Gross and Charnov to include cannibalism by satellite males, a phenomenon observed in the Miyabe char, *Salvelinus malma miyabei*. The modified equation is

$$q = \frac{h - c_1 c_2}{1 - c_1} \qquad (7.42)$$

where c_1 is the proportion of eggs eaten by satellite males, and
c_2 is the proportion of eggs eaten by a satellite male that were also fathered by that male.

The average frequency of satellite males (q) is 31% ($\pm 7.4\%$ SD, range $=23.1$–37.5%). Parameters h and c_1 were estimated as 25.5% and 6.8%, respectively. No estimate could be made for c_2 and so Maekawa and Hino considered the full range from zero to one, from which they obtained

$$0.26 < \frac{h - c_1 c_2}{1 - c_1} < 0.27 \qquad \qquad \textbf{(7.43)}$$

The addition of cannibalism has very little effect on the predicted value of q: if cannibalism is ignored the estimate is 25.5% ($= h$), which hardly differs from the range given in equation 7.43. This aside, the estimated value of q does fall within the observed range (23.1–37.5%) and is not too far removed from the observed mean (31%).

Precocial maturation is most likely to be determined by the action of many genes with environmental interaction. However, in at least one species of *Xiphophorus* size and age at maturity are determined by a sex-linked gene (Kallman et al. 1973; Kallman and Borkoski 1978; Kallman 1983; Zimmerer and Kallman 1989). Males of the species *X. nigrensis* from the river Rio Coy (Mexico) comprise four size classes, and these differ in their Y-linked *P* alleles (*s*, *I*, *II*, or *L*). All X chromosomes carry the *s* allele, which when homozygous produce small, early maturing males; the *L* allele produces large, late-maturing males, and alleles *I* and *II* produce intermediates (Zimmerer and Kallman 1989). Large males display exclusively and are significantly more successful in obtaining copulations; small males switch from display to sneaking behavior according to the size of competing males. Controlling size by using genotypes carrying the *I* allele, Zimmerer and Kallman were able to show that only small males homozygous for the *s* allele display the sneaking behavior. How the three alleles are maintained in the population has not been determined.

7.2. Variable Environments

Environmental variability can be conveniently discussed under two categories: seasonal and nonseasonal variation. Seasonal environments vary temporally in a predictable manner, with a period favorable for growth and reproduction, hereafter referred to as the "growing season," or simply the "season," and an unfavorable period through which one or more stages pass in a quiescent state (dormancy, hibernation, diapause, aestivation, etc.). The length of the growing season may fluctuate from year to year. Environments subject to nonseasonal fluctuations are defined as those which vary in space and/or time in a manner that may or may not be predictable but which are not seasonal.

7.2.1. Seasonal Variation

Consider an organism emerging at the beginning of the growing season: how should it allocate its energy to maximize fitness? In a seasonal environment the life history that confers the greatest fitness is that which results in the largest number of descendants entering the quiescent phase at the end of the growing season. The simplest case is that in which the organism has only a single generation per year (univoltine or annual), and there is no interannual variation in the length of the growing season, designated as ω. Fitness is maximized by maximization of the integral

$$\int_0^\omega l(x)m(x)\ dx \tag{7.44}$$

The end of the growing season, ω, is the same as maximum life-span and hence the above equation is equivalent to maximizing R_0 when life-span has a maximum (i.e., the integral is not taken to infinity).

The question of the optimal timing of reproduction for the above scenario was first considered mathematically by Cohen (1971). His analysis deals with the problem faced by an annual plant of allocating energy to somatic tissue and seeds so as to maximize final seed yield. Assumptions of the model are:

1. Assimilation of energy is proportional to leaf weight

$$A(x + 1) = c_1 c_2 W(x) \tag{7.45}$$

    where $A(x + 1)$ is the amount assimilated at age (=time) x,
        $W(x)$ is leaf weight at age x,
        c_1 is the proportion of the vegetative body which is leaf weight,
        c_2 is the net assimilation rate per unit leaf weight.

2. A constant fraction, c_3, of assimilated energy is devoted to seed production. Leaf weight at age $x + 1$ is thus

$$W(x + 1) = W(x) + A(x) - c_3 A(x) \tag{7.46}$$

3. Seed production, $\Delta S(x)$, during the interval, x to $x + 1$, is proportional to leaf weight

$$\Delta S(x) = W(x)c_2 c_3 \tag{7.47}$$

4. At the end of the season some fraction, c_4, of leaf weight can be reutilized to make seeds.

Given a fixed growing season the optimal allocation pattern under the conditions outlined above is to switch from allocating 100% of surplus energy to growth to 100% seed production—i.e., $c_3 = 0$ or 1 (Cohen 1971; Vincent and Pulliam 1980; Ziolko and Kozlowski 1983; Kozlowski and Wiegert 1986). The optimal age at switching, α, is (Cohen 1971)

$$\alpha = \omega + \frac{1}{c_2}\left(c_4 - \frac{1}{c_1}\right) \qquad (7.48)$$

The optimal age at reproduction decreases as the proportion of soma devoted to leaves (c_1) increases, and assimilation rate (c_2) increases. Selection favors a delay in maturation with an increase in the length of the growing season (ω) or an increase in the amount of leaf matter that can be reutilized for seed production at the end of the season (c_4). The first conclusion, on the surface, would appear to suggest that an organism should devote all its soma to leaves. This is not possible because a certain fraction of the soma must obviously be devoted to support structures such as roots and stems. There are undoubtedly trade-offs in how the soma is divided: this issue could be addressed by further refinement of Cohen's model.

The analysis also suggests the absurd conclusion that as the season length becomes infinitely long the plant should delay reproduction indefinitely. Vincent and Pulliam (1980) reached the same conclusion utilizing a slightly different mathematical approach. The reason for this strange result is that neither model includes mortality. A similar result is found for a plant that switches to reproductive allocation at age α but actually only produces its seeds in a single burst, i.e., is semelparous. Kozlowski and Wiegert (1986) showed that for this case the optimal age at which to switch from vegetative to reproductive allocation, a (not α for reasons given below), is given by

$$a = \omega + \frac{c_4 - 1}{\dfrac{dP}{dx}} \qquad (7.49)$$

where P is production of surplus energy (i.e., energy available for either growth or reproduction). Equation 7.49 is identical to equation 7.48 except that the rate of production is a function of total plant weight ($dP/dx = f(W(x))$). Cohen (1971) assumed that only a fraction, c_1, of the plant contributes to production and that that production rate is constant ($dP/dx = c_2$). The end of the growing season enters as a variable because the plant is constrained to actually reproduce at the end of the growing season, i.e., $\alpha = \omega$. More realistically, we wish to optimize final seed production with the restriction that the production of seeds occurs before the end of the growing season ($\alpha < \omega$). For a constant "adult" mortality rate, M (adult age being defined as ages greater than a, the age at which surplus energy is shunted into reproductive tissues), the optimal age at reproduction is

$$\alpha = a + \frac{1}{M} - \frac{c_4 W(a)}{f(W(a))} \qquad (7.50)$$

If α, calculated from equation 7.50, is greater than ω the plant should reproduce as close to the end of the growing season as possible. Note that now the end of growing season does not enter into the equation, but "adult" mortality rate does. As mortality increases, the period devoted to reproduction decreases. Mortality enters the model as a stage-specific phenomenon, and as it increases, selection favors a reduction in the period of time over which mortality acts. Maturity occurs at age a but reproduction at age α; thus if age at maturity, a, stays constant, increases in mortality favor a decrease in α.

The iteroparous case with mortality was considered by Cohen (1976), Ziolko and Kozlowski (1983), and Kozlowski and Wiegert (1986, 1987). The general findings are that delayed maturity and large adult size are favored by

1. A high rate of somatic growth
2. A high percentage increase in reproductive rate with increases in body size
3. A high life expectancy (for annual species) or high expected number of productive days (for a perennial species) at maturity
4. Life expectancy increasing with body size

These predictions have not been tested, but they are similar to the predictions given in Table 7.3. The relative mortality of eggs or seeds in diapause versus adult mortality may also influence the timing of reproduction (Schaal and Leverich 1984). A high mortality rate of dormant seeds or eggs during the summer months selects for a delay in maturation and the production of more offspring later in the season.

For a multivoline phenology, analysis of the optimal age at reproduction must also consider the optimal number of generations per season. Many invertebrates and some plants have several generations per season, the unfavorable portion of the year being passed over in a state of dormancy. Thus in addition to the interactions between size, development time, fecundity, and survival that are important in determining the optimal age at maturity in a continuous environment, there is the additional constraint that at the end of the season offspring must be in the appropriate stage for overwintering (or oversummering in some hot, dry areas).

For a season length of fixed duration fitness will be maximized by utilizing the entire available period; furthermore, for a semelparous species all generations will be the same length (Roff 1980). If the season length is ω and the age at maturity α, the number of generations, n, will be $n = \omega/\alpha$. However, n must be an integer, and therefore, both the phenology and the optimal age at maturity are constrained. To obtain α we proceed by first computing the optimal age for $R_0^{\omega/\alpha}$, without regard to the constraint that ω/α must be an integer: α is that value which satisfies the relationship (Fig. 7.12)

$$\alpha = \frac{R_0(\alpha)\log_e R_0(\alpha)}{R_0'(\alpha)} \qquad \text{where} \qquad R_0(\alpha) = l(\alpha)m(\alpha) \qquad (7.51)$$

Since the number of generations must be integral we examine the two closest integers to the value of ω/α obtained from the above equation, say n and $n + 1$. The annual rates of increase, $R_y(n)$, $R_y(n + 1)$, generated by these two phenologies are

$$R_y(n) = R_0(\alpha)^n$$

$$R_y(n + 1) = R_0(\alpha)^{n+1} \qquad (7.52)$$

Finally, we compute the annual rate of increase for the case $R_y(*) = R_0(\alpha^*)^m$, where α^* is the optimal age at reproduction in a continuous environment and m is the integral portion of ω/α^*. The optimal age at reproduction is that value from among the three "candidates" which gives the highest annual rate of increase.

As season length increases, the optimal age and size at maturity first increases, and the fitness of the phenology comprising $n + 1$ generations per year increases more rapidly than the phenology of n generations. Ultimately, the two phenologies, n and $n + 1$ generations, are equal in fitness. Thereafter, the latter phenology will be favored. Because fitness is maximized when the total season is utilized the shift from n to $n + 1$ generations will be accompanied by a decrease in one or more of the generations. This shift will be manifested as a change in both α and body size. The repetition of the foregoing sequence as season length increases will generate a saw-tooth cline in α and body size. This pattern is illustrated for a hypothetical organism in Fig. 7.13. Because the length of the growing season decreases with altitude and latitude, body size and age at maturity (calculated in

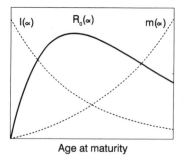

Age at maturity

Figure 7.12. Per generation fitness, $R_0(\alpha)$ for a semelparous organism in which fecundity, $2m(\alpha)$, increases with age at maturity, α. The probability of surviving to reproduce, $l(\alpha)$, decreases even if the mortality rate is constant (i.e., $e^{-M\alpha}$ decreases as α increases). Fitness is defined by the function $R_0(\alpha) = l(\alpha)m(\alpha)$.

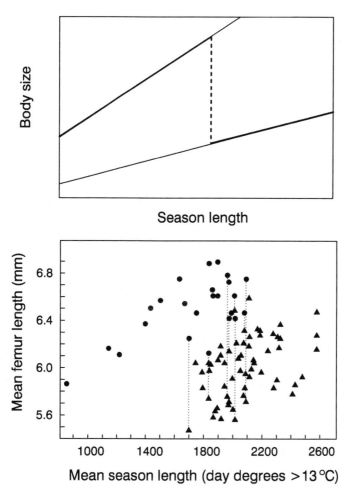

Figure 7.13. Upper panel: "Sawtooth" cline predicted as an organism shifts from a univoltine to a bivoltine phenology (based on Fig. 3 of Roff 1980). Thin lines show the optimal body size for each phenology as the season length increases. Bold line shows the phenology with the highest fitness, and the dotted line shows the point at which both phenologies have equal fitness. Lower panel: Observed pattern of variation in body size in macropterous males of the cricket, *Allonemobius fasciatus*, in relation to season length. Circles represent univoltine populations sampled in September; triangles, bivoltine populations sampled in either July or September. Symbols connected by dotted lines show populations sampled in July (circles) and September (triangles). From Mousseau and Roff (1989).

degree days not chronological time) are predicted to decrease in general with altitude and latitude, and at the transition from one phenology to another there will be a marked "sawtooth" in size and age at maturity. Such patterns have been observed in a variety of cricket species (Masaki 1967, 1973, 1978a,b; Mousseau and Roff 1989; Fig. 7.13). Although these observations demonstrate the "sawtooth" phenomenon associated with a shift in voltinism, the data are insufficient to examine the relative fitnesses of the two phenologies.

In the real world the length of the growing season at any particular location will vary from year to year. Depending on the particular form of the $l(x)m(x)$ function, selection may favor variable generation numbers and/or variable development time (Roff 1983a). If the generation time is very short relative to the length of the growing season, fitness will be maximized by the addition of an extra generation in years with longer-than-average growing seasons. Contrariwise, variation in development time will be favored in those species in which the generation length is long relative to the season length, i.e., univoltine and bivoltine species.

How can an organism such as an insect use environment cues to adjust its development time? A simple rule is "add an instar if the number of degree days exceeds Y on Julian date X." This rule is not beyond the physical capabilities of an insect since degree days can be measured indirectly by size achieved and Julian data by photoperiod: the rule then becomes "add an instar if size exceeds Z when the photoperiod is W." The red-legged grasshopper, *Melanoplus femurrubrum*, appears to follow this type of rule. A useful feature of this species is that the number of instars can be estimated from the number of antennal segments in the adult (Bellinger and Pienkowski 1987). The number of instars increases with rearing temperature and in the field varies both temporally and geographically. This variation is correlated with environmental conditions, the number of instars being correlated with the number of degree days achieved in May. Thus, in warmer years or at warmer locales the grasshopper increases the number of instars and presumably its final size. (The latter has not been verified.) It remains to be demonstrated that this norm of reaction is optimal and that there is genetic variation for the trait.

Use of photoperiod and temperature to modulate life history events is a general phenomenon in both the animal and plant kingdom. Variation among habitats may require different responses that, due to morphological or physiological restrictions, cannot be met by a single norm of reaction. This may lead to the maintenance of genetic variation for norms of reaction or in the extreme case to speciation. Jago (1973, p. 192), in a review of savanna grasshoppers, noted that "in 99 percent of the cases, there is a different subspecies or related species," on either side of the equator, and suggested that this is a consequence of changing photoperiods and seasons;

"Species keyed to decreasing day-length north of the equator experience difficulty in adjusting maturation and timing of hatching emergence to rains that are 6 months out of phase across the equator." The same problem may also be experienced by populations that straddle the transition zone between different phenologies. Because climatic variables are relatively easy to measure, and because shifts in phenologies are common, this particular problem is ideal for the analysis of the relative roles of genetic variation and phenotypic plasticity in natural populations.

For organisms that live more than a single season the timing of reproduction each year may be critical. The onset of breeding typically coincides with the onset of favorable weather and food conditions. However, female deer mice, *Peromyscus maniculatus*, do not always follow this rule. Some females attempt to breed before conditions are favorable. Only 25% of females that attempted to breed in early spring survived a 6-week period whereas 69% of nonreproductive females survived (Fairbairn 1977). Females breeding early in the season do not represent simply the tail of the breeding distribution, for the early breeding episode was separated from the main breeding period by at least 4 weeks, during which no females were lactating (Britton 1966; Fairbairn 1977). In contrast to the pattern displayed by females, males continue to be in breeding condition throughout this time, though the proportion breeding during the early phase is small.

Two advantages to breeding early are, first, that the female can breed twice in the season and, second, that the young of the first brood may also be able to breed in their first year rather than waiting until the next year. To address this question Fairbairn (1977) estimated fitness by computing the number of genes at the end of the main breeding season that are descended from either an early breeding female or a late-breeding female. This approach takes into account the fact that offspring of early breeding females that themselves breed leave offspring (grandchildren of the early breeding female) that carry only one-half of the genes of the early breeding female. The reproductive success of late-breeding females is

$$R_l = \left(S_l n_l \sum_{i=1}^{5} P(i) \right) + \frac{1}{2} B_l n_l \left(S_l n_l \sum_{i=1}^{5} P(i) \right)$$

$$= \left(S_l n_l \sum_{i=1}^{5} P(i) \right) \left(1 + \frac{1}{2} B_l n_l \right) \tag{7.53}$$

The term in the first set of parentheses is the number of offspring produced by a late-breeding female and that in the second is the number of offspring produced by the offspring of this female divided by one-half to account for the decrease in the number of genes descended from the original female. Factors taken into account are:

S_1: The probability that a female that does not breed early will survive to the main breeding period; $S_l = 0.51$.

n_1: the number of juveniles per litter born in the late peak of breeding that recruit into the population; $n_l = 2.81$.

$P(i)$: the probability of a female having at least i litters during the late peak of breeding. The maximum number of litters is five, and the probabilities for $i = 0$ to five are 0.35, 0.65, 0.24, 0.09, 0.04, 0.02.

B_1: the probability that a juvenile from the late peak of breeding will breed during the summer of its birth; $0.07 < B_l < 0.18$.

Reproductive success of an early breeding female is given by

$$R_e = S_e n_e + \frac{1}{2} B_e n_1 S_e n_e + \frac{S_e}{S_l} R_l \qquad (7.54)$$

The first term is the number of offspring per litter from the early breeding period that recruit into the population; the second is the expected number of grandchildren resulting from these offspring; and the third is the relative success of an early breeding female in the late-breeding peak. Definitions of parameters are:

S_e: 4-week survival rate of females lactating during the early peak of breeding; $S_e = 0.18$.

n_e: number of juveniles per litter born in the early breeding period that recruit into the population; $n_e = 4.33$. The higher number that can recruit into the population from the early breeding females compared to offspring born in the late peak is due to the low frequency of breeding males in the population at this time. Breeding males are very aggressive to juveniles and probably account for the higher mortalities of juveniles born in the late peak (Fairbairn 1977).

B_e: probability that a juvenile from the early peak of breeding will breed during the summer of its birth; $0.33 < B_e < 0.60$.

Substituting these values, the estimated ranges for the reproductive success of early and late-breeding females is

$$1.64 < R_l < 1.87$$
$$1.72 < R_e < 2.10 \qquad (7.55)$$

Therefore, the hypothesis that the fitnesses of the two patterns of reproduction are equal cannot be rejected. The potential advantages of more children and grandchildren combined with the good survival of offspring born in the early breeding period compensate for the high mortality of early breeding females.

Why does there exist a period between the early and late-breeding periods when no female attempts to breed? Fairbairn (1977, p. 868) suggested the following reason:

> A female that breeds early has the potential advantage of having more children and grandchildren within that summer than females that do not breed early. This is a time-dependent advantage which decreases as the later peak approaches. Further, as the late peak approaches and the proportion of adult males breeding increases, litter survival declines. Thus, there will be a period when the advantages of early breeding have diminished, and litter survival is poor. During this period, females that become pregnant will have little success. They will do better to breed later, when their offspring will have a better chance of survival. Thus, we can expect a bimodal distribution of pregnancies.

If the survival of young were to increase, this bimodality should disappear. Fairbairn cited two studies from the Gulf Islands (British Columbia, Canada) that fulfill this prediction. On one island, Samuel Island, males are significantly less aggressive than on the mainland and juveniles are recruited into the population throughout the breeding season (Sullivan 1976). As predicted, there is no bimodality in breeding. On Santurna Island the survival rates were found to be intermediate between those on the mainland and Samuel Island. There is no bimodality, but many females begin breeding before a high proportion of males have entered breeding condition.

7.2.2. Nonseasonal Variation

Environments may be variable in space and/or time. Variation in time will be more significant in modulating changes in the life history than variation in space (Roff 1978: section 3.1.1.2). This can be demonstrated as follows. Let $\lambda(\alpha)$ be the average finite rate of increase for individuals maturing at age α. Population size at time t, $N(t)$, is given by

$$N(t) = N(0)\lambda(\alpha)^t \tag{7.56}$$

Dividing time into discrete increments equation 7.56 can be written as

$$N(t) = N(0) \prod_{i=1}^{t} \lambda(\alpha, i) \tag{7.57}$$

where $\lambda(\alpha, i)$ is the finite rate of increase in time interval i. Combining equations 7.56 and 7.57,

$$\lambda(\alpha) = \sqrt[t]{\prod_{i=1}^{t} \lambda(\alpha, i)} \qquad (7.58)$$

The mean finite rate of increase is equal to the geometric mean of the finite rates of increase per time unit. Low values of λ will have an inordinately large impact on the overall average. This is intuitively reasonable since if mortality is 100% the overall average will be zero, but the rate of increase in any single time unit can be no greater than the product of the largest fecundity.

Now consider a spatially heterogeneous environment. If patches are selected at random, the finite rate of increase is

$$\lambda(\alpha) = \sum_{j=1}^{n} f(j)\lambda(\alpha, j) \qquad (7.59)$$

where $f(j)$ is the frequency of type j patches, n is the number of different types of patches, and $\lambda(\alpha, j)$ is the finite rate of increase in patch j. The average finite rate of increase is the arithmetic mean of the finite rates of increase per patch. Unlike the case of temporal variation, very low values will not have an inordinately large effect.

These results are intuitively clear: temporal changes in conditions affect the whole population and will therefore exert a significant effect, but extreme spatial variation affects only part of the population and hence will have proportionally less effect.

If patches vary only in space there would be considerable selection for an organism to remain within the patch and evolve to match the requirements of that patch. It has been hypothesized that the evolution of flightlessness and parthenogenesis are partly a response to habitats that vary in space but not time (reviewed in Roff 1991a). More generally, patches will vary in both time and space. Furthermore, patches will not persist indefinitely, forcing organisms to migrate either occasionally or every generation. What effect will this have on the optimal age at first reproduction and body size?

In insects, and invertebrates in general, migration has evolved in response to the ephemeral nature of the sites in which larvae and/or adults reside (Southwood 1962; Harrison 1980; Dingle 1985). Because sites have an uncertain existence there should be strong selection for rapid development. This could be achieved by early eclosion, resulting in a small adult, or by an increased growth rate. The cost of migration, in terms of fecundity, decreases with body size (Roff 1977, 1991b) and thus selection should act to increase growth rate even though this may have detrimental consequences such as increased mortality resulting from increased foraging.

Consequently, selection for migration should result in insects that grow rapidly and eclose at a large size relative to nonmigrants (Fairbairn 1984). This hypothesis is supported by intra- and interspecific comparisons and by selection experiments (Table 7.8).

Migration in fish differs from that of insects in that migration routes are more predictable and consist of either an annual or lifetime return journey, corresponding to the dictionary definition (*Concise Oxford Dictionary* 1964): "to come and go within a lifetime either once or periodically (e.g., seasonally)." Therefore, there is not the premium on rapid growth at a particular site that is predicted to be found among insects, in which migration is a movement between sites whose duration and/or location are unpredictable. Migratory fish are able to take advantage of highly productive areas that last for only part of the year or are suitable only for a particular period of development such as the larval stage. This is well illustrated by the reversal of the saltwater/freshwater migration routes in the temperate and tropical regions: in the temperate regions, productivity of the oceans exceeds that of freshwater environments and anadromy predominates, while in the tropics the productivity relationship is reversed and catadromy is most frequent (Gross et al. 1988).

As with insects, the loss in potential fecundity of a female migrant fish decreases with its body size (Roff 1988). Therefore, migrants should be larger than nonmigrants. Given the potential growth advantages accruing

Table 7.8. Summary of studies used to test the hypothesis that among insects, migratory forms will be larger (LS) and grow more rapidly (GR) than nonmigrant forms

Type of comparison	Subject	Hypothesis		Reference
		LS	GR	
Within populations	*Drosophila melanogaster*	Y	—	Roff 1977
	Oncopeltus fasciatus	Y	Y	Palmer and Dingle 1986; Dingle and Evans 1987
Between populations	*Limnoporus notabilis*	Y	Y	Fairbairn 1984
Between species	*Oncopeltus*	Y	—	Dingle et al. 1980
	Chironomidae	Y	Y	McLachlan 1983
	Anisoptera	N	—	Roff 1991b
	Papilionoidea	Y	Y	Roff 1991b

For detailed analyses see Roff (1991).

to a migratory species due to enhanced resources the growth rate of migrants is predicted to be greater than that of nonmigrants. However, in the temperate regions spawning generally occurs only once each year; as a consequence, the required increase in size of migrants is likely to lead to a delay in the age at maturity, even though their growth rates are higher than nonmigrants. The hypothesis that migrants will be larger than nonmigrants is supported by analysis at several taxonomic levels (Table 7.9). Analysis of fish species from the North Atlantic also supports the hypothesis that migrants will mature later than nonmigrants but grow more rapidly, as does a comparison between migratory and landlocked alewifes, *Alosa pseudoharengus*, and migratory and nonmigratory sticklebacks, *Gasterosteus aculeatus* (Table 7.9).

As described above, temporal variation in habitat quality will exert a strong influence on the timing of life history events. An organism might persist by adopting a highly conservative, "bet-hedging" life history, or respond to the changing conditions in such a manner as to maximize its fitness under the changing conditions (norm of reaction). Both responses may be adaptive but the latter will generally lead to higher fitness since

Table 7.9. Summary of studies used to test the hypotheses that among fish migratory forms will be larger (LS) and grow more rapidly (GR) than nonmigrant forms

Type of comparison	Subject	Hypothesis		Reference
		LS	GR	
Within populations	*Osmerus mordax*	Y	—	McKenzie 1964
Between populations	*Salmo salar*	Y	—	Schaffer and Elston 1975; Thorp and Mitchell 1981
	Alosa pseudoharengus	Y	Y	Scott and Crossman 1973
	Gasterosteus aculeatus	Y	Y	Hagen 1967; Snyder and Dingle 1989, 1990; Snyder 1991a
Between species	Gadiformes	Y	—	Roff 1988
	Marines teleosts	Y	Y	Roff 1988

For detailed analyses see Roff (1991b).

the response does not have to be tailored to the worse-case scenario. Lack of predictability in future conditions will favor the evolution of bet-hedging phenologies.

The norm of reaction between age and size at maturity that confers the highest fitness in a variable environment can be estimated by simply asking how the optimal values of these variables respond to variation in another variable, just as is done in Table 7.3. Stearns and Koella (1986, p. 894) asked the question, "How should an organism encountering an unavoidable stress that results in slower growth alter its age at maturity to keep fitness as high as possible despite the constraints imposed by slower growth?" To address this question they modified the quality-of-young model such that mortality and growth are related. The model components are:

1. Fecundity increases with body size, which increases with age,

$$W(x) = c_1(1 - c_2e^{-kx})$$

$$m(x) = c_3W(x) + c_4 \qquad (7.60)$$

   where $W(x)$ is size at age x, k is the parameter determining the rate at which the asymptotic size is approached, and c_1-c_4 are constants.

2. Adult mortality rate, M_a, decreases with growth rate according to the function

$$M_a = \frac{c_5}{k^{c_6}} \qquad (7.61)$$

3. Mortality rate of juveniles, M_j, is a function of the component of adult mortality, c_5, plus an amount that decreases with age at maturity, α, and growth rate, k

$$M_i = c_5 + \frac{c_7}{\alpha^{c_8}k^{c_9}} \qquad (7.62)$$

There are a large number of parameters in this model. The functional relationships between growth rate and mortality rate are not based on observation; therefore, for the results to be generally applicable the model must be representative of a broader class of models. As previously discussed with respect to the salamander model, there is a variety of possible biological interpretations for these equations, suggesting that the results will not be entirely idiosyncratic. Stearns and Koella assumed that fitness is maximized by maximizing r. As with the optimal age and size at maturity, the optimal norm of reaction depends upon the particular fitness metric chosen (David Berrigan, unpublished ms). Keeping these caveats in mind, what norms of reaction are predicted?

Stearns and Koella analyzed four cases, the predicted norms of reaction between age and size at maturity for these cases being illustrated in Fig. 7.14:

1. Mortality rate and growth rate independent ($c_6 = 0$, $c_9 = 0$): decreases in growth rate and k favor an increase in the age at maturity and a decrease in the size at maturity.

2. Juvenile mortality rate increases as growth rate increases ($c_6 = 0$, $c_9 > 0$): As with the previous case, decreases in k favor an increase in the age at maturity and a decrease in the size at maturity. The actual norm of reaction may have the same shape as case 1 or be sigmoidal.

3. Adult mortality rate increases as growth rate decreases ($c_6 > 0$, $c_9 = 0$): the optimum norm of reaction has the same shape as case 1 but is reversed, size at maturity and age at maturity being positively related.

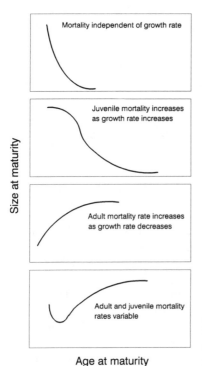

Figure 7.14. Predicted norms of reaction between age and size at maturity for a modified version of the quality-of-young model. After Stearns and Koella (1986).

4. Adult and juvenile rate variable ($c_6 > 0$, $c_9 > 0$): the relationship between age and size at maturity is "keel-shaped." The portion of the curve in which size at maturity is an increasing function of age at maturity only occurs when adult mortality rates exceed juvenile mortality rates by a factor of 100 or more (Stearns and Koella 1986). Thus, generally the trajectory is determined by mortality rates of juveniles—a conclusion reached previously (section 7.1.2).

These predictions are qualitative: testing them is difficult because it is not clear what alternate hypotheses predict the same patterns. Testing requires that the biological assumptions underlying the models be verified. This problem can be illustrated with one of the four examples Stearns and Koella present in support of the predicted norms of reaction. Poorly nourished women mature 3–4 years later and at a smaller size than well-fed women. This is consistent with juvenile mortality being a function of age at maturity and growth rate. A delay in the age of reproduction is favored in an undernourished woman for two reasons: first, she probably has an increased likelihood of dying as a result of childbearing, and second, her ability to nourish and care for her offspring is diminished. Provided these costs diminish with age it will be advantageous to delay reproduction. This qualitative result does not require a detailed model. However, an important component of the model is the assumption that if the woman delayed reproduction she would eventually reach the same size as the well-nourished woman. If this is the case, the model is valuable because, although it is intuitively obvious that she should delay reproduction, it is not clear by how much she should delay. In fact, it is very possible that undernourishment also decreases final size. (This will depend upon the extent and duration of undernourishment.)

In his study of the size and age of maturation in fish, Alm (1959) noted a norm of reaction not specified by the four cases enumerated by Stearns and Koella. This norm of reaction is dome-shaped, the optimal length at maturity first increasing with the optimal age at maturity but eventually decreasing. To examine the factors that could generate such a reaction norm Perrin and Rubin (1990) used the allocation model of Roff (1983b), described in section 6.1.1 According to this model change in length with age in immature fish is linear,

$$L(x + 1) = L(x) + c \qquad (7.63)$$

where $L(x)$ is length at age x, and c is the annual increment in length. Growth in mature fish depends upon the proportion of surplus energy devoted to reproduction, this proportion being measured as G—the ratio of gonad weight to somatic weight. Length at age for a mature fish is described by the equation

$$L(x + 1) = \frac{L(x) + c}{(1 + G)^{1/3}} \tag{7.64}$$

Perrin and Rubin assumed a stationary population and the four types of variation between annual survival and growth increment, c (Fig. 7.15):

1. Variation in survival only. The optimal size at maturity increases linearly with age.

2. Variation in growth increment only. The optimal size at maturity is a convex function.

3. Annual survival and growth increment are positively correlated. The shape of the optimal normal of reaction is the same as case 2.

4. Annual survival and growth increment are negatively correlated. This produces a concave, or dome-shaped, norm of reaction. A critical feature of the Perrin and Rubin model is the assumption of a fixed life-span. Thus fish growing very slowly must reproduce at a reduced size because of the proximity of the end of life. Empirical tests of this model remain to be undertaken.

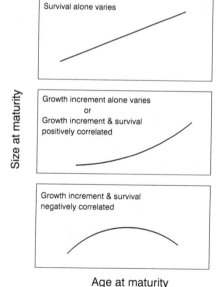

Figure 7.15. Predicted norms of reaction between age and size at maturity for the fish model in which life-span is fixed.

7.3. Predicting Genotypic Versus Phenotypic Changes

It is important to distinguish between models that deal with differences between genotypes from models designed to examine phenotypic variation within a genotype. Norms of reaction models fall into the second category, while models outlined in section 7.1 addressed the former problem, though they may be used to address the latter. The difference between the two approaches can be illustrated with two examples: age and size at maturity in *Drosophila melanogaster*, and age and size at maturity in fish.

For any particular genotype of *D. melanogaster* the growth curve is asymptotic (Fig. 7.16), but across genotypes the asymptotic size may vary, and hence we obtain an allometric relationship between development time

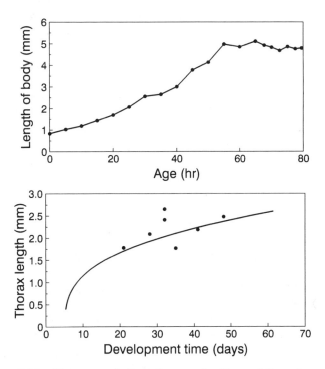

Figure 7.16. Upper panel: Growth curve for *Drosophila melanogaster* (data from Alpatov 1929). Lower panel: Solid line shows the predicted relationship between development time and adult body size in *Drosophila melanogaster*. Symbols show the observed body size and development time for some Hawaiian *Drosophila* (development times from Carson et al. 1970; thorax lengths from Kambysellis and Heed 1971).

and body size, the latter increasing monotonically with development time (Fig. 7.16). The *Drosophila* model presented in section 7.1 is based on variation across genotypes, while the model for the optimal norm of reaction in *D. melanogaster* examined by Stearns and Koella is based on variation within a genotype. Therefore, the two models are fundamentally different in perspective and design, and consequently consist of different functional components. The norms-of-reaction model asks the question, How should a genotype respond under varying environmental conditions? The *Drosophila* model asks, What genotype will be most favored under a particular set of conditions? The two answers need not be the same. First let us consider the optimal response of a particular genotype. Stearns and Koella assume an asymptotic growth function (equation 7.60) and make juvenile mortality a function of growth rate and age at maturity (equation 7.62), the latter assumption based on the premise that "parents that are older and larger when they first reproduce make offspring of higher quality with better chances of surviving to maturity" (Stearns and Koella 1986, p. 895). As stated before, while this assumption may apply to many endotherms and some ectotherms it is unlikely to be a general rule for ectotherms. Nevertheless the function may have merit although its biological interpretation does not.

In *D. melanogaster* (Bakker 1959) as with many insects (see chapter 5) successful metamorphosis requires that the larvae exceed a critical size threshold. But this critical size may not be the size at which a well-fed larva will necessarily pupate. The observation that larvae fed on reduced rations or poor-quality food induces larvae to pupate at a size lower than that achieved under optimal conditions (Sang 1949; Chiang and Hodson 1950; David et al. 1970) suggests that a larva monitors both its present size and its rate of change in size: if the rate of change is low the optimal response may be to pupate soon after the critical minimum threshold but below the "preferred" size. Mortality rate of larvae under suboptimal conditions should be higher than that under optimal conditions, both because individuals may fail to meet maintenance requirements and because it is likely that the greater the difference between the size at metamorphosis under optimal conditions and that actually attained the greater the mortality of pupae. Bakker (1959) demonstrated the latter phenomenon by removal of larvae from food after varying ages. Thus, within a given set of environmental conditions the mortality rate, averaged over the entire larval period, should decline with the age at metamorphosis, a phenomenon that can give rise to the quality-of-young model (though the two biological mechanisms are very different). Across environments the average mortality rate will increase with the severity of the environment.

A study by Sang (1949) showed that the instantaneous larval mortality rate of *D. melanogaster* larvae *increases* dramatically with larval density, from 0.01 (99% survival per day) to 0.1 (90% survival per day), leading to a very large change in the proportion of larvae that successfully meta-morphose. However, the mortality function used by Stearns and Koella predicts that the instantaneous mortality rate will *decline* from approximately 0.33 to 0.21 (data from Table 4 and text of Stearns and Koella 1986). The mortality function is therefore not a correct description of reality. Whether it is possible to obtain the observed pattern by a manipulation of the parameter values remains to be demonstrated. The statement (Stearns and Koella 1986, p. 907) that the difference in age and size at maturity under crowded conditions "can be explained as a life-history adaptation to minimize the loss of fitness imposed by slower growth" does not yet have demonstrated support. An empirically derived relationship between density, growth rate, and mortality rate for different sizes at pupation is required. This is experimentally feasible by using the technique of Bakker (1959) of simply removing larvae from their food source at different ages/sizes.

What age/size at maturity would be expected if *D. melanogaster* were maintained under crowded conditions? To answer this question we must look across genotypes. The appropriate model in this case is the *Drosophila* model with R_0 as the fitness measure. This model predicts a variety of responses depending upon how growth is altered by crowding (Table 7.3): adult size is most likely to decline but development time shows a very variable response. As crowding is most likely to increase c_6 (growth coefficient) or c_7 (growth exponent) more than c_8 (the time for the eggs to hatch and time spent as a pupa), selection under crowded conditions should lead to a decrease in the age at pupation. Scramble competition as experienced by *D. melanogaster* (Bakker 1961) will favor individuals that can pupate on reduced rations. If rations fluctuate, norms of reaction will evolve, but under continuously low food selection we might expect that genetically "small" flies able to pupate with a higher success rate than phenotypically small flies that are genotypically "large" will be at an advantage.

Shorrocks (1970) maintained 14 wild-type populations of *D. melanogaster* for a period of 25 months transferring each population to a new cage every 14 days but otherwise not culling the populations. The populations fluctuated markedly from less than 10 to greater than 450 adults. Flies drawn from peak populations produced fewer offspring than those from nonpeak populations. This difference persisted in the F_2, suggesting genetic differences between the peak and nonpeak populations. The low fecundity of the offspring from peak populations is consistent with the hypothesis that high densities selected genetically small flies.

A fly that reaches its critical weight for pupation sooner than its competitors is at an advantage. This can be achieved by being genetically small or by accumulating resources at a faster rate than other larvae. Feeding rate is genetically variable in *D. melanogaster*, and genetically fast-feeding larvae attain larger weights than either slow-feeding or control flies (Burnet et al. 1977). Fast-feeding flies have the highest survival rates under conditions of scramble competition (Bakker 1961; Burnet et al. 1977). Populations of *D. melanogaster* maintained under high or low densities for over 120 generations showed changes in competitive ability; flies from the high density populations outcompeted those from the low density (Mueller 1988b). This competitive ability is a result of increased feeding rate (Joshi and Mueller 1988). Thus these flies are not sacrificing size for the ability to persist under crowded conditions; rather they are increasing their food intake (Mueller 1988a). A pertinent question is, Why has not selection maximized feeding rate? That it has not done so implies that there are costs that favor an intermediate rate under uncrowded conditions. Compared to slow feeders, fast-feeding larvae pass food more quickly (Burnet et al. 1977) and are more active (Sewell et al. 1975). This could increase vulnerability to predators, but more significantly it may be correlated with increased metabolic rate and reduced efficiency of assimilation (Scribner and Slansky 1981). In the wild these effects may mitigate against fast-feeding larvae if food is scarce for reasons other than high density.

Clearly the models for the evolution of age and size at maturity in *D. melanogaster* have considerable room for improvement. One model (Roff 1981a) correctly predicts the size at maturity of *D. melanogaster* under uncrowded conditions and correctly predicts the changes in life history associated with changes in body size of different species. But this model does not consider the effect of feeding rate—a phenomenon that appears to be important in crowded populations—and hence cannot at present be used to predict the optimal size and age at maturity of flies under persistently crowded conditions. Given further data on the trade-offs involved the model can be modified for this purpose. The model of Stearns and Koella (1986) predicts the observed norm of reaction for *D. melanogaster* reared under suboptimal conditions but is inadequate because it predicts that under low food availability the mortality rate of larvae will decrease, which is contrary to observation. Thus the model does not account for the changes in the age and size at maturity under fluctuating conditions. The results are, however, promising and suggest what experiments are required to refine the model.

We now consider the problem of predicting phenotypic and genotypic variation in age and size at maturity in fish. Two cases are examined: first, the general pattern between age and size at maturity, and second, the phenomenon of stunting. Both among species and populations of fish there is a consistent pattern of age and size at maturity being positively correlated (Fig. 7.17). Based on the norms-of-reaction model we might hypothesize

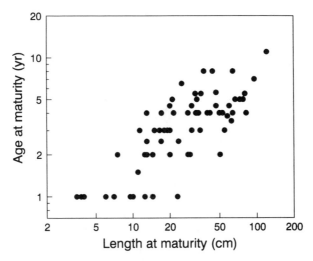

Figure 7.17. Age at maturity as a function of length at maturity in a variety of fish species. Data from Baltz (1984), Hart (1973), Hutchings and Morris (1985), Ni (1978), Oostuizen and Daan (1974), Powles (1958), Roff (1981b), and Wheeler (1969).

that this is a consequence of adult mortality rate being negatively correlated to growth rate (Fig. 7.14). In fact, the opposite is true (Roff 1984a; Fig. 7.18). The problem here is that we are examining variation across genotypes (species) and cannot assume that asymptotic length is either fixed or uncorrelated with other parameters. Indeed, asymptotic length, L_∞, and the growth parameter, k, are negatively correlated (Fig. 7.19). How this can generate a positive correlation between the age and size at maturity is illustrated by the fish model described earlier. According to this model the age, α, and length, $L(\alpha)$, at maturity are given by

$$\alpha = \frac{1}{k} \log_e\left(\frac{3k}{M} + 1\right) \tag{7.65}$$

$$L(\alpha) = L_\infty\left(\frac{1}{1 + \dfrac{M}{3k}}\right) \tag{7.66}$$

If k alone is increased (i.e., a higher growth rate), age at maturity is decreased and length at maturity increased, just as in the norms of reaction

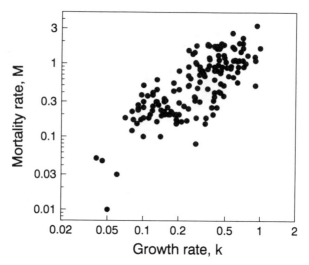

Figure 7.18. Mortality rate, M, versus growth rate, k, in a wide range of fish species ($r = 0.76$, $n = 158$, $P < 0.001$; data from Pauly 1980).

model of Stearns and Koella. For example, suppose $k = 0.1$ and $M = 0.17$; the optimal age at maturity, α, is 10.2 years and the length at maturity is $0.64L_\infty$: increasing k to 2.0 decreases α to 1.8 years and increases $L(\alpha)$ to $0.97L_\infty$. However, this pattern can be reversed if k is negatively correlated to L_∞. First, consider the effect of the positive correlation between growth rate and mortality ($M = 1.93k^{1.044}$, Fig. 7.18). Using the same values of k as in the previous example we have: $k = 0.1$, $M = 0.17$, $\alpha = 10.2$, $L(\alpha) = 0.64\,L_\infty$, and $k = 2.0$, $M = 3.98$, $\alpha = 0.46$, $L(\alpha) = 0.60L_\infty$. The optimal age at maturity is drastically reduced but the length at maturity is barely affected. Now let us introduce the negative correlation between L_∞ and k ($L_\infty = 19.55k^{-0.644}$, Fig. 7.19): the ages at maturity are unaffected but the lengths at maturity are now 86.1 cm versus 12.5 cm. Not only is there now a strong positive correlation between the optimal age and length at maturity, but the values so obtained closely correspond to those actually observed (Fig. 7.17).

Why should k and L_∞ be correlated? The reason is that these parameters are statistical descriptors of the growth curve and do not have independent biological interpretations. Asymptotic length is a function of growth rate in the absence of reproduction, the amount of surplus energy allocated to reproduction, and the age at which reproduction begins. The parameter k is largely a function of the latter two factors: the earlier a females matures or the more energy she allocates to reproduction the higher the value of k and the lower the value of L_∞.

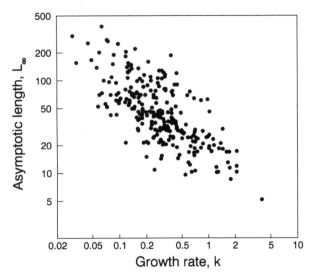

Figure 7.19. Asymptotic length, L_∞, versus growth rate, k, for fish species from 16 families. There is a highly significant regression between L_∞ and k ($r = -0.71$, $n = 260$, $P < 0.001$), and significant effects attributable to taxonomic family (Roff 1991a).

Within a local area, fish from different water bodies, most typically discrete bodies such as ponds and lakes, frequently show marked differences in growth, in some cases to such an extent that particular populations are said to be "stunted." Because of the continuous nature of changes in growth rate no unambiguous definition of stunting can be made: rather it is recognized as an extreme at one end of a continuum. However, while in some cases there is continuous gradation in growth between normal and stunted populations (e.g., populations of white sucker, *Catostomus commersoni*, Beamish 1973), the more usual situation is for most populations to lie closer to the "normal" end of the spectrum and for the greatly retarded growth of the stunted populations to be easily recognized (Diana 1987). For example, landlocked Atlantic salmon, *Salmo salar*, frequently become stunted, adult lengths being as small as 20–25 cm (Leggett and Power 1969), while the more typical anadromous form ranges in size from 70 to 100 cm (Baum and Meister 1971).

Stunting, or dwarfism, is common in landlocked forms of Salmonidae, Osmeridae, and Clupeidae (Wheeler 1969; Hart 1973; Carlander 1969; Scott and Crossman 1973; Balon 1980), and within various strictly freshwater species, particularly within the Centrachidae (Carlander 1977), and the genera *Perca* (Alm 1946; Heath 1986) and *Tilapia* (Iles 1973). In many populations reduced growth is probably a consequence of a reduced food supply, in some cases due to high population density (Table 7.10). Indirect evidence for the importance of crowding is an increase in growth observed following reduction in population size of "slow-growing" (i.e., not necessarily regarded as stunted) populations (Table 7.11). Interpretation of such data is somewhat problematical because other species were also removed (and in some cases added), and hence effects of competition or predation cannot always be excluded.

The immediate change in growth rate resulting from a reduction in population density indicates that the response is a norm of reaction rather than a change due to changes in gene frequency. Further evidence for a norm of reaction being the basis of stunting is the failure to find differences in

Table 7.10. Factors cited as contributing to stunting in freshwater fish

Family	Species	1	2	3	Reference
Clupeidae	*Alosa pseudoharengus* (alewife)		X		1
Salmonidae	*Salmo salar* (salmon)		X		2
	Salvelinus alpinus (char)		X		3
	Salvelinus willughbii (char)			X	4
	Salvelinus fontinalis (trout)	X	X		5
	Salvelinus namaycush (trout)		X		6
Cichlidae	*Tilapia mossambica*		X		7
	Tilapia nilotica			X	8
Percidae	*Perca flavescens* (perch)	X	X	X	9
	Perca fluviatilis (perch)	X	X	X	10
Centrarchidae	*Ambloplites rupestris* (bass)	X			11
	Lepomis cyanellus (sunfish)	X	X		12
	Lepomis gibbosus (sunfish)	X	X		13
	Lepomis macrochirus (sunfish)	X	X		14
	Pomoxis nigromaculatus (crappie)	X			15
Cyprinidae	*Leuciscus leuciscus* (dace)	X			16
	Rutilus rutilus (roach)	X			17

Factors are: 1 = overcrowding; 2 = low food availability; 3 = change in survival rate.
Modified from Heath (1986).

References: 1. Walton 1983; 2. Leggett and Power 1969; Barbour et al. 1979; 3. Hindar and Jonsson 1982; Nordeng 1983; 4. Frost and Kipling 1980; 5. Wydoski and Cooper 1966; Reimers 1979; 6. Donald and Alger 1986; 7. Bowen 1979; 8. Iles 1973; 9. Eschmeyer 1936; Echo 1954; Grimaldi and Leduc 1973; 10. Alm 1946; Deelder 1951; LeCren 1958; Williams 1967; Shafi and Maitland 1971; Rask 1983; Hansson 1985; 11. Beckman 1940; 12. Bailey and Lagler 1937; Mannes and Jester 1980; 13. Bailey and Lagler 1937; Beaulieu et al. 1979; 14. Bailey and Lagler 1937; 15. Hanson et al. 1983; 16. Williams 1967; 17. Burrough and Kennedy 1979; Linfield 1979.

Table 7.11. Effect of a reduction in population size on growth rate in fish

Species	% removed	Growth increased ?	Reference
Drosoma cepedianum	10	No	Sandoz 1956
Promoxis annularis	29	Yes	Rutledge and Barron 1972
Dorosoma cepedianum	31	Yes	Carter 1956
Lepomis macrochirus	20–100[a]	Yes	Hooper et al. 1964
Lepomis macrochirus	71	Yes	Cooper et al. 1971
Lepomis macrochirus	90	Yes	Jenkins 1956

[a]15 lakes; increase in growth dependent on percentage of population removed (Hooper et al. 1964).

growth rate between sunfish and perch from normal and stunted populations when grown in a common environment (Heath and Roff 1987).

By definition, a reduction in growth results when the annual increment in length is decreased. Unless the allocation to reproduction is changed dramatically this will result in a change in asymptotic length (Roff 1983a). But differences in k will only be manifested if there are significant changes in the allocation to reproduction and the age at maturity. Thus variation in L_∞ should be much greater than variation in k, a prediction that differs from the assumption in the norms of reaction analysis of Stearns and Koella. Two observations support this hypothesis. First, changes in population density alleviated stunting in the roach, *Rutilus rutilus* (Burrough and Kennedy 1979), resulting in a dramatic increase in L_∞ from 15.7 cm to 25.8 cm but no change in k (0.73 to 0.72). Second, the asymptotic length of stunted perch, *Perca flavescens*, in Lac Hertel (Quebec, Canada) is 17.5 cm (data from Grimaldi and Leduc 1973) and that of normal perch from nearby Lake Memphremagog is 33.0 cm (data from Nakashima and Leggett 1975); in contrast, the respective values of k differ very little, that of the stunted population being 0.26 and that of the normal, 0.22.

The life-span of fish in stunted populations is typically less than that of fish from normal populations (Fig. 7.20); whether this is a consequence of the stunted fish being under stress is not known. To maximize fitness a fish should decrease its age and length at maturity with an increase in mortality (equation 7.65). Decreases in the asymptotic length further favor a reduction in the length at maturity (equation 7.66). Age at maturity, α, of fish from stunted populations is generally equal to or less than that of fish from normal populations from within the same geographic area (Fig. 7.20). Changes in α are frequently rather small, which is not unexpected given that α is related to M via a logarithmic transformation.

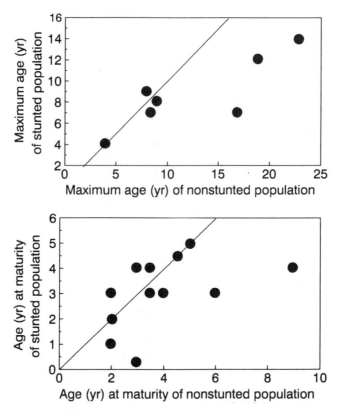

Figure 7.20. Upper panel: Maximum age of fish from stunted and nonstunted populations within the same geographical area. Lower panel: Age at maturity of fish from stunted and non-stunted populations within the same geographical area. Species and data sources: *Alosa pseudoharengus* (Walton 1983); *Salmo salar* (Alm 1959; Leggett and Power 1969; Scott and Crossman 1973); *Salvelinus alpinus* (Jonsson and Hindar 1982; Sparholt 1985); *Salvelinus fontinalis* (Rabe 1957; Wydoski and Cooper 1966; Carlander 1969); *Coregonus sardinella* (Mann and McCart 1981); *Catostomus commersoni* (Beamish 1973); *Perca fluviatilis* (Alm 1946; LeCren 1958; Williams 1967; Rask 1983); and *Lepomis gibbosus* (Scott and Crossman 1973; Beaulieu et al. 1979).

In some species (e.g., *Alosa pseudoharengus* and *Lepomis gibbosus*, age at maturity of females from the stunted population is greater than that of females from normal stocks. A possible reason for this discrepancy is that the foregoing model does not explicitly consider how the allocation to reproduction should vary. Obviously if selection favors a change in allocation that reduces k the optimal age at maturity will increase. For this to happen the allocation should decrease. Is there any evidence that stunted fish decrease their allocation to reproduction? Data on the allocation of energy to reproduction by stunted fish are scarce and equivocal. Stunted brook trout, *Salvelinus fontinalis*, within infertile streams of Pennsylvania produce only one-half the number of eggs of normal brook trout of equivalent size from the same geographic area. (See Fig. 12 of Wydoski and Cooper 1966.) But since the surplus energy sequestered by the stunted fish is less than that of the normal fish the proportion of surplus energy allocated to reproduction of stunted trout could be either greater or less than that of normal fish. Landlocked Atlantic salmon, *Salmo salar*, from Flatwater Pond (Newfoundland, Canada), produce an average of only 153 eggs; but this number is almost exactly that predicted (144) from the fecundity/length function of nonstunted, anadromous Atlantic salmon from Maine (equation from Baum and Meister 1971). Since the stunted salmon accumulate less surplus energy than the anadromous form, these data indicate that the landlocked form is probably allocating a greater fraction of its annual energy income to reproduction than the normal morph. This will increase k and favor a reduction in the age at maturity. As predicted, the age of maturity of the landlocked form is less than the average age at maturity of the anadromous form (3 year versus 4 year, respectively, Leggett and Power 1969; Scott and Crossman 1973).

Analysis of changes in the optimal age at maturity in fish stunted by a reduction in food supply predicts that, in general, age at maturity will decrease, albeit by not a great deal. This prediction is upheld but the presence of contrary observations indicates the need for more detailed analyses, both empirical and theoretical.

Stunting in *Tilapia* is not caused by a decrease in food supply since it occurs in artificial ponds where food is abundant (Iles 1973; Dadzie and Wangila 1980). In Lake Albert (Africa) stunted *T. nilotica* were found in isolated lagoons in which predation was higher than in the adjoining lake— the instantaneous mortality rate for the former being 3.37 and for the latter, 0.3 (Iles 1973). Values of k for the stunted and normal populations were 2.8 and 0.5, respectively. From equation 7.65 the predicted age at maturity for normal *T. nilotica* is 3.6 years and for the stunted form it is 0.44 year. These predictions correspond well with the observation that "stunted *Tilapia* are precocious breeders and can mature in 3 months or less (Chimmits 1955), whereas tropical lake populations require from 2 to 4 years" (Iles 1973, p. 251). While equation 7.65 correctly predicts the norm of reaction in *Tilapia*, it is unsatisfactory because it makes use of the composite parameter k, and the biological processes underlying it—namely, the effect of changes in allocation on growth and mortality—are largely "swept under the rug." The important issue of how an organism should allocate its surplus energy to growth and reproduction is the subject of the next chapter.

7.4. Summary

If the rate of increase is high, selection will act more strongly on development time, and hence age at maturity, than fecundity. But as r approaches zero (stationary population) the relative fitness of changes in fecundity and development time decline and at some point it is reversed.

The optimal age and size at maturity in females can frequently be analyzed assuming frequency-independent selection. A variety of models (McLaren 1966; Roff 1981a, 1984a; Stearns and Crandall 1981; Kusano 1982; Sterans and Koella 1986; Kachi and Hirose 1985), based on trade-offs between fecundity, development time, and survival give accurate predictions of age and/or size at maturity in insects, amphibians, reptiles, fish, and plants. Despite apparent differences in the mathematical formulations of these models their qualitative predictions are the same. (See, for example, Table 7.3.) The most serious deficiency of these models is that the trade-off functions are not well enough understood.

The general finding of the above models is that mortality prior to maturation typically is far more important in determining the optimal age at first reproduction than mortality after maturity, a result that is intuitively reasonable since it obviously matters more to get a chance to breed once than to breed again. A critical factor is whether mortality rate is stage- or age-specific. The former type occurs when there are specific developmental stages such as larval and adult, as in holometabolus insects and amphibia. High mortality rates in a later stage may increase the time spent in earlier stages, thereby producing larger, more fecund females entering the high-mortality stage. But this option is not available when mortality rate is age-specific, and the optimal response is to mature early. The latter prediction has been verified with studies on guppies and *Daphnia*. Though the mortality regime is critical, so also is the age/size-specific reproduction rate. The important trade-off in the models considered is that between increased fecundity resulting from a delay in maturation versus the increased mortality incurred.

An alternative approach to mechanistic-based models is to aggregate life history traits into a few compound parameters and attempt to predict or measure the pattern of variation (Charnov and Berrigan 1990, 1991a,b). Preliminary investigations are encouraging inasmuch as broad patterns appear evident, but the problem of comparing patterns predicted on the basis of optimality criteria with suitable null models has not been resolved.

In males, selection is likely to be most often frequency-dependent; analysis, therefore, requires a game-theoretic approach. Satellite behavior is a common phenomenon among male animals. Estimation of the relative fitnesses of the satellite and territorial behaviors in salmon and sunfish supports the hypothesis that the two behaviors are maintained by frequency-dependent selection, the fitnessess being equal at equilibrium.

In a seasonal environment delayed maturity and large size are favored in a univoltine organism by a high growth rate, a large rate of increase in fecundity with increases in body size, a high life expectancy, or a mortality rate that declines with body size. These predictions are similar to those obtained for a nonseasonal, deterministic environment (Table 7.3). Organisms with multivoltine phenologies are constrained by the requirement that they be at the appropriate stage for passing through the inclement seasons. As a consequence, body size and development times are predicted to vary in a sawtooth fashion with season length. Such patterns have been observed. Organisms that live for several years face the problem of when to initiate breeding each year. The advantages of breeding early and hence having offspring breed in the same year favor early breeding, but poor weather reducing survival favors a later start. The interaction of these two factors, in conjunction with an increasing frequency of aggressive males, produces a bimodality in the age at first breeding in *Peromyscus maniculatus* (Fairbairn 1977).

In a variable but nonseasonal environment temporal fluctuations act more strongly on age and size at maturity than spatial variation. Migration is one adaptation to variable environments. From bioenergetic and biomechanical arguments migrant insects and fish are predicted to be larger and grow more rapidly than nonmigrants. This hypothesis is supported by analyses at several taxonomic levels. A second mode of adaptation to a variable environment is to change the age schedule of growth, survival, maturity, and fecundity. Certain changes may be imposed upon the organism, such as, for example, when growth rate slows as a result of decreased ration. Given such changes, how should an organism change those life history components over which it can practice at least some control? Specifically, how should its age and size at maturity vary? Stearns and Koella (1986) and Perrin and Ruben (1990) have examined this question using a variety of models. While a range of optimal norms of reaction can be produced, lack of adequate biological data upon which to base the underlying model components impede adequate testing.

Great care should be taken determining the scope of a particular model: does it apply just to norms of reaction within a particular population, does it apply primarily to the analysis of variation across species, or is appropriate for both questions? Three case studies illustrate how the analysis of phenotypic versus genotypic variation can be undertaken.

8

Reproductive Effort

Maturation is an "all or none" phenomenon in the sense that it can be defined as the production of the first offspring or first attempted mating (not necessarily successful). But how many offspring are produced or how much effort is put into securing matings or defending young is quantitative. Thus an organism is faced with two "decisions": should it breed, and how much effort should it expend? The former was the subject of the last chapter, the latter is the focus of the present. Six questions are addressed:

1. Under what circumstances will fitness be maximized by the devotion of so much effort in the first episode of reproduction that death ensues (semelparity)?

2. If iteroparity is favored, under what circumstances will an organism partition surplus energy into both growth and reproduction?

3. How should reproductive effort vary with age?

4. For some organisms reproductive effort includes investment in offspring after birth (parental care). Under what circumstances will parental care be favored, and how much should be provided?

5. Males frequently compete for the opportunity to mate with females. Phenotypic differences between males may favor the evolution of alternative behaviors, particularly territorial versus satellite behavior. What factors favor the adoption of a particular behavior?

6. The amount of energy devoted to current reproduction clearly depends upon the prospects for future reproduction. Environmental variation modulates expected future output and hence plays a role in the evolution of the age schedule of reproduction. How important is such variation?

8.1. Reproductive Investment and Breeding More Than Once

In a seminal paper for life history theory Cole (1954, p. 118) made the following proposition. *"For an annual species, the absolute gain in intrinsic population growth that can be achieved by changing to the perennial repro- ductive habit would be exactly equivalent to adding one more individual to the average litter size"* (his italics). Thus, an annual species with a clutch size of 101 would increase in numbers as fast as a perennial that produces 100 young every year forever. There is obviously something amiss with this result, for perennials are common. While there is good evidence that survival and reproduction are negatively correlated (chapter 6) it seems highly unlikely that perennial species are committing so much energy to reproduction that they cannot produce one more offspring. The result derives from Cole's failure to consider the consequences of juvenile and adult survival rates. The importance of mortality was recognized by Gadgil and Bossert (1970), who, however, mistakenly attributed Cole's result to the absence of mortality per se. This error was noted by Bryant (1971), who derived Cole's result with mortality present. A correct solution to the paradox was given by Charnov and Schaffer (1973).

Consider two species, one annual and one perennial, both breeding at age 1 and producing m_a and m_p female offspring per year, respectively. The proportion surviving the first year (juvenile survival) is s and the annual adult survival of the perennial is S. The annual population grows according to

$$N(t + 1) = m_a s N(t) \qquad (8.1)$$

and the perennial grows according to

$$N(t + 1) = m_p s N(t) + S N(t) \qquad (8.2)$$

where the first term represents the number of recruits to the population and the second the surviving adult population. To find the annual rates of increase, we divide throughout by $N(t)$, ($\lambda = N(t + 1)/N(t)$). Letting the two rates for annual and perennial be λ_a and λ_p, respectively, we have

$$\lambda_a = m_a s \qquad (8.3)$$

$$\lambda_p = m_p s + S \qquad (8.4)$$

The two patterns of reproduction confer equal fitnesses when $\lambda_a = \lambda_p$

$$m_a s = m_p s + S$$

$$m_a = m_p + \frac{S}{s} \qquad (8.5)$$

Cole considered the case in which $s = S = 1$ and Bryant that in which $s = S < 1$. In both cases the result is

$$m_a = m_p + 1 \qquad (8.6)$$

As discussed in chapter 5, survival rates of juveniles are generally lower than those of adults and hence in general S/s will be greater than one. Thus Cole's result applies only to a very special circumstance.

Charlesworth (1980) extended the model of Charnov and Schaffer to the more general case of semelparous and iteroparous organisms with an arbitrary age at first reproduction. This formulation is instructive in the present context: the two life histories have equal fitness when

$$\frac{m_i}{m_s} = 1 - Se^{-r}$$

$$= 1 - S/\lambda \qquad (8.7)$$

where i and s refer to iteroparity and semelparity, respectively. The ratio of iteroparous to semelparous reproduction decreases with decreasing λ and adult survival (Fig. 8.1). High adult survival and a low rate of increase will tend to shift the selective advantage to iteroparity since the reproductive allocation of a semelparous organism must be large to equal that of iteroparous. For example, when adult survival is 0.1 and the per-generation rate of increase is 2, a semelparous organism must produce only 1.05, (1/ 0.95) times the number of progeny of an iteroparous species with the same population parameters; but when adult survival is 0.9 and the population is stationary ($\lambda = 1$) the semelparous life history requires a productivity 10 times that of the iteroparous to have equal fitness.

Though the majority of plants are either annual or perennial, some are biennial, setting seed only after two season of growth. Hart (1977) addressed the question of why this strategy is relatively uncommon. He approached the problem in the same manner as Charnov and Schaffer (1973), deriving the three rates of increase

$$\lambda_a = s m_a \qquad (8.8)$$

$$\lambda_p = S + s m_p \qquad (8.9)$$

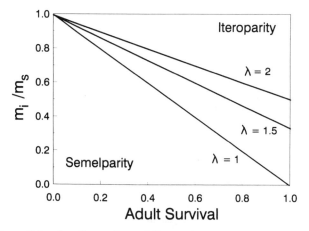

Figure 8.1. Isoclines of equal fitness for semelparous and iteroparous life histories, given particular values of λ, fecundities, and annual survival rate. Combination lying above the lines favor iteroparity; below the lines selection favors semelparity.

$$\lambda_b = \sqrt{sSm_b} \tag{8.10}$$

where the subscripts refer to annual, perennial, and biennial, respectively. Consider first the requirement for the fitness of the biennial to equal the perennial: equating equations 8.9 and 8.10 and rearranging, we arrive at

$$m_b = \frac{S}{s} + 2m_p + \frac{s}{S}m_p^2 \tag{8.11}$$

A biennial life history will be favored over the perennial when the value on the left exceeds that on the right. The relative productivity of the biennial (m_b/m_p) is a convex function of the ratio of first-year survival to "adult survival" $(s/S,$ Fig. 8.2), with a minimum at four (obtained easily by differentiation): for the biennial mode of reproduction to have greater fitness than the perennial it must produce more than four times as many seeds. A similar analysis of the biennial versus annual life histories indicates that the biennial must produce greater than twice as many seeds as the annual to be favored. Is this possible? Seed production is typically larger in biennials than either annuals or perennials (Table 8.1), possibly because biennials have longer than annuals to store energy and grow larger but do not have to devote as much energy as perennials to support structures. On average, biennials produce four to five times as many seeds as annuals or perennials, which roughly matches the minimum requirement for them to

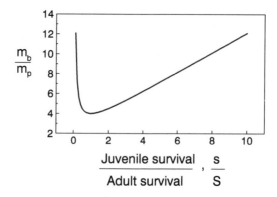

$\dfrac{m_b}{m_p}$

Juvenile survival , $\dfrac{s}{S}$
Adult survival

Figure 8.2. Plot of the required fecundity of a biennial relative to a perennial (m_b/m_p) at which the fitnesses of the two life histories will be equal. The required fecundity varies with the ratio of juvenile to adult survival (s/S).

Table 8.1. The distribution of annuals, biennials, and perennials among different habitat types, and the seed production of biennials relative to annuals and perennials

	Percentage of each type			Ratio of biennial over	
Community type	Annuals	Biennials	Perennials	Annual	Perennial
Closed	0	2.3	97.7	—	3.8
Woodland gaps	5.6	44.4	50.0	24.0	1.9
Semiopen	26.1	26.1	47.8	0.2	0.3
Mud and shingle shore	52.2	17.4	30.4	2.5	2.1
Open	92.3	5.1	2.6	1.3	13.5
Mean ratios based on means of seeds =				4.5	4.9

Modified from Hart (1977).

have equal fitness. Without estimates of survival rates the precise conditions cannot be stipulated. The number of seeds required to make the biennial life history equal in fitness to the annual and perennial habits is least at intermediate values of s/S. Hart (1977, p. 795) suggested that this condition will be most frequently met "in situations where survival is likely for at least a few years, e.g., midsuccessional stages or forest openings." This is supported by the data presented in Table 8.1, inasmuch as annuals predominate in permanently open habitats and perennials in closed habitats. More detailed studies of life history parameters are required to confirm this hypothesis.

Bell (1976) and Young (1981) examined the general question of when breeding more than once will be favored. The resulting formulation "leads to no very neat generalizations" (Bell 1976, p. 64). Despite this, the simpler models do indicate that the selective advantage of switching from semelparity to iteroparity depends upon the costs of increased reproductive effort. Semelparity may arise because the amount that must be allocated to reproduction so drains the organism's resources that its death is certain. In some cases the organism may have little choice in the matter. For example, semelparity and iteroparity among various anadromous fish species may be, in part, a consequence of difficulties of migration. This may account for Pacific salmon, *Oncorhynchus* spp., being semelparous, while Atlantic salmon, *Salmo salar*, are iteroparous, though Schaffer (1979b) disputes this, suggesting that differences in physiology or phylogenetic constraints are responsible. Difficulties of migration may determine not only whether a fish is semelparous or iteroparous but also the degree of iteroparity. For example, the proportion of repeat spawners in sea-run migrant trout, *Salmo trutta*, increases with passable river length and water discharge, which l'Abée-Lund et al. (1989) attributed in part to more effective passive expulsion of spent adults in larger rivers. Further, Leggett and Carscadden (1978) hypothesized that the relative allocation of energy to migration may account for a latitudinal gradient in the proportion of repeat spawners in American shad, *Alosa sapidissima*. The important point to draw from these examples is that semelparity or a significant reduction in potential life-span may be an unavoidable consequence of reproduction. If an individual female has little or no chance of surviving after the breeding episode, she has no reason to conserve resources. On the other hand, death might be avoidable if the organism devoted less energy to reproduction. As discussed in chapter 5, reproduction is likely to increase the likelihood of dying either because the act of reproduction places the organism directly at risk or because it increases its susceptibility to death in the period subsequent to reproduction. These arguments suggest that, in general, iteroparous species will allocate fewer resources to reproduction than semelparous species.

Because no two organisms have exactly the same life histories this prediction cannot readily be applied to a comparison of just two species, one semelparous and one iteroparous. However, the hypothesis can be tested by using a group of semelparous species and a group of iteroparous species, provided there is enough taxonomic diversity within the two groups to randomize differences between species within groups. (If one group comprised only a single genus and the second group another, different, genus, differences between the groups might be due to effects ascribable to being in the same genus rather than being semelparous or iteroparous.)

The best test of this prediction is provided by data on 40 species of annual and perennial grasses, comprising a large number of different genera (Wilson and Thompson 1989). If we consider our unit of time to be a single year, annuals can be termed semelparous and perennials iteroparous. A further subdivision is possible within annuals, for some reproduce only once and are, therefore, semelparous within any time scale, while others flower repeatedly throughout the summer and, hence, are iteroparous with respect to annuals that flower only once, but semelparous with respect to perennials. In the grasses examined by Wilson and Thompson (1989) flowering occurred both synchronously and over a period of weeks, and in the latter case inflorescences were harvested as they matured. However, only the mean total weight of mature inflorescences is presented, and, therefore, I shall restrict the analysis to a consideration of annual versus perennial. There is no overlap in the distributions of reproductive effort (= % of plant weight comprising mature inflorescence) of annuals and perennials: annual grasses range from 41% to 66% and perennials from 0.2% to 31% (Fig. 8.3). The two groups are significantly different, with annuals having, as predicted, a higher reproductive effort than perennials. Annual grasses are larger than perennials, and thus the difference might be a consequence of size differences rather than differences in phenology. However, though reproductive effort declines with plant height, the regression for annuals lies above that of perennials, indicating that there are differences associated both with plant size and with reproductive mode. (Fig. 8.3: Strictly speaking, a regression of reproductive effort on plant height is not legitimate since the regression is basically Y/X versus X, where Y is inflorescence weight and X is plant weight. The correct procedure is to do a covariance analysis of the allometric relationship, $\log(Y) = a + b\log(X)$, with reproductive mode introduced as a categorical variable. In the present case this cannot be done since the independent variable is height, not weight. The difference between the two strategies is so clear that though a correct statistical level cannot be assigned to the difference, the acceptance of a significant difference is warranted).

Pitelka (1977) reported a similar finding for three species of lupine, the annual species, *Lupinus nanus*, having an expenditure of 61%, while the two perennial species, *L. variicolor* and *L. arboreus*, expended only 18% and 20%, respectively. However, for the reasons given above this cannot be viewed as an adequate test of the hypothesis.

Calow (1978) obtained support for the hypothesis using semelparous and iteroparous gastropods (Table 8.2). Calow's measure of reproductive effort is a type of gonadosomatic index, which he termed an "indirect index of effort," IIE:

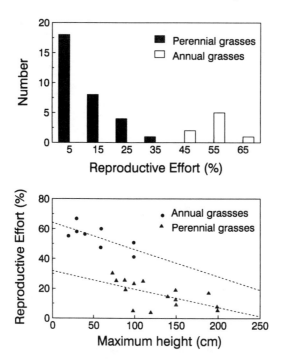

Figure 8.3. Top panel: Frequency distributions of reproductive effort in perennial and annual grasses. (See text for definition of RE.) Bottom panel: Regression of reproductive effort on maximum potential height. Redrawn from Wilson and Thompson (1989).

$$IIE = \frac{\text{No. of eggs per breeding season} \times \text{Egg volume}}{\text{Volume of parent}} \qquad (8.12)$$

Though the sample sizes are small, the differences are statistically significant (Mann-Whitney U-test, $U = 15$, $P < 0.025$, one-tailed test), the same result being obtained if we minimize the difference between modes of reproduction by taking the lower values of the range for the semelparous species, *Lymnaea peregra* and *Ancylus fluviatilis*, and the upper values of the three semelparous species.

Calow and Woollhead (1977) estimated reproductive effort over a 70-day period for two iteroparous and one semelparous species of freshwater triclads (flatworms). They computed reproductive effort in a variety of ways, based on number and energy content of young relative to the parent (Table 8.3), and provide sufficient data to construct a gonadosomatic index

Table 8.2. A comparison of reproductive effort in semelparous and iteroparous gastropods

Species	Reproductive effort
Semelparous species	
Lymnaea peregra	3.05 (1.23–4.88)
Physa fontinalis	2.17
Physa gyrina	2.01
Planorbis contortus	1.18
Ancylus fluviatilis	1.89 (1.42–2.36)
Mean = 2.06 SE = 0.30	
Iteroparous species	
Lymnaea stagnalis	0.43 (0.34–0.51)
Lymnaea palustris	0.19 (0.03–0.35)
Bithynia tentaculata	0.21 (0.08–0.33)
Mean = 0.28 SE = 0.08	

See text for measure of reproductive effort. Where range is shown in parentheses, the figure displayed is the median.

Adapted from Calow (1978).

Table 8.3. Estimates of reproductive effort over 70 days for three species of triclad, *Dendrocoelum lacteum*, *Polycelis tenuis*, and *Dugesia lugubris*

Measure of reproductive effort	*D. lacteum* annual	*P. tenuis* perennial	*D. lugubris* perennial
No. young per parent	103.7	53.2	61.1
No. young per unit wt.	11.2	9.5	5.6
Energy (J) in young per parent	249.0	13.3	15.3
Energy (J) in young per unit wt.	26.8	2.4	1.3
Wt. of young per unit wt.	0.95	0.10	0.17

From Calow and Woollhead (1977).

for the three species. All measures of reproductive effort agree that the semelparous species, *Dendrocoelum lacteum*, has a higher reproductive effort than the two iteroparous species. But there is no consistent pattern between the two iteroparous species, perhaps due to the two species having very similar efforts. Because of the small number of species used, the results, though consistent with the hypothesis, must be viewed with caution.

The sum of evidence presented above does suggest that iteroparous organisms expend less energy per period of reproduction than semelparous species.

8.2. Optimal Reproductive Effort and Growth

Under what circumstances will it profit an organism to continue growing after maturity? This question has been addressed by Schaffer (1974a), Taylor et al. (1974), Pianka and Parker (1975), and León (1976). A graphical analysis has been discussed in chapter 4 (section 4.2.3). The following, based on León's analysis, is an alternative approach. (The analysis by Taylor et al. [1974] contains a mathematical error that invalidates the numerical analysis; see León [1976, p. 317] for a discussion of this.) León (1976) considered the question with respect to the following simple model:

1. Natural selection maximizes r.
2. Mortality rate, M, is constant: $l(x) = e^{-Mx}$, where x is age.
3. Fecundity is a function, f, of the difference between the surplus energy and the energy allocated to growth: defining this difference as reproductive effort, E, than fecundity $= f(E)$.
4. Surplus energy is a function, g, of the weight, W, of the organism, $g(W)$.
5. The rate of change in weight is proportional to the amount of energy allocated to growth. Without loss of generality the constant of proportionality can be scaled to one, and hence $dW/dt = G$, where G is the allocation to growth. This is the control equation; i.e., changes in G govern changes in weight and fecundity and hence the value of r.

Converting these assumptions into mathematical form we obtain the characteristic equation

$$\int_0^\infty e^{-(M+r)x} f(E)\, dx = 1 \qquad \text{where } E = g(W) - G \qquad \textbf{(8.13)}$$

We must now describe the relationship between fecundity and reproductive effort, $(E = g(W) - G)$. Four possibilities that encompass the likely range of relationships are (Fig. 8.4):

1. Fecundity increases linearly with reproductive effort.
2. Fecundity is a convex function of reproductive effort.
3. Fecundity is a concave function of reproductive effort.
4. Fecundity is a sigmoidal function of reproductive effort.

The next step is to find the age schedule of reproductive effort that maximizes r. This can be done using either dynamic programming or Pon-

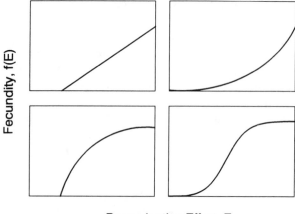

Reproductive Effort, E

Figure 8.4. Four plausible relationships between reproductive effort and fecundity. The signs of the first derivative, f', and second derivative, f'', of these function are as follows:
Top left panel: $f' > 0, f'' = 0$
Top right panel: $f' > 0, f'' > 0$
Bottom left panel: $f' > 0, f'' < 0$
Bottom right panel: $f' > 0$, initially $f'' > 0$, but after inflection $f'' < 0$

tryagin's Maximum Principle. In the present case the latter method is the easiest route to take. Maximizing r is equivalent to maximizing the Hamiltonian of equation 8.13,

$$H = e^{-(M+r)x}f(E) + p(x)G \qquad (8.14)$$

where $p(x)$ is a function of x, to be determined. For a given weight W, the maximal value of H with respect to G is found by differentiating H with respect to G,

$$\frac{\partial H}{\partial G} = e^{-(M+r)x}\frac{\partial f(E)}{\partial G} + p(x) \qquad (8.15)$$

Equation 8.15 is maximized with respect to H when $\partial H/\partial G = 0$,

$$p(x) = -e^{-(M+r)x}\frac{\partial f(E)}{\partial G} \qquad (8.16)$$

Recalling that $f(E) = f(g(W) - G)$, we expand the partial derivative on the right hand side using the chain rule (see the appendix),

$$\frac{\partial f(E)}{\partial G} = \frac{\partial f(g(W) - G)}{\partial G}$$

$$= \frac{\partial f(g(W) - G)}{\partial(g(W) - G)} \frac{\partial(g(W) - G)}{\partial G}$$

$$= \frac{\partial f(g(W) - G)}{\partial(g(W) - G)} (-1)$$

$$= -\frac{\partial f(E)}{\partial E} \qquad (8.17)$$

The derivative on the right-hand side is known since we have defined it as one of four possibilities (Fig. 8.4). Substituting the right-hand side in equation 8.16, and then $p(x)$ into equation 8.14, gives

$$H = e^{-(M+r)x}f(E) + \frac{\partial f(E)}{\partial E} e^{-(M+r)x}G^*$$

$$= e^{-(M+r)x}\left[f(E) + \frac{\partial f(E)}{\partial E} G^* \right] \qquad (8.18)$$

where G^* is the value of G at which H is maximized. However, what we really want is the value of E that maximizes H. To obtain this we must go through another round of differentiation: differentiating H with respect to E gives

$$\frac{\partial H}{\partial E} = \frac{\partial f(E)}{\partial E} + \frac{\partial^2 f(E)}{\partial E^2} G^* \qquad (8.19)$$

and hence $\partial H/\partial RE = 0$ when

$$\frac{\partial f(E)}{\partial E} = -\frac{\partial^2 f(E)}{\partial E^2} G^* \qquad (8.20)$$

The sign of both terms in equation 8.20 are known by inspection of Fig. 8.4, and hence we can determine if the equality holds. If it does not, there can be no turning point and hence no maximum: the optimum pattern of allocation is to switch from allocating 100% of surplus energy to growth to allocating all surplus energy to reproduction. Equation 8.20 cannot hold when $f(E)$ is either a linear or convex function of reproductive effort, E (Fig. 8.4). If fecundity is a concave or sigmoidal function of E, the above equality can hold and hence the organism should, after maturation, simultaneously allocate energy to both growth and reproduction. Using the same protocol as outlined above it can be shown that both growth and reproduction will be sustained if mortality is a concave or sigmoidal function of reproductive effort.

To my knowledge there is no experimental analysis of the functional relationship between reproductive effort and survival or fecundity in a natural situation. There are a number of studies that have attempted to manipulate reproductive effort in animals, particularly birds (see chapter 6), but the majority of these have only been capable of addressing the question of whether mortality increased, or future fecundity decreased with reproductive effort; the data are simply too crude to permit the fitting of any meaningful functional relationship. Koufopanou and Bell (1984) demonstrated that in an aquarium the relationship between survival from predation by guppies and egg number in the cladoceran, *Daphnia pulex*, is more or less linear. However, what might happen under more natural conditions awaits investigation.

An alternate method of study is to assume a particular functional relationship and see if it produces a reasonable fit to observed data. This approach was adopted by Kitahara et al. (1987) in their analysis of age-specific growth, reproduction, and survival schedules of two fish species. Instead of using r as the measure of fitness, they chose R_0, the expected lifetime reproductive success,

$$R_0 = \sum_{x=1}^{\omega} l(x)m(x) \tag{8.21}$$

where ω is the maximal age, $l(x)$ is survival to age x, and $m(x)$ is the number of female births at age x. Fecundity was assumed proportional to surplus energy,

$$m(x) = c_1 u(x) g(W, x) \tag{8.22}$$

where c_1 is a constant,
      $u(x)$ is the proportion of surplus energy allocated to reproduction at age x,
      $g(W, x)$ is the surplus energy of a fish of weight W and age x.
Like León (1976), Kitahara et al. (1987) assumed growth to be a simple linear function of surplus energy and allocation to reproduction, such that weight at age $x + 1$, $W(x + 1)$, is given by

$$W(x + 1) = W(x) + [1 - u(x)] g(W, x) \qquad \textbf{(8.23)}$$

i.e., weight at age $x + 1$ is equal to the weight at age x plus the amount of surplus energy not allocated to reproduction. Following Ware (1980) they assumed surplus energy to be allometrically related to weight,

$$g(W, x) = c_2 W^{c_3} \qquad \textbf{(8.24)}$$

where c_2 and c_3 are constants. Equation 8.24 holds for Atlantic cod, *Gadus morhua*, and Atlantic herring, *Clupea harengus*, the two species analyzed. The mortality schedule is rather complicated: the year x to $x + 1$ is divided into two parts, the first comprising the period from the spawning season to the beginning of gonad development and the second the remainder. Survival in the first period is assumed to be a consequence of postspawning stress in year x and in the second part to be due to prespawning stress in year $x + 1$. This model leads to some difficulties in defining $l(x)$ because the effect in year $x + 1$ is now split between an effect due to year x and an effect due to year $x + 1$. For this reason they defined the survival function in terms of a new function, $L(x)$. In their application of the model Kitahara et al. make some simplifying assumptions that boil down to assuming survival in year x, in terms of this new function $L(x)$, to be a function of reproductive effort in the previous "year." Given the arbitrary nature of the survival function it seems rather pointless to adopt such a tedious method of relating survival to reproductive effort: I see no compelling reason not to simply assume that mortality in year x is due to reproductive effort in year x (or year $x - 1$ if one believes that mortality occurs after reproduction).

Unlike León (1976), who worked with the instantaneous rate of mortality, Kitahara et al. used the annual survival rate. In both cases a concave relationship between instantaneous mortality rate or survival and reproductive effort is required to produce allocation to both growth and reproduction. To ensure a graded pattern of allocation to growth and reproduction, Kitahara et al. adopted the simplest arbitrary concave function, a parabola, relating survival and reproductive effort,

$$L(x) = [c_4 - c_5 u(x)^2] L(x - 1) \qquad \textbf{(8.25)}$$

where c_4 and c_5 are constants. Accordingly, the expected lifetime reproductive success is

$$R_0 = \sum_{x=1}^{\omega - 1} [c_4 - c_5 u(x)^2] L(x) c_1 c_2 W(x)^{c_3} \qquad \textbf{(8.26)}$$

The Hamiltonian for the above equation is not a pretty sight and application of the model proved rather difficult. The two constants c_1 and c_2 were estimated independently but c_3 and c_4 had to be obtained by fitting the model to the same data required to verify the model. As stressed earlier (see chapter 4, section 4.5.1), with several "free" parameters a model is not fully constrained and hence any conclusions concerning its adequacy must be viewed with caution. With this caveat in mind we may note that the fit between prediction and observation for growth and reproductive effort is very good for the two species analyzed (Fig. 8.5), but predicted survival rates are less than those estimated by traditional methods (which themselves may be wrong).

The model of Kitahara et al. (1987) is very similar to that of Myers and Doyle (1983; see chapter 5, section 5.2.1), who also used the same fish stocks in their analysis. They obtained good fits using two different models, which must make us pause before applauding the relatively good fits ob-

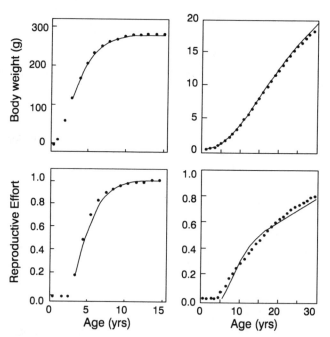

Figure 8.5. Predicted (lines) and observed (circles) age-specific changes in body weight and reproductive effort (proportion of surplus energy devoted to reproduction) for herring (left panels) and cod (right panels) stocks from the Southern Gulf of St. Lawrence. Redrawn from Kitahara et al. (1987).

tained by Kitahara et al. The problem is that we analyze the product $l(x)m(x)$ and not the component functions: unless we can specify the values of all parameters without recourse to the data we wish to predict, or are sure of the shape of the functional relationships, we may be left with a wide range of possible survival and fecundity functions that generate more or less the same $l(x)m(x)$ function. In spite of this criticism, the attempts of Kitahara et al. (1987) and Myers and Doyle (1983) are important because they demonstrate that the theoretical analyses of León and others are consistent with the real world. What we require is an experimental analysis of the relationship between reproductive effort and survival that can be applied to a field situation: at present no such analyses exist.

The above analyses do not explicitly distinguish how growth and reproduction are distributed throughout the year. Kozlowski and Uchmanski (1987) examined the situation of a perennial species which grows for part of the year and then switches entirely to reproduction for the remainder of the year, this pattern being repeated each year. Using a dynamic programming approach and assuming that

1. Selection maximizes the expected lifetime reproductive success
2. Mortality rate is constant
3. Fecundity is size dependent
4. Growth rate is size dependent

they developed a formula for determining the optimal time of switching in any given year. This prediction is not directly tested: rather the model is fitted to growth curves based on annual increments in size. It is not clear how specific the model actually is. For example, suppose we assume simply that some fraction of surplus energy is devoted to reproduction in a particular year, would we obtain an equally good fit to data? The model seems unduly complicated given the available data.

As with the previous studies there are two "free" parameters that must be estimated from the data. Good fits were obtained for the gastropod, *Conus pennaceus*, and the Iceland scallop, *Chlamys islandica*. For the Iceland scallop, Vahl (1981) reports the age-specific allocation of energy to growth and reproduction: assuming the fraction of energy devoted to reproduction is equal to the fraction of the year devoted to reproduction, Kozlowski and Uchmanski calculated the age-specific changes in the proportion of the growing season used for reproduction. The model and data agree inasmuch as reproductive effort increases with age, but the model consistently underestimates reproductive effort. Again, these results may not be inconsistent with a simpler model that ignored the switching and considered only the fraction of surplus energy devoted to reproduction.

For an annual plant growing in an environment in which the season is of fixed length the optimal allocation pattern of surplus energy is "bang bang": the plant switches entirely from growth to reproduction at some fixed time (Cohen 1971). But if season length varies in a stochastic manner there may be a period in which the optimal pattern is for both growth and reproduction to occur simultaneously (Cohen 1971; King and Roughgarden 1982). The reason for this can be explained verbally as follows: suppose for some season lengths, S_1 and S_2, the optimal switching times are t_1 and t_2, respectively, with $S_1 > S_2$ and $t_1 > S_2$—i.e., the optimal switching time in the first environment (t_1, S_1) is later than the length of the second season (S_2). Obviously if both seasons occur a plant must switch no later than t_2 or it will leave no progeny when seasons of length S_1 occur. However, when seasons of length S_1 do occur, switching at so early a date as t_2 may have very low fitness. A compromise response that can confer a higher fitness than switching entirely to reproduction is to begin producing some propagules at t_2 but continuing growing at least until t_1 if time permits. Mathematically, this result can be proven as follows: first, since the plant is annual, the appropriate measure of fitness in a deterministic environment is the number of seeds that a single plant produces by the end of the year (all seeds are assumed to be alike), which since the generation is 1 year is also equivalent to the expected lifetime reproductive success, R_0. Letting the fraction of surplus energy diverted to seeds at age x be $u(x)$, we find the optimal allocation by differentiating R_0 with respect to $u(x)$,

$$\frac{\partial R_0}{\partial u(x)} = Z \tag{8.27}$$

where Z is a complicated function that does not contain $u(x)$. (See Cohen 1971.) Since $\partial R_0/\partial u(x)$ is independent of $u(x)$, there is no maximum and the optimal strategy is $u(x) = 0$ or $u(x) = 1$, i.e., a "bang bang" pattern. Vincent and Pulliam (1980) arrived at the same result using optimal-control theory. If season length is variable the appropriate measure of fitness is the mean of the logarithm of R_0 (see chapter 3),

$$\text{Fitness} = w = \frac{\sum_{i=1}^{n} \log_e R_0(i)}{n} \tag{8.28}$$

Differentiating,

$$\frac{\partial w}{\partial u(x)} = \frac{1}{n} \sum_{i=1}^{n} \frac{Z(i)}{R_0(i)} \tag{8.29}$$

Now $Z(i)$ is independent of $u(x)$, but $R_0(x)$ is a function of $u(x)$. (As $u(x)$ varies then so must the number of seeds left at the end of the year.) To evaluate $\partial w/\partial u(x)$ we put the right-hand side over a common denominator, which will bring $R_0(i)$ into the numerator. A maximum occurs when $\partial w/\partial u(x) = 0$, which is when the numerator equals zero, since by virtue of its relationship with $R_0(i)$, $u(x)$ is now in the numerator, it is possible for the numerator to equal zero. Hence, a graded allocation pattern may be favored.

Another reason for a gradual transition from growth to reproduction is that the reproductive system may initially be too small to absorb all the available energy. For example, many species breed annually and release all their eggs at the one time: there may simply not be the physical space to store all the eggs that could be produced given the available energy, and hence the excess should be channeled into growth, thereby increasing the potential fecundity in the next year. Kozlowski and Ziolko (1988) demonstrated that such a design constraint could result in a graded response but did not provide any evidence that this actually occurs. Bertness (1981a) reports a situation in which mechanical constraints can impede growth and result in the females shunting energy into reproduction. The organism in question is a hermit crab, *Clibanarius albidigitus*. Females which because of inadequate shell size could not grow reproduced significantly more than those provided with a large range in shell sizes. Furthermore, the addition of shells into natural populations decreased the incidence of reproduction and increased the rate of growth (Bertness 1981a).

8.3. Reproductive Effort and Age

The analyses of Cohen (1971) and Kozlowski and Uchmanski (1987) indicate that for those particular models reproductive effort should increase with age—i.e., the organism should devote an increasing fraction of its surplus energy to reproduction as it ages. Based on special cases, Williams (1966) and Gadgil and Bossert (1970) also concluded that reproductive effort should increase with age. That this is not necessarily a universally correct conclusion can be demonstrated with two examples. Suppose mortality varies with reproductive effort as a step function (Fig. 8.6) and growth is determinate; the optimal reproductive effort will lie just to the left of the step. Now if the step moves to the left as the organism ages the optimal reproductive effort will decline with age (Fig. 8.6).

The second example is taken from Fagen (1972). The assumptions of this model are:

1. Natural selection maximizes r.
2. The organism lives for three time intervals.

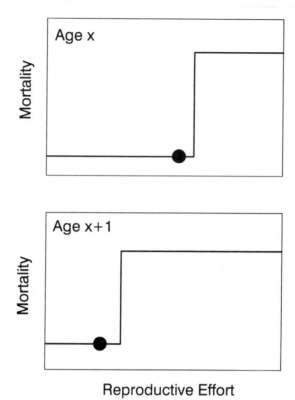

Figure 8.6. Hypothetical relationship between mortality and reproductive effort at two ages, x and $x + 1$. Above some threshold value of reproductive effort the organism does not survive to reproduce. Therefore, the optimal reproductive effort (circle) lies to the left of the "step." The threshold for total reproductive failure declines with age and hence so does reproductive effort.

3. Survival is a linear function of reproductive effort, $1 - E(x)$, ($x = 1, 2,$ or 3).

4. Fecundity is proportional to weight times a quadratic function of reproductive effort at age x:

$$\text{Fecundity} = W(x) (2E - E^2) \qquad (8.30)$$

5. Weight at age $x + 1$ is equal to weight at age x plus an increment, $c(x)$, times the proportion of energy not allocated to reproduction, $1 - E(x)$:

$$W(x + 1) = W(x) + c(x) (1 - E(x)) \qquad (8.31)$$

The maximum incremental increase for age 2 is 0, meaning that regardless of reproductive effort no change in size can occur between ages 1 and 2.

The above model produces an age schedule of reproduction in which the optimal reproductive efforts are 0.300, 0.175, and 1 for ages 1, 2, and 3, respectively: reproductive effort is least at age 2. Growth rate decreases and then increases; size therefore plateaus for a period. Fagen notes that male African elephants (Laws 1969), certain seals, and toothed whales (Laws 1970) exhibit this type of growth. This, however, is not a particularly good argument for the model unless it can be demonstrated that the decrease in growth rate is not a consequence of reproductive effort.

These two counterexamples demonstrate that reproductive effort will not inevitably increase with age. But it must be admitted that these models are quite contrived, and in most instances we might well expect to find reproductive effort increasing with age.

Charlesworth and León (1976) tackled the problem analytically using a model proposed by Schaffer (1974a). Their model is constructed as follows:

1. The probability of survival from age x to $x + 1$ is a decreasing function of reproductive effort where reproductive effort is expressed as a proportion of surplus energy: let this function be denoted by $S(E,x)$, where E is reproductive effort ($0 < E < 1$). Both survival and reproductive effort may vary with age x.

2. Fecundity (female births) at age x is equal to size at age x, $W(x)$, times an age-specific coefficient, $b(E,x)$, that increases with reproductive effort:

$$m(x) = b(E,x) \, W(x) \tag{8.32}$$

3. Size at age x is a product of reproductive effort at the previous age and size:

$$
\begin{aligned}
W(x) &= (\text{function of } E \text{ at age } x - 1) \, W(x - 1) \\
&= g(E, x - 1) \, W(x - 1) \\
&= g(E, x - 1) \, g(E, x - 2) \, W(x - 2) \\
&= g(E, x - 1) \, g(E, x - 2) \, g(E, x - 3) \ldots W(0) \\
&= \prod_{i=0}^{x-1} g(E, i) \, W(0) \tag{8.33}
\end{aligned}
$$

The rate of increase in growth decreases with reproductive effort—

i.e., $g(E, x)$ is a decreasing function of E. Because both survival, $S(E, x)$, and growth rate, $g(E, x)$, are decreasing functions of reproductive effort, their product (which occurs in the characteristic equation) is also a decreasing function. Therefore, we can simply define a single function that is equal to their product,

$$P(E, x) = S(E, x) \, g(E, x) \qquad (8.34)$$

The characteristic equation can thus be written as

$$\sum_{x=0}^{\omega} e^{-r(x+1)} L(E, x) \, b(E, x) = 1 \qquad (8.35)$$

where

$$L(E, x) = W(0) \prod_{i=0}^{x-1} P(E, i) \qquad (8.36)$$

In order that both growth and reproduction can be favored after maturity (i.e., the optimal reproductive effort is greater than 0 but less than 1), P must be a concave function of b (Fig. 8.7).

Based on this model Charlesworth and León (1976) showed that, given an optimal reproductive effort which is intermediate and both ages are at a stable evolutionary equilibrium, then a *sufficient* condition for reproductive effort to increase from age $x - 1$ to x is

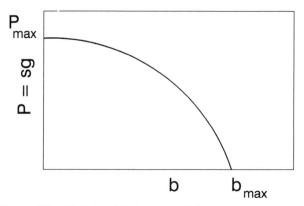

Figure 8.7. Relationship between P ($=$ survival $\times$ growth rate) and the fecundity coefficient (b) for which natural selection favors both growth and reproduction.

$$\log_e P_{\max} \geq r^* \tag{8.37}$$

where $P_{\max}$ is the value of $P(E, x)$ at zero reproductive effort (Fig. 8.7), and the rate of increase is at its global maximum r^*. Recall that $P(E, x)$ is the product of the survival and increment in growth: thus an increase in reproductive effort will be favored if r is not too large and growth and/or survival are high at low levels of reproductive effort. As r approaches zero (a stable population size) the above inequality becomes

$$P_{\max} \geq 1 \tag{8.38}$$

and the likelihood of reproductive effort increasing with age increases. In a population controlled by density-dependent phenomena, reproductive effort is likely to decrease with age if a population is subjected to a high-density-dependent mortality or reduction of individual growth rate (Charlesworth and León, 1976, p. 455).

Since the above inequality is a *sufficient* condition only, optimal reproductive effort might still increase with age even if the inequality is not satisfied. The *necessary* condition for reproductive effort to increase with age is

$$\log_e \left(\frac{1}{2} W(0) \, TANb_{\max} \right) \geq r^* \tag{8.39}$$

where TAN is the absolute value of the tangent at which the fecundity coefficient is greatest, $b_{\max}$ (i.e., when $P = 0$, Fig. 8.7). While it may be relatively easy to estimate the relationship between growth rate and the fecundity coefficient (b), that between survival and b poses formidable technical difficulties that have yet to be satisfactorily solved for any species. Therefore, calculating the necessary condition for reproductive effort to increase with age is not simple. It may be possible, however, to reverse the procedure and ask what minimal survival rate is required to satisfy the inequality, given that reproductive effort is known to increase with age.

Although we cannot estimate the values required for the necessary condition, it is possible to at least roughly estimate $P_{\max}$ for a number of fish stocks. For most stocks it is not unreasonable to assume that population growth rate in an unexploited population is close to zero, and hence as a first approximation the sufficient condition for reproductive effort to increase with age can be estimated from equation 8.38.

Ware (1980) estimated surplus energy and allocation to reproduction in the southern Gulf of St. Lawrence cod stock. Annual survival of fish from this stock is estimated at 86% (Myers and Doyle 1983). Using these two pieces of information we can construct the relationships between P_{max} and age and between reproductive effort and age. P_{max} declines with age but is always greater than 1, and hence reproductive effort should increase with age, which indeed it does (Fig. 8.8).

For most fish such detailed information as available for the above cod stock is not at hand, but it is possible to predict at least the response for the first years after maturation by considering the growth rate of immature fish. In the year prior to maturation, energy is being diverted into gonad production, and hence the appropriate year in which to estimate the growth rate is the year 2 years prior to maturation. I estimated the product of survival times growth rate for 23 species (31 stocks) using the data given in Table 1 of Roff (1984a): in all but three cases the product exceeds unity (Fig. 8.9). This suggests that, in general, reproductive effort should increase with age in fish. For one species in this data set, *Pleuronectes platessa*, the allocation of energy into eggs and soma has been directly measured and, as predicted, it increases substantially with age (Rijnsdorp et al. 1983). Analysis in other species, both fish and other taxa, must rely upon indirect measures of reproductive effort.

Reproductive effort is typically measured in terms of annual productivity or, more frequently, clutch weight relative to total body weight (the gonadosomatic index, GSI—see chapter 5). How do we interpret age-related changes in the gonadosomatic index? For fish the energy available for somatic growth is known to be an allometric function of body weight (Parker and Larkin 1959; Ware 1975a, 1978),

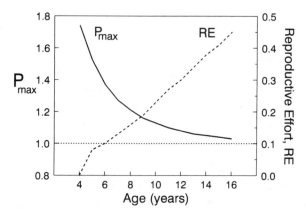

Figure 8.8. Estimated values of P_{max} and reproductive effort (*RE* = proportion of surplus energy devoted to reproduction) for the Southern Gulf of St. Lawrence cod. Data from Ware (1980).

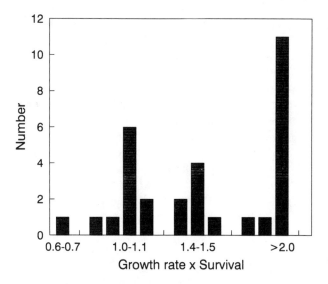

Figure 8.9. Frequency distribution of estimated values of P (= survival × growth rate) for 31 species and stocks of fish. From Roff (1991a).

$$\text{Surplus energy} = c_1 W^{c_2} \qquad \textbf{(8.40)}$$

where c_1 and c_2 are constants characteristic of a species and population. Many other animals (e.g., crustacea, molluscs, and lizards) show the same form of growth function as fish, and thus equation 8.40 may be a generally applicable description of the relationship between surplus energy and size in indeterminate growers. From equation 8.40 we can infer that if an organism allocates a constant fraction of its surplus energy to reproduction, clutch mass should vary allometrically with body weight, with an exponent of c_2. In nine species of fish the value of c_2 ranges from 0.46 to 1.1, with a mean of 0.81, and only two greater than 1 (Myers and Doyle 1983). Across species, regressions of log(clutch mass) on log(body mass) give similar exponents (spiders 0.84; hover flies 0.95; mammals 0.80–0.84; birds 0.74; reptiles 0.88; frogs 0.90; salamanders 0.64; aquatic poikilotherms 0.92; data from appendix VIIa of Peters 1983; and Table 4 of Reiss 1985).

If an organism allocates a constant fraction of its surplus energy to reproduction, clutch mass should increase with weight, but with an exponent less than 1: the gonadosomatic index should therefore decrease with size. Since size and age are closely connected we predict, therefore, that an increase in reproductive effort with age will be indicated by either an increase in the gonadosomatic index with age *or an index that remains constant with age*. Given the uncertainty in the value of the exponent and statistical problems of determining slopes, good evidence of an increasing reproductive effort is provided only by those cases in which the index increases significantly. In some instances the relationship between gonadosomatic index and age has been estimated by examining the allometric relationship between clutch mass and body mass, an exponent greater than 1 (where body weight is the independent variable) indicating an increased reproductive effort with size, and hence age.

Whether one uses data on energy budgets or relative clutch mass, among a wide range of organisms reproductive effort generally increases with age: increases have been observed in 30 species; no change has been detected in 2 species; and in only 1 case, the harp seal, is there sufficient evidence to suggest that reproductive effort may decline (Table 8.4). Harp seals feed their pups for about 13 days, the pups growing at the rate of 2 kg per day (Kovacs and Lavigne 1985). During this period the females do not feed, and all lactation costs are met from stored reserves (Kovacs et al. 1990). Young female harp seals use 36% of their stored energy for milk production, middle-aged females use 29%, and old females 25% (Stewart 1986). The actual amount transferred remains the same, the difference being due to younger females beginning with smaller reserves.

A disturbing feature of the reproductive-effort estimates for the two molluscs, *Viviparus georgianus* and *Corbicula manilensis*, is that while the reproductive efforts estimated from the energy budgets increase with age, the reproductive efforts calculated from the ratio of the amount of carbon channeled into reproduction versus the amount contained within the average adult female shows a decline with age. (See Table 2 in Browne and Russell-Hunter 1978.) Similarly, reproductive effort for various gastropod species shows an increase with age when estimated from energy budgets but no change or a decrease when calculated on the basis of reproductive production per time unit over the somatic energy content at the beginning of the time unit over which reproduction is measured. (See Figs. 2 and 4 in Hughes and Roberts 1980.) The second of the two measures in each of these two studies in roughly equivalent to the gonadosomatic index, and in the first case the two measures are *positively* correlated when comparisons are made across species. (See chapter 5, section 5.2.3.) These results suggest some caution in the interpretation of indirect measures of reproductive effort that do not vary substantially.

The probability that reproductive effort will increase with age increases as r approaches zero (equations 8.37, 8.38). The tropical palm, *Astrocaryum mexicanum*, is very long-lived (> 100 years) and shows a consistent increase in reproductive effort with age (Piñero et al. 1982). Though rates of increase cannot be estimated they are undoubtedly very low. Most of the molluscs listed in Table 8.4 are also relatively long-lived, the Iceland scallop reaching 26 years (Vahl 1981), winkles up to 16 years (Hughes and

Table 8.4. Reproductive effort, RE, and age in a variety of organisms

Common name	Scientific name	RE	RE and age	Reference
Palm	*Astrocaryum mexicanum*	B	+	1
Iceland scallop	*Chlamys islandica*	B	+	2
Mollusc	3 species	B	+[a]	3
Mollusc	7 species	B[b]	+	4
Mussel	*Anodonta piscinalis*	B[c]	+	5
Insect	*Gargaphia solani*	O	+	6
Fish	9 species	GSI[d]	+	7
Slimy sculpin	*Cottus cognatus*	GSI[e]	+	8
Fish	4 species	GSI	?[f]	7
Black goby	*Gobius niger*	O	+	9
Worm snake	*Carphophis vermis*	GSI	+	10
Lizard	*Sceloporus jarrovi*	GSI	+[g]	11
Lizard	3 species	B	?[h]	12
Lizard	2 species	GSI	0	13
Common lizard	*Lacerta vivipara*	GSI	?[f]	14
Anole lizard	*Anolis limifrons*	B	+	15
California gull	*Larus californicus*	O	+	16
Glaucous-winged gull	*Larus glaucescens*	O	0	17
Harp seal	*Phoca groenlandica*	B	−	18
Red deer	*Cervus elaphus*	O	+	19

+, RE increases with age; 0, no change with age; −, RE decreases with age.

References: 1. Piñero et al. 1982; 2. Vahl 1981; 3. Browne and Russell-Hunter 1978; 4. Hughes and Roberts 1980; 5. Haukioja and Hakala 1978; 6. Tallamy 1982; 7. Roff 1983b; 8. Mousseau et al. 1987; 9. Magnhagen 1990; 10. Clark 1970; 11. Tinkle and Hadley 1973; Pianka and Parker 1975; 12. Tinkle and Hadley 1975; 13. Pianka and Parker 1975; 14. Avery 1975; 15. Andrews 1979; 16. Pugesek 1981; Pugesek and Diem 1990; 17. Reid 1988; 18; Stewart 1986; 19. Clutton-Brock 1984.

B = reproductive effort calculations based on an energy budget
GSI = reproductive effort based on gonadosomatic index
O = reproductive effort calculated otherwise (see below for details)

Species not explicitly cited in table (listed according to reference number):

3. *Viviparus georgianus, Corbicula manilensis, Patinopecten yessoensis*;

4. *Littorina rudis, L. nigrolineata, L. littorea, Nerita versicolor, N. tessellata, Fissurella barbadensis, Thais lapillus, Lacuna pallidua, L. vincta.*

7. *Hippoglossoides platessoides, Pleuronectes platessa, Clupea h. pallasi, Lesueurigobius freiesii, Leuciscus leuciscus, Squalius cephalus, Hysterocarpus traski, Gobio gobio, Cobitis taenia.*

7. *Platichethys flesus, Noturus miurus, Perca fluviatilis, Rutilus rutilus, Esox lucius.*

Table 8.4.   (Continued)

12. *Sceloporus jarrovi, S. graciosus, Uta stansburiana.*

13. *Crotaphytus wizlizeni, Masticophis taeniatus.*

Notes on "Other" methods of estimating reproductive efforts (listed according to reference number):

6. Vigor with which female defended successive broods.

9. The propensity of males to build nests in the presence of predators increases with age.

16. RE based on feeding rate and territorial defense.

17. RE based on relative changes in mass of parents during breeding.

19. RE based on calf condition relative to hind condition.

Footnotes:

[a]Reproductive effort based on energy budget increased with age but reproductive effort based on a "GSI" measure decreased. (See text for discussion.)

[b]Reproductive effort calculated in two ways. That based on production giving an increasing RE with age, while that based on weight gives no change. (See text for further discussion.)

[c]Reproductive effort for mussels of a given length was calculated as

$$RE = (W_g - W_r + W_{nr})/W_{nr}$$

W_g: mean weight of all glochidia within a female (in this group the eggs are brooded in the gill chamber where they develop through to the veliger stage, called glochidia),
W_r: mean weight of reproductive females (including glochidia),
W_{nr}: mean weight of nonreproductive females.

[d]Analysis based on the relationship between gonad weight and body weight.

[e]Both males and females show increasing RE with age.

[f]The exponent in the allometric regression of gonad weight on body weight does not differ significantly from one.

[g]The original analysis of Tinkle and Hadley (1973) showed no increase, but a later analysis by Pianka and Parker (1975) demonstrated a significant correlation.

[h]For *S. jarrovi* the REs based on total annual production are 10, 11 and 13% which Tinkle and Hadley (1973) take as indicative of increasing effort. Given the large errors that must be associated with the budget calculations a difference of only 1% is probably not significant. Furthermore, if only surplus energy is used to calculate the RE, the values are 57.4%, 38.2%, and 62.6%, which clearly does not indicate increasing effort. But see footnote g.

Roberts 1980), and the mussel, *Anodonta piscinalis*, 3–6 years (Haukioja and Hakala 1978). Though the early mortality of these molluscs may be very high—for example, the larval survival of Iceland scallop is estimated to be only 0.00005% (Vahl 1981)—the long life-span of the adults suggests a generally low rate of increase.

Among the vertebrate ectotherms listed in Table 8.4 life-spans range widely, from the long-lived American plaice (20–30 years) to lizards with life-spans of about 4 years. Life tables have been constructed for *Sceloporus jarrovi*, one of the lizard species included in Table 8.4, and two related species, *Sceloporus undulatus* and *Sceloporus poinsetti*: estimates of R_0

range from 0.61 to 1.23 (Tinkle and Ballinger 1972; Ballinger 1973), indicating a low rate of increase. Rates of increase of the endotherms are also probably quite low.

The observed increase of reproductive effort with age is consistent with theory. However, more accurate estimates of age schedules of reproductive effort and experimental manipulations of effort are required to adequately test the model.

8.4. Reproductive Effort and Parental Care

The subject of the evolution of parental care is very large and amply reviewed in a recent book by Clutton-Brock (1991). Parental care is shown by a wide range of organisms, but is is particularly common among vertebrates (Table 8.5). Which sex participates in parental care varies among taxa, with fish showing the greatest variety (Gross and Sargent 1985). It clearly will profit a parent to care for its offspring if this increases the survival rate of its offspring without incurring excessive costs such as increased mortality of the parents, or missed breeding opportunities. Clutton-Brock (1991, p. 101) suggested that parental care in ectotherms might "be

Table 8.5. A summary of parental care and its correlates

Group	Sex most commonly caring	Conditions under which parental care most frequently found
Invertebrates	Female	Eggs or young are clumped in time and space, and commonly associated with physically harsh or biotically dangerous habitats
Fish	Male	Egg guarding most common. Most frequent in fresh or inland waters. Hypothesized that this association a consequence of increased predation risk on eggs and limited spawning sites. Egg size large and fecundity low
Anurans	Male	Egg guarding most common. Associated with terrestrial reproduction. Egg size large and fecundity low
Reptiles	Females	Defense of nest site
Birds	Both	Care found in most species
Mammals	Female	Some form of care found in all species

Information extracted from Clutton-Brock 1991.

expected where environmental conditions are harsh, predation is heavy, or competition for resources is intense—or where the costs to the parent of providing care are reduced." A general survey of the occurrence of parental care among ectotherms supports this hypothesis (Table 8.5) but adequate statistical analysis is lacking. Parental care is found in practically all endotherms, and is undoubtedly necessary given the relatively undeveloped or vulnerable condition of neonatal birds and mammals. Despite the relatively large literature on parental care there are no quantitative analyses that address the problem of the amount of care either sex should provide. Whether any care should be provided and when a parent should desert its offspring have received some attention. Mathematically these two questions are essentially the same since failure to provide any care is simply desertion at time 0.

Trivers (1972) considered the subject of desertion and sexual selection and concluded that (p. 146) "At any point in time the individual whose cumulative investment is exceeded by his partner's is theoretically tempted to desert, especially if the disparity is large." The proposition that selection is a function of past investment has been criticized by Dawkins and Carlisle (1976) and Boucher (1977), who pointed out that it is future investment and returns that are important. The hypothesis that selection acts on past rather than future investment has been dubbed the Concorde fallacy after the rationalization given by certain politicians for continued investment in the development of the supersonic aircraft, the Concorde (Dawkins 1976). Triver's method of analysis stands in most cases because the two measures are likely to be highly correlated (Sargent and Gross 1985). However, the possibility of reaching the wrong conclusion is sufficient grounds for not using this approach. The appropriate formulation of the problem is in terms of game theory (Dawkins and Carlisle 1976). Maynard Smith (1977) developed some simple models based on the game-theoretic method but empirical investigations are largely lacking. Artificial reduction of primary broods caused male Brewer's blackbirds, *Euphagu cyanocephalus*, to increase their effort in their secondary broods but did not result in increased allocation in mating attempts (Patterson et al. 1980). Problems of correctly defining the appropriate model and estimation of parameter values make interpretation of these results difficult.

A rather easier subject to analyze is abandonment when only a single parent is involved. Consider, for example a single parent defending its offspring. The vigor with which it defends its young should be a function of how close to independence the young are, and future prospects for reproduction. Male three-spine stickleback, *Gasterosteus aculeatus*, guard eggs and young, keeping the eggs aerated and defending them against predators such as other sticklebacks, other species of fish, and invertebrate predators. The fitness value of a brood increases with the number and age of eggs, the latter not because of the amount invested but because of the reduction in the amount still required to be invested. Therefore, males should defend broods more vigorously as age or number of eggs increases. Pressley (1981) tested this prediction using a dummy of a potential predator of both eggs and parent, the prickly sculpin, *Cottus asper*. In accord with theory, males that remained within the nest area and attacked the sculpin had more and older eggs than those which deserted.

Desertion may occur not only in response to a risk to the parent but also because it leads to an overall increase in reproductive output. The probability of desertion of their clutch by the blue-winged teal, *Anas discors*, is a direct function of reduction in clutch size (Armstrong and Robertson 1988). Such desertion may be adaptive because females can increase their fitness by producing another, larger, clutch. Female grizzly bears give birth to up to four young and have been observed to abandon single young. Tait (1980) put forward the thesis that such abandonment may have a selective advantage. A female could nurse a lone cub for the 2 years required to reach independence or abandon the cub, mate again the same year, and produce a new brood of perhaps three or four young. In this framework it is obvious that there will exist conditions under which desertion will be the optimal behavior. Tait was able show that for realistic parameter values abandonment could be optimal. The model may also be applicable to black bears and polar bears (Bunnell and Tait 1981). Mock and Parker (1986) advanced the same hypothesis for high desertion rates of singleton broods by egrets and great blue herons. The theory is attractive and warrants further empirical study.

In contrast to the above study, Townsend's analysis of brood caring in the neotropical frog, *Eleutherodactylus coqui*, showed that desertion is never profitable and that the male should remain with the eggs until hatching (Townsend 1986). Male *E. coqui* provide parental care in elevated, terrestrial nests, protecting their eggs from predators and desiccation. Eggs of this species undergo direct development, the young completing metamorphosis and hatching as tiny mobile frogs. The male parent remains with the clutch until hatching and for several days thereafter (Townsend et al. 1984). To analyze the problem Townsend used the marginal-value theorem (see chapter 4, section 4.2.4), the benefit curve being the probability of hatching following abandonment at any given age and the cost of care-giving being the time required to remate. For desertion to be advantageous the benefit curve must be concave at some point. (See Fig. 4.7.) However, the probability of hatching is a convex function of time of desertion (Townsend 1986), and hence there is no intermediate point at which desertion is favored: fitness is maximized by remaining with the young at least until hatching. (Consequences of desertion beyond this age were not empirically investigated.)

From a parent's perspective the best situation is to get someone else to raise your young. This both reduces the mortality risks attendant with parental care and potentially increases brood size. Thus animals able to practice brood parasitism should be able to direct their reproductive effort into the production rather than the rearing of young. This hypothesis can be tested with data on cuckoos and bees. Like their nesting counterparts, parasitic cuckoos produce eggs in clutches. Payne (1974) found the average clutch size for eight parasitic species to be 3.48, compared to 2.82 for 39 nesting species. This difference is statistically significant ($P = 0.0495$, Mann-Whitney U-test, one-tailed); furthermore, the estimated clutch size for parasitic cuckoos was probably underestimated and the difference in clutch size greater than the 0.66 observed (Payne 1974). But there is a highly significant correlation between egg size and body size, with parasitic cuckoos laying smaller eggs for a given body weight than nesting cuckoos (Fig. 8.10). Computations of reproductive effort should take this variation

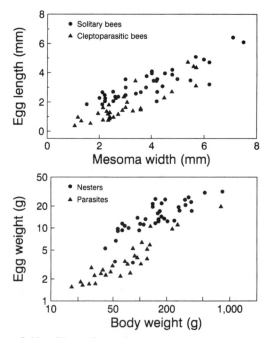

Figure 8.10. Plots of egg size against body size for parasitic and nonparasitic bees and cuckoos. Upper panel redrawn from Alexander and Rozen (1987), lower plot from Payne (1974). Note that both parasitic cuckoos and cleptoparasitic bees typically lay smaller eggs than nonparasitic species.

into account and may influence the results given above. (Unfortunately Payne does not provide sufficient detail to do this analysis.) A more relevant statistic than clutch size is the number of eggs laid per season. Parasitic cuckoos lay about 18 eggs per season (range 16–22, $n = 8$, data from Table 1 of Payne 1974). The number of clutches laid by nesting cuckoos is not well known but is probably only 1 or 2 (Payne 1974), giving a maximum output of between 4 and 8 eggs per season (data from Fig. 1 of Payne 1974). The difference in seasonal output of parasitic and nesting cuckoos is sufficiently great that, even without considering egg size and with the uncertainties attached to the two estimates, it is reasonable to conclude that parasitic cuckoos are investing more heavily in egg production than nesting cuckoos. The egg production of other parasitic birds has not been so extensively studied but they also appear to have higher-than-average values (Payne 1965).

The second case study examined the reproductive output of solitary versus cleptoparasitic bees. Alexander and Rozen (1987) observed that cleptoparasitic bees in all families tend to have a larger number of mature oocytes in their ovaries than do solitary bees (Table 8.6). This increased output is accomplished by an increased number of oocytes per ovariole, and in some species also by an increased number of ovarioles (Table 8.6). Increasing oocytes per ovarioles should be accompanied by a decrease in oocyte size, which is indeed the case (Fig. 8.10). Payne (1974) suggested that one reason for the small size of parasitic cuckoo eggs is that their hosts lay small eggs. Another contributing factor suggested by comparison with cleptoparasitic bees is a trade-off between egg size and egg number.

8.5. Reproductive Effort and Male Fitness

Much of the above discussion has concerned, explicitly or implicitly, the reproductive expenditure of females. While males may not invest as heavily as females in gonads (Trivers 1972; but see Dewsbury 1982 for evidence that sperm limitation can occur), they may invest very heavily in display and other reproductive behaviors that carry a cost. Such costs appear to have led in many instances to several distinct behaviors, the most frequently observed being that of a territorial or displaying male and a satellite or sneaker male. (See Table 7.7 for a survey of species in which satellite males have been observed.) In the first case the male actively attracts the female by maintaining a territory and/or displaying; in the second case the male acts as a satellite of the "focal" male and attempts to intercept the female attracted to the latter. In some cases, (e.g., *Hyla*) the satellites appear not to intercept the females, but simply wait for the focal male to leave its territory or cease calling (Fellers 1975, 1979a).

Table 8.6. A comparison of mature oocytes in parastic versus solitary bees

Category	Total number of mature oocytes	Mature oocytes per ovariole	Sample size
Solitary bees	2.17	0.34	26
Cleptoparasitic bees: all species	7.18	0.86	31
Cleptoparasitic bees: characteristic number of ovarioles	5.62	0.90	14
Cleptoparasitic bees: increased number of ovarioles	8.46	0.84	17

Cleptoparasitic bees have been divided into species which have increased numbers of ovarioles compared to the characteristic number of that family, and those with the characteristic number.

Modified from Alexander and Rozen (1987).

Statistical comparisons (Mann-Whitney U-tests; see Alexander and Rosen for details):
  Total number of mature oocytes:
    Solitary bees versus all cleptoparasitic bees, $P < 0.001$
    Solitary bees versus cleptoparasitic bees with the characteristic number of ovarioles, $P < 0.001$
    Between two categories of cleptoparasitic bees, $P < 0.05$
  Mature oocytes per ovariole:
    Solitary bees versus all cleptoparasitic bees, $P < 0.01$
    Between two categories of cleptoparasitic bees, $P = 0.84$

Since satellite males do not invest energy in calling or displaying, behaviors that can be energetically very expensive (see chapter 5, section 5.2.3) their reproductive effort will be less than that of the focal male. However, this reduction in investment is probably offset by a reduced success in the number of successful copulations. The satellite strategy might also be one that is "the best of a bad lot": males that for reasons outside of their control cannot compete with other males (e.g., they may be very small) might adopt this behavior as being that which produces the highest probability of obtaining copulations, even though this probability is still less than that of the focal males.

Callers and satellite males are reported to occur in the cricket species, *Gryllus integer*: the satellite male in this case does not necessarily take up station beside a calling male but wanders in search of a female (Cade 1979, personal communication). The potential advantages and disadvantages of calling in this species are well understood. Females are attracted to males by virtue of the male's call. But there is a cost to calling. First, calling is energetically expensive (see Table 5.9), and second, the parasitic tachinid fly, *Euphasiopteryx ochracea*, locates its host, *G. integer*, and other orthopteran species, by orienting to the sound of the male's call (Cade 1975). The amount of calling declines rapidly in parasitized males and death occurs prematurely (Cade 1984). Noncalling males are less likely to be parasitized than calling males: Cade (1975) found that 9 of 11 calling crickets were parasitized whereas only 1 of 17 noncalling males had been parasitized, a difference that is highly significant ($\chi^2 = 16.8$, $P < 0.002$). A noncalling male can still obtain copulations by acting as a satellite to a calling male and intercepting incoming females, or by simply searching for them. Satellite behavior is clearly frequency-dependent since satellites can only exist if there are calling males. A quantitative analysis using a game-theoretic approach requires more detailed information on the relative success rates of calling and noncalling males. Such data are presently lacking.

Satellite behavior can result from a variety of causes:

1. There may be two distinct phenotypes within the population. This does not imply a simple Mendelian inheritance, i.e., a single-locus mechanism, since the expression of the behavior could be a consequence of a threshold response to some continuously distributed factor such as hormone titer.

2. An individual might switch from one mode to another as it gets older and/or bigger, as, for example, in the case of bluegill sunfish described in chapter 7 (section 7.1.3).

3. An individual might switch to a satellite mode after it has depleted a given fraction of its energy resources.

4. An individual might switch to a satellite mode in response to the call of a neighbor. This last possibility has been studied in some detail by Arak (1983a, 1988) with respect to calling behavior in the natterjack toad, *Bufo calamita*.

Male natterjack toads, like most frogs and toads, gather at a pond in the breeding season and attempt to attract females by repeated calling. As with other anurans (Table 7.7), some individuals do not call but attempt to intercept females attracted to calling males. The behavior adopted by an individual is not fixed, but contingent on characteristics of the individual and its neighbor. Multiple regression analysis indicated that the probability that a male will adopt the satellite behavior is primarily a function of the pitch of its call—the weaker the call the greater the probability of becoming a satellite—and secondarily a function of the size of the toad (Arak 1988). The probability that a calling male is parasitized by a satellite male is also significantly correlated with call intensity ($r = 0.93$, $P < 0.001$, $n = 14$), and males with satellites produced louder calls than males without satellites (Arak 1988). These data "suggest that males adopt a conditional strategy.

Small males with weak calls were most likely to be satellites, and tended to associate with the largest and loudest subset of callers" (Arak 1988, p. 419).

The above results can be interpreted on the basis of male-male interactions over territorial control and the attractiveness of a male to a female. Calling males maintain territories by fighting with intruding males and neighboring males. The outcome of a contest is correlated with size, the larger male typically winning, a phenomenon observed in a variety of anurans (Table 5.2). Experimental manipulations have shown that males use the pitch of their rival to assess its size and fighting ability (Arak 1983a), a finding reported for other anurans (Ramer et al. 1983; Given 1987). Furthermore, females preferentially orient to males with louder calls (Arak 1988). Thus a small male next to a large male may have little chance of (1) maintaining its territory against the incursion of its neighbor and (2) attracting a female when a more attractive male is in the adjoining territory. Under these conditions the optimal response may be to conserve energy, remain in the vicinity of the large male (but out of sight to avoid being attacked), and intercept females as they approach the large calling male. Further support for this hypothesis is provided by the context in which a male was observed to switch modes. On six occasions the male stopped calling and became a satellite on a larger caller in a neighboring territory, and on 21 occasions a calling male, supplanted from his calling site by a larger male, became a satellite to the winner (Arak 1988).

A calling male has two responses to an approaching female—waiting or dashing. The first response entails simply waiting for the female to initiate physical contact: this is the response most likely when the caller is alone, and is also more successful under these conditions than dashing. (All 17 cases of the male waiting resulted in copulation but only 6 of 9 cases did so when the male dashed toward the female.) Dashing involves the male lunging at the female as she approaches: this increases the likelihood that the female will abandon her approach but increases the frequency of successful copulations when a satellite is present. (The territorial male obtained no copulations when it waited with a satellite present, the latter obtaining 6 copulations out of 8 attempts. In contrast, dashing resulted in territorial males obtaining 10 copulations, the satellites obtained 2, and in 4 cases the females retreated.) That callers more frequently "dash" in the presence of satellites indicates that at some point the calling male detects the presence of the satellite and responds to the potential threat. The total number of observed encounters is low but the data suggest that satellite males are relatively successful, the number of successful copulations obtained by callers in the presence of satellites being not significantly different from that of the satellites (10 versus 8, ns, binomial test). The population studied by Arak comprised 67% callers and 33% satellites. Callers obtained

80.5% of the matings (33 of 41) and satellites 19.5%. Though the success rate of callers was higher than that of satellites the success of the latter was by no means insignificant. More importantly, a male that becomes a satellite is as likely to obtain copulations as its neighbor, while if it remains a caller it may have a reduced success.

Playback experiments using synthetic calls of high and low intensity showed that large males continued calling regardless of the intensity of the synthetic song. But all small males tested switched to being satellites with the high-intensity song and one-half switched when hearing the low-intensity song. On the basis of these observations Arak constructed three models to predict the cue required to make a toad switch behaviors: the omniscient strategy, the local-assessment strategy, and the naive strategy. The first model assumes that males know the attractiveness of all males in the population and the location of all satellites. This model is biologically unreasonable and does not give satisfactory predictions. The third model assumes that a male can assess his own attractiveness but not that of his neighbor. This assumption is not supported by observation, and the model does not give accurate predictions. The second model is predicated on the assumptions that

1. A male cannot detect the presence of a satellite, at least while calling. Given that males will attack satellites that they detect, this is a reasonable assumption.

2. A male is capable of assessing only the attractiveness of it nearest calling neighbor.

The analysis proceeds as follows:

1. The probability that a given male, x, attracts a female depends upon his attractiveness relative to all other calling males. This is computed as

$$P(x) = \frac{a(x)}{A(N)} \tag{8.41}$$

where $P(x)$ is the probability of caller x attracting a given female,
$a(x)$ is the "attractiveness" of caller x, measured in terms of call intensity,
$A(N)$ is the summed attractiveness of all N callers in the population

$$A(N) = \sum_{i=1}^{N} a(i) \tag{8.42}$$

This basic formulation was tested by estimating the seasonal mating success of a male and the intensity of its call. Nightly mating success was estimated by multiplying the probability of mating for a given male by the number of females arriving at the pond on that night. Seasonal mating success was obtained by summing nightly success over all nights the male was present. There is very good agreement between observed and predicted seasonal mating success (Fig. 8.11) and no evidence of bias (i.e., slope not significantly different from one and no trend for points to lie unequally about the 1:1 line).

2. The impact of satellites is now considered. Data on success rates of satellites and callers suggest that contest for an incoming female is of the scramble type, from which we get

$$p(x) = \frac{1}{1 + n(x)} \qquad \text{(8.43)}$$

where $p(x)$ is the probability of calling male x obtaining the copulation,

$n(x)$ is the number of noncalling (satellite) males around male x. The probability of obtaining the copulation, $C(x)$, is the product

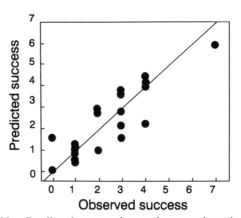

Figure 8.11. Predicted versus observed seasonal mating success of male natterjack toads. Predicted values obtained from equation 8.41. Solid line is 1:1 reference line. The fitted regression is $y = 0.99x + 0.02$ ($r = 0.89$, $n = 21$, $P < 0.001$). Redrawn from Arak (1988).

of the probability of attracting a female and the probability of mating with her, $C(x) = P(x)p(x)$,

$$C(x) = \frac{a(x)}{A(N)(1 + n(x))} \tag{8.44}$$

Since satellites have the same success rate once females are approaching the calling male, the mating success of any given satellite, say male y, about male x is equal to that of male x, i.e., $C(x) = C(y)$.

Now consider the consequences of a change in behavior by either the caller or satellite. First, suppose male x stops calling and becomes a satellite on another caller, z. The mating probability of caller z and also male x is

$$C(z) = \frac{a(z)}{(A(N) - a(x))(2 + n(z))} \tag{8.45}$$

Note that there are now $2 + n(z)$ males competing for a female and that the total attractiveness of the population has decreased by $a(x)$. The two have equal fitnesses when $C(x) = C(z)$,

$$\frac{a(z)}{(A(N) - a(x))(2 + n(z))} = \frac{a(x)}{A(N)(1 + n(x))}$$

$$\frac{a(z)}{a(x)} = \left(\frac{A(N) - a(x)}{A(N)}\right)\left(\frac{2 + n(z)}{1 + n(x)}\right) \tag{8.46}$$

It has previously been demonstrated that males assess their neighbors in the same manner as do females. Therefore, Arak assumed that the probability that satellite male y is associated with male x is the same as the probability that a female will be attracted to male x, $P(x)$. This is a logical manner for a satellite to act. If the total number of noncalling males is T, the calling male's estimate of the number of satellites, $Est(n(x))$, is

$$Est(n(x)) = T\frac{a(x)}{A(N)} \tag{8.47}$$

To obtain the point at which a male switches we substitute the estimated number of satellites for the actual numbers in equation 8.46,

$$\frac{a(z)}{a(x)} = \left(\frac{A(N) - a(x)}{A(N)}\right)\left(\frac{2 + T\dfrac{a(z)}{A(N)}}{1 + T\dfrac{a(x)}{A(N)}}\right) \tag{8.48}$$

Arak assumed $a(x)$ to be much smaller than $A(N)$; so $A(N) - a(x) \approx A(N)$. Assuming this and rearranging equation 8.48 gives

$$\frac{a(z)}{a(x)} = 2 \tag{8.49}$$

which can be interpreted as "switch to being a satellite if your nearest-neighbor's call is twice as intense as yours; otherwise call."

The predicted switching line and observed behavioral modes are shown in Fig. 8.12: toads to the right of the line should call, those to the left

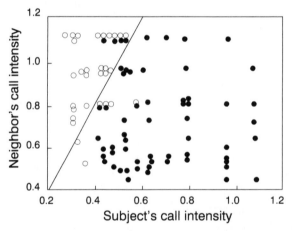

Figure 8.12. Test of Arak's model for satellite behavior in the natterjack toad. The subject's call intensity (N/m^2) is plotted against the call intensity of its nearest neighbor. The subject should adopt the satellite behavior whenever its neighbor's call is twice as loud as its own, such combinations occurring to the left of the diagonal line. Filled circles indicate calling males and open circles, satellite males (thus four males were callers when predicted to be satellites and six males predicted to be satellites were callers). Redrawn from Arak (1988).

should not. Callers and satellites are clearly not randomly distributed. The model makes the correct prediction for 24 out of 30 (80.0%) satellites and 56 out of 60 (93.3%) callers. In total "88.9% of all males adopted the tactic that was predicted to maximize fitness within the constraints of the information available to them" (Arak 1988, p. 426). This is better than expected by chance and supports the assumptions underlying the model. Further support for the decision rule predicted by the model could be obtained using playback experiments.

The essential element in the optimal reproductive effort of males described thus far is that one behavior is contingent on the presence of another (satellite and caller). But contingent behaviors may be frequency-independent. An example of this is provided by the mating activities of male scorpion flies of the genus *Panorpia*, studied by Thornhill (1975, 1978, 1979a,b,c, 1980c,d, 1981a,b).

Scorpion flies belong to the order Mecoptera; the genus *Panorpa* contains the largest number (47) of species in North America. The life histories of all species are very similar and there may be both intra- and interspecific competition for food (Thornhill 1980c). Dead insects are the principal food, and scorpion flies have the dangerous habit of feeding on insects caught in spider webs (Thornhill 1975, 1978): dangerous, because although they have evolved defenses against spiders, including a noxious oral effluent that is dabbed on the spider, mortality rates from web-building spiders is high—65% of observed predation mortality being due to these predators (Thornhill 1981b). Male scorpion flies use one of three behaviors to obtain a mate; the first two involve a nuptial gift:

1. The male locates a dead insect, and after feeding briefly emits a pheromone to attract females. If the female accepts the "gift" the male mates while the female feeds. Defense of the dead insect against other males can be strenuous and wings and legs may be torn in the contest.

2. The male secretes a salivary mass and emits a pheromone. Again, if the female accepts the offering the male mates while she feeds. Salivary glands are highly developed for the production of a nuptial gift, comprising between 14% and 28% of the dry body weight (Fig. 8.13). As with dead insects, salivary masses are defended against other males.

3. The male forcibly mates with a female, no nuptial gift being offered. That this is not an uncommon opportunistic method is evidenced by the presence of specialized abdominal clamps designed to hold the female during forced copulation (Thornhill 1980a, 1981a).

Females approach males with offerings but attempt to escape from males without offerings. Laboratory experiments with *P. latipennis* have demonstrated that females prefer males offering dead insects to those offering salivary masses. In these experiments 7.4% of males without offerings were able to obtain a copulation (all by force), while 56% of males with salivary masses obtained copulations, compared to 89% of males possessing a dead cricket (Thornhill 1981b, all differences statistically significant). Experiments with two Mexican species of *Panorpa* produced similar results (Thornhill 1979a). This choice by the female is adaptive: females mating with males that provide a dead arthropod lay significantly more eggs than females mating with males that provide a salivary mass: females that were "raped" lay very few eggs (Thornhill 1981b). There is a second manner in which nuptial feeding may be beneficial for the female; it alleviates the need to search for food and hence reduces the risk of being caught by predators such as spiders.

Although a male scorpion fly has the highest probability of attracting a female if he is able to present a dead insect as a nuptial gift, searching for dead insects costs energy and may be risky; the searching male may fall prey to spiders. Secreting a salivary mass might decrease the risk from predators but requires energy, and such males are only "second best" from the female's point of view. In the absence of dead insects males will shift to this "second-best" behavior. The absence of dead insects may be a result of other males commandeering them or of simple scarcity. As with the natterjack toad, larger males have an advantage and can more successfully defend a nuptial gift. Contests over food items occur not only within species but also across species, and here we also find that the spoils go the biggest (Fig. 8.13). Furthermore, allocation to salivary glands decreases with dominance rank (Fig. 8.13), suggesting that smaller species, being unable to compete against the larger, allocate more resources to the second-best response of producing a salivary mass for a nuptial gift (Thornhill 1981b).

8.6. Reproductive Effort in a Variable Environment

Thus far we have assumed that the environmental conditions remain constant over time. However, conditions are generally temporally variable. Though there is obviously an autocorrelation between present, past, and future environmental conditions, analyses of reproductive effort in variable environments have assumed at least implicitly that variations are stochastic and uncorrelated. The problem was first tackled by Murphy (1968), who proposed the following hypothesis (p. 392):

> Evolutionary pressure for long life, late maturity, and many reproductions may be generated either by an environment in which density-in-

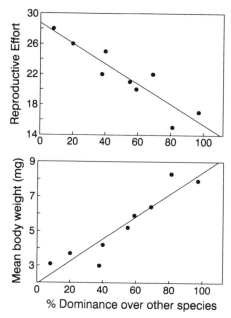

Figure 8.13. Reproductive effort (measured as % total dry body weight that is salivary gland) and mean body weight (mg dry wt.) of various species of *Panorpa* regressed against % dominance of each species over the others (computed as % heterospecific interactions won). Data from Table 4 of Thornhill (1981b).

dependent factors cause wide variation in the survival of pre-reproductives or by an environment that is biologically inhospitable to pre-reproductives because of intense competition with the reproductives. Conversely, either high or variable adult mortality will tend to generate evolutionary pressure toward early reproduction, high fecundity, and few reproductions, or only one reproduction.

The logic of this hypothesis can be illustrated with a very simple example. Consider two species (or genotypes)—one that is semelparous producing m_s female offspring after 1 year and a second that is iteroparous, living 2 years, and producing m_i female offspring each year. Now let us assume that the fitness of the semelparous type is greater that of the iteroparous (i.e., $\lambda_s > \lambda_i$). In a deterministic environment the semelparous form will eventually prevail. But suppose that the environment consists of "good" and "bad" years, the survival of offspring in the "bad" years being zero. Even if the probability of a "bad" year occurring is very small (e.g., one in 1,000) the demise of the semelparous type is certain, while the persistence time of the iteroparous type may be extremely long (e.g., if the probability is 1 in 1,000, the probability of 2 bad years occurring in a row is 1 in 10^6). Thus in a variable environment iteroparity will be favored.

The argument advanced above can be extended to the iteroparous condition itself: increased variability in reproductive success should favor an age schedule of reproductive effort that is extended so that, to follow the advice of an old adage, "not all eggs are placed in one basket." Murphy demonstrated the selective advantage of variation in the degree of iteroparity with a simulation model comprising two clones in which the number of recruits is governed by a Ricker recruitment function:

$$R_1 = P_1 e^{\frac{c_1 - P_1 - c_2 P_2}{c_3}}$$

$$R_2 = P_2 e^{\frac{c_4 - P_2 - c_5 P_1}{c_6}} \qquad (8.50)$$

where R_i is the number of recruits in clone i, P_i is the number of parents in clone i and c_i is a constant. Parameter values were selected to approximately conform to species such as anchovies and sardines. Values of the interaction coefficients c_2 and c_5 were set such that in a deterministic environment the clones coexisted, the long-lived clone at a population size of 2,364 and the short-lived clone at 6,757. Murphy introduced random variation in reproductive success, though the exact details of how this was done are not clear. Following the introduction of random variation the population sizes of the two clones switched—the longer-lived clone maintaining a higher population size than the short-lived clone; but over the 230 time units the simulations were run neither clone went extinct.

Murphy also examined the effects of variation when iteroparity is determined by a single-locus, two-allele mechanism. The AA homozygote was defined as the short-lived form, the aa homozygote as the long-lived, and Aa as intermediate. Relative fecundities, designated as m(AA), m(Aa), and m(aa), were set at m(AA) = 3m(aa) and m(Aa) = 2m(aa). Without environmental variation allele A became fixed. Variation comprised "good" and "bad" years, either in random or alternating sequence. In both cases allele a became fixed.

The results of Murphy's analysis were criticized by Hairston et al. (1970) on the grounds that they depended on strong density dependence. They produced a similar genetic model in which the population showed a gradual increase in size over time and no density-dependent effects. In a variable environment the more iteroparous genotype predominated in the population as in Murphy's simulations, but they claim that the result "was more a function of the mean rather than the variance in survivorship" (Hairston et al. 1970, p. 687). This proposition is not proven and is certainly incorrect since, as discussed in chapter 3 (3.1.1.2) and below, fitness in a variable environment is a function of both the mean and the variance. The important point is that in both Murphy's simulations and those of Hairston et al., iteroparity was favored in a variable environment.

A more formal analysis of the impact of variation in survival on the optimal reproductive effort was given by Schaffer (1974b). Recall (section 8.1) that for a perennial species

$$\lambda = ms + S \qquad (8.51)$$

where m is female offspring per year, s is the proportion surviving the first year, and S is the annual probability of survival thereafter. For simplicity, subscripts have been dropped. Schaffer combined fecundity and survival-to-first-reproduction into a single value which he called "effective fecundity," $B = ms$. Years are assumed to be "good" or "bad" with equal frequency; therefore, the mean geometric rate of increase is given by

$$\overline{\lambda}^2 = \lambda_g \lambda_b \qquad (8.52)$$

where g and b stand for "good" and "bad" years, respectively. The good and bad years can be mathematically represented by $B(1 + d)$ and $B(1 - d)$, where d is the departure from the overall average. Thus

$$\lambda_g = B(1 + d) + S$$

$$\lambda_b = B(1 - d) + S$$

$$\overline{\lambda}^2 = [B(1 + d) + S]\,[B(1 - d) + S]$$

$$= (B + S)^2 - d^2 B^2 \qquad (8.53)$$

To find the reproductive effort (E) that maximizes fitness we differentiate equation 8.53 with respect to E and find the value of E at which the differential is equal to zero,

$$\frac{\partial \overline{\lambda}^2}{\partial E} = 2(B + S)\frac{\partial B}{\partial E} - 2Bd^2\frac{\partial B}{\partial E} + 2(B + S)\frac{\partial S}{\partial E}$$

$$= 0 \quad \text{when} \quad \frac{\partial B}{\partial E}\left(1 - \frac{Bd^2}{B + S}\right) = -\frac{\partial S}{\partial E} \qquad (8.54)$$

For selection to favor an intermediate reproductive effort, B and S must be concave functions of reproductive effort (Fig. 8.14; for the justification of this, see section 8.2). Increasing values of d, which correspond to increasing variability in reproductive success (juvenile mortality or fecundity), lead to a decrease in the optimal reproductive effort (Fig. 8.14). A similar analysis can be applied to variation in adult survival, S, with the intuitively reasonable result that increasing adult survival favors greater

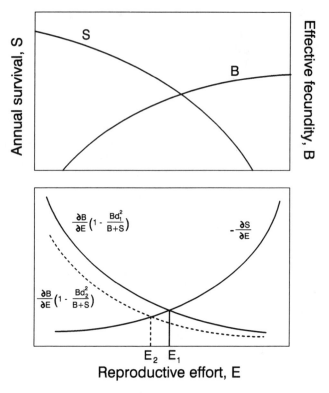

Figure 8.14. Upper panel: Hypothesized relationship of effective fecundity, B, and annual survival, S, versus reproductive effort. Lower panel: The first derivatives of the functions shown in upper panel. That for fecundity has been multiplied by the fraction $1 - Bd^2/(B + S)$. The optimal reproductive effort occurs where the two curves intersect. (See equation 8.54.) As the parameter d, measuring the degree of departure from the average, increases from d_1 to d_2, the optimal reproductive effort decreases from E_1 to E_2.

reproductive effort. Charlesworth (1980) used a similar approach to demonstrate that iteroparity will be favored in environments with high juvenile survival and semelparity will be favored when adult survival is variable. But, he cautions,

> The weakness of these results is that they are strictly valid only for the case when the population spends a long time in each environment . . . no adequate theory has yet been developed for the dynamics of selection in age-structured populations, when the time period of the environmental

change is short compared with the time needed to stablise age-structure.
(Charlesworth 1980, p. 229)

In an age-structured population the realized growth rate of a population
is less than the growth rate calculated from the mean Leslie matrix associated with the population growth history (Goodman 1984). More importantly, Goodman was able to demonstrate that this discrepancy increased
with increasing environmental variation but decreased with iteroparity,
suggesting that the analyses of Schaffer (1974b) and Charlesworth (1980)
are at least qualitatively correct. As is intuitively obvious, analyses have
shown that the benefit of iteroparity is that it smooths out the variability
in the long-term growth rate (Goodman 1984; Bulmer 1985b; Orzack and
Tuljapurkar 1989).

The analysis of Hairston et al. (1970) suggested that polymorphisms may
be favored rather than a particular allele being fixed. This possibility was
examined by Giesel (1974), who considered genotypes that differed in their
age schedules of reproduction, and by Bulmer (1985b), who used an ESS
approach (i.e., noninterbreeding genotypes). Giesel (1974) assumed a single-locus, two-allele model, the three genotypes differing in the shape of
their $l(x)m(x)$ functions. Though the number of simulations was limited,
the results suggest that under some circumstances the geometric increase
of a polymorphic population may be greater than that of a monomorphic
population comprising individuals with $l(x)m(x)$ curves equal to that of
one of the three genotypes. The reason for this is that different genotypes
are favored under different conditions, and the overall result is a population
that has a variable degree of iteroparity. However, since the simulations
were run for only 300 cycles it is not clear if the polymorphisms were stable.

Bulmer (1985b) extended the analyses of Charnov and Schaffer (1973)
to the case in which population size is constrained by density-dependent
factors. He considered two models, the lottery model and the exponential
model. In the former, each plant, annual or perennial, has an equal chance
of occupying the available spaces. The exponential model was based on a
Ricker recruitment function. (See equation 8.50.) Coexistence is not possible in the lottery model but is possible in the exponential model; but the
general message is as found by other workers: variability in juvenile survival
rates favors perennials while variability in adult survival rates favors annuals.

Given the considerable interest in the importance of variable environments, it is surprising to find that attempts to test predictions are very few.
There have been three attempts to test the predictions using interspecific
comparisons: Murphy (1968), Mann and Mills (1979), and Roff (1981b).
Murphy (1968) computed the correlation between reproductive life-span
and brood-strength variation (measured as the ratio of maximum brood

strength to minimum brood strength) in five species of schooling, plankton-feeding fish of the order Clupeiformes (sardines and herrings). The correlation is remarkably high ($r = 0.975$) but is based on an inaccurate estimate of brood strength in the Peruvian anchovy, *Engraulis ringens*. The estimate of brood-strength variation in this species was made during years when there was not an El Niño event—a climatic phenomenon that can cause drastic fluctuations in the anchovy population. Further data on the anchovy population dynamics indicate that instead of the twofold variation estimated by Murphy the ratio is at least 11.9 (Roff 1981b). The correlation between reproduction life-span and brood-strength variation is now not significant ($r = 0.776$, $0.05 < P < 0.1$).

Mann and Mills (1979) attempted an analysis similar to that of Murphy (1968), increasing the number of data points to 18 (15 fish species, 2 species comprising 2 and 3 stocks). They noted that "despite the tendency of data to be biased by the length of the study period, the points generally fit the pattern predicted by Murphy" (Mann and Mills 1979, p. 165–166). However, the correlation is not even close to being significant (Fig. 8.15; $r = 0.03$, $n = 18$, $P > 0.50$)! Roff (1981b) examined the hypothesis using data from a single order, Pleuronectiformes, the flatfishes. Two measures of brood strength were used: the ratio of maximum to minimum brood size and the coefficient of variation in brood strength. Neither measure is sig-

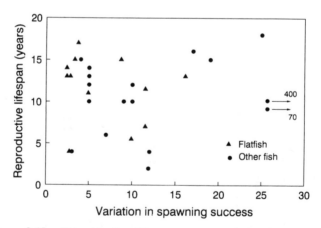

Figure 8.15. Reproductive life-span versus variation in spawning success (largest over smallest brood strengths) for various fish species. Data from Mann and Mills (1979) and Roff (1981b). The estimated reproductive life-span of plaice, which occurs in both data sets, is taken from Mann and Mills. Brood strength variation of anchovy and plaice from Roff (1981b).

nificantly correlated with reproductive life-span ($r = 0.245$, $r = 0.072$, $n = 13$, respectively). Combining the data from all three studies also produces an insignificant correlation ($r = 0.062$, $n = 29$, Fig. 8.15). Although, as noted by Mann and Mills, and as illustrated by the anchovy data, the estimate of brood-strength variation may depend upon the number of years of data (see also Pimm and Redfern, 1988, for an analysis of this phenomenon with respect to population fluctuations), the complete lack of a correlation suggests that brood-strength variation does not play an important role in the evolution of the degree of iteroparity when measured across species.

Haukioja and Hakala (1978) report a more successful finding in a study of variation within a single species, the mollusc, *Anodonta piscinalis*. In this species they found a highly significant positive correlation between reproductive life-span and variation in juvenile survival (Spearman rank correlation $= 0.823$, $P < 0.001$, $n = 13$). But this result may be a statistical artifact. Reproductive life-span (RLS) was calculated as $1 + t_{50}$, where t_{50} is the time (in years) in which the number of females decreases by half. The rate of decline was estimated from the survivorship curve, i.e., the relationship of $\log_e$ (Numbers) on age (see chapter 5); thus RLS was estimated as

$$RLS = 1 + \frac{\log_e(N) - \log_e\left(\dfrac{N}{2}\right)}{M} \tag{8.55}$$

where N is the initial number of females and M is the slope of the survivorship curve (equal to the instantaneous rate of mortality). But the survivorship curve was also used to estimate the variation in juvenile survival, estimated as $1 - r$, where r is the correlation between $\log_e$(Numbers) and age. The slope of a regression equation is equal to the product of the regression coefficient and the ratio of the standard deviation of the independent variable to the standard deviation of the dependent variable (slope $= rR$, where R is the aforementioned ratio). In the present case the slope of the regression is M and thus the relationship between reproductive life-span (RLS) and variation in juvenile survival (V) is

$$RLS = 1 + \frac{\log_e N - \log_e\left(\dfrac{N}{2}\right)}{Rr}$$

$$= 1 + \frac{\log_e N - \log_e\left(\dfrac{N}{2}\right)}{R(1 - V)} \tag{8.56}$$

Providing R is not correlated with r, reproductive life-span will necessarily increase with the variance in juvenile survival as estimated by Haukioja and Hakala. To assign a correct probability to the observed regression we must know what correlation is expected if the relationship is entirely due to the algebraic relationship between the two variables. This could be estimated using Monte Carlo simulation, but insufficient data are given in the paper to undertake such an exercise. Without further data on the possible significance of the nonindependence of RLS and V, these data cannot be accepted as supporting Murphy's hypothesis.

8.7. Summary

Cole's paradoxical claim, that for a semelparous species with a female offspring production of m, the intrinsic rate of increase can be made equivalent to that of an iteroparous species with the same birth rate simply by increasing female offspring production to $m + 1$, is true only if juvenile and adult survival rates are the same. High adult survival and low rate of increase favor iteroparity.

If death after reproduction is certain a female will have no reason to conserve resources. Similarly, if death can be averted by the reduction in reproductive effort, it may be advantageous for a female to divert less than maximal surplus energy into reproduction and hence survive to breed again. These arguments suggest the hypothesis that reproductive effort will be greater in semelparous than iteroparous organisms. Comparison of the reproductive efforts of semelparous and iteroparous grasses, lupines, gastropods, and triclads supports this hypothesis.

Surplus energy may be diverted entirely into reproduction after maturity or into both further growth and reproduction. The latter pattern is favored when fecundity or mortality is a concave or sigmoidal function of reproductive effort. Temporal variation may also select for a graded allocation pattern in annual species. A third circumstance in which surplus energy may be shunted into both growth and reproduction is when mechanical constraints limit fecundity, and this limitation can be relieved by further growth. Conversely, if growth is inhibited, surplus energy may be directed into reproduction. The hermit crab, *Clibanarius albidigitus*, exemplifies this scenario, growth being restricted due to a lack of large shells: when presented with large shells, reproduction diminished and growth increased (Bertness 1981a).

Various authors have hypothesized that reproductive effort will increase with age. Two examples, show that this hypothesis is not universally true, though it may still be generally true. Charlesworth and León (1976) produced a model for which they derived both necessary and sufficient conditions for reproductive effort to increase with age. This model is supported by an analysis of the reproductive patterns of fish. A general survey of animals and one plant indicates that reproductive effort does generally increase with age, but insufficient data exist to test for conformity with the conditions of the Charlesworth and León model.

Reproductive effort comprises, in some animals, effort spent in mating, reproduction (production of gonads and young), and parental care. There is an optimal balance between these three components. Parental care may have arisen in ectotherms in response to environments in which unguarded offspring have very high mortality rates. It may be advantageous for one or both parents to desert their offspring. The probability of desertion depends upon the amount of further effort required to bring the offspring to independence rather than the amount already invested. In practice these two may be highly correlated. Tests with the three-spined stickleback are in accord with theory: the vigor with which males defend their nests increases with the age and number of eggs contained therein.

If offspring require parental care, offspring productivity can be increased by dumping them on other individuals, members of either the same or different species. As predicted, the fecundity of parasitic cuckoos and bees is greater than in nonparasitic species.

Males frequently expend a significant amount of energy in attracting females. Differences in the attractiveness of males, or costs associated with sexual advertisement, may lead to the evolution of alternative modes of behavior, most typically territorial and satellite behavior. These two types of males may be genetically distinct, or the behaviors may be an expression of phenotypic plasticity, and contingent on relative age, size, or condition. The expression of alternative mating behaviors is often frequency-dependent, for which a game-theoretic approach is appropriate. Analyses of calling in crickets and toads and the repertoire of mating behaviors in scorpion flies demonstrate the diversity and ubiquity of alternative patterns of allocation of energy in obtaining mates.

In a temporally variable environment selection is postulated to favor a greater conservatism in the allocation of reproductive effort. While this hypothesis is clearly true in extreme conditions (e.g., semelparity versus iteroparity) it is not clear what role environmental variation plays in shaping the degree of iteroparity. Interspecific comparisons of the degree of iteroparity with temporal variation in brood size show no relationship. An intraspecific comparison using the mollusc, *Anodonta piscinalis*, is in conformity with the hypothesis but the result may be a statistical artifact.

9

Clutch Size

The previous chapter dealt with how much effort an individual should invest in reproduction. One component of this investment is the number of offspring produced at each reproductive episode. The present chapter focuses on this component of fitness, and the following chapter discusses the trade-off between the number and size of offspring.

A fundamental constraint on clutch size resides in mechanical factors. In the simplest case there is a limit to how many offspring can be squeezed inside the body of a female. But there can also be more subtle effects: increases in clutch size may impede maneuverability or speed and thus reduce a female's ability to escape predators. A female carrying an enlarged clutch might also be more visible to predators. Studies invoking mechanical factors as being instrumental in the evolution of clutch size are reviewed in the first part of this chapter.

Much of the research into how selection has molded clutch size has been strongly taxon-oriented although the models so generated have greater applicability. Two groups have been the prime focus of research—birds and insects. Hypotheses resulting from work on birds have been moved more or less intact to small mammals, which of course are very close in their general life-style. Insects typically have a very different life-style, showing no parental care and distributing their eggs among several sites. This introduces a complexity not present in the analysis of clutch size in endotherms. Nevertheless the research on birds has played an important role in shaping the theoretical approach to the analysis of the optimal clutch size in insects. Because of the peculiarities of the mathematical frameworks (particularly those dealing with insects) it is easiest to discuss the various avenues of investigation within the two general taxonomic frameworks—birds and mammals, and insects. But I cannot emphasize too strongly that this does not reflect a strict differentiation of processes between the taxa.

The most influential analysis of the evolution of clutch size in endotherms is that by Lack, whose hypothesis has formed the starting point of most analyses since its publication in 1947. Lack believed selection favors that clutch size producing the most offspring, the most productive clutch size, often referred to as the Lack value. Following his championship of this hypothesis have come numerous studies suggesting that, on both empirical and theoretical grounds, the optimum clutch size is less than the Lack value. Lack's hypothesis has been challenged by the observation that clutch sizes below the Lack value will be favored when there exist (1) trade-offs between clutch size and adult survival, and/or clutch size and future fecundity; (2) inherent variation in clutch size; (3) environmental variation; or (4) likelihood of nest failure.

In the second part of this chapter I review the evidence that in birds and mammals life history traits such as size at fledging, survival, and future reproduction are a function of clutch size both for the parents and their offspring. I examine Lack's hypothesis and find that there are sufficient grounds to be sceptical of the the hypothesis. The four factors, cited above, that are most likely to favor a reduced clutch size are next detailed. Unfortunately, while the theoretical bases for these influences have been well worked out there is still a lack of empirical support for their general importance.

Most of the research on the evolution of clutch size in endotherms has adopted a demographic perspective, examining the interactions between clutch size and the two components of the characteristic equation, survival and fecundity. In 1965 a paper was published by T. A. Wilson that adopted an entirely different perspective. This analysis, based on the concept of entropy minimization, has received very little notice despite the fact that its predictions appear to be much better than expected by chance. In section 9.2.2 I present this model, recast in terms more in keeping with the concept of selection acting on the individual.

As with all life history traits, clutch size is under continual selection. The relative ease with which clutches can be manipulated and their fates can be followed make them them an ideal subject for the analysis of natural selection in the wild. Field studies have produced the paradoxical finding that though there is additive genetic variance for clutch size, and selection favors an increased clutch size, there is no response. Section 9.2.3 examines two models proposed to account for this lack of change.

Lack's arguments have stimulated much of the research on the optimal clutch size in insects. While endotherms care for their young, insects typically scatter their eggs and pay them no further interest. An insect therefore faces the problem of selecting the optimal division of its eggs into clutches—a decision that will be modulated by a range of factors not impinging on the reproductive decisions of most endotherms. Competition between larvae within a single site will reduce their individual fitness by reducing their survival and/or their future fecundity. This, of course is the same problem faced by the endotherm; but an invertebrate, because it does not generally show parental care, can divide its offspring into several separate clutches and thereby reduce the adverse effects of competition. A division into clutches of single eggs might appear at first glance to be the optimal pattern, but a variety of factors favor larger clutches. Analyses have proceeded within one of two frameworks. In the first and simplest case only a single female deposits her eggs at any given site (section 9.3.1); this greatly simplifies analysis and is a reasonable assumption in many cases. In birds and mammals increasing clutch size has been considered to be generally negatively density-dependent, but in some invertebrates there occurs what is termed the Allee effect, in which survival increases with density. This phenomenon, discussed in section 9.3.1.4, has been little studied with respect to the optimal clutch size. Considerably more attention has been given to the more usual case in which survival and/or future fecundity decreases with density. If a female survived to lay her full complement of eggs she could minimize the competition between her offspring by laying only one egg per site. However, once there is the possibility that she will not survive to the next site, the optimal clutch size will typically increase (section 9.3.1.1). A priori it might seem that increases in patch quality would favor larger clutches, but in section 9.3.1.2 I show that this is not always the case. Although egg clumping may increase competition between larvae it may also reduce the risk of parasitism. The theoretical basis for this hypothesis, explained in section 9.3.1.3, is drawn from research on the advantages of the convoy system during the Second World War—an example of how research in a very different field can be relevant to biology.

The second mode of analysis (section 9.3.2) assumes that several females can lay their eggs at a given site. This is a potentially more complex problem and relatively few studies have attempted an analysis. In one case the situation is simplified by assuming only one oviposition episode per female. This analysis shows that the optimal clutch size depends critically upon the mathematical equation relating survival, fecundity, and larval density. Therefore, because of the absence of empirical studies on this particular relationship, no general conclusions can be presently drawn. In a second study the complexity of the analysis is handled by simulation modeling and shows very clearly that interactions between individuals can be a potent evolutionary force. Finally a study is presented in which the oviposition rules are derived for a particular species, the bean weevil.

9.1. Mechanical Factors in Evolution of Clutch Size

Mechanical factors can potentially influence clutch size in two ways: by increasing the load to be carried and hence reducing locomotory ability, or by changing the shape of the organism. In the flesh fly, *Neobellieria bullata*, lift production during takeoff is seriously affected by the number of eggs the female is carrying (Berrigan 1991). Reduction in lift production may increase susceptibility to predation, and hence there may be an optimal clutch size determined by the trade-off between clutch number and predator escape. A female bloated with eggs may be considerably more visible to a predator, a phenomenon that may account, in part, for the frequent finding that pregnant females are more likely than nonreproductives to be taken by a predator. (See section 6.2.2.)

Mechanical factors should be relatively easy to measure, and it is rather curious that so little attention has been paid to them. The importance of mechanical constraints is well illustrated by research in reptiles, where considerable interest has been shown in the mechanical problems accompanying increases in clutch size and concomitant increases in mortality. I present three examples in which the mechanical constraints imposed by a particular life-style restrict the evolution of clutch size. These three are arboreality, crevice dwelling, and foraging mode.

9.1.1. Clutch Size and Climbing Constraints

Anoline lizards are characterized by a clutch size of one egg; as a consequence clutch weight relative to body size is extremely low in this genus compared to a more typical iguanid genus such as *Sceloporus* (Fig. 9.1). The low clutch size is partly compensated for by multiple broods—a pattern possible in the tropical environment typical of the genus *Anolis*—but one that may not be available to their temperate ecological counterparts in the New World, *Sceloporus*, which lay larger clutches but only one to three clutches per season (Tinkle 1972, 1973; Tinkle and Ballinger 1972; Ballinger 1973). Anoles are highly arboreal lizards; Andrews and Rand (1974) spec-

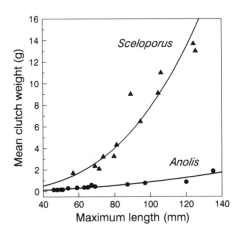

Figure 9.1. Relationship between mean clutch weight and maximum length in two genera of lizards. Solid lines show regression lines fitted after back-transformation from log/log regressions. Data from Andrews and Rand (1974).

ulate that this mode of activity may place a mechanical constraint on clutch weight. They argue as follows (pp. 1323–1324):

> The shape of the curves in Fig. [9.1] suggests that for *Sceloporus* clutch weight is set by female size as a result of the limitation to the volume capacity of the female, whereas for *Anolis* the limiting factor is somehow limited to area. A characteristic of the genus *Anolis* is the possession of expanded subdigital lamellae or adhesive toe pads. These are adaptively associated with the arboreal habits of most *Anolis* species. Collette (1961) showed that the most arboreal of the anole species he studied had the greatest number of lamellae of the widest pads and that the larger species had the greater numbers of lamellae than the smaller ones. Lamellae number and body size are associated in this way because body weight increases by the cube of the linear dimensions while lamellae area increases only by the square. The loading capacity of a female anole may limit the amount of additional weight she can carry and still climb effectively. This limit on additional weight may in part explain both the relatively low clutch weight and single egg clutch in *Anolis* and the relative decline in the size of the hatchling (clutch weight) as species size increases. Other arboreal lizards with adhesive toe pads, e.g., geckonids and some scincids, also have low clutch numbers (one or two eggs). In contrast, arboreal lizards without toe pads, e.g., *Polychrus*, *Iguana* and *Chamelaeleo*, have a large clutch number. Thus, the association between the specialized climbing habits of anoles and their low clutch number or weight is a partial explanation for their small clutch.

An important component of the argument of Andrews and Rand concerns the difference in the relationship between clutch mass and female length of the two lizard genera, but curiously, they do not verify their "visual impression." The hypothesis is that in *Anolis* mean clutch weight will increase as the square of female length while in *Sceloporus* mean clutch weight will increase as the cube of female length. A linear regression on these data after logarithmic transformation supports the hypothesis (Fig. 9.1), the slope for *Anolis* being 2.11 (SE $= 0.184$, $r = 0.957$, $n = 14$, $P < 0.001$) and that for *Sceloporus* being 2.97 (SE $= 0.269$, $r = 0.961$, $n = 12$, $P < 0.001$).

9.1.2. Clutch Size and Constraints of Crevice Living

Living in trees may not be the only ecological niche that limits clutch size. Vitt (1981) argued that use of crevices to avoid predators such as snakes may have played an important role in the evolution of body and egg morphology and clutch size in the lizard, *Platynotus semitaeniatus*. To illustrate the peculiarities of this species Vitt compared its morphology and ecology to the "morphologically and reproductively typical sympatric iguanid lizard, *Tropidurus torquatus*" (Vitt 1981, p. 507). *P. semitaeniatus* has a "pancake" morphology, which enables it to enter very narrow crevices while *T. torquatus* is cylindrical, which prevents it, like most other lizards, from entering such crevices (Fig. 9.2). Unlike the vast majority of other iguanids, *P. semitaeniatus* lays clutches of 2 eggs (excluding the anoles, only 4 of 114 species of iguanids are reported to lay clutch sizes of 2). The eggs of *P. semitaeniatus* are also dissimilar from other iguanids in being more elongate (Fig. 9.2). Finally the relative clutch mass of *P. semitaeniatus*

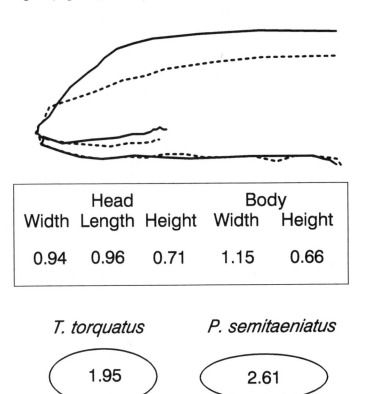

Head			Body	
Width	Length	Height	Width	Height
0.94	0.96	0.71	1.15	0.66

T. torquatus   1.95   *P. semitaeniatus*   2.61

Figure 9.2. Upper diagram shows silhouettes of the lizard *Tropidorus torquatus* (solid line) and the crevice-dwelling lizard *Platynotus semitaeniatus* (dashed line). Sizes standardized to equal snout–tympanum lengths. Modified from Fig. 1 of Vitt (1981). Central box: Relative ratios (*P. semitaeniatus/T. torquatus*) of body components in the two species. Ratios calculated after first dividing by snout–vent length for each species. Data from Table 1 of Vitt (1981). Note that the crevice-dwelling *P. semitaeniatus* differs from the more typical *T. torquatus* in being more flattened (two height ratios). Lower figure: Relative size and shape of the eggs of the two species. The ratios shown are length/width. Modified from Vitt (1981).

is significantly lower than that of *T. torquatus* (means of 0.195 and 0.258, respectively; $P < 0.01$).

Because it is difficult to experimentally manipulate the system, tests of Vitt's hypothesis must rely upon a comparative analysis. His hypothesis predicts that the same morphological modifications and clutch size should occur in unrelated genera provided they occupy the same ecological niche. It is therefore significant that crevice-dwelling lizard species of the African cordylid genus *Platysaurus* show modifications similar to those in *P. semitaeniatus*: the head and body are dorsoventrally compressed, clutch size is usually two, and the eggs are elongate. (Ratios of egg length to width for the *Platysaurus*, *P. semitaeniatus*, and *T. torquatus* are 2.13, 2.61, and 1.95, respectively; Vitt 1981.)

9.1.3. Clutch Size and Foraging Mode

A predator that actively searches for its prey is under two constraints to maintain a low clutch mass: first, the female may be impeded from catching its prey and, second, the female may be more vulnerable to predators because its escape speed is reduced. Gravid females of the Australian agamid species *Amphibolurus nuchalis* and the common lizard, *Lacerta vivipara*, show a significant reduction in running speed (Garland 1985; Van Damme et al. 1989), and in the former species preliminary data suggest reduced endurance (Garland and Else 1987). Shine (1980) demonstrated that in a laboratory setting gravid females of some Australian skinks are more likely than male skinks to fall prey to snakes. However, direct field evidence for a relationship between loading and vulnerability to predators is wanting. Field data on the stomach contents of horned adders (*Bitis caudalis*) suggest that widely foraging lacertids are more vulnerable than sedentary species (Huey and Pianka 1981), but this does not demonstrate that loading per se changes vulnerability. Bauwens and Thoen (1981) pointed out that a lizard can shift its foraging mode, supporting this hypothesis with data on the European lacertid, *Lacerta vivipara*. Gravid females of this species have reduced sprint speeds (Van Damme et al. 1989) but increase their reliance on crypsis when gravid (Bauwens and Thoen 1981).

Vitt and Congdon (1978) extended the concept of mechanical constraints by suggesting that lizards could be classified into two foraging groups, "sit-and-wait" predators versus "widely foraging" predators, and that lizards in the former category will experience a reduced "loading" constraint and hence will have larger relative clutch masses. McLaughlin (1989) examined the concept of a dichotomy in search patterns in both birds and lizards and found that species can be classified as "mobile" or "sedentary" on the basis of the frequency of moves, the former type moving approximately 7 to 10 times as often as the latter. The prediction of Vitt and

Congdon (1978) that the relative clutch mass of sit-and-wait lizards will be greater than that of widely foraging species is supported by data on relative clutch mass in 130 species of lizards (Fig. 9.3). But an important caveat must be appended: there is a high correlation between foraging mode and family (Table 9.1), and thus phylogeny cannot be excluded as a confounding factor. Of particular interest is the observation that the relative clutch mass of geckos, which are sit-and-wait predators, more closely resembles that of widely foraging lizards than other sit-and-wait foragers (Fig. 9.4; Vitt and Price 1982). This follows the prediction of Andrews and Rand (1974, see above) that arboreal lizards using adhesive disks will have relatively low clutch weights.

A detailed field study of the relationship between foraging type and relative clutch mass in four species of lizard was undertaken by Magnusson et al. (1985). Three of these species—*Cnemidophorus lemniscatus*, *Ameiva*

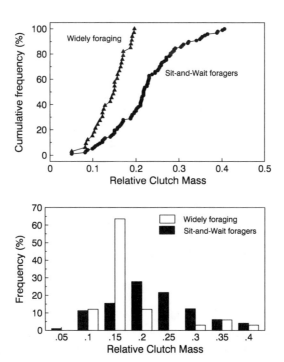

Figure 9.3. Distributions of the relative clutch masses of lizards classified according to foraging mode. Data from Huey and Pianka (1981), Vitt and Price (1982), and Annan'eva and Shammakov (1986). For widely foraging lizards, $n = 33$; for sit-and-wait lizards, $n = 97$.

Table 9.1.   Mean relative clutch masses according to foraging type and taxonomic family

Family	Sit-and-wait			Widely foraging		
	Mean	SE	*n*	Mean	SE	*n*
Iguanidae	0.25	0.008	61	—	—	0
Agamidae	0.23	0.018	10	—	—	0
Xantusiidae	0.18	0.034	2	—	—	0
Gekkonidae	0.13	0.010	21	—	—	0
Anguinidae	0.28	—	1	—	—	0
Lacertidae	0.20	0.016	2	0.16	0.008	9
Scincidae	—	—	0	0.31	0.036	5
Teiidae	—	—	0	0.15	0.007	19

Data are from Huey and Pianka (1981), Vitt and Price (1982), and Anan'eva and Shammakov (1986).

ameiva, and *Kentropyx striatus*—are widely foraging species; the last— *Anolis auratus*—is a sit-and-wait predator. The first three exhibit different rates of activity as measured by movements per hour, mean speed of movement, and area used (Fig. 9.5), demonstrating that within categories there is a continuum of rates of activity. As expected, *A. auratus* shows the lowest level of activity. The relative clutch masses of the three actively foraging species vary in the opposite direction to foraging rate (Fig. 9.5), as predicted by the loading-constraints hypothesis. The anole has a very low clutch mass, but this is typical of anoles and, as discussed above, may be a consequence of an arboreal habit.

If clutch mass impedes the movement of lizards, it should also have an effect on locomotor performance in snakes. Gravid female garter snakes, *Thamnophis marcianus*, show a reduced locomotor performance relative to nongravid females, and this effect increases with relative clutch mass (Seigel et al. 1987). But another factor that may be important is the mode of reproduction, viviparity versus oviparity. The risk of being preyed upon will be a function of the relative clutch mass and the period over which this is maintained, and hence viviparous snakes may be at greater risk than oviparous snakes (Tinkle and Gibbons 1977). Seigel and Fitch (1984) examined the relationship between relative clutch mass, reproductive mode, method of predator escape ("flee" versus "crypsis"), prey location ("active foraging" versus "sit-and-wait"), and prey capture ("pursue" versus "ambush"). They reported a significant effect of reproductive mode (Fig. 9.6), but not foraging mode or escape behavior. The lack of statistical significance may be a consequence of misclassification of the behavior of snakes since these are more difficult to observe than lizards, or it may mean that

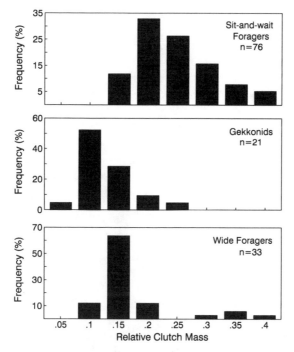

Figure 9.4. Distributions of relative clutch masses of lizards divided into three categories: sit-and-wait foragers (excluding gekkonids), gekkonids, and wide foragers. Data sources as given in Fig. 9.3.

snakes do not fit easily into dichotomous groups (Seigel and Fitch 1984). Further field studies are required to assess these possibilities.

Seigel and Fitch (1984) reported two interesting differences in the relative clutch masses of snakes and lizards: first, snakes have higher relative clutch masses than lizards, and second, unlike snakes, there is no significant difference in relative clutch mass between viviparous and oviparous lizards. They suggested that the first result may be a consequence of higher energetic costs of locomotion in lizards and that the second may be a consequence of most viviparous lizards being sit-and-wait foragers.

A further "twist" on the mechanical-constraints hypothesis in snakes was suggested by Shine (1988) in an analysis of relative clutch masses in terrestrial and aquatic snakes. Biomechanical arguments predict that aquatic locomotion in snakes will be best achieved by a laterally compressed body form, as is observed in a number of aquatic groups (e.g., hydrophiids, laticaudids, acrochordids, estuarine natricine colubrids). On this basis Shine

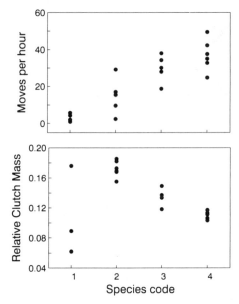

Figure 9.5. Upper panel: Measure of activity in four species of lizard. 1, *Anolis auratus*; 2. *Kentropyx striatus*; 3, *Cnemidophorus lemniscatus*; 4, *Ameiva ameiva*. Lower panel: Relative clutch masses in the four species. Figures redrawn from Magnusson et al. (1985).

predicted that aquatic snakes would have lower relatively clutch masses than terrestrial snakes, for in the former group eggs will tend to make the body more cylindrical and hence less hydrodynamically efficient. One method of reducing the mechanical problems so induced is to place the eggs at a relatively anterior part of the body (Shine 1988). Aquatic snakes do indeed have lower relative clutch masses than terrestrial snakes, and while the position of the anterior portion of the clutch is not shifted forward, the total length occupied by the clutch is less, and thus the center of gravity of the clutch as a whole is shifted anteriorly as predicted.

9.2. Hypotheses Generated by Considerations of Clutch Size in Endotherms

Birds and mammals expend reproductive effort in producing a clutch of young and, in most cases, further effort in rearing them to an age at which they can forage for themselves. A priori we might expect that the maximum clutch size will be determined by the ability of parents to supply food to their offspring and maintain their own condition. To investigate this hypothesis Bryant and Westerterp (1983) empirically estimated the net energy balance for house martins, *Delichon urbica*, by rearing broods of different sizes. Under either favorable or unfavorable conditions the parents have a positive net energy balance for broods sizes from one to five, but thereafter they must operate under a negative balance. Thus broods of six and

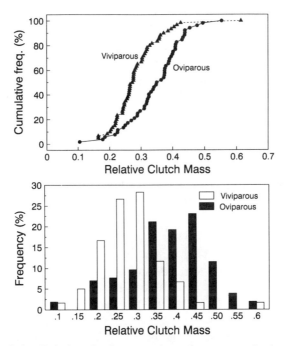

Figure 9.6. Relative clutch masses in snakes, categorized according to reproductive mode. Data from Seigel and Fitch (1984) and Seigel et al. (1986). For oviperous snakes, $n = 52$; for viviparous, $n = 60$.

above would be undernourished unless the parent subsidized chick intake by drawing on its own body reserves. Such a subsidy does not appear to be a normal pattern (Bryant 1979), and hence the above analysis predicts that the maximum brood size should be five, which indeed it is (Bryant 1975). To further test this prediction Bryant and Westerterp increased some broods to six or seven young. Survival of young dropped from 94%–96% in unmanipulated clutches to 88.1% (brood size = 6) and 92% (brood size = 7), and the average weight in enlarged broods declined significantly. Further, the frequency of very light young that would have poor survival chances increased significantly (Bryant and Westerterp 1983).

The breeding period of most organisms is generally restricted to a specific period in the year when the climate and food conditions are favorable for the rearing of young. This limits the maximum number of clutches that can be raised and may place an upper limit on clutch size. The rationale underlying the latter hypothesis is the observation that in some broods the energetic costs of rearing offspring increase with brood size, thereby increasing the period of recuperation the parent requires before starting a new brood (Smith and Roff 1980; Tinbergen 1987). Consequently, increasing brood size may decrease the total number of broods that can be raised in a season and reduce the total annual production of offspring. The possible importance of the length of the breeding season can be explored using the model of Smith and Roff (1980). The details of the model, as applied to the great tit, are as follows (specific references for the relevant data are given in Smith and Roff):

1. In the great tit, as with many bird species, only one egg is laid per day and incubation does not start until the penultimate egg is laid; thus the time (in days) required to raise a single clutch of great tits is $46.3 + x$ where x is clutch size.

2. Interbrood interval increases linearly with clutch size according to the relationship $1.11x - 2.2$.

3. Taking into account the time required from laying to fledging the total time required to raise n broods of size x, $T(x,n)$ is

$$T(x,n) = (n - 1)(29.1 + 2.11x) + (46.3 + x) \qquad \textbf{(9.1)}$$

The first term represents the time required to raise the first $n - 1$ broods and the second term of time required to raise the final brood. The highest number that can be raised given the observed season length of 125 days is two clutches of 15 eggs per clutch. If variation in season length is taken into account the optimal breeding pattern is two clutches of 14 eggs per clutch (Smith and Roff 1980). These values are the same as the maximum observed clutch size of 14–15 (Boyce and Perrins 1987). Thus, in the great tit, time and energy constraints appear to limit maximum clutch size. Additional factors acting within the breeding season will reduce the optimum clutch size below this ceiling. (The observed yearly averages are, in fact, 8.0–12.3, Lack 1966.) It is to an analysis of these factors that we now turn.

9.2.1. Optimal Clutch Size: A Demographic Perspective

Moreau (1944), after examining comparative data on clutch size in birds, came to the view that clutch size and number of clutches per season are the result of natural selection acting to maximize the rate of increase. He noted (p. 309) that

> it is far from certain that the bigger clutch is always more to the 'good of the species'—which I take to be the maintenance or increase of the population. As we have proved . . . the amount of food brought to the nest does not increase in proportion to the number of young; so that the

members of B/4 (brood size of 4) might have a significantly better start in life than members of B/5 . . . a greater abundance of helpless or inexperienced young may induce, not a proportionately, but a disproportionately, greater attention from predators; and in a climate that is uncertain the effects of a bad season might be more disastrous on bigger broods that were adapted in size to the food supply of the best seasons.

Furthermore, a large clutch may stress the parents and decrease their expected production:

It is possible, for example, that B/5, at least in some circumstances, might put a significantly greater strain on the parents than B/4, so that they were prevented from raising a larger total number as the product of the smaller broods in the same season; or that a succession of B/5 would so shorten the reproductive lives of the parents that their total of off-spring, produced in smaller, less exacting broods, would be greater.

Evidence from manipulations of clutch or brood size in birds indicates that larger clutches or broods lead to increased number of fledglings (Lessels 1986; Ydenberg and Bertrand 1989; Dijkstra et al. 1990; Tables 9.2 and 9.3). However, we must consider not only the number fledged but also their subsequent survival, which has not been well studied. Nevertheless, some general patterns emerge from Table 9.2. First, although the enlarged clutches typically produce an increased number of fledglings, most studies also demonstrate that mass at fledging and survival to fledging are reduced (Table 9.3). In six of 12 studies the later survival of the fledglings from enlarged broods was reduced. Given the difficulties of detecting a change in survivorship (see chapter 6, section 6.2.2) this percentage suggests that a decrease in survival is a typical phenomenon. Enlarged clutches produce more offspring but these are typically underweight at fledging (Table 9.3), which probably reduces their chances of survival, at least if they fledge late in the season and conditions during the winter months are more severe than average. This hypothesis is further supported by a significant association between mortality from hatching to fledging (S_1) and mass at fledging (M, Table 9.3, $\chi^2 = 5.21$, $P < 0.05$). It is reasonable to suppose that such detrimental effects will continue for a period following fledging.

In assessing the fitness increments associated with enlarged clutches we must also take into account the survival of parents and size of future clutches. Experimental manipulations of clutch size have shown that survival and/or future brood sizes can be reduced in birds. (See Tables 6.5 and 6.8.) In only a few cases have all, or at least the majority of, components, relating to optimal clutch size been evaluated (Table 9.4). Of eight studies that attempted to compare the effects of enlarged clutches on both parents and offspring, five have sample sizes less than 100. As

Table 9.2. Effects of artificially increasing clutch or brood size on the number of fledglings (N), the body mass of fledglings (M), the survival to fledging (S_1) and survival of offspring until fall or next breeding season (S_2). Also shown is the typical clutch size (C) and the size of the manipulated clutches (Exp)

Species	C	Exp	N	M	S_1	S_2	Reference
Diomedea immutabilis (albatross)	1	2	−	−	−	?	1
Oceanodroma castro (stormpetrel)	1	2	−	?	−	?	2
Oceanodroma leucorrha (stormpetrel)	1	2	?	−	−	?	3
Puffinus puffinus (shearwater)	1	2	−	−	−	−	4
Puffinus puffinus (shearwater)	1	2	+	−	0	0	5
Puffinus tenuirostris (shearwater)	1	2	?	−	−	?	6
Sula sula (gannet)	1	2	−	?	−	?	7
Sula bassana (gannet)	1	2	+	?	−	?	8
Sula bassana (gannet)	1	2	+	−	0	?	9
Sula capensis (gannet)	1	2	+	−	?	−	10
Sula dactylatra (booby)	2	2[a]	?	?	−	?	11
Sula leucogaster (booby)	2	2[a]	?	?	−	?	11
Creagrus furcatus (gull)	1	2	+	0	0	0	12
Fratercula arctica (puffin)	1	2	+	?	−	?	13
Fratercula arctica (puffin)	1	2	−	?	−	?	14
F. cirrhata, F. corniculata	1	2	+	−	?	?	15
Alca torda (razorbill)	1	2	+	−	−	?	16
Alca torda (razorbill)	1	2	+	−	−	?	17
Cepphus grylle (guillemot)	1–2	+1	+	0	0	?	18
Cepphus columba (guillemot)	2	3	+	−	−	?	19
Stercorarius longicaudus (skua)	2	3	+[b]	?	0	?	20
Rissa tridactyla (kittiwake)	2	4	+	−	0	?	21
Larus argentatus (gull)	3	±2	+	0	0	?	22
Larus glaucescens (gull)	3	4–6	+	−	?	?	23
Larus glaucescens (gull)	3	±1–3	+	?	0	0	24
Larus fuscus (gull)	3	+1–3	+[c]	0	0	?	25
Phalacrocorax auritus (cormorant)	4	5–8	+	?	−	?	26
Phalacrocorax pelagicus (cormorant)	4	5–7	−	?	−	?	26
Necrosyrtes monachus (vulture)	1	2	0	−	−	?	27
Accipiter rufiventris (sparrowhawk)	3	+2	?	−	−	?	28
Falco tinnunculus (kestrel)	5	±2	+	−	−	0	29
Aegolius funereus (owl)	5–6	±1	+	0	0	?	30
Columba palumbus (woodpigeon)	2	1, 3	+	−	−	0	31
Apus apus (swift)	2–3	4	−	?	−	?	32
Aerodramus spodiopygius (swiftlet)	1–2	1, 3	0	−	−	?	33
Delichon urbica (martin)	2–5	5–6	+	0	0	?	34
Delichon urbica (martin)	2–5	5–7	+	−	−	?	35
Iridoprocne bicolor (swallow)	5–7	+2	+	−	0	0	36
Troglodytes aedon (wren)	4–8	>+1	+	0	0	0	37
Ficedula hypoleuca (flycatcher)	5–7	9	0	−	−	?	38[d]
Ficedula hypoleuca (flycatcher)	5–7	±1	+	−	0	0	39
Ficedula albicollis (flycatcher)	6–7	±1–2	+	?	?	−	40
Turdus pilaris (fieldfare)	5	±1–2	+	0	0	?	41

Table 9.2. (*Continued*)

Species	C	Exp	N	M	S_1	S_2	Reference
Parus major (tit)	5–13	±4–5	+	–	–	–	42
Parus major (tit)	5–13	5–13[e]	+	–	–	–	43
Parus major (tit)	5–13	±4–5	+	–	?	?	44
Parus caeruleus (tit)	3–15	3–15[e]	+	–	0	?	45
Plectrophenax nivalis (bunting)	3–5	4–8	+	–	0	?	46
Agelaius phoeniceus (blackbird)	3–4	5–6	+	–	–	?	47
Pyrrhula pyrrhula (bullfinch)	4–5	6–7	–	?	–	–	48
Passer domesticus (sparrow)	3–5	+1	+	–	–	?	49
Passer domesticus (sparrow)	3–5	±2	+	–	0	?	50
Quelea quelea (dioch)	3	±1–2	0	0	–	?	51
Sturnus vulgaris (starling)	3–7	3–10	+	–	?	?	52
Sturnus cinerareus (starling)	3–7	9	+	?	?	?	53
Pica pica (magpie)	5–8	±1–2	–	0	–	?	54
Corvus corone (crow)	2–6	±1–2	+	0	–	?	55
Corvus frugilegus (rook)	4–5	4	+	?	?	?	56
Branta canadensis (goose)	2–7	2–7[f]	+	0	–	?	57
Calidris pusilla (sandpiper)	4	5	–	0	0	?	58

A " + " indicates a positive relationship, a " – " a negative relationship, a "0" no significant effect, and "?" not measured.

References: 1. Rice and Kenyon 1962; 2. Harris 1969; 3. Lack 1966; 4. Harris 1966; Perrins et al. 1973; 5. Perrins et al. 1973; 6. Norman and Gottish 1969; 7. Nelson 1966a,b; 8. Nelson 1964; 9. Wanless 1984; 10. Jarvis 1974; 11. Dorwood 1962; 12. Harris 1970; 13. Corkhill 1973; 14. Nettleship 1972; 15 Wehle 1983; 16. Lloyd 1977; 17. Plumb 1965; 18. Asbirk 1979; 19. Koelink 1972; 20. Andersson 1976; 21. Lack 1966; 22. Haymes and Morris 1977; 23. Ward 1973; 24. Vermeer 1963; 25. Harris and Plumb 1965; 26. Robertson 1971; 27. Mundy and Cook 1975; 28. Simmons, R. 1986; 29. Dijkstra et al. 1990; 30. Korpimäki 1987, 1988; 31. Murton et al. 1974; 32. Perrins 1964; 33. Tarburton 1987; 34. Bryant 1975; 35. Bryant and Westerterp 1983; 36. DeSteven 1980; 37. Finke et al. 1987; 38. Askenmo 1977; 39. Von Haartman 1954, 1967; 40. Gustafsson and Sutherland 1988; 41. Slagsvold 1982a; 42. Smith et al. 1989; 43. Tinbergen 1987, Tinbergen and Boerlijst 1990; 44. Lindén 1988; 45. Nur 1984a,b, 1986, 1988a,b; 46. Hussell 1972; 47. Cronmiller and Thompson 1980; 48. Lack 1966; 49. Schifferli 1978; 50. Hegner and Wingfield 1987; 51. Ward 1965; 52. Crossner 1977; 53. Kuroda 1959; 54. Högstedt 1980; 55. Loman 1980; 56. Røskaft 1985; 57. Lessells 1986; 58. Sanfriel 1975.

Notes:

[a]One chick added to nests with only one chick. The smaller of the pair in all cases died within a few weeks.

[b]Results are for added chicks, but preliminary data suggests that birds cannot incubate two eggs.

[c]Parents cannot protect young in bad weather which can lead to catastophic mortality of enlarged broods.

[d]Fecundity of offspring also reduced.

[e]For each experimental group three clutches of equal size chosen; one clutch was halved and the chicks removed placed with the second nest, the third serving as a control.

[f]Random assignment of clutches.

Table 9.3.   Summary statistics from Table 9.2

Response	N	M	S_1	S_2
−	10	31	33	6
0	4	13	18	8
+	41	0	0	0

	Mass at fledging, M	
Number at fledging, N	−	0
−	2	2
0	3	1
+	23	10

	Mass at fledging, M	
Survival to fledging, S_1	0	−
0	9	8
−	4	18

N = number at fledging: M = mass at fledging: S_1 = survival to fledging: S_2 = survival to fall or next breeding season.

discussed in chapter 6 (section 6.2.2) such sample sizes are inadequate to correctly assess the impact of enlarged clutches on survival rates. In the three studies in which sample sizes are greater than 100 there are adverse effects of clutch enlargement on both parents and offspring.

Studies of natural variation in litter size among small mammals raised in the laboratory show, with increasing litter size, a reduced weight at birth and weaning, a lower growth rate, and a lower survival to weaning (Table 9.5). In addition, time to weaning and maturity may be increased (Cameron 1973; Lackey 1976; Konig and Martl 1987), and fertility of adults decreased (Machin and Page 1973). The few studies of free-ranging mammals report similar findings (Millar 1973; Morris 1986; Boutin et al. 1988). Among white-tailed deer (Ransom 1967) and rabbits (Poole 1960) some rather scanty evidence indicates that prenatal losses increase with litter size. Because in most of the small-mammal studies variation in litter size was natural and not the result of manipulation, it is not possible to test the hypothesis that a female could raise more young than she actually bore. In the majority of bird studies the total number of fledglings in the artificially enlarged clutches was increased above that of control clutches (Tables 9.2, 9.3). Since the later survival of offspring and parents is generally unknown, the selective advantage or disadvantage of the enlarged clutch cannot be estimated. In at least some cases the future survival of parents is reduced.

Table 9.4. Summary of clutch enlargement experiments in which effects on both offspring and parents have been measured

Species	n	Offspring				Parents			Reference
		N	M_0	S_1	S_2	M_p	S_p	C_p	
House sparrow	19	+	−	0	?	0	0	−	1
Swallow tailed gull	30	+	0	0	?	0	0	?	2
Tree swallow	41	+	−	0	0	0	0	0	3
Tengmalm's owl	65	+	0	0	?	0	0	0	4
Kestrel	72	+	−	−	0	−	−	0	5
Pied flycatcher	111	0	−	−	?	−	−	?	6
Great tit	147	+	−	−	−	0	0	−	7
Blue tit	216	+	?	0	0	−	−	0	8

Studies arranged in ascending order of sample size (n) used to detect differences in adult survival. N = number fledging; M_0 = mass of fledgling; S_1 = survival to fledging; S_2 = survival of fledgling until fall or next breeding season; M_p = mass of parent; S_p = survival of parent; C_p = subsequent clutch size of parent.

References 1. Hegner and Wingfield 1987; 2. Harris 1970; 3. DeSteven 1980; 4. Korpimäki 1988; 5. Dijkstra et al. 1990; 6. Askenmo 1977, 1979; 7. Tinbergen 1987; Tinbergen and Boerlijst 1990; and unpublished data cited by Dijkstra et al. 1990; 8. Nur 1984ab, 1986, 1988a,b.

(See Table 6.5.) These data support Moreau's general hypothesis that offspring and/or parents from larger broods may be at a disadvantage, but they do not address the more important question of what the optimal brood size should be. It is to this question that we now turn.

9.2.1.1. Lack's Hypothesis

In a seminal paper, Lack (1947) clearly laid out the problem and his hypothesis:

> if clutch-size is inherited *and if other things are equal* [my italics], those individuals laying larger clutches will come to predominate in the population over those laying smaller clutches. It is easy to see why a species which normally lays four eggs and raises four young should not, instead, lay only three eggs. The difficult problem is to discover why such a species should not normally lay five eggs. I believe that, *in nidicolous species, the average clutch-size is ultimately determined by the average maximum number of young which the parents can successfully raise in the region and season in question* [*Lack's italics*], i.e. that natural selection eliminates a disproportionately large number of young in those clutches which are higher than the average, through the inability of the parents to get enough food for their young, so that some or all of the brood die before

Table 9.5. Effects of litter size on mass at birth (M_b), mass at weaning (M_w), growth rate (G), and survival to weaning (S) in small mammals

Species	M_b	M_w	G	S	Reference
Peromyscus leucopus	−	0	?	?	1
Peromyscus leucopus	0	−	−	−	2[a]
Peromyscus maniculatus	−	?	?	−	3
Peromyscus maniculatus	?	0	0	?	4
Peromyscus melanocarpus	−	−	−	?	5
Peromyscus yucatanicus	−	−	−	?	6
Peromyscus polinotus	−	−	−	−	7[b]
Mus musculus	?	−	−	?	8[a]
Mus musculus	−	−	−	?	9
Mus musculus	?	−	?	?	10[a]
Mus domesticus	?	−	−	?	11
Sigmodon hispidus	0	−	−	−	12
Sigmodon hispidus	?	?	−	?	13
Sigmodon hispidus	−	?	?	?	14
Microtus californicus	0	−	?	−	15
Microtus pennsylvanicus	0	?	0	?	16
Microtus ochrogaster	0	?	?	?	17
Phenacomys longicaudus	−	−	−	?	18
Dicrostonyx groenlandicus	0	−	−	?	19
Neotoma lepida	?	0	−	−	20
Clethrionomys gapperi	0	0	−	?	21
Cavia porcellus	−	−	−	−	22
Dipodomys deserti	?	0	0	?	23

Except where indicated data based on laboratory rearings using natural variation in litter size. A " − " indicates a negative relationship, a "0" no significant effect, and "?" not measured.

References: 1. Millar 1975, 1978; Hill 1972; 2. Fleming and Rauscher 1978; 3. Myers and Master 1983; 4. Linzey 1970; 5. Rickart 1977; 6. Lackey 1976; 7. Kaufman and Kaufman 1978; 8. Smith and McManus 1975; 9. Parkes 1926; 10. Machin and Page 1973; 11. Konig and Markl 1987; 12. Randolph et al. 1977; 13. Mattingly and McClure 1987; 14. Kilgore 1970; 15. Krohne 1981; 16. Innes and Millar 1979; 17. Kaufman and Kaufman 1987; 18. Hamilton 1962; 19. Hasler and Banks 1975; 20. Cameron 1973; 21 Innes and Millar 1979; 22. Wright and Eaton 1929 (cited from Mountford 1968; Lack 1948); 23. Butterworth 1961.

Notes:

[a]Litter size manipulated.

[b]Data comprise both natural variation and artificially induced variation in litter size.

or soon after fledging, with the result that few or no descendants are left
with their parent's propensity to lay a larger clutch. (Lack 1947, p. 319)

Lack considered that this hypothesis also applied to nidifugous birds with
the possible exception of some gulls where the number of eggs might be
constrained by the size of the brood patch (Lack 1947, p. 326). In a later
paper (Lack 1948) he extended the hypothesis to mammals. The emphasis
of Lack's hypothesis is on the survival of offspring, not parents: this is
made clear in his book, *The Natural Regulation of Animal Numbers*, first
published in 1954, in which he states (p. 22 of the 1967 edition): "in most
birds clutch-size has been evolved through natural selection to correspond
with the largest number of young for which the parents can on average
find enough food."

Lack's hypothesis assumes that the only important interactions are negative density-dependent interactions between siblings within a clutch. It is
not a hypothesis restricted to birds and mammals and does not require
that parental care be given after the eggs are laid. In fact it will be strictly
true when the parent lays a single isolated clutch and dies immediately
afterward, for the female then must only consider the effects of interactions
between siblings. (See section 9.3.1.1 for a detailed mathematical treatment
of this.) If more than one female deposits her clutch at a particular site
the optimal clutch size will be a function of the number of females per site
(section 9.3.2.1).

Lack's hypothesis predicts that the most productive clutch should also
be the most frequent clutch observed. Klomp (1970), in his review of the
hypothesis, noted that the most productive brood size was similar to the
most frequent in 10 species but larger in 11. A present survey of the data
from manipulation experiments gives equally equivocal results: while the
number of fledged chicks typically increases with clutch size (Table 9.3),
decreased fledging weight suggests that offspring mortality may also increase with clutch size. But even if the survival of chicks is not affected
there remains a second problem with Lack's hypothesis. The problem, as
pointed out by Williams (1966) and Klomp (1970), is that Lack's hypothesis
ignores adult survival. Moreau (1944) in the quotation given above (section
9.2.1) clearly outlines the possible importance of adult survival. Lack had
not overlooked the question of adult survival in relation to clutch size: he
addressed the issue in his 1966 book, *Population Studies of Birds*, and with
respect to the great tit concluded that adult survival is not compromised
by increased clutch size. But we now have evidence that this may not be
a typical finding (chapter 6, Table 6.5).

9.2.1.2.   Effects of Trade-Off Between Clutch Size and Parental Traits

It is intuitively clear that if parental survival declines with clutch size,
the optimal clutch will be smaller than the most productive clutch. Since

a reduced future fecundity is mathematically equivalent to a decreased survival the same argument extends to the effect of clutch size on future fecundity. The effect of reduced survival was formally demonstrated by Charnov and Krebs (1974) with the following simple model. (For a more detailed theoretical model relating fecundity and survival in birds see Ricklefs 1977b.) Recall (chapter 8, equations 8.1–8.4) that if survival rates are constant and reproduction occurs at age 1, the rate of increase of an iteroparous population is given by

$$\lambda = S(x) + xs(x) \qquad (9.2)$$

where $s(x)$ is the survival rate for the clutch for the first year and declines with clutch size x,

$S(x)$ is the yearly survival rate of parents, which also declines with clutch size. As noted above, $S(x)$ could equally well apply to future fecundity. Further note that the survival of parents does not require that the parental care be provided after birth: it might equally well apply to provisioning provided before birth, as occurs in solitary bees and dung beetles. The increase in mortality might even be a result of the stress of producing more eggs irrespective of any parental care. The particular model does assume that the interval between reproductive episodes is equal to the time to maturity, but this is a convenient mathematical detail that does not affect the general conclusion.

What is the most productive clutch size? Letting this be Y we have (Fig. 9.7)

$$Y = xs(x) \qquad (9.3)$$

To find the most productive clutch size, x^*, we differentiate with respect to x (dY/dx) and set the differential equal to zero (upper right panel, Fig. 9.7). To find the optimal clutch size we differentiate λ with respect to x and set this differential equal to zero

$$\frac{d\lambda}{dx} = \frac{dS(x)}{dx} + \frac{dY}{dx}$$

$$= 0 \qquad \text{when} \quad \frac{dS(x)}{dx} = -\frac{dY}{dx} \qquad (9.4)$$

Now note that

1. $dS(x)/dx$ is negative for all x (lower left panel, Fig. 9.7).
2. dY/dx is positive when x is less than x^* and negative when x is greater than x^* (upper right panel, Fig. 9.7).

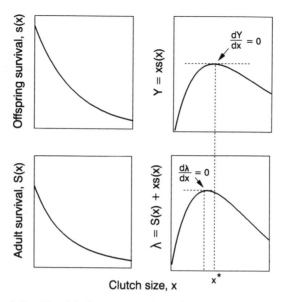

Figure 9.7. Graphical representation of the mathematical argument that if parental survival or future fecundity is a function of clutch size the optimal clutch size will be less than the most productive clutch size. See text for details.

Thus equation 9.4 will only be satisfied when x is less than x^*. A negative relationship between clutch size and parental survival and/or future fecundity favors an optimal clutch size below that which produces the most offspring. The Lack value will be the optimal clutch size when the only important factors are negative density-dependent interactions within the clutch: the incorporation of additional factors in the life history will almost certainly reduce the optimal clutch size. Two such factors are decreased adult survival and fecundity; three others, described below, are inherent variation in clutch size, environmental variation, and nest failure.

9.2.1.3. Effect of Inherent Variability in Clutch Size

The above solution to the observation that the most frequent clutch size is less than the most productive one is based on the premise that each genotype produces a fixed clutch size. Mountford (1968) proposed a solution based on the alternative premise that any given genotype will inevitably produce a range in clutch sizes even in a constant environment. Under this hypothesis the optimal clutch size is due to an interaction between the frequency distribution of clutch size and the survival rate of offspring. Suppose each genotype produces some frequency distribution,

$f(x)$, of litter sizes (top panel, Fig. 9.8), genotypes differing in their most frequent clutch size, x_f. Survival to independence is assumed to be a monotonically decreasing function, $s(x)$ of litter size (middle panel, Fig. 9.8). The number surviving from a clutch of size x is $xs(x)$ (bottom panel Fig. 9.8), and the expected number of offspring from a clutch size of x from a particular genotype is $f(x)xs(x)$. The expected number of surviving offspring from a genotype is $\Sigma f(x)xs(x)$. For the particular genotype shown in Fig. 9.8 the most frequent clutch size is 5, from which 4.88 individuals survive; the most productive clutch size is 7, giving 5.74 surviving offspring. In total this genotype produces, on average, 4.66 surviving offspring. Fig. 9.9 shows the expected offspring production for different values of x_f, assuming that the frequency distribution remains the same, simply being shifted to the right or left. The optimal clutch size is 5, from which 4.66 offspring can be expected, taking into account that genotypes producing this clutch size as their most frequent also produce a range in clutch sizes. (See upper

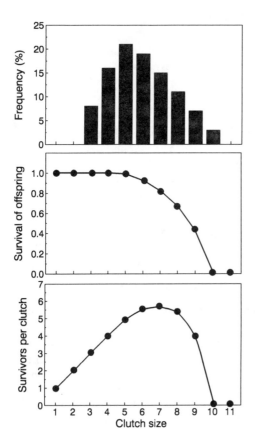

Figure 9.8. Litter-size statistics for laboratory guinea pigs (from Table 1 of Mountford 1968). Upper panel: Frequency distribution, $f(x)$, of litter sizes. Middle panel: Probability of survival to weaning, $s(x)$. Lower panel: Mean number of survivors per litter, $xs(x)$.

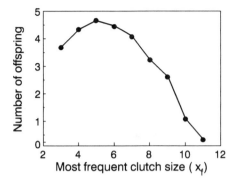

Figure 9.9. The expected number of offspring as a function of the most frequent clutch size. The distribution of clutch sizes for a given genotype is assumed to have the same shape as that shown in Fig. 9.8, but shifted right or left according to its modal clutch size.

panel Fig. 9.8.) The most productive clutch size (i.e., that clutch from which the most number of offspring survive) is 7 (bottom panel, Fig. 9.8), but genotypes which have their most frequent clutch size (x_f) at 7 will produce in total only 4.09 young. Depending on the particular distributions and survival functions chosen, the optimal clutch size may be equal to, larger than, or smaller than the most productive. The important message is that if genotypes produce a range of clutch sizes the optimal one cannot be estimated without taking this fact into consideration.

9.2.1.4. Influence of Environmental Variation

Another reason why the optimal clutch size may not be the most productive clutch size is that there may be year-to-year variation in the most productive clutch size. If a female cannot predict whether a year is going to be a "good" or "bad" year, selection may favor an optimal clutch size that is less than the most productive clutch size in an "average" year if larger clutches suffer very high mortalities in "bad" years. In a variable environment the appropriate measure of fitness is not the arithmetic mean clutch size but the geometric mean, which is weighted very strongly by years of low productivity. Boyce and Perrins (1987) used the geometric mean as the measure of fitness in their analysis of the influence of annual fluctuations in success rate in the great tit population of Wytham Wood (Oxfordshire, UK). In this population the average clutch size is 8.53 whereas the overall average number of offspring surviving 1 year is highest for females laying a clutch of 12 (Fig. 9.10). Can this difference be accounted for by variability in productivity across years?

Given that the mean generation time of great tits is 1 year, the number of offspring produced by a phenotype that matures and produces a clutch of size x in year t is

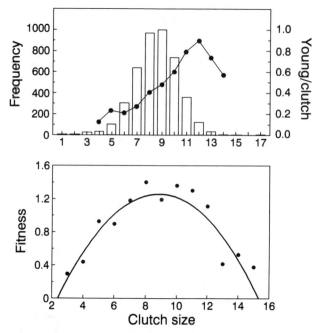

Figure 9.10. Upper panel: Frequency distribution of clutch size for 4,489 great tits in Wytham Wood over the years 1960–1982. Also shown is the number of young per clutch size that survived at least 1 year. Lower panel: Geometric mean relative fitness as a function of clutch size for the years 1960–1982. A quadratic fit through the data points indicates an optimum clutch size of about nine, corresponding quite closely to the observed mean clutch size of 8.53. Figures redrawn from Boyce and Perrins (1987).

$$R_0(x,t) = N(x,t) + \sum_{i=t+1}^{t+5} \left(N(x',i) \prod_{j=t}^{i-1} S(j) \right) \qquad (9.5)$$

where $R_0(x,t)$ is lifetime productivity of a phenotype producing a clutch of size x in year t,

$N(x,t)$ is number of young surviving per clutch of size x in year t,

$N(x',i)$ is number of young surviving per clutch of size x' in year i. A female that lays a clutch of size x in year t will not necessarily lay the same-sized clutch in subsequent years. For all years subsequent to t clutch size is determined by the repeatability of the trait, R: hence,

$$x' = R(x - \hat{x}) + \hat{x} \qquad (9.6)$$

where $\hat{x}$ is the mean clutch size (Bulmer 1985a). Clutch-size repeatability for tits within Wytham Wood is 0.51 (Perrins and Jones 1974). Across years, the parameter x refers not to a particular clutch size per se but to a particular phenotype: the actual clutch size laid is determined by equation 9.6. Following Boyce and Perrins, I shall refer to this phenotype as "clutch-size phenotype x." At evolutionary equilibrium $x = \hat{x}$, and the clutch size laid is x.

$S(j)$ is the probability of an adult female surviving from year j to year $j + 1$; the product is thus the probability of surviving to year i.

The summation is taken from $t + 1$ to $t + 5$ on the assumption that an adult female lives at most 5 years. Clutch size is assumed not to influence adult survival. (The data on this are equivocal: Boyce and Perrins make the conservative assumption of no effect.) Geometric mean fitness of clutch-size phenotype x, $G(x)$, is given by

$$G(x) = \left(\prod_{t=1}^{T} \sum \frac{R(x,t)}{\hat{R}(t)} \right)^{1/T} \tag{9.7}$$

where T is number of years considered and $\hat{R}(t)$ is the mean expected lifetime productivity of all clutch-size phenotypes in year t. The fitness of different clutch-size phenotypes was computed for the years 1960 to 1982 (Fig. 9.10). A quadratic fit to the data gives an optimal clutch size of 9.01, which is not significantly different from the observed clutch size of 8.53 ± 1.79. A clutch size of nine also corresponds to the observed modal clutch size (Fig. 9.10). These results support the hypothesis that in the great tit population of Wytham Woods the optimal clutch size is determined by year-to-year variation in the functional relationship between survival of young and clutch size. This does not mean that the optimal clutch size will be the same for all individuals within the population. As demonstrated by the experiments of Högstedt (1980) on magpies, and less convincingly, by those of Pettifor et al. (1988) on great tits (discussed in detail in 9.2.3), the optimal clutch size may vary according to parental quality. The analysis of Boyce and Perrins averages across these differences.

Millar (1973) observed that although pikas, *Ochotona princeps*, produce three young, only two typically survive, and suggested that the third offspring might be an "insurance policy" against the loss of a young. Provided the initial cost of producing the third offspring is not too high the female may also benefit if the season is particularly favorable and she can rear all three. Thus the production of excess young may be an adaptive response to a variable environment. Brood reduction has been most commonly reported in bird species, generally being a result of starvation in birds other than raptors (Howe 1976, 1978; O'Connor 1978; Horsfall 1984; Mock et al. 1990). Siblicide is most frequent among raptors (Stinson 1979) and birds nesting in large colonies (Mock et al. 1990).

O'Connor (1978) analyzed the circumstances under which brood reduction should occur: his analysis suggested that selection should favor first siblicide, followed by infanticide, and finally suicide. It is difficult to conceive how suicidal behavior might evolve, and none has been detected, though O'Connor's theory did predict it in some species. However, the theory was successful in predicting the occurrence of brood reduction (predicted in 17 cases, observed in 16; predicted not to occur in 6 cases and observed not to occur in 5). But the question of whether the birds would have had higher fitness by laying a smaller clutch was not addressed, as is true for other theoretical treatments of siblicide (Stinson 1970; Mock and Parker 1986; Parker et al. 1989; Godfray and Harper 1990). The basic lack of field data presently impedes such extensions (Mock et al. 1990). Analysis of siblicide in egrets and great blue herons failed to show an increase in fitness with brood reduction (Mock and Parker 1986).

9.2.1.5. Effects of Nest Failure

It has been postulated that the likelihood of nesting failure due to predation on nests or inclement weather may limit clutch size (Lack 1954, 1968; Cody 1966, 1971; Ricklefs 1969; Klomp 1970; Perrins 1977; Slagsvold 1982b). Losses due to these sources can be significant both during the egg and nestling stages (Table 9.6). Among 63 species of New World passerines the proportion of offspring lost to nest predators ranges from 10% to 90% (Kulesza 1990).

Predation may increase on larger clutches for two reasons: first, larger broods may be noisier and more likely to attract predators, and second, larger broods require more frequent feeding trips to and from the nest and hence may be likely to be traced by an observant predator. Based on

Table 9.6. Causes of mortality (% losses) in the young of six passerine species

Source of mortality	Eggs	Hatched young
Hatching failure	12.0	—
Cowbird parasitism	2.4	0.4
Nest-site competition	4.2	1.1
Adult death	3.5	—
Desertion	6.7	2.1
Predation	54.9	65.8
Weather	6.2	6.9
Other	10.2	17.5

Table modified from Table 1 of Ricklefs (1969).

different models, Ricklefs (1977a) and Perrins (1977) concluded that predation rates would have to be unrealistically high to significantly affect the optimum clutch size. However, these models do not take into account the possibility that predation results in the complete destruction of the brood and that clutch size may affect the ability to renest after a nest failure. Milonoff (1989) examined a model incorporating these considerations and found that significant effects on clutch size are obtainable with reasonable rates of predation. Since a bird under these circumstances must reserve resources for a possible future brood, the optimum clutch size will be less than the most productive calculated on the basis of a single brood.

Although renesting may occur more quickly for smaller clutch sizes (Slagsvold 1984), the evidence that clutch size is limited by nest failure is equivocal (Slagsvold 1982b; McLaughlin and Montgomerie 1989; Kulesza 1990). Using species as the basic datum, Kulesza (1990) found no correlation between clutch size and the proportion of offspring taken by nest predators among 63 New World passerines. A significant negative correlation was obtained when the species were grouped by genus ($n = 50$, $P < 0.05$), supporting the hypothesis that species suffering high rates of nest predation will lay smaller clutches. In this analysis we are faced with the problem of inferring causation from correlation; it remains to be demonstrated experimentally that increased clutch size leads to higher mortality rates from nest predation.

Predation rates are generally greater on open nests on the ground than on those off the ground; therefore, if predation rates limit clutch size, by the arguments advanced above, the average clutch size of ground-nesting birds will be less than that of off-ground nesters (Martin 1988). In fact, among the wood warblers the reverse is true: ground nesters have larger clutches than off-ground nesters (Martin 1988). Martin considered various factors that might override the effects of nest predation (adult mortality, thermoregulatory costs, food availability, nest size) but the data are too meager to draw any conclusions. Slagsvold (1982a, 1989a,b) paid particular attention to the possible importance of nest size. His research (Slagsvold 1982a, 1989a) suggests that an enlarged nest may relieve crowding of nestlings and enhance survival but decrease hatching success. Martin (1988) proposed this to be the most parsimonious explanation for the difference between ground and off-ground clutch sizes. He posited (p. 907) that "nest size may be a greater constraint for off-ground nesters because overcrowding in off-ground nests causes chicks to be pushed out of the nest and onto the ground." The data compiled by Kulesza (1990) are also important in demonstrating the significance of nest structure: for given values of nest predation and latitude, small-pensile nesters lay smaller clutches than open-cup nesters, which lay smaller clutches than domed nesters.

9.2.2. *Optimal Litter Size: An Energetic Perspective*

The models discussed thus far in this chapter have been derived from considerations of the effect of clutch size on the life history traits, survival, and reproduction. This is not the only perspective available. Wilson (1965, p. 373) attempted to predict the optimal litter size for mammals based upon the premise that "parameters of biological systems are chosen so that the entropy production is a minimum." His arguments can be recast in terms of individual selection: selection will favor those genotypes that maximize the amount of energy that is converted into offspring production. The basic model, restated in an individual selection framework, is as follows:

1. The population is assumed to be stationary: each female produces two offspring that themselves survive to reproduce.
2. The rate at which energy, E, is accumulated by the immature animal is given by the von Bertalanffy equation.

$$E = E_{max}(1 - e^{-kt}) \qquad (9.8)$$

where E_{max} is the maximum rate, k is a constant, and t is age. At maturity the animal accumulates energy at the maximum rate E_{max}. This equation is not explicitly justified but since weight increases roughly in this manner and metabolism also scales approximately with weight it is a reasonable approximation. (See, for example, Kooijman 1986.) The constant k was estimated by Wilson from the change in weight with age.

3. Mortality rate, M, is constant; hence the probability of being alive at age t, $l(t)$, is $l(t) = e^{-Mt}$. The average rate of energy production is approximately equal to

$$E^* = \int_0^\infty -\frac{dl(t)}{dt} E\,dt$$

$$= \frac{E_{max}}{1 + \dfrac{M}{k}} \qquad (9.9)$$

4. The total accumulation of energy to the age at maturity, α, is

$$E_\alpha = \int_0^\alpha E_{max}(1 - e^{-kt})e^{-Mt}dt$$

$$= E_{max}\left(\frac{1 - e^{-M\alpha}}{M} - \frac{1 - e^{-\alpha(M+k)}}{M + k}\right) \qquad (9.10)$$

5. The interval between litters is T, during which time the female accumulates E_{max} amount of energy per unit time, to be distributed among the x individuals within the litter. Since two offspring survive to reproduce the investment in these replacements is $C = 2E_{max}T/x$.

6. Since the energy accumulated to maturity is a fixed cost, the minimum cost, C, of producing the two replacements is $E_\alpha + 2E_{max}T/x$.

A plausible measure of fitness is the ratio E^*/C, the rate of production relative to the minimum cost. For given values of M, k, T, and E_{max}, natural selection should favor a clutch size that maximizes E^*/C. Wilson (1965) tested this prediction using data on seven species of mammal ranging from moose to meadow mice: the correlation between prediction and observation is extremely good ($r = 0.99$, $n = 7$, $P < 0.001$, slope $= 1.076$, intercept $= -0.16$). Two species that did not fit the predictions are mentioned but the relevant statistics are not given and hence the overall fit cannot be computed. I am not entirely convinced by the logic of Wilson's model or my own interpretation, but it does give very accurate predictions and therefore the approach deserves more study than it has apparently received since the paper was first published.

9.2.3. A Genetic Paradox: Clutch Size and Selection

A number of long-term studies have documented an overall positive selection differential for increased clutch size but no systematic change in this trait (Noordwijk et al. 1980, 1981; Nur 1984b; Boyce and Perrins 1987; Rockwell et al. 1987; Gibbs 1988; Fig. 9.11). Selection differentials, however, though generally positive, are not typically statistically different from zero, and correction for other confounding factors such as laying date may considerably reduce the estimated differential (Fig. 9.11). If a positive and

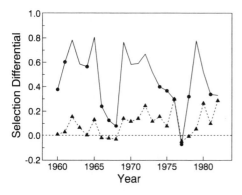

Figure 9.11. Selection differentials (solid line) on unmanipulated clutch size of the great tit for the years 1960–1982 at Wytham Wood. Unmarked points are significantly different from zero. Correction for laying data (triangles) considerably reduces the selection differential, and none are significantly different from zero. Data from Boyce and Perrins (1987).

persistent selection differential is real, how can it not result in a change in clutch size? Obviously if the heritability of clutch size is zero no response will be expected: this is the case for Darwin's ground finch, *Geospiza fortis* (Gibbs 1988), but it is not the case for other studies, where heritabilities are quite high (Table 9.7). Price and Liou (1989) suggested that a persistent selection differential for increased clutch size without a corresponding response may arise because variation in clutch size reflects variation in nutritional status, not genetic variation. In a number of studies the ability to produce and rear offspring has been demonstrated to be related to the nutritional status of the parents or the quality of the environment, the general observation being that if conditions are poor, females lay smaller clutches. (For a review of data to 1980 see Drent and Daan 1980; further data in Högstedt 1980, 1981; Korpimäki 1987; Nur 1988b.) Price and Liou postulated that variation in clutch size may reflect variation in nutritional status of females that is not heritable. Under these conditions better-nourished females will produce larger clutches, but this tendency will not be passed to their offspring. Although there may exist significant heritability for clutch size there will be no evolutionary response since the variation in clutch size occurring in the wild reflects phenotypic, not genetic, variation. This hypothesis has also been suggested for the evolution of breeding date in birds (Price et al. 1988) and seed size in wild barley (Giles 1990). If the above hypothesis is correct, heritability estimated from variation across years should be essentially zero. In this regard it is significant that heritabilities are typically calculated after correction for such effects, and hence the supposed paradox is really a paradox generated by the way in which heritability is being estimated.

To demonstrate that variation in clutch size reflects the ability of a female (or couple) to raise the particular clutch size laid requires that clutch sizes be manipulated by artificially increasing and decreasing the number of eggs or chicks. (Experiments in which clutch size is manipulated by randomly assigning clutch sizes to females eliminate the effect that is of present

Table 9.7. Heritability (h^2) estimates and repeatibilities (R) of clutch size in birds

Species	h^2	SE	R	Reference
Great tit	0.48	0.05	0.51	Perrins and Jones 1974
Great tit	0.37	0.12	0.45	Noordwijk et al. 1980
Starling	0.33	0.02	—	Flux and Flux 1982
Collared flycatcher	0.32	0.15	—	Gustaffson 1986
Darwin's ground finch	0.03	0.03	0.08	Gibbs 1988
Lesser snow goose	0.20	0.08	0.26	Findlay and Cooke 1983, 1987

interest.) A large experiment of this type was carried out by Pettifor et al. (1988) on the great tit, *Parus major*. Their analysis shows that a female could maximize the number of offspring recruited into the population by increasing its clutch size by about one egg regardless of the initial clutch size (Fig. 9.12). Therefore, females that begin with 9 eggs are better able to rear 9 offspring than a female that lays only 5 eggs, though both birds might do better to enlarge their clutches by one. (These results do not, as claimed by Pettifor et al., "demonstrate unambiguously that the optimal clutch size is *n* for birds that choose to lay *n* eggs.") Pettifor et al. (1988) could not detect an effect of clutch size on subsequent survival of adults, though other studies on great tits have done so (Nur 1988b; unpublished data of Tinbergen reported in Dijkstra et al. 1990).

The hypothesis of Price and Liou (1989) postulates that selection acting to increase clutch size is unsuccessful because the expressed variation in clutch size is phenotypic rather than genetic. Cooke et al. (1990) propose a mechanism that makes the opposite assumption and also leads to no change in clutch size. They postulate that the environment is changing in such a manner as to negate the selective advantage of larger clutches. Suppose that success in rearing offspring depends on nutritional status, which itself depends on the quality of the territory a bird can defend. While selection may favor increased clutch size, this can only be achieved by increasing the quality of the territory. Therefore, selection acts upon both clutch size and some behavioral component that relates to territory quality. Selection for, say, increased aggressiveness, will result in a better territory

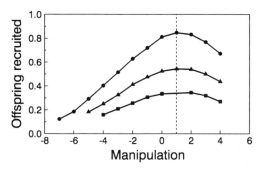

Figure 9.12. Estimated average number of offspring recruited per nest in the great tit as a function of original clutch size, cs (circles, cs = 9; triangles, cs = 7; squares cs = 5) and manipulation score. Note that offspring recruited is maximized by the addition of an extra young to the clutch. Redrawn from Pettifor et al. (1988).

only if a limited number of individuals in the population change their behavior. If, as is likely to happen, the general level of aggressiveness rises, no individual will be "better off" and hence each bird may, on average, be able only to defend the same territory as before; hence there can be no response in clutch size since the nutritional status of the bird remains the same.

This process can be described mathematically as follows: suppose that clutch size, x, is the product of three factors

$$x = FTe \tag{9.11}$$

where F is the number of eggs per unit area, determined by the foraging efficiency of the parent,

T is the parental territory size,

e is a random environmental component.

Territory size is determined by a large number of factors, which Cooke et al. lump together under the general term, A = "aggressiveness." The amount of territory a particular bird can sequester is dependent upon the aggressiveness of this bird relative to the mean aggressiveness ($\hat{A}$), and the number of birds, N, that have territories. Letting the total area available be a, the value of T is given by

$$T = \left(\frac{a}{N}\right)\left(\frac{A}{\hat{A}}\right) \tag{9.12}$$

The two component traits, F and A, can be decomposed into their genetic and environmental components, $F = G_F E_F$ and $T = G_T E_T$, and hence

$$x = GE$$
$$\text{where} \quad G = G_F G_T$$
$$\text{and } E = \frac{E_F E_T ea}{N\hat{A}} \tag{9.13}$$

In a constant environment the change in mean phenotype in response to selection, R, is (chapter 2)

$$R = h^2 S \tag{9.14}$$

where h^2 is the heritability of the trait and S is the selection differential (the difference between the means of the selected and unselected populations). If the environment is changing with time, the above equation is modified to

$$R = h^2S + \Delta\hat{E} \qquad (9.15)$$

where $\Delta\hat{E}$ is the change in the mean environmental component over one generation. Now even if the environmental components E_F and E_T do not change, the overall value of E will change if $\hat{E}$ changes, as might occur if there is a change in one of the behavioral components. If $\hat{A}$ increases, $\hat{E}$ will decrease and hence $\Delta\hat{E}$ will be negative, and may negate any change in clutch size resulting from a positive selection differential, S. The difficulty with this hypothesis is that is requires that the two components be fairly well balanced to prevent any significant response. The most important message from the analysis of Cooke et al. is that responses may depend upon how the environment, both biotic and abiotic, is changing in concert with selection for increasing clutch size.

9.3. Hypotheses Generated by Consideration of Clutch Size in Insects

This section is divided into two broad divisions based upon whether competition within a site is between siblings only or between the offspring of several females. The basic difference between the models developed for insects and those for endotherms is that the insect models typically assume that a female lays several clutches distributed over different sites. Because the oviposition of a sequence of egg batches entails a sequence of decisions, dynamic programming is frequently a very useful method of analysis. Dynamic programming is not required for the simple models discussed here; the reader is referred to Iwasa et al. (1984), Mangel (1987), and Mangel and Clark (1988) for examples utilizing this method.

For the analyses discussed in this section to be valid there must exist the potential for the pattern of oviposition to evolve. Genetic variation for habitat preference has been demonstrated in a number of insects (reviewed by Hedrick 1986), and such preferences can lead to different patterns of egg dispersion, but genetic variation for oviposition behavior per se has been little studied. Females of the seed beetle, *Callosobruchus maculatus*, differ genetically in their disposition to lay eggs randomly or uniformly (Messina 1989). The inheritance of this trait appears to be polygenic, but due to largely nonadditive genes with dominance for uniform egg laying. The egg-laying behavior of female cabbage white butterflies, *Pieris rapae*, from Canada, Australia, and the United Kingdom, differ, and common garden experiments with females from the last two countries indicate that this difference has a genetic basis (Jones 1987). Females from the UK and Canada lay their eggs in more clumped distributions than Australian females. Jones and Ives (1979) postulated that the difference between the Canadian and Australian strains is a consequence of differences in the two habitats. In Vancouver (Canada) densities are low and overcrowding is not a major problem for larvae. But weather conditions frequently disfavor flight, and hence search time may be limiting. Similar conditions prevail in the UK (Jones 1987). In contrast, in Canberra (Australia) local overcrowding is frequent but the weather is generally favorable for flight. Therefore, Australian females will do better to distribute their eggs widely so as to minimize overcrowding of their own offspring (Jones 1987).

9.3.1. Analyses Assuming One Female per Site

9.3.1.1. Clutch Size and Survival Between Oviposition Episodes

Insects face a somewhat different problem than birds and mammals for the female insect generally shows no parental care and must select both the appropriate site (host) and the appropriate clumping of eggs to ensure the maximum number of offspring. This decision involves a number of considerations. First, too many offspring laid on a particular host could lead to overcrowding and increased mortality. Though in some insects an Allee effect, in which mortality is reduced by aggregation, is observed (Taylor 1937; Sang 1950; Ghent 1960; DeLoach and Rabb 1972; Seymour 1974; Berger 1989; see section 9.3.1.4.), in general, increased crowding causes a decline in growth rate, survival rate, and adult size (Fig. 9.13). If crowding were the only consideration it would clearly profit a female to lay her eggs in clutch sizes that resulted in no competition among larvae. However, the lower the clutch size the more hosts that must be found, and hence the greater the search time required. Increasing search time will undoubtedly lead to increased mortality between egg-laying episodes, and hence a lower overall fecundity. Courtney (1984) tested this hypothesis indirectly by comparing the realized fecundity of insects that lay their eggs singly with insects that lay their eggs in batches. As predicted, batch-layers realize more of their potential fecundity than single-layers (Fig. 9.14).

Weis et al. (1983) made a detailed study of clutch-size variation in a natural population of the gall midge, *Asteromyia carbonifera*. Their analysis illustrates the potential importance of survival between hosts in the evolution of clutch size. Larvae of *A. carbonifera* induce ellipsoid, biconvex, blisterlike galls on the leaves of *Solidago canadensis*. Each gall typically contains one to five larvae, though as many as 12 have been recorded (Fig. 9.15). Larval mortality and adult longevity are independent of clutch size, but adult body weight declines with increasing clutch size. This decline in body size reduces fitness since fecundity is reduced. However, percent parasitism declines with clutch size. (A theoretical reason for this is discussed in section 9.3.1.3.) With respect to a single oviposition, the inter-

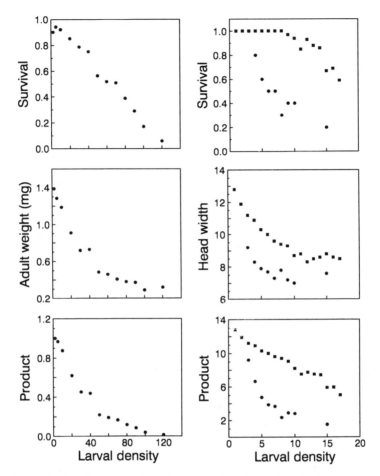

Figure 9.13. Survival and adult size as functions of larval density in two species of insect. Fecundity increases with adult size and hence an index of expected fecundity per larva can be formed by multiplying adult size by the probability of survival to the adult stage (bottom panels). Left panels: *Drosophila melanogaster* (data from Sang 1949). Right panels: The egg parasite, *Trichogramma embryophagum* (data from Klomp and Teerink 1967). Food quality is a direct function of the size of the host egg: squares, *Bupalus* eggs (egg weight = 0.26 mg); circles, *Anagesta* eggs (egg weight = 0.028 mg). Head width is given in micron units (1 unit = 0.0247 mm).

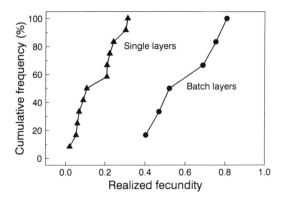

Figure 9.14. Cumulative frequency plots of realized fecundity (= fraction of maximum eggs available that are laid) for insects that lay in batches (circles) and insects that lay their eggs singly (triangles). Note that there is no overlap between the two types. Data from Courtney (1984).

action between fecundity and survival with clutch size results in clutches of size one being the most productive (Fig. 9.15). The most productive clutch size has been termed "the Lack clutch size" since it is the size Lack considered in birds and mammals to be that favored by selection (Charnov and Skinner 1984). It must, however, be remembered that in invertebrates the productivity of a clutch is a function of both number and size of offspring: throughout the present discussion productivity per clutch is used in this sense; i.e., it is the expected number of eggs that will be produced by the larvae when they themselves become adults. A clutch size of one in *A. carbonifera* is not only the most productive but also the most frequent clutch size, though the average is 1.88 (Fig. 9.15). Overall fitness depends not on the productivity of individual clutches but the total productivity of the female (the total number of eggs produced by her offspring), which depends upon the mortality rate suffered by females between ovipositions. With a survival probability between egg-laying episodes of 99% the optimal clutch size is two, and increases to four when the survival probability is decreased to 95%. Weis et al. (1983) could not estimate this critical parameter and hence their excellent study cannot, unfortunately, predict the optimal clutch size, though it does demonstrate that any such attempt must take into account the age schedule of survival of the ovipositing female.

Intuitively it is reasonable to suppose that as host density decreases search time will decrease and hence so will mortality. The effect of host density on the optimal clutch size has been formally developed by Parker

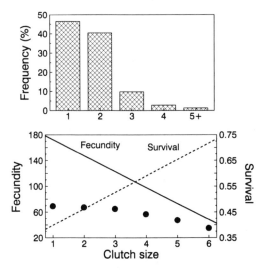

Figure 9.15. Upper panel: Frequency distribution of clutch size in the gall making insect *Asteromyia carbonifera.* Modified from Weis et al. (1983). Lower panel: Regressions of fecundity and survival as functions of clutch size in *A. carbonifera.* Circles show the expected productivity of females (read off right axis) laying clutches of the specified size.

and Courtney (1984) and Skinner (1985) by using the marginal-value theorem (Charnov and Skinner 1984, 1985). To maximize her fitness a female must maximize not the number and size of offspring emerging from a single host but her productivity per unit time. Although mortality may increase with clutch size, productivity per clutch will typically be a concave function since initially the product of clutch size, future fecundity of larvae, and survival increase with clutch size (Fig. 9.16). In the case of birds and mammals, reduced survival of offspring, reduced survival/fecundity of parents, and year-to-year variation in the relationship between survival rate and clutch size may all contribute to an optimal clutch size that is below the most productive clutch size (Lack clutch size). In the case of insects, search time favors a reduction in clutch size below that which is most productive on a per-clutch basis. To demonstrate this we first convert clutch size into oviposition time, i.e., oviposition time = clutch size × time to produce and lay an egg. For convenience we can scale time such that 1 time unit is equivalent to the time required to produce and lay one egg, in which case oviposition time and clutch size, x, are the same magnitude,

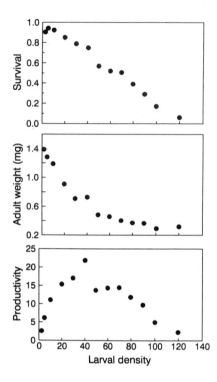

Figure 9.16. A typical response of larval survival and final weight with initial larval density in insects. Data shown are for *Drosophila melanogaster* (data from Sang 1949). Given that fecundity is proportional to weight, an index of the overall productivity (bottom panel) is obtained from the formula Productivity = Adult weight × Survival × Larval density.

and we can simply refer to clutch size rather than oviposition time. We now proceed in the usual manner dictated by the marginal-value approach (see chapter 4, section 4.2.4): the optimal clutch size is that at which a line drawn from the point $-t_s$, where t_s is search time, is tangential to the curve of per clutch productivity on clutch size (Fig. 9.17). It is immediately apparent that this value will be less than that which maximizes productivity per clutch (i.e., the Lack clutch size). This can be mathematically demonstrated as follows: let $f(x)$ be the number of offspring produced from a clutch of size x. The most productive (Lack) clutch size with respect to a single host is that clutch size at which the derivative of $f(x)$ with respect to x, $f'(x)$, is equal to zero. But fitness, w, is maximized by maximizing the productivity of a female per unit time,

$$w = \frac{f(x)}{x + t_s} \qquad (9.16)$$

Differentiating w with respect to x gives

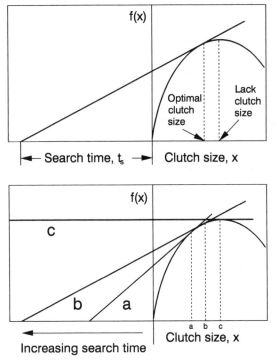

Figure 9.17. Graphical calculation of the optimal clutch size when an insect must search for several oviposition sites. See text for method of calculation. $f(x)$ is the productivity for a clutch of size x. Upper panel: The optimal clutch size is less than the most productive (= Lack) clutch size. Lower panel: Effect of increasing search time (a < b < c) on the optimal clutch size. In the limit, when search time becomes infinite (line c), the optimal clutch size is the Lack clutch size.

$$\frac{dw}{dx} = \frac{f'(x)}{x + t_s} - \frac{f(x)}{(x + t_s)^2}$$

$$= 0 \quad \text{when} \quad f'(x) = \frac{f(x)}{x + t_s} \qquad \textbf{(9.17)}$$

All the terms on the right-hand side are positive and hence the optimal clutch size must occur when $f'(x)$ is positive, which is to the left of the clutch size at which $f'(x) = 0$, i.e., at a lower clutch size.

As search time increases, the optimal clutch size increases (Fig. 9.17, see also Waage and Godfray [1985] for an alternate approach that makes the same prediction), and in the limit ($t_s = \infty$) the optimal and most productive clutch sizes converge. This makes intuitive sense: if search time is infinite the female has only a single chance of laying, and hence she should lay the Lack clutch size. Various models also suggest that if the female has a fixed number of eggs, the optimal clutch size declines as the number of eggs laid increases (Parker and Courtney 1984; Iwasa et al. 1984; Waage and Godfray 1985).

The prediction that clutch size will increase as host density declines either because the absolute density of hosts declines or the density of the insect increases (both leading to increased search time) is supported by two studies on parasitoid wasps (Fig. 9.18) and the bean beetle, *Callosobruchus maculatus* (Mitchell 1975).

9.3.1.2. Clutch Size and Host Quality

Skinner (1985) suggested that for a constant search time clutch size will increase with host quality. This prediction is not universally true but depends upon the particular manner in which host quality is entered into the per-clutch production function, $f(x)$. Skinner proposed the relationship

$$f(x) = c_1\left(1 - \frac{x}{v}\right)^{c_2} \tag{9.18}$$

where v is host quality and c_1 and c_2 are constants, the latter defining the shape of the relationship ($c_2 < 1$ concave, $c_2 = 1$ linear, $c_2 > 1$ convex). As host quality, v, increases, $f(x)$ increases as required. The optimal clutch size is (Skinner 1985; the equation has been slightly modified to conform with the present definitions)

$$x = -c_3 t_s + \left(c_3^2 t_s^2 + \frac{vs}{c_2}\right)^{1/2} \quad \text{where} \quad c_3 = \frac{1 + c_2}{2c_2} \tag{9.19}$$

From equation 9.19 it is clear that clutch size, x, increases with host quality, v.

But the above relationship between survival and host quality is not the only possible relationship. Suppose the per-clutch production function is defined as

$$f(x) = vg(x) \tag{9.20}$$

where the function $g(x)$ is defined with respect to some reference host quality. This production function simply states that fitness increases linearly with host quality, a not-unreasonable model since a doubling of host mass

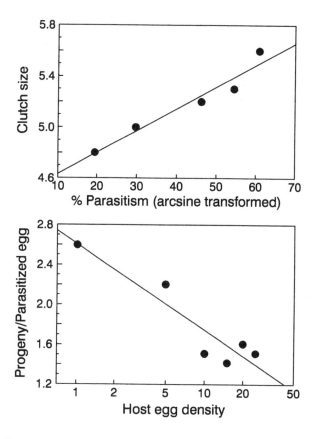

Figure 9.18. Effect of increasing search time on clutch size. Upper panel: *Trichogramma kalkae* parasitizing eggs of the dipteran *Diopsis macrophthalma*. As the proportion parasitized increases so search time increases and clutch size, in accord with prediction, increases ($r = 0.97$, $n = 5$, $P < 0.005$). Data from Feijen and Schulten (1981). Lower panel: *Trichogramma brevicapillum* laying on eggs of *Vanessa* sp. Single females were provided with a specific density of eggs for a period of 8 hours. The number of progeny per parasitized egg (assumed to reflect clutch size) declines significantly ($r = -0.93$, $n = 6$, $P < 0.05$) as the number of host eggs increases (i.e., search time is diminished). Data from Pak and Oatman (1982).

might reasonably be assumed to lead to a halving of mortality rate. Rearrangement of equation 9.17 gives the optimal oviposition time,

$$x = \frac{f(x)}{f'(x)} - t_s \tag{9.21}$$

Substituting the modified function gives

$$x = \frac{vg(x)}{vg'(x)} - t_s$$

$$= \frac{g(x)}{g'(x)} - t_s \tag{9.22}$$

In this model, changes in host quality have no effect on the optimal clutch size. It is also easy to graphically construct functions in which the per-clutch productivity increases with host quality but the optimal clutch size actually declines (Fig. 9.19). Thus, although Skinner's prediction may be true for the particular model he adopted it is not universally true. However, a number of studies have demonstrated that insects increase their clutch size with host size (Fig. 9.20; Salt 1961; Purrington and Uleman 1972; Godfray 1986). Further research is required to assess the incidence with which clutch size increases with host quality and the actual relationship between productivity in a single host and host quality.

9.3.1.3. Clutch Size and Convoy Hypothesis

Escape from parasites and predators has been put forward as a factor in the evolution of egg clumping (Stamp 1980). The data of Weis et al. (1983) on the gall midge, *A. carbonifera* (discussed in 9.3.1.1), support this hypothesis, percent parasitism declining with clutch size. The advan-

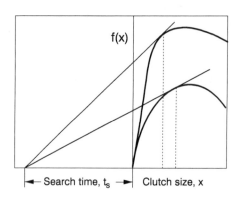

Figure 9.19. Graphical illustration of a decline in optimal clutch size with increased host quality. $f(x)$ is the per-clutch productivity.

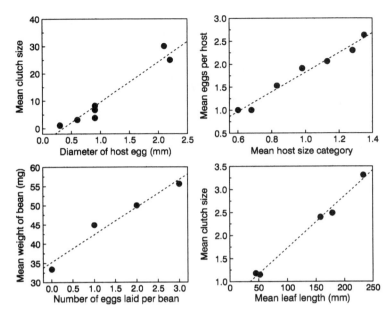

Figure 9.20. Four cases in which clutch size increases with host quality. Upper left panel: Parasitic wasp *Trichogramma embryophagum* laying on eggs of different host species ($r = 0.97$, $n = 8$, $P < 0.001$; data from Klomp and Teerink 1962, 1967). Upper right panel: Ectoparasitic wasp *Aphytis melinus* laying on various sizes of one of its hosts, the oleander scale, *Aspidiotus nerii* ($r = 0.99$, $n = 7$, $P < 0.001$). Size categories of hosts are < 0.59, $0.6–0.75$, $0.76–0.9$, $0.91–1.05$, $1.06–1.2$, $1.21–1.35$, >1.35 mm. Similar results were obtained for *A. lingnanensis* laying on oleander scale, and both species laying (separately) on California red scale, *Aonidiella aurantii* (data from Luck et al. 1982). Lower left panel: Beetle, *Callosobruchus maculatus*, laying on its host, mung beans, ($r = 0.98$, $n = 4$, $P < 0.05$). Because of the experimental design the data have been classified according to the number of eggs rather than bean weight (i.e., axes reversed). Data from Mitchell (1975). Lower right panel: Leaf-mining fly, *Pegomya nigritarsus*, laying on different species of *Rumex* ($r = 0.995$, $n = 5$, $P < 0.0.001$). In one species, *Rumex acetosa*, a second leaf-mining fly, *P. bicolor*, accounted for 30% of eggs laid but the two species cannot be separated in the larval stage. Data from Godfray (1986).

tages of aggregation have been much discussed in reference to fish schooling (Brock and Riffenburgh 1959; Olson 1964; Cushing and Harden Jones 1968; Partridge 1982; Pitcher 1986), and the theoretical framework established can be applied to the present hypothesis. With respect to the present problem the analysis of Olson (1964) is satisfactory. This analysis derives from a paper by Koopman (1956) examining the advantages of running ships in convoy during the Second World War. The probability of finding a given object, P_1, in an area A by a parasite (or predator) moving at random with constant speed V for a time t is

$$P_1 = 1 - e^{\frac{-2rVt}{A}} \qquad (9.23)$$

where r is the sight range of the parasite. Now let R be the radius of an aggregation (e.g., a gall that is increased in size because it contains more than one larva). In search theory $2r$ is regarded as the width of the swept path. If the visual range to an aggregation is practically the same as that to an individual (this is true for *Asteromyia carbonifera* where gall size does not increase much with clutch size), the equivalent swept path is $2(r + R)$. The probability of finding the aggregation is

$$P_2 = 1 - e^{-\frac{2(r+R)Vt}{A}} \qquad (9.24)$$

If the parasitic female lays eggs on m of the x larvae in a gall the probability of a larva being parasitized is $P_1 m/x$. The potential advantage of aggregation thus depends upon the rate at which the aggregation becomes more visible versus the number of individuals within the aggregation that become victims if the aggregation is located. To illustrate under what circumstances multiple larvae may be advantageous we can substitute some plausible figures. From the data of Weis et al. the probability of locating a gall containing a single larva is approximately 0.6. Letting the sight range of the parasite, r, be unity, the value of $2Vt/A$ is 0.92. Gall size is proportional to the number of larvae within it, but according to Weis et al. (1983, p. 693) the range "is not sufficiently great to make large clutches more easily discovered." For the purposes of illustration let us assume that R, the radius of the gall with two larvae, is $1.25r$, which is still a significant increase in the visibility of the enlarged gall. The probability of a parasite finding this gall is $P_2 = 1 - e^{-(2.25 \times 0.92)} = 0.87$. Letting the proportion of larvae parasitized in a multiple gall be P_3, it will be advantageous for a female to clump her larvae whenever $P_1 > P_2 P_3$: in the present example this requires that $P_3 < 0.6/0.87 = 0.69$. Therefore, if the parasitic wasp on average locates only 69% of the larvae within a multiple gall, selection will favor the evolution of clutch sizes greater than one. Given that a female parasite probably cannot accurately estimate the number of larvae within the gall by its external appearance it is reasonable to assume, as indeed is found, that the female will in many cases leave the gall without laying eggs on all the larvae. The foregoing numerical example illustrates that the hypothesis of egg clumping to alleviate losses from parasites or predators is reasonable. Only further behavioral analysis can establish the selective differential generated by this phenomenon.

9.3.1.4. Clutch Size and Allee Effect

If the fitness of individual larvae decreases with each additional larvae (e.g., Fig. 9.13), the optimal clutch size, given no searching costs and an unlimited resource, will be a single individual per site. But in some insects the survival of larvae is enhanced by aggregation, a phenomenon known as the Allee effect. This effect can come about because an aggregation of larvae can more effectively overcome a host's defense system (Taylor 1937; DeLoach and Rabb 1972; Kalin and Knerer 1977). Regardless of how such an effect is generated it will clearly favor clutch sizes in excess of one. Godfray (1986) examined the consequences of this phenomenon in the leaf-mining fly, *Pegomya nigritarsis*.

P. nigritarsis lays its eggs on the leaves of various species of *Rumex*, into which the newly hatched larvae mine and within which they remain until ready to pupate, when they drop to the ground and pupate in the litter layer or soil. Godfray (1986) studied the life history in a single *Rumex* species, *R. obtusifolius*. Mortality can be divided into two categories (Godfray 1986): "larval death" and endoparasitic attack. The first category, larval death, includes death due to a number of causes that cannot be separated: starvation in early instars, invertebrate predation, parasitoid host feeding, and interference from other larvae. The proportion of larvae surviving this category decreases with clutch size (Fig. 9.21).

Attacks by endoparasites generally occur when the larva is in the first instar. The parasite egg hatches, but the larva remains as a first instar until the fly pupates, at which time the parasite completes development, causing the death of the fly. Only one parasite emerges per host. The proportion of larvae surviving attacks by endoparasites is a concave function of clutch size (Fig. 9.21). It is not known why such a relationship should occur. But the consequence is that the overall mortality rate is also a concave function of clutch size (Fig. 9.21).

The adult size of *P. nigritarsus* is independent of clutch size and thus there is probably no fecundity cost to increased larval density (at least within the range one to nine larvae per leaf). Therefore, the optimal clutch size will be greater than one even if there are no costs associated with

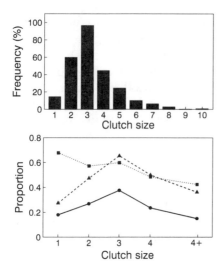

Figure 9.21. Upper panel: Frequency distribution of clutch sizes of the leaf-mining fly, *Pegomya nigritarsus*, laying on one of its hosts, *Rumex obtusifolius*. Lower panel: Survival rates of *P. nigritarsus* on *R. obtusifolius*. Squares: proportion of larvae surviving "larval death"; triangles: proportion of larvae surviving attacks by endoparasites; circles: proportion of larvae surviving all sources of mortality. Figures modified from Godfray (1986).

searching and egg production, or resources are not limiting. Indeed, for a leaf miner the amount of resources available is likely to be very large and the time to move from one suitable oviposition site to another is probably very small. Under these conditions the optimal clutch size is that at which larval survival is greatest (since fecundity does not vary with larval density). If resources are very scarce or survival between ovipositions is very high the optimal clutch size is that at which the product of clutch size × survival is greatest (the Lack clutch size). For *P. nigritarsis* the optimal clutch size under either scenario is the same, three eggs per clutch (Fig. 9.21). This predicted value corresponds to the most frequently observed in the leaves of *R. obtusifolius* (Fig. 9.21).

9.3.2. Analyses Based on Assumption of Several Females per Site

9.3.2.1. A Single Oviposition per Female

Thus far, analysis has been predicated on the assumption that only a single female lays on a patch. Relaxation of this assumption produces some interesting results. Ives (1989) examined three functions that describe the expected per capita production (survival × the adult fecundity of females, or survival × the mating success of males) of larvae emerging from a patch in which N eggs have been laid. Letting the per capita production function be designated $p(N)$ the three models are

$$p_1(N) = c_1(1 - a_1N) \qquad N < \frac{1}{a_1}$$

$$= 0 \qquad N \geq \frac{1}{a_1} \qquad \textbf{(9.25)}$$

$$p_2(N) = c_2 e^{-a_2 N} \tag{9.26}$$

$$p_3(N) = c_3(1 + a_3 N^{-a_4}) \tag{9.27}$$

The constants c_1, c_2, and c_3 are simply constants of proportionality; the constant a_i defines the shape and rate of decline in per capita productivity with N. All constants are greater than 0, except for a_4, which is greater that one. Examples of these three curves are shown in Fig. 9.22. (Examples for various insects can be found in Table 1 of Ives.)

At evolutionary equilibrium all females lay the same clutch size. Consider the simple case of only one oviposition per female and f females laying at a given site. The expected per capita production of larvae from a female laying a clutch of size x, given that all other females do the same, is $p_i(xf)$, ($i = 1, 2$ or 3), and hence the fitness of the female, $w(x)$, is

$$w(x) = xp(xf) \tag{9.28}$$

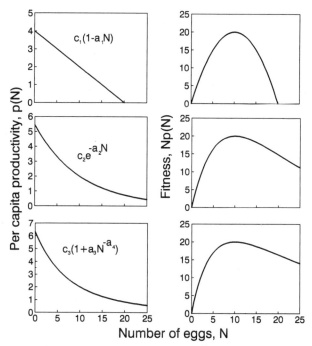

Figure 9.22. Graphical illustration of the three production functions suggested by Ives (1989). Modified from Ives (1989). Panels on the left show the per capita productivity, those on the right the per patch productivity.

Differentiating $w(x)$ with respect to x and finding that value of x at which $dw(x)/dx$ is equal to zero gives the optimal clutch sizes (x_1, x_2, x_3) for the three fitness functions:

$$x_1 = \frac{1}{a_1(f + 1)} \tag{9.29}$$

$$x_2 = \frac{1}{a_2} \tag{9.30}$$

$$x_3 = \frac{1}{a_3(a_4 - f)} \quad \text{for} \quad f < a_4 \tag{9.31}$$

These three results are very different from each other: in the first case *the optimal clutch size decreases as the number of ovipositing females increases* (see also Parker and Begon 1986), but in the second case *the optimal clutch size is independent of the number of ovipositing females.* (Smith and Lessels [1985] obtain the same result for a somewhat different model.) In the third model *the optimal clutch size increases as the number of ovipositing females increases.* (When $f > a_4$ no evolutionarily stable equilibrium is possible.) This result, and that previously given for the optimal clutch size in relation to host quality, emphasize that *caution* must be exercised in inferring general properties from specific models.

9.3.2.2. Multiple Ovipositions by Females and Evolution of Egg Dispersion

Monro (1967) looked at the issue of egg distribution from a perspective of population persistence. He suggested that selection may favor patterns of clumping that preserve food refuges and hence prevent populations from overexploiting their resource and becoming extinct. This group selection hypothesis has not received much attention, and no data in support of it have been produced. Nevertheless, the consequences of egg clumping on population dynamics are important and may play a role in the evolution of individual oviposition behavior. Myers (1976) explored the consequences of different patterns of egg deposition on population persistence and stability using simulation models, allowing each female multiple clutches. In these models mortality of larvae resulted from overcrowding and dispersal from an overexploited patch. Population persistence increased with increased aggregation of eggs, and in the absence of dispersal mean population size declined, but when dispersal was included in the model, population size increased. Similar results were obtained by Jong (1979), Kuno (1983, 1988), Itô et al. (1982), Ives and May (1985), and Yamamura (1989).

But can a mutant that lays its eggs contagiously spread in a population, or must we rely upon group selection? To address this question Myers incorporated two clones into her model—clutch size of females from clone R being random and clutch size of females from clone C being contagious. These models demonstrated that, depending on the particular combination of parameters, clone C can be competitively superior even if the total fecundity of its females is less than that of females from clone R (Myers 1976). Thus selection can favor females that lay their eggs contagiously, i.e., larger clutch sizes.

The models of Ives (1989) and Myers (1976) clearly illustrate the complexities that can occur when multiple females lay on the same site. One of the most important findings of the latter study is that we should pay considerable attention not only to the mean clutch size but also to the size distribution of clutches. Females do not typically have knowledge of the distribution of egg batches in the habitat. Therefore, the decision on the number of eggs to be laid at a particular location must be made using some "rule of thumb." Mitchell (1975) attempted to derive the egg oviposition rules used by the bean weevil, *Callosobruchus maculatus*.

The eggs of the bean weevil are laid on the outside of beans; the larvae burrow into the bean upon hatching and remains there until ready to eclose into the adult form. To obtain differences in host density Mitchell (1975) exposed 10 g of beans to 200 females for three different time periods. With increasing density (i.e., increasing exposure time) the mean number of eggs per bean increased (0.2, 1.1, and 1.8, respectively). Females did not lay their eggs at random: the distributions were more uniform than expected by a Poisson model and larger numbers of eggs were laid on larger beans. Among beans containing only one egg there was a significant positive regression between survival rate and bean weight. For beans containing two or three eggs no significant correlation was found.

In the best of all worlds a female *C. maculatus* would lay no more than one egg on each bean, since survivorship is highest at this density (regardless of how individual larval survival rates are estimated), and would select beans *in decreasing sequence of size*. Thus in the experiment yielding 0.2 eggs per bean the most fit behavior would be to lay the 46 eggs laid in this experiment in the 46 largest beans. If the number of eggs exceeds the number of beans the second egg should be laid on beans *in ascending order of bean weight*. The reason for this is that the survival rate of two larvae is independent of bean size (at least over the range in bean size examined). The survival rate of a single larva on the smallest bean is approximately 50% compared to 46% for two larvae, while for a single larva on the largest bean it is approximately 72%: therefore, the expected yield from laying two larvae on the smallest bean is hardly changed, while there is a great decrease in yield if the second egg is laid on the largest bean. The rule for laying the second egg presupposes that the first egg belongs to the present female: if it could not, because the female were laying her first egg or the number of beans or beetle density were very large, the female should lay her egg on the largest bean since that gives the highest larval survival.

Note that the above analysis does not address the question of whether the second egg should be laid at all, but begins with the proposition that a specific number of eggs will be laid. Based on this assumption, Mitchell computed the yield expected for five different oviposition behaviors (Table 9.8). As a fraction of the best yield possible, a random distribution of eggs does not perform well, the ratio ranging from 0.57 to 0.77; in contrast the observed distribution achieves a ratio from 0.93 to 0.94. One obvious assumption in the foregoing analysis that is unlikely to be correct is that a female can evaluate and rank all the eggs at her disposal. Without perfect knowledge she cannot achieve the best distribution of eggs; nevertheless the observed distribution comes impressively close to this goal.

The oviposition behavior giving the highest total yield is not expected under the assumption of individual selection (unless the females are very closely related), and hence the fact that the observed distribution comes close to this yield might be attributed to group selection. More likely it is

Table 9.8. A comparison of patterns of oviposition behavior in the bean weevil, *Callosobruchus maculatus*

Distribution	Relative fittness at densities (eggs per bean) of		
	0.2	1.1	1.8
Poisson distribution	0.77	0.57	0.77
Observed frequency distribution	0.93	0.94	0.94
Observed distribution with no weight discrimination by females	0.89	0.94	0.94
Uniform distribution on beans ranked by weight with the largest beans used first	1	0.93	0.93
As above but with the second egg added to lighter beans (the best pattern)	No 2nd egg on beans	1	1

Fitness is measured by the number of offspring produced by each pattern of oviposition relative to the best possible (last row; see text for a justification of this).

Modified from Mitchell (1975).

a consequence of the pattern favored by individual selection being very similar. In fact, if a female simply selects the largest beans (pattern 4 in Table 9.8), a behavior certainly favored by individual selection, she achieves a relative productivity very close to the highest (1, 0.93, 0.93), and very similar to the productivity obtained by using the observed distribution (0.93, 0.94, 0.94).

A female weevil obviously cannot inspect all the beans in her "universe" and hence she must use some "rule of thumb" to determine whether she lays an egg on the bean presently under inspection. Mitchell proposed that females use a simple decision rule based on the size of the previous bean encountered and the number of eggs laid upon it, relative to the size and egg complement of the current bean (Fig. 9.23). The model does a good job of predicting the frequency distribution of eggs per bean but is relatively poor at predicting the size of beans selected (i.e., the weight of beans containing 0, 1, 2, or 3 eggs) and slightly underestimates the yield of weevils (Fig. 9.23). The reason for the latter two discrepancies is that the simulated distribution is more uniform (contrary to the opposite claim by Mitchell) than that actually observed. The reduced selectivity of the weevils in the simulation leads to the mean size of beans in each egg class (zero to three) being less variable than in the observed population. Despite this short-coming the model does illustrate how the oviposition behavior of the weevil is likely to be accomplished by a very simple decision rule.

9.4. Summary

Mechanical constraints may limit clutch size. This factor has formed the core of several studies on the evolution of clutch size in reptiles. The low clutch size in anolis lizards has been attributed to the restriction on body mass of climbing using expanded subdigital or adhesive toe pads. A similar restriction could account for the relatively low clutch mass of geckos. Lizards that live in very narrow crevices cannot substantially increase their body dimensions when gravid. Predictably these lizards have low clutch masses and eggs that are more elongate than ecologically comparable lizards that do not use very narrow crevices to avoid predators. Because of reduced efficiency in prey pursuit and predator avoidance, lizards that actively forage for their prey are hypothesized to have relatively lower clutch masses than sedentary species (Vitt and Congdon 1978). This hypothesis is supported by the available data, but a high correlation between foraging mode and family confounds the analysis. Seigel and Fitch (1984) found no relationship between foraging mode and relative clutch mass in snakes but a higher ratio in oviparous than viviparous snakes. The lack of a correlation with foraging mode they suggest may be due to problems of

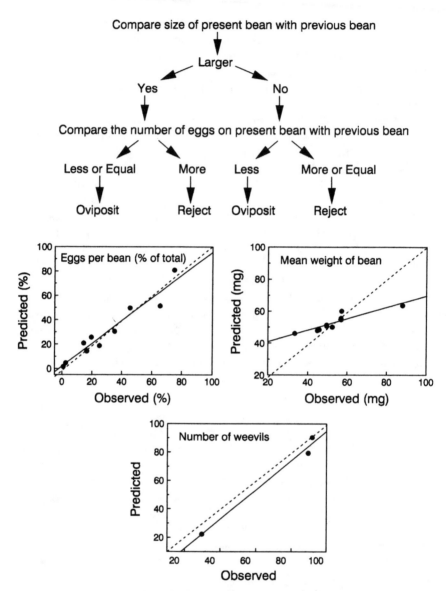

Figure 9.23. Upper section: Flow diagram of proposed decision tree for the bean weevil, *Callosobruchus maculatus*, laying on mung beans. Lower panels: Predicted and observed relationships. Dotted lines shows 1:1 reference line, solid lines the fitted regressions. Data from Mitchell (1975).

classifying behavior in snakes. Viviparous snakes must carry their clutches longer than oviparous snakes and this may increase their vulnerability to predators and hence favor a reduced clutch mass.

Lack proposed that in mammals and birds selection favors that clutch size which maximizes the number of offspring. This hypothesis does not take into account effects of increasing clutch size on parental survival or future fecundity, which have been shown to be decreasing functions of clutch size in some birds (Tables 6.5, 6.8). With such a decreasing function the optimal clutch size will be less than the Lack value. An alternate mechanism favoring an optimal clutch size less than the most productive was proposed by Mountford (1968). If a genotype produces, for whatever reason, a range in clutch sizes, the optimal clutch size may be greater or less than the most productive. A third factor causing the optimal clutch size to be less than the Lack value is environmental variability. The appropriate measure of fitness in a temporally variable environment is the geometric mean, which is strongly influenced by the frequency of low values. The effect of this is to favor a conservative clutch size. Based on a model incorporating temporal variability Boyce and Perrins (1987) successfully predicted clutch size in the great tit. Finally, if the probability of complete clutch loss through predation at the nest or inclement weather is high, a low clutch size may be optimal because it permits rapid renesting. Data in support of this hypothesis are equivocal, possibly due in part to confounding influences of nest size and structure.

Wilson (1965) addressed the question of clutch size in mammals by using the proposition that selection minimizes entropy. His arguments are recast in the framework of individual selection. Selection is hypothesized to maximize the ratio of the rate of production of energy relative to the minimum energetic cost of producing replacement offspring. Predictions from Wilson's models are significantly better than expected by chance.

Clutch size in a number of birds has a high heritability and selection differentials in wild populations are frequently positive, but clutch size does not change. A plausible explanation is that variation in clutch size is due to phenotypic variation in female condition that is uncorrelated with genetic variation. Heritabilities are typically computed after correction for year-to-year variation and hence eliminate this phenotypic variation. The apparent paradox is thus primarily a consequence of the frame of reference within which heritability is estimated. Cooke et al. (1990) proposed that heritabilities could be high, selection differentials positive, but clutch size show no response because behavioral components necessary to increase clutch size are also evolving. Consequently, in each generation a genotype may show increased ability to sequester resources relative to the previous generation, but its ability relative to members of the present generation is the same because the sequestering ability of the average individual within the population has increased.

Analysis of the optimal clutch size in insects and other organisms that disperse their eggs in several clutches must consider both density effects within a patch (host) and the probability of the female surviving the interval between ovipositions on different patches. The latter factor causes the optimal clutch size to be less than that which maximizes per clutch productivity. Increasing host density decreases travel time between hosts, increases survival, and hence increases the optimal clutch size. Theory predicts that as travel time increases, the difference between the optimal clutch size and the Lack clutch size decreases. Evidence from three insects supports this prediction.

Skinner (1985) conjectured that optimal clutch size increased with host quality. Two counterexamples are given demonstrating that this is not universally true. However, a number of studies have shown that in some insects clutch size is positively correlated with host size. Other factors that can favor increased clutch size are predators that do not destroy all larvae within a patch (the convoy hypothesis), and the Allee effect, whereby survival is greatest at an intermediate density.

When several females lay on the same host the optimal clutch size may be positively, negatively, or unrelated to the number of females ovipositing. This results and that demonstrated for host quality emphasize the care that must be exercised in using specific models to infer general patterns. Using simulation models Myers (1976) demonstrated that aggregative patterns of egg dispersion can evolve by individual selection.

10

Offspring Size

Size at birth is undoubtedly an important fitness component. A priori it is reasonable to suppose that offspring would benefit most from being born as close to adult size as possible. But this clearly does not occur: many organisms are, relative to their adult size, born both very small and underdeveloped. From a life history perspective the obvious hypothesis is that there is a trade-off between offspring size, offspring quality, and offspring number. The following sections explore this hypothesis and its ramifications. Section 10.1 reviews the evidence that propagule size influences fitness, showing that, indeed, larger size increases some fitness components. But increases in propagule size may be constrained by mechanical or physiological factors and reduce overall fitness by reducing the total number of offspring produced. These considerations are explored in section 10.2. The remainder of the chapter examines how propagule size evolves in different environments, constant or variable. Studies in the former category can be divided into the following five areas:

1. It is intuitively obvious that the larger the size at birth the shorter the time to maturity, all other things being equal. Thus there is a trade-off between size, fecundity, and time to maturity. The ramifications of this trade-off are explored in section 10.3.1.

2. Size at birth is one factor determining the time to maturity; another is the rate of development. Section 10.3.2 shows that the optimal propagule size is a function of the rate of development.

3. It has been observed that egg size varies with the size, condition, and age of the female. Size and condition can be handled by the same algebraic formulations. Two different models are presented in section 10.3.3 that each predict an increase in egg size with female size or condition.

4. Egg size frequently decreases with the age of the female in invertebrates but increases in vertebrates: reasons for this different response are discussed in section 10.3.4.

5. The final section in this category examines several models that have implicated the evolution of egg size with the evolution of parental care.

Research on the effect of a variable environment on the evolution of propagule size has been much less. Two problems have received significant attention. First, there have been a variety of hypotheses put forward to explain the observation that propagule size often varies in a systematic fashion during the year. As the phenomenon is not consistent across taxa it is unlikely that a single explanation will suffice. Section 10.4.1 explores possible hypotheses. The second problem concerning environmental variation deals not with seasonal variation in propagule size but variation within a single clutch. Simulations show that such variation is unlikely to be promoted by spatial heterogeneity but that temporal heterogeneity can be important.

10.1. Effect of Size on Fitness Components of the Young

There is considerable evidence that in plants large seeds have a selective advantage (Table 10.1): percent germination and percent emergence both increase with seed size; large seeds give rise to large seedlings (see reviews

Table 10.1. Summary of studies examining the consequences of variation in seed size

Factor	+	NS	−
Percentage germination	21	1	0
Percentage emergence	8	0	0
Germination rate	9	2	7
Initial growth rate of seedlings	5	0	0
Seedling survival	4	1	0
Competitive advantage	4	1	0

A study may comprise more than a single species. The reference list comprises all those authorities consulted but not all studies contained within the table, some being taken from the studies cited. +, a positive correlation between seed size and factor; 0, NS, no significant relationship; −, negative correlation.

References searched: Black 1958; Cavers and Harper 1966; Harper et al. 1970; Schaal 1980; Howell 1981; Thompson 1981; Weis 1982; Pitelka et al. 1983; Zimmerman and Weis 1983; Gross 1984; Hendrix 1984; Schaal and Leverich 1984; Stanton 1984; Marshall 1986; McGinley et al. 1987; Winn 1988; Kalisz 1989.

in Willson 1972; and McGinley et al. 1987), which either grow faster than smaller seedlings or maintain their initial advantage (at least for a short period), survive better, and have a competitive advantage under crowded conditions. Studies showing an increase in germination rate with seed size are approximately equalled by studies showing the reverse. The initial advantage accruing to large seeds does not generally persist; but this is not surprising since the absolute difference in seed size is very small compared to the final size of the plant. However, increased germination rate and survival should be themselves sufficient to favor an increasing seed size.

Effect of propagule size on fitness components has been less explored in animals than plants, but the same general pattern emerges (Table 10.2). Proportion hatching increases with egg size but there is no relationship between time to hatch and egg size. Size at hatching is positively correlated to egg size in snails (Bondesen 1940; Orton and Sibly 1900), insects (Campbell 1962), fish (Marshall 1953; Blaxter and Hempel 1963; Bagenal 1969; Mann and Mills 1979; Kazakov 1981; Zastrow et al. 1990), amphibians (Ferguson and Brockman 1980; Kaplan 1980; Berven 1982; Crump 1984b; Ferguson and Fox 1984; Sinervo 1990), turtles (Congdon et al. 1983), and birds (Schifferli 1973; Pinkowski 1975; Ricklefs et al. 1978; O'Connor 1979; Horsfall 1984; Soler 1988), and this initial size advantage probably accounts for the increased survival rate, the increased resistance to starvation, and the maintenance of a larger size at age (at least for the period immediately following hatching). As with plants the initial differences in size may not be maintained, but this does not negate the increased overall fitness of offspring from large eggs. Differences in survival rate are also not generally found to persist until maturity, but this may reflect the lack of power of the statistical tests resulting from the small sample size of animals or plants surviving to older ages.

Despite the selective advantage of large propagule size there is not an evident increase in egg size, a fact suggesting that either large egg size is not always favored and/or there are constraints or trade-offs that prevent the continuous increase in size, or favor an intermediate size. Stabilizing selection on egg size has been demonstrated in *Drosophila melanogaster* (Curtsinger 1976a,b; Roff 1976), chickens (Lerner 1951; Lerner and Gunns 1952), and on birth weight in humans (Karn and Penrose 1951; Van Valen and Mellin 1967). However, such data cannot be taken as unambiguous evidence that in these cases large propagule size per se is disadvantageous: if selection for other reasons has favored an intermediate propagule size then the small and large eggs in *D. melaogaster* and chickens or small and large babies in humans may represent "errors" in physiology, resulting in offspring with reduced viability for reasons unconnected with their size. Studies of the proximal causes of mortality in these cases may help resolve this problem.

Table 10.2. Effects of increasing propagule size on various fitness components in animals

Group	+	NS	−	Reference
	Hatching success			
Lepidoptera	2	0	1	1–3
Fish	2	0	0	4–5
Birds	5	3	0	6–13
Total	9	3	1	
	Time to hatch			
Fish	0	6	0	1–6
Amphibians	0	2	0	7–8
Total	0	8	0	
	Survival			
Insects	0	1	0	1
Fish	6	1	0	2–8
Reptiles and Amphibians	3	0	0	9–11
Birds	10	0	0	12–21
Mammals	1	0	0	22
Total	20	2	0	
	Starvation time			
Invertebrates	3	0	0	1–3
Fish	4	1	0	4–8
Total	7	1	0	
	Growth			
Insects	3	0	0	1–3
Fish	11	1	0	4–15
Amphibians	4	1	0	16–20
Birds	5	1	0	21–26
Mammals	2	1	0	27–29
Total	25	4	0	

+, positive correlation between egg size and component; NS, correlation not significant; −, negative correlation.

References (numbers refer to separate species: two or more species within a single publication are designated by multiple numbers, while studies of the same species appear under the same number):

Hatching success: 1. Harvey 1977; 2. Richards and Myers 1980; 3. Marshall 1990; 4. Gall 1974; Pitman 1979; 5. Zastrow et al. 1990; 6. Dawson 1972; 7. Murton et al. 1974; 8. Koivunen et al. 1975; 9. Ryder 1975; 10. O'Connor 1979; 11. Quinn and Morris 1986; 12. Moss et al. 1981.

Time to hatch: 1. De Ciechomski 1966; 2. Blaxter and Hempel 1963; 3. Marsh 1986; 4–5. Beacham et al. 1985; 6. Knufsen and Tilseth 1985; 7. Kaplan 1980; 8. Crump 1984.

Survival: 1. Harvey 1977; 2. Blaxter and Hempel 1963; 3. Kirpichnikov 1966; 4. Bagenal 1969; 5. Fowler 1972; 6. Pitmana 1979; 7. Henrich 1988; 8. Zastrow et al. 1990; 9. Derickson 1976; 10. Ferguson and Fox 1984; 11. Semilitsch and Gibbons 1990; 12. Skoglund et al. 1952; 13–14. Nisbet 1973; 15. Schifferli 1973; 16. Howe 1976; 17. Lundberg and Vaiisanen 1978;

Table 10.2. (*Continued*)

18. O'Connor 1979; 19 Moss et al. 1981; 20. Tiainen et al. 1989; 21 Quinn and Morris 1986; 22. Myers and Master 1983.

Starvation time: 1. Campbell 1962; 2. Tessier and Consolatti 1989; 3. Solbreck et al. 1989; 4. Blaxter and Hempel 1963; 5. De Ciechomski 1966; 6. Bagenal 1969; 7. Marsh 1986; 8. Rana 1985.

Growth: 1. Campbell 1962; 2. Steinswascher 1984; 3. Larsson 1989; 4. Magnusson 1962; 5. Blaxter and Hempel 1963; 6. Kirpichnikov 1966; 7. De Ciechomski 1966; 8. Fowler 1972; 9. Gall 1974; Pitman 1979; 10–11. Beacham et al. 1985; 12. Marsh 1986; 13. Rana 1985; 14. Knutsen and Tilseth 1985; 15. Monteleone and Houde 1990; Zastrow et al. 1990; 16–18. Kaplan 1980; 19. Kaplan 1985; 20. Newman 1988; 21. Skoglund et al. 1952; 22. Schifferli 1973; 23. Murton et al. 1974; 24. Davis 1975; 25–26. Nisbet 1978; 27 Whitehead et al. 1990; 28. Rickart 1977; 29. Millar 1978; 30. Myers and Master 1983.

Another approach to decoupling the possible correlation between the mother's condition and egg size is to manipulate the size of eggs directly. This has been successfully done in sea urchins (Sinervo and McEdward 1988) and the lizard *Sceloporus occidentalis* (Sinervo 1990; Sinervo and Huey 1990). Hatchling size in *S. occidentalis* was manipulated by removal of varying amounts of yolk in the egg, resulting in a twofold variation in hatchling weight. This range in hatchling size spans the range found among populations from Washington to California. Analysis of data from four different population showed that

1. Incubation time increased with egg mass.
2. Mass-specific growth rate ($\ln[\text{mass}_{\text{time }2}/\text{mass}_{\text{time }1}]/[\text{time }2 - \text{time }1]$) decreased with hatchling mass.
3. Spring speed of hatchlings increased with hatchling mass.

The same patterns were found in the hatchlings that had been experimentally reduced in size, demonstrating that some of the differences between populations can be explained by differences in egg size alone.

Sinervo and McEdward (1988) experimentally reduced egg size in the two sea urchins *Strongylocentrotus droebachiensis* and *S. purpuratus* by isolating blastomeres from embryos. The two species are approximately the same size as adults but *S. purpuratus* produces an egg that is six times smaller in volume than the egg of *S. droebachiensis*. Larvae from the eggs of *S. purpuratus* are less elaborate in morphology and develop more slowly than the larvae of *S. droebachiensis*. When the eggs of the latter species were experimentally reduced the larvae displayed the morphology and early developmental trajectory characteristic of the larvae of *S. purpuratus*. There were no effects on later development: the adult size remained the same for all treatments. Thus differences between these two species in the early development of the larva are in large measure a simple consequence of differences in egg size.

These two studies by Sinervo strongly suggest that differences in egg size may have profound effects on early development and survival. Both studies indicate that being born large is advantageous; in the remainder of this chapter we shall examine circumstances that limit egg size.

10.2. Optimal Propagule Size

10.2.1. *Constraints on Propagule Size*

There are clearly physical limits to how large a propagule can be: for example, a hummingbird cannot lay a goose-sized egg. In vertebrates other than fish an obvious limiting factor is the size of the pelvis through which the egg or neonate must pass. Following a suggestion by Tucker et al. (1978), Congdon et al. (1983) and Congdon and Gibbons (1987) noted that in some turtles, egg size increases with female body size and postulated that selection favors a larger egg size than can pass through the pelvic aperture, and hence a female will maximize her fitness by maximizing the size of her eggs as she gets larger. In two species, *Chrysemys picta* and *Deirochelys reticularia*, the size of the pelvic opening increases at the same rate as egg width, while in *Pseudemys scripta* egg width varies little with female size. Long and Rose (1989, p. 316) tested "the hypothesis that natural selection favors larger pelvic girdles in female turtles to facilitate the passage of eggs" by comparing the relative sizes of pelvic girdles in conspecific males and females of the three species. *Gopherus berlandieri*, *Kinosternon flavescens*, and *Terrapene ornata*. In all three species, after correcting for body size, mean pelvic canal apertures were larger in females than males. While these observations indicate that pelvic aperture must increase in synchrony with egg width, they are insufficient grounds for accepting the hypothesis of constraint since it has not been demonstrated that a smaller, younger turtle could not have a larger pelvic opening. This is important because a correlation between female size and egg size has been found in a wide range of taxa (Table 10.3), some of which (e.g., fish and insects) are probably not constrained by mechanical factors.

Bears are unusual in giving birth to cubs that are far smaller than predicted from the allometric relationship between litter weight and maternal weight (Fig. 10.1). Ramsey and Dunbrack (1986, p. 735) advanced the hypothesis that the small size and immature condition of bear neonates constitute "an adaptive response to physiological constraints associated with supporting fetal development while in a state of hibernation and without access to food." During hibernation the fasting female mammal obtains energy by the catabolisis of free fatty acids. Although free fatty

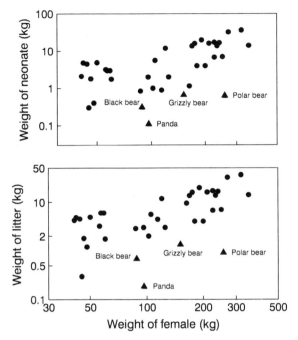

Figure 10.1. Neonate and clutch weights of bears in relation to adult weight, compared to other mammals of comparable body size. The relative litter masses of bears are substantially below those predicted by the overall allometric regression (equation 5.31). For bears the relative litter masses range from 0.18% to 0.87%, while the predicted range is 3.1% to 3.8%. Data from Leitch et al. (1959), Millar (1981), and Schaller (1985).

acids can pass through the placenta in some species, in vitro experiments indicate the fetal tissues cannot utilize these to meet their energy requirements. Two possible solutions to this problem are (Ramsey and Dunbrack 1986)

1. For the female to convert the maternal fat stores into ketone bodies for use by the fetus. Whether a fetus could in fact sustain and grow by this method has not been determined, but the levels circulating in the mother's system may interfere with the physiological adaptations required for survival under conditions of hibernation.

2. For the fat stores of the mother to be made available to her offspring by incorporating them into her milk. In contrast to fetuses, neonates can utilize free fatty acids. The solution apparently

Table 10.3. Propagule size in relation to the size of the female

Species	Rel	Reference
Pastinaca sativa (plant)	+	Hendrix 1979, 1984
Pseudocalanus sp. (copepod)	+	McLaren 1965; Hart and McLaren 1978
Bosmina longirostris (cladoceran)	+	Kerfoot 1974
Balanus balanus (barnacle)	+	Barnes and Barnes 1965
Armadillidium vulgare (isopod)	0	Lawlor 1976
Trichogramma embryophagum (wasp)	+	Klomp and Teerink 1967
Aedes aegypti (mosquito)	+	Steinwascher 1984
Colletes cunicularius (wasp)	+	Larsson 1990
Bembix rostrata (wasp)	+	Larsson 1990
Wasps 5spp	+	O'Neil and Skinner 1990
Graphosoma lineatum (bug)	+	Larsson 1989
Parapediasia teterrella (moth)	+	Marshall 1990
Chilo partellus (moth)	+	Berger 1989
Speyeria mormonia (butterfly)	+	Boggs 1986
Pieris rapae (butterfly)	−	Jones et al. 1982
Dacus 2 species	0	Fitt 1990
Drosophila melanogaster (fruit fly)	0	Warren 1924; David and Legay 1977
Drosophila melanogaster (fruit fly)	+	Eigenbrodt and Zahl 1939
Stenurella melanura (beetle)	+	Larsson and Kustrall 1990
Salmo gairdneri (Rainbow trout)	0	Scott 1962
Salmo salar (Atlantic salmon)	+	Pope et al. 1961; Kazakov 1981
Etheostoma 2 species (darter)	0	Hubbs et al. 1968
Gadus morhua (cod)	0	Oosthuizen and Daan 1974
Ariid catfish, 5 species	+	Coates 1988
Salvelinus alpinus (Arctic charr)	+	Jonsson and Hindar 1982
Oncorhynchus kisutch (coho salmon)	+	Fleming and Gross 1990
Cymatogaster aggregata (shiner perch)	+	Wilson and Milleman 1969
Morone saxatilis (striped bass)	+	Zastrow et al. 1990
Seriphus politus (queenfish)	+	DeMartini 1991
Rana sylvatica (wood frog)	+	Berven 1982
Rana temporaria (common frog)	+	Ryser 1988
Ambystoma talpoideum (salamander)	+	Semilitsch 1985
Hynobius nigrescens (salamander)	+	Takahashi and Iwasawa 1989
Urosaurus ornatus (lizard)	+	Dunham 1982
Cnemidorphorus (lizard) 1 sp.	+	Schall 1978
Cnemidorphorus 4 spp.	0	Schall 1978

Table 10.3. (*Continued*)

Species	Rel	Reference
Gekkonidae (lizard) 7 spp.	0	Vitt 1986; How et al. 1986; Selcer 1990
Deirochelys reticularia (turtle)	+	Congdon et al. 1983
Pseudemy scripta (turtle)	+	Congdon and Gibbons 1987
Chrysemys picta (turtle)	+	Tucker et al. 1978; Congdon and Gibbons 1987
Kinosternon flavescens (turtle)	+	Long 1986; Long and Rose 1989
Gopherus berlandieri (turtle)	+	Long and Rose 1989
Terrapene carolina (turtle)	+	Tucker et al. 1978
Terrapene ornata (turtle)	+	Long and Rose 1989
Wading birds 3 species	+	Väisänen et al. 1972; Jönsson 1987
Wading birds 2 species	0	Väisänen et al. 1972
Peromyscus polinotus (deermouse)	+	Kaufman and Kaufman 1987
Peromyscus leucopus (deermouse)	+	Millar 1978

Summary			
Group	+	0	−
Plants	1	0	0
Invertebrates	18	3	1
Vertebrates	29	17	0
Total	48	20	1

All analyses refer to variation within a population. Rel = size of propagule in relation to an increase in body size of female; + = increase, 0 = no effect, − = decrease.

taken by the female bear is to give birth at a very early stage of fetal development and supply further nourishment via suckling, thereby being able to utilize her fat reserves jointly for her own maintenance and for development of the neonate.

If this hypothesis is correct, bears that do not hibernate or fast for extended periods during pregnancy should produce young that are closer in size to those produced by other mammals of a similar size. Giant pandas (*Ailuropoda melanoleuca*) are closely related to modern bears, and give birth to one or two small young, only one of which typically survives (Schaller 1985). Unlike temperate bears, female giant pandas neither hibernate nor fast for long periods; therefore, the exceptionally low weight of neonates is unexpected. Ramsey and Dunbrack (1986, 1987) invoke the possibility of phylogenetic constraints to explain this anomalous pattern.

Without more detailed information on the evolutionary origins of denning and speciation within bears this hypothesis cannot be addressed.

10.2.2. Trade-Off Between Propagule Size and Fecundity

The most commonly cited reason for egg size not to increase continually is that there is a negative relationship between the number and size of eggs. This assumption was made by Smith and Fretwell (1974) in their pioneering analysis of the evolution of optimal clutch number and propagule size, though they provided no evidence for such a trade-off. Similarly, Vance (1973a,b) made the same assumption, without verification, in his analysis of the evolution of egg size in planktonic organisms. The majority of analyses since these studies either cite as authorities for this assumption these two studies (e.g., Brockelman 1975; McGinley et al. 1987; Sargent et al. 1987; Sibly and Monk 1987; McGinley and Charnov 1988) or make the assumption without citing either authority or evidence (e.g., Ware 1977; Winkler and Wallin 1987; Sibly et al. 1988). Verification of the assumption that increasing the number of propagules decreases fecundity is essential because it is critical to the analysis of the optimal combination of size and number of propagules. Such a relationship could arise because of mechanical constraints or a limited supply of resources. Featherwing beetles (Ptiliidae) are the smallest-known insects, measuring less than 1 mm. However, the egg produced by females of the genus *Bambara* is about 0.32 mm, or nearly one-half the length of the adult female (Dybas 1976). It is, therefore, not surprising that a female produces only one egg at a time. The common observation that fecundity increases with body size (chapter 5, section 5.2.2) suggests that egg production is in some way limited and that an increase in egg size would probably cause a reduction in fecundity. That the exponent in the allometric equation of log(fecundity) on log(length) is frequently within the range two to three further suggests that size may be causally connected: the number of eggs may be limited by total body volume or by cross-sectional area.

A variety of studies have addressed the hypothesis of a trade-off between egg size and clutch size or fecundity (Table 10.4). Tests have been based on both interspecific and intraspecific comparisons, the latter involving either interpopulation or intrapopulation comparisons. All tests are weak because they are based on unmanipulated observations and hence cases in which the predicted correlation is not observed may reflect correlated differences associated with the two traits. For example, females laying small eggs may be physiologically stressed and both egg size and fecundity may be pathologically small: this may account for the persistent failure to find a trade-off in birds (Ojanen et al. 1978). Tests between species and different populations within the same species are also suspect because selective

Table 10.4. Studies reporting a phenotypic correlation between clutch size and propagule size

Taxon	Trade-off?	Reference
Comparison between species		
Flatfish	Yes	Roff 1982
Darters	Yes	Paine 1990
Salamanders	Yes	Kaplan and Salthe 1979
Hylidae	Yes	Crump and Kaplan 1979
Solidago sp. (plant)	Yes	Werner 1979
Nematodes	Yes	Zullini and Pagani 1989
Gammarus sp.	Yes	Kölding and Fenchel 1981
Insects	Yes	Berrigan 1991
Turtles	Yes	Elgar and Heaphy 1989
Waterfowl	No	Rohwer 1988
Comparison between populations		
Clupea harengus (herring)	Yes	Blaxter 1969
Oncorhynchus kisutch (salmon)	Yes	Fleming and Gross 1990
Hynobius nigrescens (salamander)	Yes	Takahashi and Iwasawa 1988a
Lacerta lepida (lizard)	Yes	Castilla and Bauwens 1989
Waterfowl (17 species)	No	Rohwer 1988
Comparison within a population		
Brassica napus (plant)	Yes	Olsson 1960
Brassica campestris (plant)	Yes	Olsson 1960
Vicia fabae (bean)	Yes	Hodgson and Blackman 1956
Simocephalus exspinosus (cladoceran)	Yes	Agar 1914
Mesocyclops edax (copepod)	Yes	Allan 1984
Armadillidium vulgare (isopod)	Yes	Lawlor 1976a
Gasterosteus aculeatus (stickleback)	Yes	Snyder 1990, 1991b
Ambystoma tigrinum (salamander)	Yes	Kaplan and Salthe 1979
Ambystoma spp. (2 species)	No	Kaplan and Salthe 1979
Uta stansburiana (lizard)	Yes	Nussbaum and Diller 1976; Ferguson and Fox 1984
Sceloporus occidentalis (lizard)	Yes	Sinervo 1990
Waterfowl (3 species)	No	Rohwer and Eisenhauer 1989
Passerines (5 species)	No	Ojanen et al. 1978
Sturnus vulgaris (starling)	Neg[a]	Ojanen et al. 1978
Spinus tristis (goldfinch)	Neg[a]	Holcomb 1969
Troglodytes aedon (wren)	No	Kendeigh et al. 1956
Lagopus lagopus (red grouse)	No	Moss et al. 1981
Parus major (great tit)	Yes	Jones 1973
Corvus monedula (jackdaw)	No	Soler 1988
Larus argentatus (herring gull)	Yes	Gordi and Herrera 1983
Homo sapiens (human)	Yes	McKeown et al. 1976

For data on small mammals see Table 9.5. Under the trade-off hypothesis the correlation will be negative.

[a]In these species the correlation is negative; i.e., larger clutches have larger eggs.

regimes may be very different and hence patterns of allocation to reproduction could vary widely. Nevertheless, these criticisms notwithstanding, excluding birds, there are clearly significantly more cases in which a trade-off has been observed than not (from Table 10.4: 23+, 2 NS; and for small mammals, from Table 9.5: 9+, 7 NS), giving overall support to the hypothesis.

Propagule size may be under direct control of the female or may be a consequence of competition between the propagules themselves for a limited resource supply. In the former case egg size will be fixed, and experimentally aborting some of the embryos should not produce changes in the size of the eggs or neonates. On the other hand, if there is sibling competition within the mother, abortion should lead to an increased initial size of offspring. Manipulation of the number of propagules is most easily done in plants. Removal of flowers of early developing seeds leads to an increase in seed size in *Vicia fabae* (Hodgson and Blackman 1956), *Phaseolus vulgaris* (Adams 1967), *Lupinus luteus* (Steveninck 1967), wheat (Bingham 1967), soybeans (Smith and Bass 1972), and *Zea mays* (Dyar 1975). Sinervo and Licht (unpublished work cited in Bernado 1991) have tested the fixed-resources hypothesis, using the lizard, *Uta stansburiana*, by experimentally ablating follicles early in vitellogenesis after the state at which further follicles can be recruited. The result was an increase in egg size. Both the experiments on plants and those by Sinervo and Licht support the hypothesis that a female allocates a fixed amount of energy to the developing propagules. However, if the concept of reproductive investment is taken to include parental care the situation is not so clear; birds may increase their effort in experimentally increased brood sizes or may continue to forage and supply food at the same rate. Similarly, among small mammals rate of food intake may increase with litter size, reflecting the increased demands of lactation.

From an evolutionary perspective we require not only that there be a phenotypic correlation between egg size and egg number but also that this trade-off be genetically based. There is abundant evidence that egg size is genetically determined, frequently with significant additive genetic variation (Table 10.5), though maternal effects may be very significant (Roach and Wulff 1987; Mousseau and Dingle 1991). Clutch size and fecundity also frequently show significant heritabilities (Tables 9.7 and 10.6). Unfortunately there is very little information on the genetic correlation between size and number, though in the three-spined stickleback the genetic correlation is negative (-0.98 ± 0.02) as expected (Snyder 1991b).

Selection for large seed size in *Sinapis alba*, *Brassica napus*, and *B. campestris* resulted in a decreased number per pod (Olsson 1960), indicating a negative genetic correlation between size and number. Tedin (1925) obtained similar results for selection on seed size in an annual species of

Table 10.5. Evidence that propagule size is genetically variable

Species	Evidence[a]	Reference
Silene dioica (red campion plant)	CG	Thompson 1981
Bromus inermis (grass)	S	Christie and Kalton 1960
Lotus corniculatus (birdsfoot trefoil)	S	Draper and Wilsie 1965
Phaseolus vulgaris (kidney bean)	S	Allard et al. 1968
P. lunatus (lima bean)	S	Allard et al. 1968
Sorghum vulgare (sorghum)	S	Allard et al. 1968
Sorghum vulgare (sorghum)	0.6	Voigt et al. 1966
Glycine max (soybean)	0.93	Fehr and Weber 1968
Linum usitatissumum (plant)	D	Khan 1967
Brassica sp.	S	Olsson 1960
Sinapis alba	S	Olsson 1960
Cleome serrulata (bee plant)	0	Farris 1988
Lupinus texensis (lupine)	0.1	Schaal 1980
Anthoxanthum odoratum (plant)	D	Antonovics and Schmitt 1986
Phlox drummondi (phlox)	0	Schwaegerle and Levin 1990, 1991
Drosophila melanogaster (fruit fly)	CG	Oksengorn-Proust 1954; Cals-Usciati 1964; David and Legay 1977
Drosophila melanogaster (fruit fly)	S	Parsons 1964
Hyalella azteca (amphipod)	CG	Strong 1972
Poecilia reticulata (guppy)	CG	Reznick 1982; Reznick and Endler 1982; Reznick et al. 1990
Gambusia affinis (mosquito fish)	CG	Reznick 1981
Cyprinius carpio (carp)	0.24	Kirpichnikov 1981
Gasterosteus aculeatus (stickleback)	0.38	Snyder 1991b
Hyla crucifer (tree frog)	0.28	Travis et al. 1987
Domestic fowl	0.49	Kinney 1969
Domestic fowl	0.63	Van Tijen and Kuit 1970
Domestic turkey	0.42	Nestor et al. 1972
Anas platyrynchos (mallard)	0.55	Prince et al. 1970
Lagopus lagopus (red grouse)	0.70	Moss and Watson 1982
Anser caerulescens (goose)	0.53	Lessells et al. 1989
Parus major (great tit)	0.61	Noordwijk et al. 1980
Parus major (great tit)	0.72	Jones 1973
Parus major (great tit)	0.86	Ojanen et al. 1979

Where heritability estimates have not been made, the method of analysis is presented.

[a]CG, "common garden" experiments (different populations grown under identical conditions).

S, selection experiments.

D, diallel cross.

Table 10.6. Heritability estimates for fecundity or clutch size in some animal species, mostly insects

Species	h^2	SE	Reference
Apis mellifera	0.29	0.01	Hillesheim 1984
Dysdercus bimaculatus	0.27	0.09	Derr 1980
Oncopeltus fasciatus	0.24	0.11	Hegman and Dingle 1982; Palmer and Dingle 1986
Tribolium casteneum	0.33	0.04	Orozco 1976
Drosophila melanogaster	0.18	—	Robertson 1957
Drosophila melanogaster	0.04	0.04	Tantawy and El-Helw 1966; Tantawy and Rakha 1964
Drosophila melanogaster	0.35	—	Rose and Charlesworth 1981a
Drosophila simulans	0.11	0.07	Tantawy and Rakha 1964
Callosobruchus maculatus	0.10	0.11	Møller et al. 1989
Gambusia affinis (fish)	0.16	—	Busack and Gall 1983
Heterandria formosa (fish)	0.20	0.12	Henrich and Travis 1988
Gasterosteus aculeatus (fish)	0.21	0.26	Snyder 1991b
Domestic fowl	0.22	—	Kinney 1969
Domestic geese	0.34	—	Merritt 1962

Multiple estimates from a single study averaged.

Camelina: a decline in seed number per pod accompanied selection for increased seed size, but selection for both large seed and high number could be achieved by simultaneous selection for both seed size and pod size. In an attempt to select for high yield in winter barley, Nickell and Grafius (1969) selected plants on the basis of yield, kernel weight, and seed number per unit area. Instead of an increase in yield there was a significant drop from 4,271 kg/ha to 3,267 kg/ha. The reason for this decline was attributed to negative genetic correlations between the number of heads, the number of kernels per head, and kernel weight: while the number of heads per unit area increased from 100.9 to 199.5, the number of kernels per head decreased from 43.8 to 20.2 and kernel weight from 37.4 mg to 28.5 mg. As no control line was grown, these results must be interpreted with caution.

The assumption that there is, in general, a trade-off between size and number of offspring appears warranted, though in analyses of specific cases this should be tested. Smith and Fretwell (1974) began with this assumption and asked the question, How should a fixed quantity of resource be divided into propagules? They assumed

1. that the population is stationary and natural selection maximizes the expected number of offspring produced

2. that the number produced at each breeding episode does not affect the survival or future fecundity of the parent
3. that the relationship between the mean fecundity of offspring (= survival × fecundity of offspring) is a concave function of the amount invested per offspring with some minimum investment required (Fig. 10.2)

The last assumption can be relaxed to the extent that the minimum requirement for there to be an optimum size/number combination is for the rate of change of the mean fecundity of offspring to monotonically decrease beyond some effort per offspring. (See Fig. 4.5 for such an alternative curve, a sigmoidal curve.)

The fitness of a particular investment per offspring is obtained by multiplying the mean fecundity of offspring by the number of offspring produced (fecundity). To graphically locate the investment per offspring, x, at which fitness, $w(x)$, is maximized, we find the tangent to the mean offspring fecundity function for a line originating from the origin (Fig. 10.2): the value of x at this point is the optimal investment per offspring. This procedure can be analytically justified as follows:

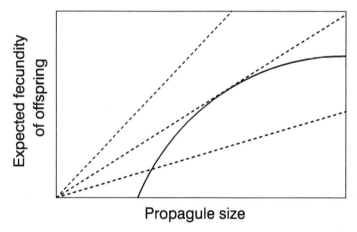

Propagule size

Figure 10.2. The model (solid line) proposed by Smith and Fretwell (1974) relating expected fecundity of offspring, $f(x)$ (=survival × fecundity), to propagule size, x (or resources invested into a propagule). Dashed lines show three fitness isoclines ($f(x) = xw(x)/I$). Only lines that intersect the curve relating expected fecundity to propagule size (solid line) are permissible (hence upper line is not possible). The largest fitness isocline (i.e., highest value of $w(x)$) is that which is tangential to the expected fecundity curve (i.e., the middle dashed line).

$$\text{Fitness} = w(x) = f(x)N(x)$$

$$= f(x)\frac{I}{x} \tag{10.1}$$

where $f(x)$ is the mean fecundity of offspring for an investment per offspring of x,

$N(x)$ is the number of offspring for an investment per offspring of x,

I is the total invested into offspring.

Note that $f(x)$ can be written as $xw(x)/I$; thus for a given value of fitness, $w(x)$, and a fixed investment I, $f(x)$ is a linear function of x passing through the origin (Fig. 10.2). Each line passing through the origin represents a fitness isocline (i.e., a particular value of $w(x)$), the value increasing with the angle subtended at the origin. The investment per offspring, x, that generates a fitness value $w(x)$ is the value of x at the intersection of the fitness isocline and the mean fecundity of offspring function, $f(x)$, (Fig. 10.2). From the above argument it is obvious that the maximal value of $w(x)$ obtains where the fitness isocline is tangent to $f(x)$.

The model of Smith and Fretwell establishes that, given a trade-off between propagule size and propagule number, there will exist a particular combination that maximizes fitness. In a constant environment there will be, in general, a single optimal size, but in an environment that is changing there will be a concurrent change in the most fit size/number combination of offspring. The following discussion will consider these two scenarios in turn.

10.3.   Optimal Propagule Size in a Constant Environment

10.3.1.   Propagule Size and Development Time

Onset of maturation or metamorphosis is frequently triggered by the attainment of a critical size. (See chapter 5, section 5.2.2.) Given a fixed size at maturation, all things being equal, the larger the size at birth (hatching or sprouting) the shorter the time required to become an adult. If the initial size at hatching is $W(0)$ and the growth rate some simple function of age, $f(x)$, then the size at age x will be $W(x) = W(0)f(x)$. Increasing the initial size by some constant factor will increase size at all ages by an equivalent amount and hence could conceivably significantly affect development time. This effect may be accentuated by correlations between the growth parameters (Kaplan 1980).

The difference in time to maturation will depend upon the particular form of the growth function: we shall consider two cases—first that in which initial size has little effect and second that in which initial size may have a very significant impact on development time. Suppose growth is linear with time,

$$W(x) = W(0) + cx \tag{10.2}$$

It is evident that if initial size is a very small fraction of the adult size, as is generally the case, variation in initial size will have an insignificant impact on the time required to reach some threshold size for maturation.
But suppose growth is exponential,

$$W(x) = W(0)e^{cx} \tag{10.3}$$

where $W(x)$ is size at age x, $W(0)$ is size at birth, assumed to be proportional to egg or seed size in those organisms that are not viviparous (see section 10.1 for references supporting this assumption), and c is a constant.

If size at hatching is increased by some fraction p, size at each age, $W(x)$, will be increased by a similar amount. The time, X, required to reach the threshold size, $W(X)$, for maturation of metamorphosis is

$$X = \frac{1}{c} \log_e \left(\frac{W(X)}{W(0)} \right) \tag{10.4}$$

and the time required the reach the same size, X_p, given an initial size of $pW(0)$ ($p > 1$), is

$$X_p = \frac{1}{c} \log_e \left(\frac{W(X)}{pW(0)} \right) \tag{10.5}$$

The ratio of X_p to X is

$$\frac{X_p}{X} = \frac{\log_e \left(\dfrac{W(X)}{pW(0)} \right)}{\log_e \left(\dfrac{W(X)}{W(0)} \right)}$$

$$= 1 - \frac{\log_e p}{\log_e \left(\dfrac{W(X)}{W(0)} \right)} \tag{10.6}$$

Reasonable figures for initial and final sizes for an insect such as a grasshopper are 0.001 g and 0.5 g, respectively. A doubling of the initial size ($p = 2$) will reduce the time to adulthood by a factor of 0.9; i.e., if hatchling size is doubled the time to reach the threshold size is 90% of its former value. But doubling of egg size will reduce fecundity by one-half, and hence selection will not necessarily favor such an increase in egg size. The optimal size will depend upon the effect of a reduction in development time on overall fitness, which can only be ascertained by incorporating considerations of egg size on other life history components into a more comprehensive model.

The growth trajectory of young isopods, *Amadillidium vulgare*, is exponential (Hubbell 1971), and this, in conjunction with initial differences in size at hatching, in large measure account for the negative correlation between mean size of ova and weight of 37-day-old young (Lawlor 1976). Lawlor used these observations to predict the optimum number of young per unit weight in female *A. vulgare*. Specifically, he assumed

1. Population size is stationary, and thus the appropriate measure of fitness is the average lifetime fecundity, R_0.

2. All offspring mature at a fixed age X.

3. The probability of surviving from hatching to reproduction is a constant, S.

4. The coefficient of exponential growth (c in equation 10.3) is uncorrelated with size at hatching. The actual form of the growth function in this analysis is not critical. The important assumption is that between two individuals the ratio of size at some given age is the same as that between their respective egg sizes.

The size of young at maturity, $W(X)$, is

$$W(X) = G(2.587 - 0.317\ F) \tag{10.7}$$

The first term on the right-hand side is the increase in size between age 37 days and maturity; the second term is the linear regression between weight at 37 days and the size-specific fecundity of the female (F). Fecundity is proportional to female weight (Lawlor 1976), and hence fitness can be equated to the total biomass, B, of female offspring that survive to reproduce (i.e., R_0 is proportional to B),

$$B = \frac{1}{2}G(2.587 - 0.317F)FW(X)S \tag{10.8}$$

To obtain the optimal size-specific fecundity we differentiate B with respect to F and find that value of F at which dB/dF equals zero,

$$\frac{dB}{dF} = \frac{1}{2}G(2.587 - 0.317F^2)W(X)S$$

$$= 0 \quad \text{when} \quad F = \frac{1}{2}\frac{2.587}{0.317} = 4.08 \text{ young per mg of female} \quad \textbf{(10.9)}$$

The optimal size of young is thus $1/4.08$ mg $= 0.245$ mg. The observed size-specific fecundity of 3.48 young/mg female, equal to a weight of 0.287 mg per young, is reasonably close to that predicted. Given the very simple model the correspondence between observation and prediction is encouraging.

Sibly and Monk (1987) examined the fitness consequences of variable egg size in two grasshopper species, *Chorthippus brunneus* and *C. parallelus*. Details of their model have been presented in chapter 4 (section 4.2.1), and only a general description is presented here. Comparisons between different populations showed that development time in these species decreases with egg weight (Fig. 4.1). Both species are univoltine and thus fitness can be equated with the total number of eggs laid by a female that survive to hatch the following spring. The trade-offs that form the core of Sibly and Monk's analysis are the trade-offs between egg size and number, and between development time and egg size. Increasing egg size decreases development time and both reduces the mortality incurred by the nymphs (simply because they spend less time in this stage) and increases the number of eggs laid by increasing the time available to lay eggs. Assuming constant rates of mortality for the various life-cycle stages, Sibly and Monk derived the equation describing the fitness isoclines between egg size and development time. As described in section 4.2.1, the optimal combination can be found exactly using the calculus, though this requires precise specification of the trade-off function of development time versus egg size. Sibly and Monk did not use this approach but used their theoretical model to predict what the slope of the trade-off function should be at the observed combinations of egg size and development time. The trade-off curves for each species considered separately are roughly linear and therefore the observed slope can be estimated by fitting a straight line through the three points for each species giving slopes of -0.106 for *C. parallelus* and -0.056 for *C. brunneus* (Sibly and Monk 1987). From the three populations the predicted slopes for *C. parellelus* are -0.024, -0.030, and -0.033, substantially different from that observed (-0.106). For *C. brunneus* the fit is closer, the predicted slopes from the three populations being -0.040, -0.059, and -0.081, compared to the observed value of -0.056. A problem with this approach is that it is very difficult to assess the magnitude of the differences. Predictions based on the egg size/development time combinations would be much easier to evaluate. Further analysis is required to test the predictive ability of the model. An important function that this analysis serves is to evaluate the relative importance of the different components. Sensitivity analysis indicates that most attention should be paid to estimating the egg-laying period, development time, and juvenile mortality rate. The analysis can also be extended to include other factors (Table 10.7); testing these predictions remains to be undertaken.

Ware (1975b, p. 2507) developed a model for optimal egg size in marine organisms based on the argument that "growth and mortality in the sea are continuous interdependent processes, and that the instantaneous mortality rate is inversely proportional to particle size regardless of whether it is an egg or larvae." Evidence presented in chapter 5 (Fig. 5.13) indicates that mortality rates in marine systems depend in large measure on size. McGurk's (1986) analysis of mortality rates of fish eggs and larvae suggest that the relationship between size and mortality differs between these two life history stages. The difference, however, is not sufficient to negate the qualitative predictions of Ware's model. The fundamental equation of this model relates the probability that an egg survives the incubation period and hatches, S_1, to egg weight, W,

$$S_1 = e^{-xc_1 W^{-c_2}} \tag{10.10}$$

Table 10.7. Effect on optimal egg size of increasing parameter values in the model of Sibly and Monk

Effect of increasing	Trade-off between egg size and			
	t_n	T	M_e	M_n
Time from laying to hatch, t_e	0	0	+	0
Duration of nymphal stage, t_n	+	+	−	+
Interclutch interval, t_I	0	0	+	0
Egg-laying period, T	−	−	+	0
Egg mortality rate, M_e	−	+	?	0
Nymphal mortality rate, M_n	+	0	0	0
Adult mortality rate, M_a	−	−	?	0

+ = increase in egg size, − = decrease, 0 = no effect, ? = depends on parameter values. It is assumed that $M_e \neq M_a$.

From Sibly and Monk 1987.

Nature of trade-offs.

t_n: As indicated in the text and Fig. 4.1.

T: Assumed that the larger the egg, the earlier the hatching date, and therefore, T increases with egg size.

M_e: Egg mortality decreases with egg size.

M_n: Nymphal mortality decreases with egg size.

where c_1 and c_2 are constants and x is incubation time. Note that this equation is virtually the same as that later derived by Peterson and Wroblewski (1984, equation 5.18). The number of hatching larvae, N_1 is equal to ovary weight divided by the weight of a single egg, multiplied by S_1. In pelagic fish the bulk of mortality occurs in the egg and larval stages during the first growing season, which may last up to 200 days (Ware 1975b). If we take the number of offspring surviving to the end of the first growing season as a fitness component and assume that effects due to initial egg size do not extend to the remainder of the life, we can compute the optimal egg size as that which maximizes $N = N_1 S_1 S_2$, where S_2 is the probability of a larvae of weight W surviving to the end of the season. Hatching weight was estimated from egg diameter assuming that the egg is spherical and hatching weight is proportional to egg volume. From considerations of survival as a function of particle size and effects of density Ware derived the probability of survival to the end of the season after hatching to be

$$S_2 = \frac{W^{c_3}}{((c_4 - c_5 N_1)(\omega - x) + W^{c_2})^{c_6}} \tag{10.11}$$

where the c_is are constants derived empirically and ω is the length of the growing season, which Ware arbitrarily set at 200 days. For any given incubation period there is an optimal egg size that maximizes the number of offspring surviving to the end of the season. Increasing incubation time favors an increasing egg size as actually observed, though the predicted egg diameters are consistently too small (Fig. 10.3). Unfortunately, Ware did not undertake a sensitivity analysis to locate those parameter values that might be responsible for the mismatch.

Kölding and Fenchel (1981) proposed essentially the same hypothesis as Ware to explain variation in egg size among five species of amphipods of the genus *Gammarus*. In their analysis Kolding and Fenchel considered the entire juvenile period, suggesting that lower temperatures produce a longer juvenile stage and hence higher mortality, simply by virtue of the increased period of exposure to risks. Selection will then favor an increase in egg size to decrease the juvenile period. From these considerations Kolding and Fenchel predicted that species active in the winter months will lay larger eggs, as observed in *Gammarus*. This hypothesis and that of Ware assume changes in temperature do not change mortality rates. But if the primary source of mortality is predation from ectotherms, mortality rates will vary with temperature, decreasing as temperature decreases. Consequently, qualitative arguments are insufficient to support the logic of the hypothesis that colder temperatures increase development time, increase mortality, and thus favor larger initial size. Support for this

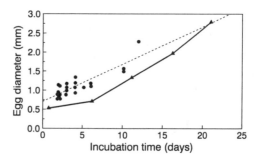

Figure 10.3. Comparison of predicted optimal egg size (triangles and solid line) and observed (dots) for 14 species of fish that spawn pelagically in the Northwest Atlantic, as a function of incubation duration. Dashed line shows the fitted regression ($y = 0.101 x + 0.67$, $r = 0.87$). Modified from Ware (1975b). Reproduced with the permission of the Minister of Supply and Services Canada, 1992.

hypothesis requires quantitative estimates of the variation in mortality rates with temperature.

Vance (1973a,b) postulated that predation and starvation may play a significant role in the evolution of egg size in marine benthic invertebrates. Three principal modes of development can be recognized among this group:

1. Planktotrophic larvae: species with a free-swimming larval stage that feeds upon smaller planktonic organisms

2. Pelagic lecithotrophic larvae: species in which the larval period is spent in the plankton but which do not feed, nutrients for development coming from stored reserves in a yolk sac

3. Nonpelagic lecithotrophic larvae: species in which an independent larval stage has been suppressed, the larvae either developing in an egg deposited in the benthos or retained within the body of the mother

Propagule size increases, in general, from pattern 1 to pattern 3. Vance proposed the following model to account for this variation: The principal assumption is that there is a fixed development time from conception to metamorphosis. This period is divided into two components: a nonfeeding stage during which stored nutrients are utilized and a feeding stage during which the larvae feed on planktonic organisms. To increase the first phase a female must increase egg size to supply the required increase in nutrients. Vance then examined the consequences of different mortality rates in the two phases. As is intuitively obvious, different modes of larval development will evolve in response to variation in the mortality schedule. This model is a specific case of Shine's (1978) "safe harbor hypothesis," which itself (as noted by Shine) can be traced to Williams (1966): natural selection should adjust the life history such that organisms minimize the time spent in life stages characterized by high mortality. Shine (1989, p. 311) hypothesized that "if the egg stage is a safe harbor (as in species with parental care), whereas juvenile life is hazardous, selection should favor an increase in egg size and thus a decrease in the duration of the high-risk juvenile phase." An important caveat that must be attached to this statement is that high mortality rates may be traded-off against other fitness components such as increased fecundity. Underwood (1974) criticized the models of Vance on the grounds that among species there is no correlation between egg size and development time, either in total or when only the length of the prefeeding period is considered. Steele (1977) countered with examples in which such correlations have been observed. But this argument is ill-founded, for interspecific comparisons are not relevant unless properly analyzed: interspecific differences in other life history components such as the size at metamorphosis may confound attempts to extract the required relationship (Strathmann 1977). A correct examination of the hypothesis requires intraspecific comparisons (Vance 1974), and even in this case care must be taken to statistically hold constant other confounding variables.

10.3.2. Propagule Size and Rate of Development

Implicit in the model of Sibly and Monk (1987) is the assumption that each nymph follows the same growth trajectory, simply starting life at different points of the curve. But development rates may differ, both between individuals and between populations. A decrease in the rate of growth favors larger propagules (Sibly and Calow 1983, 1985; Parker and Begon 1986; Sibly et al. 1988). Sibly and Calow (1983, 1985) provide a general demonstration of this proposition, but for purposes of illustration I present here the more specific model of Sibly et al. (1988). This model, graphically illustrated in Fig. 10.4 assumes

1. That developmental period is fixed.
2. The amount of resources, R, accumulated by a female increases with her size, A.
3. Since the duration of development is fixed, adult size is increased by increasing propagule size or development rate.
4. Adult size is a linear function of egg size, $A = cx$, where x is propagule size.

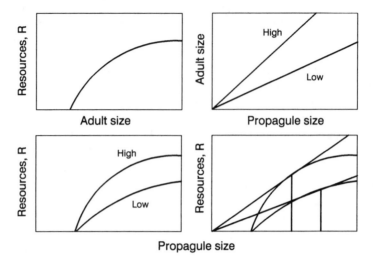

Figure 10.4. Upper left panel: Hypothetical relationship between the resources an adult gathers and its size. Increasing resources are used to increase fecundity and/or propagule size. Upper right panel: Adult size increases linearly with propagule size, the upper line showing the relationship with a high rate of development and the lower that for a low developmental rate. Lower left panel: Eliminating adult size from the above two graphs gives the relationship between the size of the propagule and its projected resource-gathering ability (i.e., fitness). Lower right panel: The optimal propagule size is obtained by drawing the line subtended at the origin that is tangent to the curve describing resource-gathering ability as a function of propagule size. Note that the optimal propagule size is reduced when development rate is high.

Eliminating adult size from the above relationships gives the relationship between resources gathered and propagule size, x.

Following the recipe of Smith and Fretwell (1974), the optimal propagule size is obtained where a line from the origin lies tangent to R (Fig. 10.4), a condition specified by

$$\frac{dR}{dx} = \frac{R}{x} \qquad (10.12)$$

Now R is a function of adult size, $f(A)$ (Fig. 10.4). Using the chain rule equation 10.12 can be rewritten

$$\frac{df(A)dA}{dA\ dx} = \frac{f(A)}{x} \tag{10.13}$$

and since from assumption 4 $(A = cx)$ $dA/dx = c$, we obtain

$$\frac{df(A)}{dA} = \frac{f(A)}{A} \tag{10.14}$$

The optimal adult size A^* satisfies equation 10.14, and does not depend upon the growth rate, c. But the optimal propagule size, x^*, is given by A^*/c (since $A = cx$), and hence as growth rate increases the optimal propagule size decreases.

Orton and Sibly (1990) attempted to test the hypothesis that propagule size and growth rate will be inversely correlated using the freshwater snail, *Theodoxus fluviatilis*. Hatchling length was estimated from egg capsule length and growth rate from the sizes of hatchlings and yearlings. Five sites were studied, two sites for 2 years and three for a single year. The correlation between capsule base diameter and growth rate was negative $(r = -0.68, n = 7, P < 0.05,$ one-tailed test), though this correlation depends upon a single extreme population. A further statistical difficulty with interpreting this result is that the multiple measures may not represent independent data points and hence the degrees of freedom may be inflated. Thus while the result is consistent with the prediction it is not sufficiently convincing to accept it as an adequate test.

Perrin (1988) examined, in the cladoceran *Simocephalus vetulus*, the response of egg size to changes in growth rate induced by changes in temperature. All animals used in the experiment were derived from a single female and hence the potential confounding effect of genetic variation between individuals was removed. Across the three temperatures used, egg size declined significantly with increasing temperature, while growth rate at birth increased (Perrin 1988). This experiment is not without its problems. First, temperature may be used as a cue by cladocerans to indicate the predator regime to which their offspring will be subjected (see section 10.4.1) and hence the response may have no causal connection with growth rate. Second, the measurement of growth rate on a chronological scale may not be appropriate for an ectotherm: a physiological scale may be the correct frame of reference. By changing the temperature all physiological processes in an ectotherm are altered and hence an apparently slow growth rate from an endothermic perspective may not be so when time is measured in degree days. (Measured in degree days, development times typically remain constant with temperature across the tolerance range of a particular population.)

Perrin (1989) further tested the hypothesis of a negative correlation between egg size and growth rate in *S. vetulus* by rearing them at different densities. Increased density was accompanied by a decrease in growth rate and an increase in offspring size. A similar change in propagule size with density was reported by Smith (1963) for the cladoceran, *Daphnia magna*. Again, though this result is consistent with the hypothesis under test, it is somewhat flawed by the complication that the adult female may have also been under stress and the response may have been due to this phenomenon and not due to the growth rate of juveniles. Despite the shortcomings of the three experiments discussed above, the overall consistency of the results does suggest that, indeed, selection favors an increased propagule size with a decreased juvenile growth rate.

10.3.3. Propagule Size, Female Condition, and Female Body Size

Egg size is frequently correlated with the ration received and size of the female. Typically, reduced ration is accompanied by a reduction in egg size (Table 10.8), while an increase in female size leads to an increase in egg size (Table 10.3). Tucker et al. (1978) and Congdon and Gibbons (1987) postulated that egg size is constrained in turtles by the size of the pelvic aperture and that the optimal egg size is larger than this aperture, favoring an increase in egg size with body size. (See section 10.2.1.) However, the ubiquity of the phenomenon and its occurrence in organisms that are not likely to be so constrained (Table 10.3) suggest that other factors may be important, at least in other taxa. Though at first sight the corre-

Table 10.8. Relation between propagule size and ration received by the female

Species	Relation	Reference
Linyphia triangularis (spider)	+	Turnbull 1962
Pardosa (4 spp.) (spider)	+	Kessler 1971
Mytilus edulis (mussel)	+	Bayne et al. 1978
Armadillidium vulgare (isopod)	−	Brody and Lawlor 1984
Aedes aegypti (mosquito)	+	Steinwascher 1984
Salmo gairdneri (trout)	NS	Scott 1962
Salmo trutta (trout)	+	Bagenal 1969
Lebistes reticularis (guppy)	NS	Hester 1964
Gasterosteus aculeatus (stickleback)	NS	Wootton 1973
Melanogrammus aeglefinus (haddock)	+	Hislop et al. 1978
Vipera berus (adder)	+	Andrén and Nilson 1983
Sterna hirundo (tern)	+	Nisbet 1978
Fulica atra (coot)	−	Horsfall 1984

+, propagule size increased with ration; −, propagule size decreased with ration; NS, no significant relationship observed.

lations with ration and female size appear unrelated, there is, as shown below, a theoretical framework within which both phenomena can be understood.

Parker and Begon (1986) analyzed the question by using as a premise the assumption that selection maximizes a female's productivity. The female spends time t foraging for resources and then lays a clutch of eggs. In the first instance we assume that the survival of offspring is related only to the provisions they receive (i.e., no competition between immatures). Let $f(x)$ be the relationship between mean fecundity of offspring (= survival $\times$ fecundity of offspring) and propagule size. As with the case of optimal clutch size (section 9.3.1.) we can write

$$\text{Productivity per unit time} = w(x,t) = \frac{g(t)}{x} \frac{f(x)}{(t + d)} \qquad \textbf{(10.15)}$$

where $g(t)$ is the amount of resources gathered by a female during time t (thus $g(t)/x$ is clutch size) and d is the minimum time required between the laying of each clutch. (For example, d could be the time required to travel from the feeding site to the oviposition site.) The function $g(t)$ may represent either ration or, if resource acquisition increases with size, female size. There is both an optimal egg size and an optimal foraging time. To find the optimal egg size we differentiate with respect to x, setting the result to zero

$$\frac{\partial w(x,t)}{\partial x} = 0 \qquad \text{when} \quad \frac{\partial f(x)}{\partial x} = \frac{f(x)}{x} \qquad \textbf{(10.16)}$$

The optimal foraging time can be found in the same manner,

$$\frac{\partial w(x,t)}{\partial t} = 0 \qquad \text{when} \quad \frac{\partial g(t)}{\partial t} = \frac{g(t)}{t + d} \qquad \textbf{(10.17)}$$

Note that the optimal egg size is independent of the optimal foraging time, t. Furthermore, since the optimal egg size is independent of the resource-gathering function, $g(x)$, the optimal egg size is independent of female size and ration. This conclusion is quite general and can be obtained, for example, from the *Drosophila* model of chapter 7. The situation changes, however, when we introduce competition into the model.

First we consider only competition between siblings. Parker and Begon (1986) assumed that competition is primarily a consequence of clutch size rather than egg size, arguing that competitive interactions occur when the larvae are significantly larger than their initial size at hatching, and hence initial provisioning is not important in determining the degree of competition. Letting the number of young in a clutch be n and the relationship between that component of the expected offspring fecundity attributable to the effect of competition be $h(n)$ we have

$$w(x,t) = \frac{nh(n)f(x)}{t + d} \tag{10.18}$$

Proceeding as above, the optimal egg size is that size at which

$$\frac{\partial f(x)}{\partial x} = \frac{f(x)}{x}\left(\frac{h(n) + n\dfrac{\partial h(n)}{\partial x}}{h(n)}\right) \tag{10.19}$$

and the optimal foraging time when

$$\frac{\partial g(t)}{\partial t} = \frac{g(t)}{t + d}\left(\frac{h(n)}{h(n) + n\dfrac{\partial h(n)}{\partial t}}\right) \tag{10.20}$$

Unlike the previous case, egg size is not independent of female size since when both egg size and foraging time are optimized equations 10.19 and 10.20 must be simultaneously satisfied, with $n = g(t)/x$.

To examine the implications of the above equations with respect to egg and female size, Parker and Begon inserted explicit functions for $f(x)$ and $h(n)$ (Fig. 10.5). Under this scenario both clutch size and egg size increase with female size or ration (Fig. 10.5). What we presently lack is a general analysis of the above model—specifically, an analysis of the conditions under which *egg size will decrease or remain constant* with ration or female size. Obviously as the influence of competition decreases so will the change in egg size with ration or female size. The foregoing model assumed that competition occurs only between siblings. Extension of the model to include competition with nonsiblings does not change the general conclusion that large females should lay large eggs (Parker and Begon 1986), and by extension that females that receive a large ration should lay large eggs.

From the above analysis we find that egg size increases with ration or body size when the expected fecundity of offspring is composed of two components: a density-independent component, $f(x)$, in which mean offspring fecundity increases with egg size, and a density-dependent component, $h(n)$, in which mean offspring fecundity decreases with the number of competing young. It is therefore surprising to find that a model in which

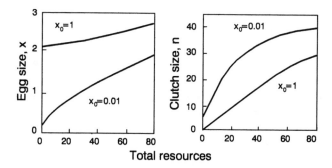

Figure 10.5. Optimal egg and clutch sizes as a function of resources available for reproduction (i.e., better female condition or larger female size). Results are for the following specific model:

$$f(x) = 1 - e^{(x - x_o)}$$
$$h(n) = 1 - 0.01n$$

where x_0 is the smallest egg size necessary for survival. Only a single clutch is laid per feeding cycle. Results shown are for two values of x_0. Redrawn from Parker and Begon (1986).

survival increases with clutch size also predicts an increasing egg size with ration or body size (McGinley 1989). This model is predicated on the assumption that survival, $s(n)$, varies with clutch size, n, according to

$$s(n) = \frac{I}{x} - c_1 \tag{10.21}$$

where I is the total amount of resources to be divided into eggs,
         x is egg size (hence clutch size, $n = I/x$),
         c_1 is a constant.
The above scenario posits that as clutch size increases, the number surviving increases. McGinley suggested that predators may be saturated by the sudden eflux of young, as occurs when hatchling sea turtles make a scramble for the sea, or when predators such as leeches are attracted to frog clutches independently of clutch size and cannot consume all eggs in large clutches. These circumstances are variants on the "convoy" hypothesis discussed in chapter 9 (section 9.3.1.3). Assuming that present investment does not affect future fecundity, and only one clutch per breeding season, fitness, $w(x)$, is simply per-clutch productivity:

$$w(x) = f(x)(\frac{I}{x} - c_1) \qquad (10.22)$$

where $f(x)$ is the relationship between mean fecundity of offspring ($=$ survival after the initial mortality $\times$ fecundity of offspring) and propagule size. To find the optimal egg size we differentiate $w(x)$ with respect to x and find that value of x at which $dw(x)/dx = 0$,

$$\frac{dw(x)}{dx} = f'(x)\left(\frac{I}{x} - c_1\right) - \frac{f(x)I}{x^2}$$

$$= 0 \qquad \text{when} \qquad I = \frac{f'(x)c_1x^2}{f'(x)x - f(x)} \qquad (10.23)$$

where for visual simplicity I have used the prime notation for the differential (i.e., $f'(x) = df(x)/dx$).

Fecundity typically increases with body size (chapter 5), and hence an increase in body size will generally lead to an increase in I. Similarly, an increase in ration will increase I. Therefore, we wish to evaluate the relationship between I and x. To do this McGinley (1989) used a specific form of $f(x)$, $f(x) = 1 - (c_2/x)^{c_3}$, where c_2 is a constant defining the minimum viable propagule size and c_3 is a constant greater than 1 that controls the rate of approach to the asymptotic value of $f(x)$. For this model egg size increases with I (i.e., ration or body size). How general this result is remains to be demonstrated.

The fact that a variety of models, based on very different premises, predict an increasing propagule size with increasing ration or body size suggests caution in interpreting observations within any particular model framework without realistic estimates of the components of the life history.

10.3.4. Propagule Size and Female Age

In addition to increasing with female size, propagule size frequently changes with age (Table 10.9). Though there are exceptions, the pattern among invertebrates is for egg size to decrease with age. This is particularly evident among lepidopterans (moths and butterflies). Moore and Singer (1987) have questioned this observation, suggesting that it may be a consequence of unnatural conditions. Among vertebrates older females frequently lay larger eggs though more varied patterns are also found. A possible explanation for the latter phenomenon is provided by the previous analysis. Recall that an increase in the amount of resources available for egg production (I) favors an increase in propagule size. Increased resources available may be a consequence of increased size or ability to sequester resources. Ectothermic vertebrates (fish, amphibians, and reptiles) gen-

Table 10.9. Effect of female age on propagule size

Species	Correlation	Reference
Drosophila melanogaster (fruit fly)	0	Warren 1924
Drosophila melanogaster (fruit fly)	+/−	Delcour 1969
Hyalophora cecropia (lepidopteran)	−	Telfer and Rutberg 1960
Porthetria dispar (lepidopteran)	−	Leonard 1970
Choristoneura fumiferana (lepidopteran)	−	Harvey 1977
Euphydras editha (lepidopteran)	−	Murphy et al. 1983
Euphydras editha (lepidopteran)	0	Moore and Singer 1987
Tyria jacobaeae (lepidopteran)	−	Richards and Myers 1980
Pieris rapae (lepidopteran)	−	Jones et al. 1982
Pararge aegeria (lepidopteran)	−	Wiklund and Persson 1983
Lasiommata megera (lepidopteran)	−	Karlsson and Wiklund 1984
Speyeria mormonia (lepidopteran)	−	Boggs 1986
Chilo partellus (lepidopteran)	0	Berger 1989
Parapediasia teterrella (lepidopteran)	0	Marshall 1990
Graphosoma lineatum (bug)	−	Larsson 1989
Salmo salar (salmon)	+	Kazakov 1981
Salmo gairdneri (trout)	+	Gall 1974; Pitman 1979
Poecilia reticulata (guppy)	+/−	Reznick et al. 1990
Cyprinius carpio (carp)	+	Moav and Wohlfarth 1974
Ambystoma talpoideum (salamander)	−/+	Semilitsch 1985
Sceloporus scalaris (lizard)	0	Newlin 1976
Melospiza melodia (sparrow)		Nice 1937
Troglodytes aedon (wren)	+	Kendeigh et al. 1956
Sterna hirundo (tern)	+	Nisbet 1978
Parus major (great tit)	0	Ojanen et al. 1979

+, size increases with age; 0, size does not change with age; −, size decreases with age; more complex patterns indicated by a slash—e.g., +/−, propagule size initially increases with age but later decreases.

erally grow throughout their lives and hence I will increase with age. Endothermic vertebrates (birds and mammals) may not grow significantly after maturity but may increase their ability to sequester resources either by virtue of their ability to defend territories or because they gain experience.

The decline in egg size with age in insects has been postulated to be a consequence of depletion of resources (Richards and Myers 1980 and references therein) though the rationale for this hypothesis is not explained. The model presented in the previous section can provide the necessary basis for the hypothesis: as resources decline the optimal egg size may also decline. Another model which also makes this prediction is that of Begon and Parker (1986). This model was constructed specifically to match the life history pattern of an insect. Model assumptions are:

1. The adult female accumulates a total reproductive reserve of I, to be divided among all eggs to be produced during the lifetime of the female.

2. The probability that a female will survive to lay the ith clutch is p_i.

3. The size of eggs in the ith clutch is x_i.

4. Clutch size is fixed at n, and the female produces a maximum of N clutches in her lifetime.

5. The expected fecundity of offspring (survival × fecundity) is an asymptotic function of egg size, $f(x)$ (Fig. 10.6).

6. Population size is stationary, making expected lifetime fecundity an appropriate measure of fitness.

The optimal propagule size is that which maximizes the expected lifetime fecundity

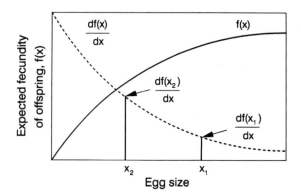

Figure 10.6. Hypothetical relationship between the expected fecundity of offspring and egg size (solid line) and its derivative (dashed line). For a female laying two clutches of eggs that are optimal in size the following inequality holds (see text):

$$\frac{df(x_1)}{dx_1} < \frac{df(x_2)}{dx_2}$$

where x_i is the size of eggs in the ith clutch. Thus, as can be seen from the above figure, $x_2 < x_1$.

$$R_0 = \sum_{i=1}^{N} p_i f(x_i) \tag{10.24}$$

subject to the constraint that

$$I = \sum_{i=1}^{N} n x_i \tag{10.25}$$

To solve equation 10.25, Begon and Parker (1986) used Lagrange's method of optimization subject to constraint. I present here a simpler method of analysis. First we shall consider an insect that produces only two clutches, for which the fitness function is

$$R_0 = p_1 n f(x_1) + p_2 n f(x_2) \tag{10.26}$$

Since $I = n(x_1 + x_2)$, the above can be rewritten

$$R_0 = n p_1 f(x_1) + n p_2 f\left(\frac{I}{n} - x_1\right) \tag{10.27}$$

Differentiating gives

$$\frac{dR_0}{dx_1} = n p_1 \frac{df(x_1)}{dx_1} + n p_2 \frac{df\left(\frac{I}{n} - x_1\right)}{dx_1}$$

$$= n p_1 \frac{df(x_1)}{dx_1} - n p_2 \frac{df(x_2)}{dx_2} \tag{10.28}$$

and $dR_0/dx_1 = 0$ when

$$p_1 \frac{df(x_1)}{dx_1} = p_2 \frac{df(x_2)}{dx_2} \tag{10.29}$$

Now since the probability of surviving to lay the second clutch (p_2) must be less than the probability of surviving to lay the first clutch (p_1), then

$$\frac{df(x_1)}{dx_1} < \frac{df(x_2)}{dx_2} \tag{10.30}$$

which requires that x_1 be larger than x_2 (Fig. 10.6)—i.e., eggs laid in the second clutch will be smaller than eggs laid in the first. We can extend this analysis to an arbitrary number of clutches by considering pairwise comparisons: in each case the preceding egg size will be greater, and hence egg size will decrease with age, as observed in a variety of insects (Table 10.9).

10.3.5. Egg Size and Evolution of Parental Care

A persistent pattern within ectotherms is for species that protect their young to produce larger and fewer offspring than related species that do not protect their offspring (reviewed in Shine 1978). Immature animals may suffer higher mortality rates both because they are undeveloped and cannot actively escape predators and because mortality rates appear to decrease with particle size (chapter 5). To increase offspring survival a female can retain her offspring until they are more fully developed (e.g., altricial versus precocial) and/or are larger. Either case will generally lead to an increase in size at birth. A female may either retain its young within its body, possibly further enhancing their survival, or simply lay larger, better-provisioned eggs; both patterns will likely lead to a reduction in the number of offspring that can be produced.

Viviparity can be considered a type of parental care, increasing offspring survival, but possibly decreasing maternal survival and most likely decreasing fecundity (Shine and Bull 1979). There are no viviparous birds, a phenomenon that Blackburn and Evans (1986) relate to the large size of the cleidoic avian egg which prevents more than a single egg being retained within the oviduct. Given this constraint, what circumstances would favor egg retention and viviparity in birds? To address this question Blackburn and Evans turned to a model proposed by Shine and Bull (1979) for the evolution of egg retention in lizards. Primary assumptions of this model are:

1. Overlapping generations with reproduction at regular intervals
2. One clutch per reproductive episode
3. Adult survival and fecundity independent of age

The expected lifetime production of daughters, R_0, is given by

$$R_0 = sm + sSm + sS^2m + \ldots = \frac{sm}{1 - S} \qquad (10.31)$$

where s is the survival to first reproduction,
m is the number of female births at each reproduction,
S is adult survival from one reproductive episode to the next.

Letting the measure of fitness be R_0 (i.e., assuming a stationary population) and the life history parameters of an alternate mode of reproduction be denoted by *, the new pattern will be favored provided $R_0^* > R_0$, which is equivalent to

$$\frac{m^*}{m} > \frac{s}{s^*} \frac{(1 - S^*)}{(1 - S)} \tag{10.32}$$

Dividing survival to first reproduction into survival to hatching, h, and survival thereafter, and making the assumption that viviparity does not affect the latter component, the above inequality becomes

$$\frac{m^*}{m} > \frac{h}{h^*} \frac{(1 - S^*)}{(1 - S)} \tag{10.33}$$

To quantify the above inequality Blackburn and Evans estimated or assumed the following values

1. From data in Lack (1967) adult survival, S, was estimated to be 0.7.
2. From data in Lack (1967) survival of the egg prior to hatching, h, was estimated to be 0.63.
3. Survival of the offspring to birth for the viviparous bird, h^*, was set at 1.
4. From anatomical considerations a viviparous bird would be limited to one egg per clutch, which, assuming a 1:1 sex ratio, gives $m^* = 0.5$.

Substitution of these values in equation 10.33 and rearranging gives the condition for viviparity to be favored as

$$S^* > 1 - \frac{0.238}{m} \tag{10.34}$$

Blackburn and Evans suggested that adult survival of a viviparous bird will be less than that of a oviparous bird but give no arguments for this assertion. Given this, and the obvious requirement that clutch size cannot be less than 1 ($m > 0.5$), the proportion of combinations of S^* and m that favor the evolution of viviparity is small (Fig. 10.7).

Under what conditions would it pay to retain eggs within the oviduct— a transitionary stage to viviparity? Letting t_h be the total time to hatching (including the time retained within the female) and t_r the period of egg

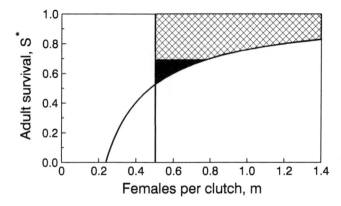

Figure 10.7. Relationship between clutch size (= 2m) of an oviparous bird and adult survival, S^*, as specified by equation 10.34. combinations less than $m = 0.5$ (left of solid vertical line) are not possible since they are less than replacement rate. Combinations lying above the curve favor the evolution of viviparity. If the survival rate of viviparous birds is less than that of oviparous birds ($S = 0.7$) the only combinations favoring viviparity are those in the small solid region. Modified from Blackburn and Evans (1986).

retention, the condition for egg retention to be favored (Blackburn and Evans 1986) is

$$\frac{m^*}{m} \left(\frac{h}{h^*}\right)^{\frac{t_r}{t_h}} \frac{(1 - S^*)}{(1 - S)} \qquad (10.35)$$

As pointed out by Dunbrack and Ramsay (1989b), this equation does not support the hypothesis that the evolution of viviparity is unlikely since if the inequality is satisfied for a very small value of t_r then it is satisfied for all larger values. This can be shown as follows: by assumption, h/h^* is less than 1 (i.e., retention of the egg increases the probability of survival to hatching), and hence

$$\left(\frac{h}{h^*}\right)^{\frac{t_r}{t_h}} \leq 1 \qquad (10.36)$$

Now suppose that for some egg retention time, say T_r, inequality 10.36 is satisfied. To examine the inequality for values of t_r greater than T_r we multiply T_r by a constant c, greater than 1. It follows immediately from inequality 10.36 that

$$\left(\left(\frac{h}{h^*}\right)^{\frac{T_r}{t_h}}\right)^{c} \leq \left(\frac{h}{h^*}\right)^{\frac{T_r}{t_h}} \tag{10.37}$$

and hence inequality 10.35 will be satisfied for all c greater than 1, meaning that selection will favor egg retention until hatching.

Thus if egg retention for a small period of time is favored, then oviparity will be evolutionarily unstable and selection will favor egg retention over the entire period. Inequality 10.35 is still unlikely to hold even for small values of t_r unless the clutch size is small, but still, as Dunbrack and Ramsay (1989b, p. 139) note, "rather than demonstrate that viviparity is unlikely to evolve in birds, this result leaves considerable doubt about why it has never appeared." Egg retention is found in cuckoos where it apparently functions to give the cuckoo hatchlings a competitive advantage by enabling them to eject the host's eggs or younger hatchlings (Payne 1974). Therefore, egg retention is possible in birds, and the failure for viviparity to evolve must be due to other factors.

A possible factor differentiating squamates and birds with respect to viviparity may be endothermy (Dunbrack and Ramsay 1989b). Eggs retained within the body probably receive less oxygen than eggs in direct contact with the atmosphere and hence their development time will be increased. This increase will be further increased with elevation in temperature. The oxygen demands of an embryo increase as the embryo develops, which could lead to an accelerating increase in development time, and select against egg retention, even when equation 10.35 is satisfied for small t_r (Dunbrack and Ramsay 1989b). This problem can be circumvented if the female reduces egg size, thereby reducing the oxygen demands of the embryo. Dunbrack and Ramsay suggest that mammals were able to take this evolutionary route because of the evolution of lactation, which enables than to nourish small, poorly developed young. Some credence is given to this hypothesis by the observation that the eggs of monotremes are extremely small relative to the egg size of a bird of the same mass: the eggs of the platypus, *Ornithorhynchus anatinus*, and echidna, *Tachyglossus aculeatus*, weigh approximately 0.1% and 0.03%, respectively, of the body mass (data from Griffiths 1979; Rismiller and Segmour 1991), while for passerines, application of equation 5.32 gives relative egg masses of 3.7% and 2.6% for birds of equivalent weight to the platypus (1,300 g) and echidna (5,000 g), respectively. (For nonpasserines equation 5.33 predicts 5.4% and 3.7%, respectively.) There is no reason to suppose that small egg size could not readily evolve in birds, but birds may lack the critical preadaption of lactation, which is necessary for them to care for small, undeveloped young. The question of why a bird could not, or does not, produce a small, precocial offspring remains to be explored.

Within various fish species there is evidence that the time to resorb the yolk sac increases with the size of the hatching larvae. (Data from intraspecific comparisons reviewed by Sargent et al. 1987; interspecific comparisons analyzed by Duarte and Alcaraz 1989.) This does not mean that selection will therefore favor smaller eggs, for larger larvae may give rise to larger juveniles which have higher survival and growth rates. Sargent et al. (1987) used these propositions to explore how the evolution of egg size and parental care in fish may be related.

The basic elements of the model of Sargent et al. are derived from the model of Smith and Fretwell (1974): fitness of a female laying an egg of size x, $w(x)$, is the product of maternal investment, I, and offspring survival divided by the investment per egg ($=$ egg size). Offspring survival, $f(x)$, increases in a sigmoidal manner with egg size, x (e.g., Fig 4.5 with egg size $=$ investment per offspring). Thus,

$$w(x) = f(x)\frac{I}{x} \tag{10.38}$$

The optimal egg size is obtained by differentiating $w(x)$ with respect to x and finding that egg size at which the differential equals zero, which is when,

$$\frac{df(x)}{dx} - \frac{f(x)}{x} = 0 \tag{10.39}$$

To specify offspring survival, Sargent et al. make the same assumption as Ware (1975b)—that the egg experiences a constant instantaneous rate of mortality. They differ, however, in specifying this mortality rate as a function of egg size. Survival from the time of laying to reproduction is given by

$$f(x) = e^{-M(x)t(x)}S_j(x) \tag{10.40}$$

where $M(x)$ is the instantaneous rate of egg mortality,
$t(x)$ is the time to yolk-sac absorption for an egg of size x,
$S_j(x)$ is juvenile stage survival, assumed to be an increasing function of egg size.

We now examine the dependence of the optimal egg size on the egg mortality rate, $M(x)$: Specifically, we are interested in whether the optimal egg size, x, should increase or decrease with decreases in egg mortality.

To answer this question it is sufficient to know the sign of the derivative of x with respect to $M(x)$. (Recall that if the slope is negative the function must decrease while if positive it must be increasing.) From rearrangement of equation 10.39 we have

$$x = \frac{f(x)}{f'(x)} \tag{10.41}$$

The derivative, dx/dM, is found by implicit differentiation (see Sargent et al. for details),

$$\frac{dx}{dM} = \frac{f(x)\dfrac{dt(x)}{dx}}{\dfrac{d^2 f(x)}{dx^2}}$$

$$= \frac{f'(x)t'(x)}{f''(x)} \tag{10.42}$$

(The prime notation is used for convenience.) Now since the time required to absorb the yolk sac, $t(x)$, increases with egg size, $t'(x)$ must be positive. At the optimal egg size the second derivative of $f(x)$ with respect to x, $f''(x)$, is negative. Therefore, the rate of change of egg mortality, M, with egg size, dx/dM, must be negative, and hence a decrease in egg mortality corresponds to an increased optimal egg size. Thus if a female can increase the survival of her offspring by showing parental care, this behavior will be selected and such females will produce larger eggs. This result, however, depends on parental care not decreasing the survival of future reproduction of the parent.

The above model is an extension of Shine's safe-harbor hypothesis discussed in section 10.3.1. Nussbaum (1985, 1987) argued that the argument may be the wrong way around: parental care may evolve because of large eggs. Nussbaum developed his hypothesis based on observations of the breeding biology of salamanders and a hypothesis put forward by Itô (1980) and Itô and Iwasa (1981). According to Itô's hypothesis the evolution of egg size is driven, in part at least, by the problems of procurability of food by the young: when food is scarce selection will favor large, developmentally advanced young. Salamander species inhabiting stream (lotic) environments lay large eggs while those breeding in ponds (lentic) produce small eggs. Nussbaum argues that in streams the most abundant larval foods are much larger than in ponds. The increased size of eggs in lotic environments is postulated to be an evolutionary response to a shift in the size spectrum of food available to the newly hatched larvae. But increases in egg size also increase incubation time and hence potentially increase mortality during this stage. Parental care may then evolve as a mechanism to reduce egg mortality. Nussbaum and Schultz (1989) refined the various models into a single model in which parental care and egg size coevolve. Tests of these different models will rely largely upon comparative analyses (Shine 1989). Detailed studies remain to be undertaken.

10.4. Optimal Propagule Size in a Variable Environment

10.4.1. Seasonal Variation and Optimal Propagule Size

In a wide variety of species there is a seasonal change in egg size (Table 10.10), but reasons for such a change are largely unknown. The observation that female isopods (*Armadillidium vulgare*) produce larger eggs when fed a reduced ration led Brody and Lawler (1984) to suggest that seasonal variation in offspring size is an adaptive response, increasing the survival

Table 10.10. Summary of studies reporting on seasonal variation, or lack of, in propagule size

| Group | Propagule size in relation to season[a] | | |
	Increases	Decreases	No change
Plants	0	13	1
Invertebrates	7	0	0
Fish	1	12	0
Reptiles	5	1[b]	8
Birds	2	6	1

[a]All entries represent separate species except for the bird *Parus major*, in which one study reported an increase with season (Perrins 1970) and another no change (Ojanen et al. 1979).

[b]In the Aldabran giant tortoise (*Geochelone gigantea*) one population shows a decline in egg size with season while another shows variation in mean egg size but no consistent trend (Swingland and Coe 1978).

References:

Plants: Cavers and Steel 1984 + references cited therein (Turner et al. 1979; Hurka and Benneway 1979; O'Toole 1982; Hermanutz and Steele 1983; Frost 1971).

Invertebrates: Green 1966; Kerfoot 1974; Kimura and Masaki 1977; Brody and Lawlor 1984; Nakasuji and Kimura 1984.

Fish: Hempel and Blaxter 1967; Blaxter 1969; Cushing 1967; Bagenal 1971; Oosthuizen and Daan 1974; Ware 1977; Daoulas and Economou 1986; DeMartini 1991.

Reptiles: Ballinger et al. 1972; Derickson 1976; Schall 1978; Swingland and Coe 1978; Nussbaum 1981; Congdon et al. 1983; Selcer 1990.

Birds: Coulson 1963; Nelson 1966b; Perrins 1970; Murton et al. 1974; Lloyd 1979; Ojanen et al. 1979; Birkhead and Nettleship 1982; Soler 1988.

chances of the offspring in times of food scarcity. However, ration has either no effect or a reversed effect in other taxa (Table 10.8); thus the hypothesis cannot be readily applied to other groups. The fall eggs of the moth *Orgyia thyellina* are significantly larger than those laid by the spring generation (Kimura and Masaki 1977). This difference may be connected with the long diapause through which the fall eggs must pass, but since diapause is not a feature of most of the organisms listed in Table 10.10, this hypothesis also cannot account for variation observed in these species. Nussbaum (1981) advanced the hypothesis that seasonal variation is a bet-hedging strategy, conditions at the end of the season being more variable than at the beginning. The theoretical basis of this hypothesis is examined in section 10.4.2.

In section 10.3.1 the hypothesis was advanced that evolutionary changes in the size of fish eggs are a consequence of variation in incubation time (Ware 1975b). This hypothesis postulates that as temperatures fall, incubation time increases but mortality rates do not, and hence size-selective mortality rates increase on eggs, favoring a larger egg size but a lower number per female. Such a mechanism predicts a decrease in egg size as the season progresses and temperatures increase, as observed in many fish species (Bagenal 1971; Table 10.10).

Another hypothesis put forward to explain seasonal variation in size of fish eggs is that the size spectrum of food available for the newly hatched larvae shifts over the season, and hence selection favors females that alter their egg size such that their larvae can take advantage of the change in the mean size of their food. The mean egg diameter of the Atlantic mackerel, *Scomber scombrus*, declines during the summer and is negatively correlated with the mean sea surface temperature and positively correlated with the mean particle size in the plankton (Ware 1977). While this is good circumstantial evidence for the food size hypothesis, it remains to be shown that the size of the larvae is optimal in relation to the size of prey. Since adult mackerel are piscivorous, the proximal cue used by the female mackerel to adjust her egg size is unlikely to be the size of the plankton but more likely is the water temperature, which correlates with particle size (Ware 1977). The "incubation time" and "food size" hypotheses are not mutually exclusive but could operate in conjunction, increasing the selective pressure on egg size.

A food effect has also been implicated in the seasonal variation in egg size of the hesperid lepidopteran *Parnara guttata*. Egg size in *P. guttata* increases during the summer in conjunction with an increase in one of its host grasses, *Imperata cylindrica*. Though the larger, later-born larvae can eat this grass, the smaller larvae produced early in the year cannot survive on it, which Nakasuji and Kimura (1984) hypothesized is due to mechanical problems associated with the toughness of the leaf. Further support for this hypothesis is provided by a highly significant correlation between leaf toughness and egg size among the Hesperidae (Nakasuji 1987).

Larger young hatch from larger eggs and hence if rates of size-selective predation shift during the year, selection will favor a shift in egg size to optimize the size and number of offspring. Kerfoot (1974) used this argument to account for shifts in egg size of the cladoceran, *Bosmina longirostris*. In this species females produce small eggs in the summer and then switch to large eggs in late fall. Winter generations produce large offspring which mature at larger sizes and then shift to producing small eggs in the spring. During the summer *Bosmina* suffer heavy predation from fish, which, however, reduce competition from larger invertebrate planktivores. In the fall, after the warm-water fishes cease feeding, *Bosmina* is subjected to predation from two invertebrates, *Cyclops* and *Chaoborus*. Predation by these two invertebrates is known to favor large size in *Bosmina* (reviewed in Kerfoot 1974). The changing pattern of heavy visual predation by fishes in the summer to predation by grasping invertebrates in the winter favors a changing pattern of size in *Bosmina*, from small to large phenotypes. This hypothesis is attractive and has received theoretical support from an analysis by Lynch (1980b), but more empirical data are required on the benefits and costs of different-sized eggs (e.g., the relationship between egg size and clutch size, that between body size and mortality under different predator regimes, and the differences in development time resulting from changes in egg size).

10.4.2. Optimal Propagule Size in a Stochastic Environment

Egg size varies with age (Table 10.9), size (Table 10.3) and ration (Table 10.8); additionally, propagules within a clutch may be variable in size. In part, such variation may be a consequence of morphological constraints, such as position on a plant (Harper et al. 1970; Silvertown 1984; McGinley et al. 1987; Roach and Wulff 1987) or position within the ovary (Telfer and Rutberg 1960; McKeown et al. 1976; Takahashi and Iwasawa 1988b). But variation may also be adaptive, and several authors have argued that variation in propagule size may be a response to environmental heterogeneity (Janzen 1977; Capinera 1979; Kaplan 1980; Crump 1981, 1984; Nussbaum 1981; Westoby 1981; Stamp and Lucas 1983; Thompson 1984). The theoretical basis for such conjectures has been explored by McGinley et al. (1987). As with many of the models discussed in this chapter, McGinley et al. begin with the Smith and Fretwell model. They extend the model by assuming that the environment consists of two types of habitats, each with a specific relationship between propagule size and expected fecundity of offspring.

In their first analysis McGinley et al. examined the consequences of random dispersal of propagules in an environment in which a proportion p of habitats comprise one type, and $1-p$ consist of a second type of habitat. After some tedious algebra, they were able to demonstrate that eggs of a single size are optimal. This result does not, however, hold when there is competition between propagules within a habitat. McGinley et al. (1987, p. 375) concluded that "the production of variable offspring sizes is favored under a very limited set of conditions." Simulations showed that under random dispersal a single egg size was always favored, but when a proportion of the propagules was able to select the most appropriate habitat variation in propagule sizes was favored. When 60% of propagules dispersed to the correct habitat 15% of the simulations indicated that a female maximizes her fitness by producing eggs of several sizes. The fraction of simulations in which a variable egg size was favored increased to 26% when 80% of the propagules dispersed correctly and to 35% when all propagules dispersed to their proper habitat. The message from these simulations is that some form of habitat choice is necessary for the evolution of variable propagule size in a spatially heterogeneous environment.

Temporal variation in habitat quality is more likely to select for variable propagule size since all individuals are subject to the same habitat and fitness is determined by the geometric mean, which is greatly influenced by cases of low fitness (e.g., see chapter 6, section 6.2.2). Kaplan and Cooper (1984) demonstrated that temporal heterogeneity can favor variable egg size when each egg size has a high fitness in one environment but a decreasing fitness with deviations from this environment. McGinley et al. (1987) criticized this model on the grounds that fitness is likely to be a monotonic function of egg size and hence a monotonic function of environmental quality. This point is supported by the data presented in the earlier sections of this chapter and makes the conclusions of Kaplan and Cooper (1984) suspect with respect to the evolution of egg size. Nevertheless, their model remains appropriate for traits that show fitness functions of the type they assume.

To evaluate the importance of temporal variation McGinley et al. constructed a simulation model. At each "year" (iteration) the simulated environment comprised only two types of habitats—"good" and "bad." The likelihood that the production of variable propagule size will be favored depends on both the difference in quality between "years" and the probability of "good" years occurring, an increase in either favoring a variable propagule size. Of the 24 combinations run, 25% resulted in selection for variation in propagule size. These results demonstrate the potential importance of temporal heterogeneity but their significance remains to be demonstrated by empirical studies.

10.5. Summary

A large size at birth increases the fitness of propagules: in plants, percent germination, percent emergence, initial growth rate, seedling survival, and competitive advantage all increase with seed size; in animals, hatching success, survival, starvation time, and growth increase with propagule size. However, propagule size may be constrained by physiological or mechanical factors. But more importantly, increases in propagule size reduce fecundity and hence do not necessarily lead to an overall increase in fitness. The trade-off between propagule size and number favors an intermediate propagule size.

Studies of the evolution of propagule size in a constant environment are divided into five categories:

1. Changes in propagule size alter the time required to achieve the critical size for maturation, but the importance of this effect depends upon the form of the growth curve. Time to maturity is little affected by initial size when growth is linear with age, but can be profoundly affected by initial size if the growth curve is exponential. The combination of a trade-off between egg size and number and between egg size and development time has been postulated to be important in determining the evolution of egg size in isopods, grasshoppers, pelagic fish, amphipods, and marine benthic invertebrates.

2. Decreases in growth rate favor an increased propagule size. Tests of this hypothesis are not definitive but are consistent with the hypothesis.

3. Propagule size has been observed in many taxa to increase with female size (Table 10.8). Several different models have addressed this question using different sets of biological assumptions. The model of Parker and Begon (1986) indicates that a positive correlation between female size (or condition) and propagule size will occur only if there is density-dependent mortality within a site. On the other hand, McGinley (1989) produced a model that showed that a positive correlation occurs when survival increases with clutch size. These two models demonstrate that resolution of empirical observation with theory depends critically upon a proper understanding of the biology of the organisms. That disparate models can give the same qualitative prediction suggests the general correlation may be a consequence of different biological mechanisms operating in the different taxa.

4. Among Lepidoptera, egg size typically declines with age, while among vertebrates the reverse is more frequently found. The latter may result from increasing size and/or experience with age. A theoretical model specifically tailored to an insect correctly predicts a decline in optimal egg size with age but has still to be tested quantitatively.

5. Among ectotherms parental care is correlated with large propagule size. The large size of the cleidoic avian egg has been hypothesized to be the factor preventing the evolution of viviparity in birds. The theoretical analysis of Backburn and Evans (1986) supporting this contention is flawed; further analysis of their model by Dunbrack and Ramsay (1989b) suggests that viviparity should have appeared. Dunbrack and Ramsay (1989b) hypothesized that physiological constraints (oxygen demands of the developing embryo) resulting from endothermy are responsible. Two methods of circumventing this problem are to lay eggs, or small, undeveloped young which are nourished by the parent. The evolution of lactation has permitted mammals to become viviparous because it allows them to nourish small, poorly developed young, which birds cannot do.

   Parental care in fishes may have evolved as a consequence of the fact that although larger eggs give rise to larger juveniles which have increased survival rates, the time from hatching to yolk absorption increases with egg size. Therefore, there are two trade-offs: egg size and number, and juvenile survival and survival during the yolk sac stage. These trade-offs are sufficient to favor the evolution of parental care and a concomitant increase in egg size (Sargent et al. 1987). It has, however, been argued that an alternative direction is for evolution to first favor large egg size because of the food regime available to larvae. But large eggs are more vulnerable and hence parental care evolves (Nussbaum 1985, 1987). Further studies are required to resolve these conflicting models, or the circumstances in which either is applicable.

Studies on the evolution of propagule size in a heterogeneous environment have focused on two issues:

1. Propagule size very often varies seasonally, but the reasons for this are largely unknown. Various hypothesis—seasonal changes in ration, overwintering mortality and egg size, seasonal change in the size spectrum or quality of prey, and seasonal variation in the size spectrum of predators—have been proposed for specific organisms. Differences among taxa in the seasonal pattern of propagule size (Table 10.10) argue against any general explanation.

2. Propagule size may vary within a clutch. Though several authors have suggested that this is an adaptation to environmental heterogeneity, both theory and tests are scarce. Conditions under which spatial heterogeneity will lead to variable propagule size are very restricted. Habitat choice by newly hatched young increases the parameter set over which variable propagule size is favored. However, as intuitively obvious, variable propagule size is very likely to evolve when conditions are temporally heterogeneous.

11

Final Thoughts

There is a continuing variation in biotic and abiotic conditions, and consequently a continuing evolution of life history characteristics. The fact that evolution continues and that any attempt to predict an observed combination of traits must necessarily make simplifying assumptions means that we cannot expect to predict *exactly* that which is observed. It is important to determine whether any discrepancy is due to one of the simplifying assumptions or results because the organism is constrained by phylogeny, lack of genetic variation, or some other factor from readily approaching the optimal combination.

The majority of analyses of life history evolution considered in this book are predicated on two assumptions: (1) selection maximizes some measure of fitness, and (2) there exist trade-offs that limit the set of possible combinations. Given these two premises we can relatively easily find the suite of trait values that will maximize fitness. If prediction and observation do not match, the next step is to modify the assumed trade-offs and seek a reasonable set of trade-offs that produce congruence between prediction and observation. The finding that the predicted trait value(s) matches the observed is evidence that we have taken into account all major factors affecting the evolution of the trait(s). However, the various trade-offs and constraints that go into a model must be verified independently; a match between prediction and observation is not evidence that the assumed biology is correct. In several cases we have found that the same phenomenon can be predicted by more than one model. The likelihood that a particular model is inappropriate even though it correctly predicts some observed phenomenon depends upon the nature of the prediction. If the prediction is very qualitative, such as an increase in egg size with age, then several models may be applicable. The choice of model will depend upon the particularities of the biology of the organism in question. It may be possible to narrow the scope of candidate models by a very general outline of life history, but even if only one model remains, a qualitative prediction must be viewed with considerable caution until sufficient data have been gathered to verify the model quantitatively. Modeling is an aid to research: it permits us to ask "what if" questions and seek hypothetical trade-offs that can then be subject to experimental verification.

When to breed, how much to commit to reproduction, the optimal division between size and number of offspring: these have been the central focus of the last half of the book. Analyses of the optimal age at first reproduction in females have shown that the critical factors are the age schedule of mortality and the cost of reproduction. Males are more difficult to model and study than females because the measure of fitness is the number of offspring sired, which typically can only be crudely estimated from the number of mates obtained and their fecundities. The schedule of mortality and costs of reproduction are also important determinants of the optimal age at maturity in males but contest for mates introduces the added complexity of frequency-dependent selection.

The impact of survival rates depends upon whether mortality is stage-, age-, or size-specific. Under stage-specific mortality rates an increase in adult mortality selects for an increase in the age at maturity, while an increase in juvenile mortality favors a decrease. Age- or size-specific mortality rates produce the opposite result. Which mortality pattern is more prevalent or important in nature remains to be investigated. If there were no cost to reproduction an organism would mature as soon as physiologically possible. But reproduction does carry costs, in terms of reduced future growth, survival, and fecundity. Though this fact has now been adequately demonstrated we lack any satisfactory estimate of the cost of reproduction under field conditions.

Understanding the evolution of reproductive effort, or its more specific component clutch size, also requires analysis of the costs of reproduction. Although Cole's paradox has been resolved we are still far from fully understanding why some organisms are semelparous and others are iteroparous. In particular, does semelparity result from extreme reproductive effort, or is death inevitable and extreme reproduction effort a response to this? Similarly, the relationship between age and reproductive effort has long been pondered, and though the circumstances under which the age schedule of reproductive effort will vary have been quite well studied, there is yet to be a field study that unites theory and observation. In this regard the problem of how to allocate energy between growth and reproduction is particularly important. Theoretical analyses have established the conditions under which allocation to both growth and reproduction is optimal, but we lack satisfactory data with which to test these models. There may also be a dichotomy in the manner in which energy is allocated to reproductive behavior within a population. Some males may allocate a relatively large amount by, for example, defending territories and/or actively attracting females, and other males may allocate little and attempt to gain copulations by lower-cost behavior such as sneaking. One of the most significant questions in this regard is the extent to which the behaviors are genetically fixed, ontogenetically controlled, or facultative.

The survival of offspring can be enhanced either by increased provisioning of the propagules or by parental care. Parental care has evolved in a wide range of taxa: female care is characteristically found in invertebrates, reptiles, and mammals; male care in fish and anurans; and biparental care in birds. Quantitative analyses of the amount of care that is optimal are lacking but the more qualitative issues of desertion and egg parasitism have received attention. The theoretical basis of desertion is now well understood but field tests are lacking.

Starting with the work of Lack, the evolution of clutch size has been much discussed. Increased clutch size may increase vulnerability to predators by impeding movement or enhancing visibility of the gravid female. This hypothesis is supported by data on invertebrates and reptiles and deserves greater study as a factor determining the optimal clutch size. Lack's hypothesis has received considerable attention, and there are both theoretical and empirical grounds for doubting its validity. Nevertheless, a proper test of the hypothesis has been impeded by an inability to measure all relevant components, particularly the effects of clutch size on subsequent survival of young and parents. Insects typically distribute their eggs in several clutches and hence a slightly different framework is required. However, a clutch size equivalent to that postulated by Lack to be optimal for birds and mammals can also be defined for insects, and as with endotherms the optimal size is predicted to be typically less than the Lack value. Much of the theoretical work can be tested both in the laboratory and field: results to date provide qualitative support for some of the simpler hypotheses (e.g., site density and quality), but more detailed quantitative tests are required. Considerable complexity can arise when several females lay on the same host and populations are subject to density-dependent regulation. This complexity has not been adequately explored either theoretically or empirically.

A clutch can consist at one extreme of many small young and at the other of a few large young. This trade-off has undoubtedly been very important in the evolution of propagule size. Factors potentially affecting propagule size are development time, development rate, female size, female condition, and female age. Correlations between propagule size and these factors are found in a variety of taxa, suggesting that several mechanisms may be responsible. It is, therefore, not surprising to find that general trends can be predicted by several biologically dissimilar models. Future research could profitably investigate the different types of life history patterns and determine for each case which theoretical model is appropriate.

The overview presented above has assumed a nonfluctuating environment. But, as shown in each chapter, the introduction of heterogeneity—spatial, temporal, seasonal, and nonseasonal—can profoundly change predictions. It is obvious that the world is variable, but the importance of this variation remains to be explored. One of the most important effects of environmental variation is selection for variation in traits. Such variation may be genetically based or a result of phenotypic plasticity. Phenotypic plasticity may itself be genetically based, and there is an increasing interest in measuring it with respect to both its phenotypic expression and its genetic architecture. While it may be a phenomenon of interest in its own right its importance in life history evolution cannot be overstressed, and future research should pay considerable attention to its impact on fitness.

Future research will likely take two directions: first, more detailed analysis of the mechanisms producing trade-offs and the construction of models to test the adequacy of these trade-offs in generating observed combinations of life history traits in particular species; second, the construction of very general models that attempt to predict the broad pattern of correlations observed among life history traits at various taxonomic levels. Both avenues of investigation can make use of the optimality approach, but the former will have to pay relatively more attention to genetic aspects, particularly with respect to the functional and genetical basis of trade-offs (a topic discussed in more detail below).

In formulating a study of life history variation there are five questions that need to be addressed:

Question 1:   To What Extent Must We Understand the Genetic Basis of Traits to Understand Their Evolution?

There is both a general and a specific answer to this question. Here I shall present the general answer; the specific answer is discussed in question 4. The "discovery" of natural selection by Darwin and Wallace not only permitted an explanation of the evolution of life histories, but more importantly, made it possible to predict which life histories are likely to occur. But in the elucidation of his theory Darwin could not provide a satisfactory model for the genetic transmission of traits. Though we now have such a mechanism, much of our understanding of the evolution of life histories has not been predicated on this knowledge. Starting from the proposition that natural selection maximizes fitness it is possible to construct models for the evolution of traits by mathematically defining fitness and the interactions between the various life history components that directly or indirectly affect the traits under study. We can therefore commence our study by initially examining trade-offs at the phenotypic level.

Question 2: What Is the Appropriate Measure of Fitness?

To answer this question we must proceed in several steps. First, to what extent are the traits under density- or frequency-dependent selection? Frequency-dependent selection is most likely to occur in males where there is competition for females, giving rise to alternative mating behaviors. (See Table 7.7.) The evolution of alternative mating behaviors, and consequently frequency-dependent selection, among males is most probable in species which form a chorus, hold a harem, are territorial, or display in a common arena. Females being the "choosy sex" will not generally be under frequency-dependent selection. Frequency-dependent selection might occur at other stages in the life history but there is little empirical evidence for this (section 2.2.4).

We must now enquire as to whether selection is density-dependent. The question of the importance of density-independent versus density-dependent factors in modulating the evolution of life history traits is not new but has still not been resolved. This question, placed in the broader perspective of population fluctuation, was very much the center of discussion during the 1950s and 60s. It appears to me that no consensus was ever reached. Interests simply shifted to other questions. Of course there is no simple answer: the real question is, How often do densities reach a level at which their effects have to be considered? Most of the theory and analysis of age and size at maturity and reproductive effort in females is based on the assumption that selection is density-independent. This generally requires making the assumption that events at one stage in life history do not affect other stages. For example, population size in fish may be regulated in large measure by density-dependent mortality in the larval stage, but provided that there is no feasible alternate life history pattern that will modify this mortality, analysis may proceed assuming density-independent selection. Studies on the evolution of clutch size and propagule size frequently assume density-dependence between individuals within a clutch or common oviposition site (chapters 9 and 10). Such an assumption can be readily assessed by a consideration of the life history. Organisms growing up in very close proximity to each other and dependent upon a common food source such as a parent are most likely to experience effects of density if food is in relatively short supply. Manipulation experiments with birds, when properly conducted, suggest that density effects can be important on the size at fledgling and future survival (chapter 9). Although eggs may be laid in a single clutch there may be no deleterious effects of clutch size if the offspring disperse immediately upon hatching: insects that lay their eggs in pods (e.g., cockroaches and grasshoppers) and many reptiles (e.g., turtles) are such examples.

In the case of density-independent selection one must select from a number of fitness measures that may be relevant. An intuitively appealing measure is the instantaneous rate of increase, r, or if the population is stationary, the expected lifetime fecundity, R_0. The demonstration that these variables are indeed maximized by selection given a particular mode of inheritance has been relatively slow in coming (Charlesworth 1980; Lande 1982), and even these demonstrations are restricted to the case of weak selection. Nevertheless, the lack of a rigorous demonstration should not detract from the use of this assumption since its "track record" suggests that under many if not most circumstances it is indeed valid. (See particularly the examples in chapters 7 and 8.) The expected lifetime fecundity, R_0, is the limiting case of selection acting on r, and therefore one might be tempted to use the latter under the assumption that conclusions will likewise apply to R_0. But, as shown in Table 7.3, this may not be the case. Unless there is clear evidence for the appropriateness of r or R_0 the analysis should be done for each measure. Analyses under the assumption that R_0 is maximized are typically far simpler than those based on r. In some cases reproductive value might be an analytically more tractable metric to use, but this is simply an alternative formulation for r or R_0.

The life cycles of many organisms are constrained by seasonal fluctuations in their environment. In this case an additional constraint that must be placed on the fitness measure is that the organism at the end of the growing season be in the appropriate stage to endure the inclement period. This restriction can have very important influences on the optimal life history. (See, for example, section 7.2.1, and the model of Sibly and Monk discussed in chapters 4 and 10.) Similarly, stochastic variation in environmental conditions can have profound effects on the optimal life history. In a spatially heterogeneous environment fitness is maximized by maximizing the arithmetic value of the finite rates of increase, while in a temporally varying environment the appropriate measure is the geometric mean of λ or R_0 (section 3.1.1.2). It is clearly easier to assume a constant environment but what is easier is not necessarily what is correct. Heterogeneity is ubiquitous and is certainly critical in the evolution of norms of reaction but it may be of much less importance in the estimation of the mean value of a trait. Therefore, the extent to which environmental heterogeneity is incorporated into a model will depend initially upon the focus of the study. If a constant environment is assumed and results do not match observation the consequences of environmental variation should be critically assessed.

In some studies it may be most convenient to work with what I have termed a local measure of fitness: a metric which when maximized leads to the maximization of the global measures r or R_0. If, for example, one assumes that increasing clutch size does not have detrimental effects on the parents, and effects on the offspring in the nest do not impact on their future age schedules of survival and reproduction beyond, say, the first year, then the optimal clutch size can be analyzed within the framework of 1 year, instead of within a full life-span. Game-theoretic models frequently use a local measure: the adoption of satellite behavior, for example, can be analyzed under the premise that selection is maximizing the rate of mating success per unit time with no effects carrying over between breeding seasons. (See section 8.5.)

Question 3: What Method of Analysis Is Appropriate?

The answer here depends in part upon the nature of the study and in part upon personal preference. A graphical representation of the various trade-offs and constraints involved with the resulting relationship between fitness and the trait(s) under examination can be very helpful in visualizing the problem. But quantitative predictions are typically best achieved by using some analytical method. For predictions involving a sequence of decisions dynamic programming may be the appropriate tool. However, use of this technique in the wrong circumstance can make a simple problem complex. (See section 4.3.) When only a single decision is required, or the sequence of decisions can be described by some simple function, dynamic programming is inappropriate. Matrix methods have been successfully employed in some analyses (section 4.4). They are particularly important in quantitative genetic analyses (e.g., section 2.1.2) but have not received a great deal of attention with respect to optimality modeling. By far the most utilized technique is that of the calculus. Problems that are analytically intractable, as are many involving spatial and temporal heterogeneity, may be solvable only by computer simulation (e.g., section 10.4.2; optimal propagule size in a stochastic environment). In adopting this approach care must be taken in judging the generality of the findings. Any computer model is a specific model; generalizations require a survey of a range of possible models, though a single model can demonstrate that a hypothesis is incorrect. In some cases a model can be made analytically tractable by simplifying assumptions; computer models can be used to test the validity of this simple model. Ecological and evolutionary models have a nasty habit of quickly becoming analytically intractable, and the use of computer modeling is likely to play an increasingly important role in analyses.

Whether one is analyzing the life history of a specific organism or undertaking a theoretical investigation great care must be taken in the examination of the generality of the findings. A mathematical equation may have a variety of biological interpretations, and therefore the model may apply to a wider range of life histories than implied by the original biological premises. For example, the function $l(x)m(x)$ contained within the characteristic equation will be triangular in shape for both triangular and asymptotic age schedules of birth (section 5.2.1). While biological generality may be obtained from a single mathematical model, this cannot be assumed a priori for there will be cases in which even qualitative findings cannot be extrapolated. Two examples of this are presented in chapter 9: the relationship between the optimal clutch size and host quality depends upon the particular form of the monotonically increasing function relating fitness to host quality (section 9.3.1); similarly the optimal clutch size when several females lay on the same host (section 9.3.2) depends critically upon the relationship between per-capita production and number of competing larvae. In the latter example the results can be entirely reversed with apparently minor changes in the function.

Attention should also be paid to the effect of quantitative variation in parameter values because these also can change the relative importance of traits. A frequently cited statement that selection will act more strongly on development time than fecundity in determining the onset of reproduction is correct only if the rate of increase is relatively high; as r approaches zero (stationary population) selection may be stronger on fecundity rather than the age at maturity (section 7.1.1).

Because a particular phenomenon is found among a range of taxa does not mean that there is a single underlying cause. (Ockham's razor may be a generally useful tool but it is not a law of nature.) This is well illustrated by studies on the evolution of propagule size. Female size and female ration are both positively correlated with propagule size for taxa as disparate as turtles and insects (Tables 10.3, 10.8), and it seems highly unlikely that a single model will apply to all cases. In fact, two models—one predicated on a decrease in survival with increasing clutch size, and another based on the opposite assumption (positive density-dependence)—both predict an increasing egg size with female size and/or condition. (See section 10.3.3.) These models obviously differ in other respects: the important point is that either or both models may be valid depending upon the specifics of the life history.

Congruence between prediction and observation is heartening but one must be sure that the model is an adequate reflection of the biology of the species concerned. Generalization to other species requires that their life histories be biologically or mathematically equivalent to the original subject.

Question 4: Should We Measure Phenotypic or Genetic Correlations?

The core to any optimality argument is the set of trade-off functions: if some trait x increases then some other trait y decreases (e.g., if egg size is increased, fecundity is decreased). These trade-offs are typically based on phenotypic correlations, and have significance only if they reflect genetic correlations. The particular genetic model chosen to describe the genetic architecture of correlations determines the extent to which the genetic correlations constrain the equilibrium values and the trajectory. For the predictions of the optimality and genetic models to be the same, the trade-offs must have a functional basis that restricts the set of possible combinations (section 3.2). The usual formulation for quantitative genetic analysis (section 2.1) is therefore inappropriate to predict the long-term effects of selection. Suppose, for example, we are interested in the evolution of fecundity: under the usual quantitative genetic model there are few conditions under which the evolution of increased fecundity is precluded, which seems an unreasonable proposition. This problem disappears if functional constraints are introduced. For example, mechanical constraints may place a limit on the maximum reproductive biomass, and hence increases in the number of offspring can only be accomplished by a diminution of their size (section 10.2.2), which may be accompanied by a decrease in survival (section 10.1).

An "unconstrained analysis" may predict the variation about the presently existing value and hence may be a useful tool with which to predict immediate responses to selection. But the problems in accurately estimating the genetic covariance matrix are so daunting (section 3.3.3) that it is debatable whether attempts to accurately measure this matrix are worthwhile. Providing one has an accurate representation of the functional constraints that limit the possible set of life history components, one can proceed with an optimality analysis without recourse to genetic considerations.

A genetic analysis is nevertheless very useful. Establishing the presence of genetic variation for traits demonstrates that such traits can evolve and so are likely to have evolved in the past. This is a fairly trivial issue since the diversity of traits is excellent prima facie evidence for evolution and genetic variation. Not trivial is the importance of genetic variation underlying phenotypic variation in traits. Phenotypic variation is predicted by life history theory, resulting particularly from environmental heterogeneity and frequency-dependent selection. But at present we do not have a clear idea of how much phenotypic variation represents the phenotypic plasticity of a single genotype versus genetic variation. In the latter case, what prevents the erosion of variance? Theoretical considerations suggest that those mechanisms promoting phenotypic variation—environmental heterogeneity and frequency-dependent selection—may also promote genetic variation (sections 2.2.4, 2.2.5). Sexual selection, being generally frequency-dependent, can be placed in the same category. Other mechanisms more strictly genetic are mutation-selection balance (section 2.2.1), heterosis (2.2.2), and antagonistic pleiotropy (2.2.3).

Since genetic variation is required for evolution to proceed, the analysis of factors favoring the retention of variation is an important issue in the study of life history evaluation. Quantitative genetic theory gives us a means, however imperfect, of tackling this problem, and the next decade should see a significant increase in our understanding.

Though it may be logistically too difficult to accurately measure the genetic variance/covariance matrix it may be useful and feasible to establish the signs of the phenotypic and genetic correlations. According to theory, with a number of functional constraints, "positive genetic correlations are mathematically possible between some pairs of components of fitness that are under constraints that imply negative trade-offs between them" (Charlesworth 1990, p. 525). But at present we have no idea how frequently such paradoxical correlations occur. We do not even have a theoretical prediction. As stressed above, a primary goal in life history analysis should be the disentangling of the mechanisms underlying phenotypic correlations. Knowledge of these mechanisms may enable the construction of the appropriate genetic model (in the manner demonstrated by Charlesworth 1990) and the prediction of at least the relative size of the parameter space over which positive genetic correlations are likely to occur. Of course in looking for mechanisms one may get trapped into an infinite regress: there will necessarily have to be some point at which a mechanism is deemed "satisfactory." In any event the search for the causal factors underlying trade-offs should lead to a better understanding of the limits of evolution.

Since genetic correlations may not reflect functional constraints we might be tempted to discount them altogether. This would be wrong since genetic correlations do provide information on genetic architecture and on the significance of the presumed trade-off in the evolution of the trait(s) under study. For evolution to occur there must be genetic variance in traits and for trade-offs to be of evolutionary importance they must be genetically correlated. Suppose we hypothesize that a trade-off exists between two traits; what experiments or observations are required to confirm the evolutionary significance of the hypothesized trade-off?

Experiments or observations of trade-offs can be divided into four categories (Reznick 1985; section 3.3): phenotypic correlations from unmanipulated situations; phenotypic correlations from experiments in which one trait has been manipulated (e.g., a comparison between virgin and mated individuals); genetic correlations between traits; correlated response to selection on one trait. Data in the first category may be unreliable in demonstrating a trade-off because correlations with other traits can obscure or even apparently reverse the appearance of the putative trade-off. For example, a trade-off between egg size and fecundity may be obscured or reversed by correlation between body size and egg size. (See Fig. 3.4.) The reason for including phenotypic correlations from unmanipulated situations is precisely that such measurements may indicate the operation of confounding influences and hence the relative importance in nature of the hypothesized trade-off. Manipulation experiments provide sound evidence for a phenotypic trade-off between the traits. Although these experiments do not directly address the issue of whether the trade-offs are of evolutionary significance, they can give insight into the functional basis of the trade-off. Manipulation experiments are therefore a crucial part of any life history study. Sib analysis or selection experiments (categories 3 and 4) demonstrate that the evolution of traits will be modulated by genetic architecture, but without information on the functional basis of the trade-off they cannot indicate the degree to which evolution in certain directions will be prevented.

Suppose that the genetic correlation is found to be in the opposite direction to that postulated on the basis of the functional constraint. Such a result indicates that other functional constraints must be operative, and further experiments are required to locate these. The presence of a negative genetic correlation between traits is definitive evidence of a trade-off. A positive correlation does not by itself permit the rejection of the hypothesis of a trade-off, but it does indicate that the functional basis of the trade-off cannot directly involve the two traits alone. To illustrate this problem consider the following model discussed by Charlesworth (1990). The hypothetical organism lives for 3 years, breeding for the first time at the end of the first year (Fig. 11.1). All potential eggs are produced prior to first reproduction and hence

$$m(1) + m(2) + m(3) = \text{const} \qquad \textbf{(11.1)}$$

where $m(x)$ is the number of female offspring produced at age x. This is the first functional constraint. Two further constraints are assumed: that

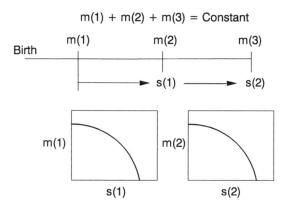

Figure 11.1. Graphical representation of the hypothetical life history discussed in the text.

there is a concave trade-off function relating fecundity at age 1 with survival, $s(1)$, from ages 1 to 2, and a similar function for age 2 (Fig. 11.1. For a discussion of the rationale for this type of trade-off see section 8.3.) The particular numerical forms chosen were

$$s(1) = 1 - m(1)^2$$
$$s(2) = 1 - m(2)^2 \qquad \textbf{(11.2)}$$

Population size is assumed fixed with density-dependence occurring in the immature stage. The appropriate fitness measure is thus lifetime reproduction

$$R_0 \propto m(1) + s(1)m(2) + s(2)m(3) \qquad \textbf{(11.3)}$$

Given the above constraints the signs of the genetic correlations between traits can be determined (see Charlesworth 1990 for the algorithm) and are presented in Table 11.1. Note first that the genetic correlations between the two functionally defined trade-offs relating fecundity and survival are both negative one. But correlations between survival from ages x to $x + 1$ and fecundity at ages other than x are positive or zero. The correlation between $m(2)$ and $m(3)$ is also zero. One might have hypothesized costs of reproduction that predicted negative trade-offs in these cases, but such trade-offs are not explicitly specified by the assumed set of functional constraints. Those that are explicitly specified do indeed produce negative genetic correlations. The set of genetic correlations does, therefore, reflect the underlying functional constraints. Whether the same matrix is possible from another set of constraints is not clear. What phenotypic correlations might we expect? These will depend upon whether traits are held constant. Suppose, for example, $m(2)$ is kept constant: in this case there will be a negative phenotypic correlation between $m(1)$ and $m(3)$, despite the genetic correlation being zero.

The above example emphasizes that experiments should be aimed at both the genetic basis of trade-offs and the mechanisms underlying their phenotypic manifestation.

Table 11.1. The signs of the genetic correlations between the life history traits discussed in the text

	$s(2)$	$m(1)$	$m(2)$	$m(3)$
$s(1)$	<0	−1	>0	>0
$s(2)$		>0	−1	0
$m(1)$			<0	<0
$m(2)$				0

Question 5: Can Variation Be Ignored?

Most life history traits show considerable phenotypic variation. Propagule size is generally the least variable. Most of the analyses discussed in this book have attempted to ignore this variation, concentrating on the prediction of some optimum value. This approach cannot produce a prediction that exactly matches observation. But this degree of accuracy is not expected since such models not only ignore parameter variation but also trivial features of the life history that might play a minor role in the evolution of the trait under study. The attempt is to get "reasonably" close to the observed value. How the term "reasonable" is defined will depend upon the nature of the study: one that is concerned with a single population will clearly wish to achieve a closer fit than one that seeks to predict the optimal life history trait across a wide taxonomic unit. The more focused the study, the more likely variability will intrude upon the analysis.

Variation is ubiquitous and there is an increasing concern for explanations of its origins and maintenance. Phenotypic variation can be the result of plasticity of a single genotype or of different genotypes within a population. Selection will clearly favor that genotype producing the most fit offspring in all environments encountered. Consequently we might expect phenotypic variation to be largely a consequence of a reaction norm. But estimates of genetic variation have shown that there is a significant amount in natural populations (chapter 1). This leads us to reject or modify our original model: either no single genotype is most fit across all environments or mutation is generating variations as fast as it is being eroded by natural selection. There is continuing debate on the role mutation plays in the maintenance of genetic variance, but in practically all studies $G \times E$ interactions have been found, suggesting that this phenomenon probably plays a significant role. The common observation of significant genotype-by-environment interactions suggests further that there are functional constraints that prevent any single genotype from achieving maximal fitness. A very fruitful area of future study is that of the functional constraints that limit the fitness of a single genotype.

Phenotypic plasticity may be adaptive or simply reflect physiological effects of no adaptive significance. Although an examination of norms of reaction may be useful, their importance is greatly diminished if their ecological significance cannot be quantified. Therefore, studies of this important question should examine traits for which fitness functions can be specified. For example, age at maturity of a univoltine species in a seasonal environment is strongly constrained by the end of the season, and hence the relative importance of maturing at different ages can be readily assessed.

An examination of norms of reaction is a promising field for study. But equally important is an examination of the extent to which adaptation is via phenotypic plasticity and the extent to which it is achieved via genetic differentiation. This will require not only experimental studies designed to determine the mode of adaptation but also the reasons why one mode has been favored over another. For example, in many insects the production of diapause eggs is controlled by a photoperiodic reaction norm: across a geographic range there may be no single norm of reaction that can accommodate the changes in thermal regime and photoperiod, and adaption will be primarily by genetic differentiation of the reaction norm (Bradford 1991).

The area of life history evolution is too broad to advocate any single approach or any narrowly defined set of problems. At one extreme we wish to explain why there are consistent patterns between life history characters across very broad taxonomic groupings, while at the other we seek to understand why it is that a particular species cannot evolve a single genotype that is the best in all possible worlds. The paradigm of natural selection is a major organizing principle but it is becoming increasingly clear that chance events may play an important role and that to attempt to ascribe adaptive significance to all phenomena is entirely wrong-headed. But much of what we see has been shaped by natural selection and hence is amenable to analysis based on the principle that fitness is being maximized. Underlying all analyses is the stated or unstated assumption that not all combinations are possible. To fully understand why trade-offs occur and how phenotypic and genetic variation is generated and maintained will require the combined efforts of workers from a broad array of biological disciplines.

Appendix: A Brief Review of Differentiation

To locate a turning point of a function we examine the first and second derivative of that function. Fig. A.1 shows the various possible configurations that may occur. Table A.1 gives the rules of differentiation for the functions used in this book. Some simple rules for the differentiation of "compound" functions follow:

A.1 The Derivative of a Sum of Functions

$$y = f(x) + g(x)$$

$$\frac{dy}{dx} = \frac{df(x)}{dx} + \frac{dg(x)}{dx}$$

Example:

$$y = ax^n + e^{bx}$$

$$\frac{dy}{dx} = anx^{n-1} + be^{bx}$$

A.2 The Chain Rule

Suppose we have $y = f(g(x))$; i.e., y is a function of x that itself can be decomposed into two or more functions; e.g., letting $u = g(x)$, we have

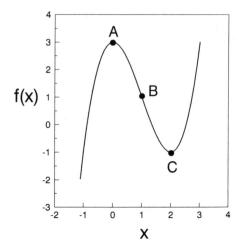

Figure A.1. Example of a curve with two turning points and an inflection. The curve changes from being concave (maximum at A) to convex (minimum at C). The actual function plotted is $f(x) = x^3 - 3x^2 + 3$. To analytically locate the turning points we differentiate with respect to x: $f'(x) = 3x^2 - 6x$. Maxima or minima occur when the first derivative equals zero, but this is an insufficient criterion since the first derivative may also be zero at a point of inflection. (At the inflection B in the above figure this is not the case.) To distinguish between turning points and points of inflection we examine the sign of the second derivative, $f''(x)$: at a maximum, $f''(x) < 0$; at a minimum $f''(x) > 0$; at an inflection point $f''(x) = 0$. The following table describes the changes in $f(x)$, $f'(x)$, and $f''(x)$ that occur in the above function. (Modified from Leithold 1969.)

Value of x	$f(x)$	$f'(x)$	$f''(x)$
$-\infty < x < 0$	$-\infty < f(x) < 3$	+	−
0 (point A)	3	0	−
$0 < x < 1$	$1 < f(x) < 3$	−	−
1 (point B)	1	−3	0
$1 < x < 2$	$-1 < f(x) < 1$	−	+
2 (point C)	−1	0	+
$2 < x < +\infty$	$-1 < f(x) < +\infty$	+	+

Table A.1. The derivatives of $y = f(x)$ and $y = f(u)$, where $u = g(x)$

Formula		Example	
$f(x)$	dy/dx	$f(x)$	dy/dx
ax^n	anx^{n-1}	$3x^5$	$15x^4$
au^n	$au^{n-1}du/dx$	$(x^5 + 4)^3$	$15(x^5 + 4)^2x^4$
$\log_e x$	$1/x$	—	—
$\log_e u$	$(du/dx)(1/u)$	$\log_e 3x^5$	$15x^4/3x^5 = 5/x$
e^x	e^x	—	—
e^u	$e^u\,du/dx$	e^{3x}	$15x^4e^{3x}$

$y = f(u)$. For example,

$$y = (1 - e^{-kx})^3$$

can be decomposed into

$$g(x) = (1 - e^{-kx}) = u$$
$$y = u^3$$

The chain rule states

$$\frac{dy}{dx} = \frac{dy}{du} \cdot \frac{du}{dx}$$

Thus for the above example

$$y = (1 - e^{-kx}) \qquad u = 1 - e^{-kx} \qquad y = u^3$$

$$\text{Hence} \qquad \frac{dy}{dx} = 3u^2 \cdot (-)(- ke^{-kx})$$

$$= 3(1 - e^{-kx})^2 ke^{-kx}$$

A.3 Differentiating the Product of Two or More Functions

Given $y = f(x)g(x)$ we differentiate by noting that

$$\frac{dy}{dx} = f(x)\frac{dg(x)}{dx} + g(x)\frac{df(x)}{dx}$$

Example

$$y = e^{bx}(1 - e^{-kx})$$

$$\frac{dy}{dx} = e^{bx}ke^{-kx} + (1 - e^{-kx})be^{bx}$$

A.4 Quotients

Suppose we have

$$y = \frac{f(x)}{g(x)}$$

We can proceed by making use of A.2 and A.3 by writing y as

$$y = f(x)g(x)^{-1} = f(x)u^{-1} \qquad \text{where} \qquad u = g(x)$$

and thus differentiation can be accomplished by

$$\frac{dy}{dx} = f(x)\left((-)u^{-2}\frac{du}{dx}\right) + g(x)^{-1}\frac{df(x)}{dx}$$

Now

$$\frac{du}{dx} = \frac{dg(x)}{dx}$$

and hence

$$\frac{dy}{dx} = f(x)\left((-)g(x)^{-2}\frac{dg(x)}{dx}\right) + g(x)^{-1}\frac{df(x)}{dx}$$

$$= \frac{-f(x)\dfrac{dg(x)}{dx} + \dfrac{df(x)}{dx}g(x)}{g(x)^2}$$

Example

$$y = \frac{ax^n}{1 - e^{-kx}} = ax^n(1 - e^{-kx})^{-1}$$

$$\frac{dy}{dx} = nax^{n-1}(1 - e^{-kx})^{-1} + ax^n(-1)(1 - e^{-kx})^{-2}ke^{-kx}$$

$$= \frac{nax^{n-1}}{1 - e^{-kx}} - \frac{ax^nke^{-kx}}{(1 - e^{-kx})^2}$$

A.5. Implicit Differentiation

Consider the equation $f(x) = h(y)$, e.g., $x^5 + 3x = 2y^3 + y^2 - 1$. This equation cannot be converted into the form y = function of x. We can make use of the chain rule on the right-hand side of the equation to give

$$\frac{df(x)}{dx} = \frac{dy}{dx} \cdot \frac{dh(y)}{dy}$$

Example:

$x^5 + 3x = 2y^3 + y^2 - 1$. Let $f(x) = x^5 + 3x$, $h(y) = 2y^3 + y^2 - 1$.

Therefore,

$$\frac{df(x)}{dx} = 5x^4 + 3$$

$$\frac{dh(y)}{dy} = 6y^2 + 2y$$

$$5x^4 + 3 = (6y^2 + 2y)\frac{dy}{dx}$$

Hence

$$\frac{dy}{dx} = \frac{5x^4 + 3}{6y^2 + 2y}$$

All of the foregoing rules can be illustrated using the fish model considered in chapter 7 and Roff (1984a). The object is to predict the optimal age at first reproduction in a semelparous species of fish assuming that the appropriate measure of fitness is r. The relevant equation to be solved is

$$\frac{e^{-\alpha(r+M)}(1 - e^{-k\alpha})^3 C}{1 - e^{-(r+M)}} = 1$$

where α is the age at first reproduction, M is the rate of mortality after the larval period, k is the growth-rate constant in the von Bertalanffy growth function, and C is a constant that comprises the product of several other constants (proportion of eggs and larvae surviving, asymptotic length cubed, and the coefficient in the fecundity/length function).

For convenience we first take natural logarithms and multiply throughout by negative one to give

$$\alpha(r + M) - 3 \log_e(1 - e^{-k\alpha}) - \log_e C + \log_e(1 - e^{-(r+M)}) = 0$$

To differentiate we proceed term by term as in A.1:

αr: This requires implicit differentiation to give

$$\frac{dr}{d\alpha} = r + \alpha \frac{dr}{d\alpha}$$

αM: This is simply

$$\frac{dr}{d\alpha} = M$$

$3 \log_e(1 - e^{-k\alpha})$: Using the chain rule we obtain

$$\frac{dr}{d\alpha} = \frac{3ke^{-k\alpha}}{1 - e^{-k\alpha}}$$

$\log_e C$: The differential of this is 0
$\log_e(1 - e^{-r+M})$: Again we use implicit differentiation:

$$\frac{dr}{d\alpha} = \frac{e^{-(r+M)}}{1 - e^{-(r+M)}} \left(\frac{dr}{d\alpha} \right)$$

Putting all this together we obtain

$$r + \alpha \frac{dr}{d\alpha} + M - \frac{3ke^{-k\alpha}}{1 - e^{-k\alpha}} - 0 + \frac{e^{-(r+M)}}{1 - e^{-(r+M)}} \left(\frac{dr}{d\alpha} \right) = 0$$

Rearranging,

$$\frac{dr}{d\alpha} \left(\alpha + \frac{e^{-(r+M)}}{1 - e^{-(r+M)}} \right) = \frac{3ke^{-k\alpha}}{1 - e^{-k\alpha}} - r - M$$

To locate the optimal age at reproduction, α, we must find that value of α at which $dr/d\alpha$ equals zero. Now, provided the term in parentheses on the left-hand side is not equal to zero when $dr/d\alpha = 0$, the required value for α is found by setting the right-hand side equal to zero; i.e., the optimal age at reproduction is that at which

$$r = \frac{3ke^{-k\alpha}}{1 - e^{-k\alpha}} - M$$

One further step is needed: we substitute the above into our original equation—actually the logarithmically transformed version is easier and is used here—to give

$$\alpha G + \log_e(1 - e^{-G}) - \log_e C - 2 \log_e(1 - e^{-k\alpha}) = 0$$

$$\text{where} \quad G = \frac{3ke^{-k\alpha}}{1 - e^{-k\alpha}}$$

which can be solved numerically to obtain α.

Glossary

Symbols

α: Age at maturity.

c_i: A constant.

λ: The finite rate of increase $= e^r$ where r is the rate of increase.

$l(x)$: Probability of survival to age x.

$m(x)$: Number of female offspring produced at age x.

$M(x)$: Instantaneous rate of mortality at age x.

$q(x)$: Probability of dying between the time interval $x, x + 1$.

r: Depending upon the context, the instantaneous rate of increase of a population or genotype. In this book it refers almost invariability to the genotype and is used as a measure of fitness. (See chapter 3, section 3.1.)

R_0: The net reproductive rate or expected lifetime fecundity. A measure of fitness when the population is stationary.

Definitions

Allee effect: Phenomenon in which larval survival in insects is increased by aggregation.

Allometry: Relationship between the size of two organisms or their parts: in functional form expressed as $y = ax^b$, where a and b are constants.

Altricial: Young that are born in an undeveloped state and are initially dependent on their parents for nourishment and protection. (See Precocial.)

Annual: Organism with a 1-year life-span. (See Univoltine, Perennial, Multivoltine).

Antagonistic pleiotropy: Phenomenon in which a gene has a positive effect on one component of fitness but a negative effect on another.

Arithmetic mean: $\Sigma x_i/n$, where x_i is observation i and n is sample size. (See Geometric Mean.)

Basal metabolic rate: Resting, thermoneutral rate of fuel consumption.

Bet hedging: Life history patterns that reduce variance in fitness.

Biennial: Organism that lives for 2 years, breeding in the second year.

Brachypterous: Short-winged.

Characteristic equation: Equation that relates population growth to the age schedules of birth and death. (See chapter 3, section 3.1.)

Cleptoparasitism: Phenomenon in which an organism (e.g., a solitary bee) lays its egg in the nest of another organism, the parasite feeding on the stored food, and frequently killing the host.

Clone: A group of genetically identical individuals.

Critical age group: That age group in which density dependence occurs. (See chapter 3, section 3.1.2.)

Degree days: A measure of the rate of development in ectotherms. One degree day is the number of degrees per day above threshold temperature (that temperature below which no development occurs).

Dominance: Occurs when an allele shows its effect in both the homozygous and heterozygous state.

Ectotherm: Organism that does not generate its own internal temperature: invertebrates, reptiles, and amphibians. (Plants might also be so classified.)

Endotherm: Organism that generates it own internal temperature: mammals and birds. (Some fish and insects have mechanisms that give them some degree of endothermy.)

Entropy: Measure of the state of disorder in a system.

Epistasis: Name given to the phenomenon in which the effect of two or more nonallelic genes in combination is not the sum of their separate effects.

ESS Evolutionarily stable strategy: A strategy, which, if adopted by individuals in a population, prevents the invasion into the population of a mutant adopting an alternate strategy.

Eutherian: Infraclass of vivaparous mammals that give birth to precocial young.

Evolutionarily stable strategy: See ESS.

Existence metabolism: Average energy metabolized over periods during which an organism maintains a constant weight.

Expected lifetime fecundity: See R_0.

Finite rate of increase: See λ.

Fully constrained model: Model in which the component parameters are estimated separately from the data being predicted. (See Partially Constrained Model.)

Game theory: In evolutionary biology, a method of analysis based on the principle that several individuals compete for some "prize" that can be equated to fitness.

Genotype: The genetic composition of an individual. (See Phenotype.)

Genotype-by-environment interaction: Occurs when the phenotypic value of a genotype is an additive function of the environment and an interaction term: hence $Y = c_1 + c_2D + c_3Dx$, where Y is the phenotypic value, D is a dummy variable ($= 0$, 1 for two genotypes), x is the environment and c_1, c_2, and c_3 are constants.

Geometric mean: $(\Pi x_i)^{(1/n)}$, where n is sample size and x_i the ith observation. (See Arithmetic Mean.)

Gonadosomatic index: Ratio or proportion of gonad weight to somatic weight. (See chapter 5, section 5.2.3.)

Group selection: Selection acting on groups rather than individuals. Thus evolution is seen to involve the differential survival of groups rather than individuals.

Hard selection: Selection favoring individuals with particular characteristics; both density- and frequency-independent. (See Soft Selection.)

Hemimetabolous: Insects showing incomplete metamorphosis, there being no obvious resting stage; examples are grasshoppers, aphids, and bugs. (See Holometabolous.)

Heritability: Measure of the proportion of phenotypic variance attributable to genetic effects. Heritability in the *broad sense* refers to all genetic sources. Heritability in the *narrow sense* refers only to the additive portion of genetic variance.

Heterosis: The phenomenon in which the heterozygote has a higher fitness than either homozygote.

Heterozygotic advantage: See Heterosis.

Holometabolous: Insects showing a complete metamorphosis; that is, having a pupal stage; examples are beetles, butterflies, and flies. (See Hemimetabolous.)

Inclusive fitness: That fraction of fitness accruing to an individual by virtue of interactions with related individuals. (See chapter 3, section 3.1.5.).

Individual selection: Selection acting to produce the differential survival and reproduction of individuals rather than groups. (See Group Selection.)

Infanticide: The killing of an offspring by its parent.

Instar: Growth in insects is accomplished by molting; development can, therefore, be divided into a series of discrete stages, called instars.

Intrinsic rate of increase: Rate of increase in an unsaturated environment.

Iteroparity: Repeat breeding. (See Semelparity.)

Jacks: Male salmon that mature precociously.

K-selection: Selection acting on the density-dependent fitness parameter, K. (See chapter 3, section 3.1.3.)

Malthusian parameter: Rate of increase for a given genotype. (See chapter 3, section 3.1.)

Migration: The movement of an organism from one habitat to another. It may or may not involve a return.

Multivoltine: A phenology comprising several generations per year. (See Annual, Univoltine, Perennial.)

Neonate: Newborn young.

Net reproductive rate: See R_0.

Nidicolous: Refers to young birds that remain in the nest and are completely dependent on the parent for a period after hatching. (See Altricial.)

Nidifugous: Refers to young birds that leave the nest shortly after hatching. (See Precocial.)

Norm of reaction: Phenotypic plasticity that is manifested as a continuous function of the variation in the environment.

Null model: The model of no effect. It is this model that statistical tests typically examine.

Overdominance: Phenomenon in which the character of the heterozygote is expressed more markedly in the phenotype than in that of either homozygote.

Ovoviparity: Egg laying.

Partially constrained model: Model in which some of the parameters are estimated using the same data as the model seeks to predict—for example, the logistic equation. (See chapter 4, section 4.5.1.)

Perennial: Organism that lives for several years. (See Multivoltine, Annual, Univoltine.)

Phenology: The sequence of events in a life history.

Phenotype: The phenotypic expression of a genotype.

Phenotypic plasticity: Phenotypic variation expressed by a single genotype in different environments.

Polymorphism: Strictly, the existence of several morphs: frequently used in biology to refer to a genetic polymorphism.

Power: The probability of rejecting a null hypothesis when it is in fact false ($= 1 - \beta$, where β is the probability of a type II error).

Pleiotropy: The multiple action of a gene.

Precocial: Young that are born in an advanced developmental state.

r-selection: Selection acting on the measure of fitness r, contrasted with selection acting on the density-dependent measure, K. (See chapter 3, section 3.1.3.)

Relative clutch mass: Ratio or proportion of clutch mass to body mass. (See chapter 5, section 5.2.3.)

Repeatability: The intraclass correlation coefficient.

Reproductive effort: Energy that is devoted to reproduction. There are several ways in which this may be defined. (See chapter 5, section 5.2.3.)

Reproductive value: Measure of the extent to which an individual of age x contributes to the ancestry of future generations. (See chapter 3, section 3.1.1.1.)

Satellite males: Males that take up station in the vicinity of a calling or displaying male. The satellite male may attempt to intercept incoming females or to replace the displaying male when it leaves.

Selection differential: Difference between the mean value of the population and the mean value of the parents that give rise to the subsequent generation.

Semelparity: Breeding once and dying; sometimes called "big bang" reproduction.

Senescence: Decline in function as a result of aging.

Sensitivity analysis: Analysis of the reaction of a model to variation in one or more of its component parameters or functions.

Siblicide: The killing of one offspring by its sibling.

Soft selection: Also called rank-order selection. Selection that favors a particular number or percentage of the population regardless of their absolute characteristics: frequency- and density-dependent. (See Hard Selection.)

Standard metabolism: See Basal Metabolic Rate.

Strategy: The terms *strategy* and *tactic* have frequently been applied to alternative methods of maximizing fitness: thus, for example, semelparity and iteroparity may be called strategies (or tactics). The two terms are, strictly, not synonyms and apply specifically to warfare, not biology. Their use in life history analysis has not always been well defined (Rothlisberg 1985; Chapleau et al. 1988), and even when defined, the differences in definitions merely muddy the waters (see, for example, Wootton 1984; Dominey 1984). I personally regard them as synonymous when used in a life history context. The terms have been used very sparingly in this book.

Surplus energy: Energy that is available for growth and/or reproduction after all maintenance costs have been accounted for.

Tactic: See Strategy.

Type I error: Rejection of a null hypothesis when it is, in fact, true.

Type II error: Failure to reject the null hypothesis when it is fact false.

Univoltine: Phenology that comprises a single generation per year.

Viviparity: Giving birth to live young. (See Ovoviparity.)

References

Adams, J., P. Greenwood, R. Pollitt and T. Yonow. 1985. Loading constraints and sexual size dimorphism in *Asellus aquaticus*. Behaviour **93**:277–287.

Adams, M.W. 1967. Basis of yield component compensation in crop plants with special reference to the field bean, *Phaseolus vulgaris*. Crop Science **7**:505–510.

Adams, P.B., 1980. Life history patterns in marine fishes and their consequences for fisheries management. Fisheries Bulletin **78**:1–12.

Agar, W.E. 1914. Experiments on inheritance in parthenogenesis. Transactions of the Royal Society of London (B) **205**:421–489.

Agduhr, E. 1939. Internal secretion and resistance to injurious factors. Acta Medica Scandinavica **99**:387–424.

Alcock, J. and T.F. Houston. 1987. Resource defence and alternative mating tactics in the Banksia bee, *Hylaeus alcyoneus* (Erichson). Ethology **76**:177–188.

Aleksiuk, M. 1977. Sources of mortality in concentrated garter snake populations. Canadian Field-Naturalist **91**:70–72.

Alerstam, T. and G. Högstedt. 1983. Regulation of reproductive success towards e^{-1} ($=37\%$) in animals with parental care. Oikos **40**:140–145.

Alerstam, T. and G. Högstedt. 1984. How important is clutch size dependent adult mortality? Oikos **43**:253–254.

Alexander, B. and J.J.G. Rozen. 1987. Ovaries, ovarioles, and oocytes in parasitic bees (Hymenoptera: Apodea). Pan-Pacific Entomologist **63**:155–164.

Allan, J.D. 1984. Life history variation in a freshwater copepod: evidence from population crosses. Evolution **38**:280–291.

Allard, R.W., S.K. Jain and P.L. Workman. 1968. The genetics of inbreeding populations. Advances in Genetics **14**:55–131.

Allee, A., E. Emerson, O. Park, T. Park and K.P. Schmidt. 1949. Principles of Animal Ecology. Saunders, Philadelphia.

Alm, G. 1946. Reasons for the occurrence of stunted fish populations (with special regard to the perch). Institute of Freshwater Research, Drottningholm. Report **25**:1–146.

Alm, G. 1959. Connection between maturity, size and age in fishes. Institute of Freshwater Research, Drottningholm. Report **40**:1–145.

Alpatov, W.W. 1929. Growth and variation of the larvae of *Drosophila melanogaster*. Journal of Experimental Zoology **52**:407–432.

Anan'eva, N.B. and S.M. Shammakov. 1986. Ecologic strategies and relative clutch mass in some species of lizard fauna in the USSR. Soviet Journal of Ecology (English translation Ekologiya) **16**:241–247.

Anderson, J.L. and R. Boonstra. 1979. Some aspects of reproduction in the vole *Microtus townsendii*. Canadian Journal of Zoology **57**:18–24.

Anderson, W.W. 1971. Genetic equilibrium and population growth under density-regulated selection. American Naturalist **105**:489–498.

Andersson, M. 1976. Clutch size in the Long-tailed Skua *Stercorarius longicaudus*: some field experiments. Ibis **118**:586–588.

Andrén, C. and G. Nilson. 1983. Reproductive tactics in an island population of adders, *Vipera berus* (L.), with a fluctuating food resource. Amphibia-Reptilia **4**:63–79.

Andrewartha, H.G. and L.C. Birch. 1954. The Distribution and Abundance of Animals. University of Chicago Press, Chicago.

Andrews, R.M. 1979. Reproductive effort of female *Anolis limifrons* (Sauria: Iguanidae). Copeia **1979**:620–626.

Andrews, R.M. and A.S. Rand. 1974. Reproductive effort in anoline lizards. Ecology **55**:1317–1327.

Ansell, A.D. 1960. Observations on predation of *Venus strialuta* (da Costa) by *Natica alderi* Forbes. Proceedings of the Malacological Society of London **34**:157–164.

Antonovics, J and J. Schmitt. 1986. Paternal and maternal effects on propagule size in *Anthoxanthum odoratum*. Oecologia **69**:277–282.

Antonovics, J. and N.C. Ellstrand. 1984. Expeirmental studies of the evolutionary significance of sexual reproduction. I. A test of the frequency-dependent selection hypothesis. Evolution **38**:103–115.

Arak, A. 1983a. Sexual selection by male-male competition in natterjack toad choruses. Nature **306**:261–262.

Arak, A. 1983b. Male-male competition and mate choice in anuran amphibians. pp181–210, *In* P. Bateson (editor), "Mate Choice," Cambridge University Press, Cambridge.

Arak, A. 1988. Callers and satellites in the natterjack toad: evolutionarily stable decision rule. Animal Behaviour **36**:416–432.

Armstrong, R.A. and M.E. Gilpin. 1977. Evolution in a time-varying environment. Science **195**:591–592.

Armstrong, T. and R.J. Robertson. 1988. Parental investment based on clutch value: nest desertion in response to partial clutch size in dabbling ducks. Animal Behaviour **36**:941–943.

Arnett, R.H. 1985. American Insects. Van Nostrand Reinhold, New York.

Arnold, J. and W.W. Anderson. 1983. Density-regulated selection in a heterogeneous environment. American Naturalist **121**:656–668.

Arnold, S.J. 1983. Morphology, performance and fitness. American Zoologist **23**:347–361.

Arnold, T.W. 1988. Life histories of North American game birds: a reanalysis. Canadian Journal of Zoology **66**:1906–1912.

Arnqvist, G. 1989. Multiple mating in a water strider: mutual benefits or intersexual conflict? Animal Behaviour **38**:749–756.

Asbirk, S. 1979. The adaptive significance of reproductive pattern in the black guillemot *Ceppus grylle*. Videnskaplige Meddelelser Dansk naturhistorisk Forening **141**:29–80.

Askenmo, C. 1977. Effects of addition and removal of nestlings on nestling weight, nestling survival and female weight loss in the pied flycatcher *Ficedula hypoleuca* (Pallas). Ornis Scandinavica **8**:1–8.

Askenmo, C. 1979. Reproductive effort and return rate of male pied flycatchers. American Naturalist **114**:748–753.

Asmussen, M.A. 1983. Density-dependent selection incorporating intraspecific competition. II. A diploid model. Genetics **103**:335–350.

Atchley, W.R. 1984. Ontogeny, timing of development, and genetic variance-covariance structure. American Naturalist **123**:519–540.

Atchley, W.R. and S. Newman. 1989. A quantitative-genetics perspective on mammalian development. American Naturalist **134**:486–512.

Auslander, D., J. Gukenheimer and G. Oster. 1978. Random evolutionarily stable strategies. Theoretical Population Biology **13**:276–293.

Avery, R.A. 1975. Clutch size and reproductive effort in the lizard *Lacerta vivapara* Jacquin. Oecologia **19**:165–174.

Axtell, R.W. 1958. Female reaction to the male call in two anurans (Amphibia). Southwestern Naturalist **3**:70–76.

Ayala, F.J. 1986. Mating, genetic variation and fitness. Accademia Nazionale dei Lincei **259**:9–23.

Ayala, J. and C.A. Campbell. 1974. Frequency dependent selection. Annual Review of Ecology and Systematics **5**:115–138.

Bagenal, T.B. 1955a. The growth rate of the long rough dab, *Hippoglossoides platessoides* (Fabr.). Journal of the Marine Biological Association, U.K. **34**:297–311.

Bagenal, T.B. 1955b. The growth rate of the long rough dab, *Hippoglossoides platessoides* (Fabr.). Journal of the Marine Biological Association,U.K. **34**:643–647.

Bagenal, T.B. 1966. A short review of fish fecundity. pp89–111, *In* S.D. Gerking (editor), "The Biological Basis of Freshwater Fish Production," Blackwell Scientific Publication, Oxford.

Bagenal, T.B. 1969. Relationship between egg size and fry survival in brown trout, *Salmo trutta* L. Journal of Fish Biology 1:349–353.

Bagenal, T.B. 1971. The interaction of the size of fish eggs, the date of spawning and the production cycle. Journal of Fish Biology 3:207–219.

Bailey, R.M. and K.F. Lagler. 1937. An analysis of hybridization in a population of stunted sunfishes in New York. Michigan Academy of Science, Arts and Letters 23:577–606.

Baker, R.J. and K.G. Briggs. 1982. Effects of plant density on the performance of 10 barley cultivars. Crop Science 22:1164–1167.

Bakker, K. 1959. Feeding period, growth, and pupation in larvae of *Drosophila melanogaster*. Entomologica Experimentalis et Applicata 2:171–186.

Bakker, K. 1961. An analysis of factors which determine success in competition for food among larvae of *Drosophila melanogaster*. Archives Neerlandaises Zoologie 14:200–281.

Ballinger, R.E. 1973. Comparative demography of two vivaparous iguanid lizards (*Sceloporus jarrovi* and *Sceloporus poinsetti*). Ecology 54:269–283.

Ballinger, R.E., E.D. Tyler and D.W. Tinkle. 1972. Reproductive ecology of a west Texas population of greater earless lizard, *Cophosaurus texanus*. American Midland Naturalist 88:419–428.

Balon, E.K. (editor). 1980. Charrs, Vol. 1 Dr. W. Junk Publishers, The Hague.

Baltz, D.M. 1984. Life history variation among female surfperches (Perciformes, Embiotocidae). Environmental Biology of Fishes 10:159–172.

Band, H.T. 1963. Genetic structure of populations. II. Viabilities and variances of heterozygotes in constant and fluctuating environments. Evolution 17:307–319.

Barbault, R. 1988. Body size, ecological constraints, and the evolution of life-history strategies. pp261–286, *In* M.K. Hecht, B. Wallace, and G.T. Prance (editors), "Evolutionary Biology 22," Plenum Press, New York.

Barbour, S.E., P.L.J. Rombough and J.J. Kerekes. 1979. A life history and ecological study of an isolated population of "dwarf" ouananiche, *Salmo salar*, from Gros Morne National Park, Newfoundland. Naturalist Canadien 106:305–311.

Barclay, H.J. and P.T. Gregory. 1981. An experimental test of models predicting life-history characteristics. American Naturalist 117:944–961.

Barclay, H.J. and P.T. Gregory. 1982. An experimental test of life history evolution using *Drosophila melanogaster* and *Hyla regilla*. American Naturalist 120:26–40.

Barlow, G.W. 1967. Social behavior of a South American leaf fish, *Polycentrus schomburgkii*, with an account of recurring pseudofemale behavior. American Midland Naturalist 78:215–234.

Barnes, H. 1962. So-called anecdysis in *Balanus balanoides* and the effect of breeding upon the growth of calcareous shell of some common barnacles. Limnology and Oceanography 7:462–473.

Barnes, H.H. and M. Barnes. 1965. Egg size, nauplius size, and their variation with local, geographic, and specific factors in some common cirripedes. Journal of Animal Ecology **34**:391–402.

Barnes, P.T. and C.C. Laurie-Ahlberg. 1986. Genetic variability of flight metabolism in *Drosophila melanogaster*. III Effects of GPDH allozymes and environmental temperature on power output. Genetics **112**:267–294.

Barton, N.H. 1986. The maintenance of polygenic variation through a balance between mutation and stabilizing selection. Genetic Research **47**:209–216.

Barton, N.H. and M. Turelli. 1987. Adaptive landscapes, genetic distance and the evolution of quantitative characters. Genetic Research **49**:157–173.

Baum, E.T. and A.L. Meister. 1971. Fecundity of Atlantic salmon (*Salmo salar*) from two Maine rivers. Journal of the Fisheries Research Board of Canada **28**:764–767.

Bauwens, D. and C. Thoen. 1981. Escape tactics and vulnerability to predation associated with reproduction in the lizard *Lacerta vivipara*. Journal of Animal Ecology **50**:733–743.

Bayne, B.L., D.L. Holland, M.N. Moore, D.M. Lowe and J. Widdows. 1978. Further studies on the effects of stress in the adult of the eggs of *Mytilus edulis*. Journal of the Marine Biological Association, U.K. **58**:825–841.

Beacham, T.D., F.C. Withler and R.B. Morley. 1985. Effect of egg size on incubation time and alevin and fry size in chum salmon (*Oncorhynchus keta*) and coho salmon (*Oncorhynchus kisutch*). Canadian Journal of Zoology **63**:847–850.

Beamish, R.J. 1973. Determination of age and growth of populations of the white sucker (*Catostomus commersoni*) exhibiting a wide range in size at maturity. Journal of the Fisheries Research Board of Canada **30**:607–616.

Beaulieu, M.A., S.U. Qadri and J.M. Hanson. 1979. Age, growth, and food habits of the pumkinseed sunfish *Lepomis gibbosus* (Linnaeus), in Lac Vert, Quebec. Naturlist Canadien **106**:547–553.

Becker, W.A. 1985. Manual of Quantitative Genetics. McNaughton and Gunn Inc., Ann Arbor.

Beckman, W.L. 1940. Increased growth rate of rock bass, *Ambloplites rupestris* (Rafinesque), following reduction in the density of the population. Transactions of the American Fisheries Society **70**:143–148.

Beebe, W. 1947. Notes on the Hercules beetle *Dynastes hercules* (Linn.), at Rancho Grande, Venezuela, with special reference to combat behavior. Zoologica **32**:109–116.

Begon, M. and G.A. Parker. 1986. Should egg size and clutch size decrease with age? Oikos **47**:293–302.

Bell, A.E. and M.J. Burris. 1973. Simultaneous selection for two correlated traits in *Tribolium*. Genetic Research **21**:24–46.

Bell, G. 1976. On breeding more than once. American Naturalist **110**:57–77.

Bell, G. 1980. The costs of reproduction and their consequences. American Naturalist **116**:45–76.

Bell, G. 1984a. Measuring the cost of reproduction. II. The correlation structure of the life tables of five freshwater invertebrates. Evolution **38**:314–326.

Bell, G. 1984b. Measuring the cost of reproduction. I. The correlation structure of the life table of a plankton rotifer. Evolution **38**:300–313.

Bell, G. 1989. A comparative method. American Naturalist **133**:553–571.

Bell, G. and V. Koufopanou 1985. The cost of reproduction. pp83–131, *In* R. Dawkins (editor),"Oxford Surveys of Evolutionary Biology," Oxford University Press, Oxford.

Bell, P.D. 1979. Acoustic attraction of herons by crickets. Journal of the New York Entomological Society **87**:126–127.

Bellinger, R.G. and R.L. Pienkowski. 1987. Development polymorphism in the red-legged grasshopper *Melanoplus femurrubrum* (DeGeer) (Orthoptera: Acridoidae). Environmental Entomology **16**:120–125.

Bengi, K. and G.A.E. Gall. 1978. Genotype-environment effects on growth and development in *Tribolium casteneum*. Journal of Heredity **69**:71–76.

Berger, A. 1989. Egg weight, batch size and fecundity of the spotted stalk borer, *Chilo partellus* in relation to weight of females and time of oviposition. Entomologica Experimentalis et Applicata **50**:199–207.

Berger, E. 1976. Heterosis and the maintenance of enzyme polymorphism. American Naturalist **110**:823–839.

Berglund, A., G. Rosenquist and I. Svensson. 1986. Mate choice, fecundity and sexual dimorphism in two pipefish species (Syngathidae). Behavioral Ecology and Sociobiology **19**:301–307.

Bernado, J. 1991. Manipulating egg size to study maternal effects on offspring traits. Trends in Ecology and Evolution **6**:1–2.

Berrigan, D. 1991. The allometry of egg size and number in insects. Oikos **60**:313–321.

Berrigan, D. 1991. Lift production in the flesh fly, *Neobellieria* (= *Sarcophaga*) *bullata Parker*. Functional Ecology **5**:448–456.

Berry, K.H. 1974. The ecology and social behavior of the chuckwalla, *Sauromalus obesus obesus* Baird. University of California Publications in Zoology **101**:1–60.

Bertness, M.D. 1981a. Pattern and plasticity in tropical hermit crab growth and reproduction. American Naturalist **117**:754–773.

Bertness, M.D. 1981b. Predation, physical stress, and the organisation of a tropical hermit crab community. Ecology **62**:411–425.

Bertness, M.D. 1981c. The influence of shell-type on hermit crab growth rate and clutch size. Crustaceana **40**:197–205.

Berven, K.A. 1981. Mate choice in the wood frog, *Rana sylvatica*. Evolution **35**:707–722.

Berven, K.A. 1982. The genetic basis of altitudinal variation in the wood frog *Rana sylvatica*. 1. An experiment analysis of life history traits. Evolution **36**:962–983.

Berven, K.A. 1987. The heritable basis of variation in larval developmental patterns within populations of the wood frog (*Rana sylvatica*). Evolution **41**:1088–1097.

Berven, K.A. and D.E. Gill. 1983. Interpreting geographic variation in life-history traits. American Zoologist **23**:85–97.

Beverton, R.J.H. 1963. Maturation, growth and mortality of clupeid and engraulid stocks in relation to fishing. Journal du Conseil Permanent International pour l'Exploration de la Mer **154**:44–67.

Beverton, R.J.H. and S.J. Holt. 1959. A review of the lifespans and mortality rates of fish in nature, and their relation to growth and other physiological characteristics. pp142–177, *In* G.E.W. Wolstenholme and M. O'Connor (editors), "CIBA Foundation Colloquia on Ageing. Volume 5," J.A. Churchill Ltd., London.

Bijlsma-Meeles, E. and R. Bijlsma. 1988. The alcohol dehydrogenase polymorphism in *Drosophila melanogaster*: fitness mesurements and predictions under conditions with no alcohol stress. Genetics **120**:743–753.

Bingham, J. 1967. Investigations on the physiology of yield in winter wheat, by comparisons of varieties and by artificial variations in grain number per ear. Journal of Agricultural Science **68**:411–422.

Birch, L.C. 1960. The genetic factor in population ecology. American Naturalist **94**:5–24.

Birkhead, T.R. and D.N. Nettleship. 1982. The adaptive significance of egg size and laying date in thick-billed murres. Ecology **63**:300–306.

Bisazza, A. and A. Marconato. 1988. Female mate choice, male-male competition and parental care in the river bullhead. *Cottus gobio* L. (Pisces, Cottidae). Animal Behaviour **36**:1352–1360.

Bisazza, A., A. Marconato and G. Marin. 1989. Male competition and female choice in *Padogobius martensi* (Pisces, Gobidae). Animal Behaviour **38**:406–413.

Black, J.N. 1958. Competition between plants of differential initial seed sizes in swards of subterranean clover (*Trifolium subterranean*) with particular reference to leaf area and light microhabitat. Australian Journal of Agricultural Research **9**:299–318.

Blackburn, D.G. and H.E. Evans. 1986. Why are there no vivaparous birds? American Naturalist **128**:165–190.

Blakley, N. and S.R. Goodner. 1978. Size-dependent timing of metamorphosis in milkweed bugs (*Oncopeltus*) and its life history implications. Biological Bulletin **155**:499–510.

Blaxter, J.H.S. 1969. Development: eggs and larvae. pp177–252, *In* W.S. Hoar and D.J. Randall (editors), "Fish Physiology 3," Academic Press, New York.

Blaxter, J.H.S. and G. Hempel. 1963. The influence of egg size on herring larvae (*Clupea harengus* L.). Journal du Conseil International Permanent pour l'Exploration de la Mer **28**:211–240.

Blueweiss, L., H. Fox, V. Kudzma, D. Nakashima, R. Peters and S. Sams. 1978. Relationships between body size and some life history parameters. Oecologia **37**:257–272.

Boag, P.T. and P.R. Grant. 1981. Intense natural selection in a population of Darwin's finches (Geospizinae) in the Galapagos. Science **214**:82–85.

Boggs, C.L. 1986. Reproductive strategies of female butterflies: variation and constraints on fecundity. Ecolological Entomology **11**:7–15.

Bohren, B.B., W.G. Hill and A. Robertson. 1966. Some observations on asymmetrical correlated responses to selection. Genetical Research **7**:44–57.

Bomze, I.M., P. Schuster and K. Sigmund. 1983. The role of Mendelian genetics in strategic models of animal behaviour. Journal of Theoretical Biology **101**:19–38.

Bondesen, P. 1940. Preliminary investigations into the development of *Neritina fluviatilis* L. in brackish and fresh water. Videnskabelige Meddelelser fra Dansk naturistorik Forening i Kjobenhavn **104**:238–318.

Bonnier, G., V.B. Jonsson and C. Rammel. 1959. Experiments on the effects of homozygosity on the rate of development in *Drosophila melanogaster*. Genetics **44**:679–704.

Borgia, G. 1980. Sexual competition in *Scatophaga stercoraria*: size and density related changes in male ability to capture females. Behaviour **75**:185–206.

Borgia, G. 1982. Experimental changes in resource structure and male density: size related differences in mating success among male *Scatophaga stercoraria*. Evolution **36**:307–315.

Boscher, J. 1981. Reproductive effort in *Allium porrum*: relation to the length of the juvenile phase. Oikos **37**:328–334.

Bostock, S.J. and R.A. Benton. 1979. The reproductive strategies of five perennial compositae. Journal of Ecology **67**:91–107.

Botkin, D.B. and R.S. Miller. 1974. Mortality rates and survival of birds. American Naturalist **108**:181–192.

Boucher, D.H. 1977. On wasting parental effort. American Naturalist **111**:786–788.

Boutin, S., R.A. Moses and M.J. Caley. 1988. The relationship between juvenile survival and litter size in wild muskrats (*Ondatra zibethicus*). Journal of Animal Ecology **57**:455–462.

Bowen, S.H. 1979. A nutritional constraint in detritivory by fishes: the stunted population of *Sarotherodon mossambicus* in Lake Sibaya, South Africa. Ecological Monogaphs **49**:17–31.

Box, J.F. 1978. R.A. Fisher: The Life of a Scientist. Wiley, New York.

Boyce, M.S. 1984. Restitution of r- and K-selection as a model of density-dependent natural selection. Annual Review of Ecology and Systematics **15**:427–447.

Boyce, M.S. and C.M. Perrins. 1987. Optimizing Great Tit clutch size in a fluctuating environment. Ecology **68**:142–153.

Bradford, M. 1991. The role of environmental hetrogeneity in the evolution of life history strategies of the striped ground cricket. Ph.D. Thesis, McGill University, Montreal, Canada.

Bradshaw, A.D. 1965. Evolutionary significance of phenotypic plasticity in plants. Advances in Genetics **13**:115–155.

Brady, R.H. 1979. Natural selection and the criteria by which a theory is judged. Systematic Zoologist **28**:600–621.

Britton, M.M. 1966. Reproductive success and survival of the young in *Peromyscus*. M. Sc. Thesis, University of British Columbia, Vancouver.

Brock, V.E. and R.H. Riffenburgh. 1959. Fish schooling: a possible factor in reducing predation. Journal du Conseil International Permanent pour l'Exploration de la Mer **XXV**:307–317.

Brockelman, W.Y. 1975. Competition, the fitness of offspring, and optimal clutch size. American Naturalist **109**:677–699.

Brodie, E.D. Jr. and D.R. Formanowicz Jr. 1983. Prey size preference of predators: differential vulnerability of larval anurans. Herpetologica **39**:67–75.

Brody, M.S. and L.R. Lawlor. 1984. Adaptive variation in offspring size in the terrestrial isopod, *Armadillidium vulgare*. Oecologia **61**:55–59.

Brooks, J.L. 1968. The effects of prey size selection by lake planktivores. Systematic Zoologist **17**:273–291.

Brooks, J.L. and S.I. Dodson. 1965. Predation, body size and composition of the plankton. Science **150**:28–35.

Brousseau, D.J. and J.A. Baglivo. 1988. Life tables for two field populations of soft-shell clam, *Mya arenaria*, (Mollusca: Pelecypoda) from Long Island Sound. Fishery Bulletin **86**:567–579.

Brown, J.C. and R.C. Lasiewski. 1972. Metabolism of weasels: the costs of being long and thin. Ecology **53**:939–943.

Brown, L.E. and J.R. Pierce. 1967. Male-male interactions and chorusing, intensities of the great plains toad, *Bufo cognatus*. Copeia **1967**:149–154.

Brown, W.D. 1990a. Size-assortative mating in the blister beetle *Lytta magister* (Coleoptera: Meloidae) is due to male and female preference for larger mates. Animal Behaviour **40**:901–909.

Brown, W.D. 1990b. Constraints on size-assortative mating in the Blister Beetle *Tegrodera aloga* (Coleoptera: Meloidea). Ethology **86**:146–160.

Browne, R.A. 1982. The costs of reproduction in brine shrimp. Ecology **63**:43–47.

Browne, R.A., S.E. Sallee, D.S. Grosch, W.O. Segreti and S.M. Parser. 1984. Partitioning genetic and environmental components of reproduction and lifespan in *Artemia*. Ecology **65**:949–960.

Browne, R.A. and W.D. Russell-Hunter. 1978. Reproductive effort in molluscs. Oecologia **37**:23–27.

Bryant, D.M. and K.R. Westerterp. 1983. Time and energy limits to brood size in house martins (*Delichon urbica*). Journal of Animal Ecology **52**:905–925.

Bryant, D.M. 1975. Breeding biology of the House Martin *Delichon urbica* in relation to insect abundance. Ibis **117**:180–216.

Bryant, D.M. 1979. Reproductive costs in the house martin (*Delichon urbica*). Journal of Animal Ecology **48**:655–675.

Bryant, E.H. 1971. Life history consequences of natural selection: Cole's result. American Naturalist **105**:75–76.

Bryant, E.H. 1976. A comment on the role of envirnomental variation in maintaining polymorphisms in natural populations. Evolution **30**:188–190.

Bucher, T.L., M.J. Ryan and G.A. Bartholomew. 1982. Oxygen consumption during resting, calling and nest building in the frog *Physalaemus pustulosus*. Physiological Zoology **55**:10–22.

Bull, J.J. and C.M. Pease. 1988. Estimating relative parental investment in sons versus daughters. Journal of Evolutionary Biology **1**:305–315.

Bull, J.J., R.C. Vogt and M.G. Bulmer. 1982. Heritability of sex ratio in turtles with environmental determination. Evolution **36**:333–341.

Bull, J.J. and R. Shine. 1979. Iteroparous animals that skip opportunities for reproduction. American Naturalist **114**:296–303.

Bulmer, M.G. 1971a. The effect of selection on genetic variability. American Naturalist **105**:201–211.

Bulmer, M.G. 1971b. Stable equilibria under the two island model. Heredity **27**:321–330.

Bulmer, M.G. 1972. The genetic variability of polygenic characters under optimizing selection, mutation and drift. Genetical Research **19**:17–25.

Bulmer, M.G. 1973. The maintenance of the genetic variability of polygenic characters by heterozygous advantage. Genetical Research.

Bulmer, M.G. 1985a. The Mathematical Theory of Quantitative Genetics. Clarendon Press, Oxford.

Bulmer, M.G. 1985b. Selection of iteroparity in a variable environment. American Naturalist **126**:63–71.

Bulmer, M.G. 1989. Maintenance of genetic variability by mutation-selection balance: a child's guide through the jungle. Genome **31**:761–767.

Bulmer, M.G. and C.M. Perrins. 1973. Mortality in the great tit *Parus major*. Ibis **115**:277–281.

Bumpus, H.C. 1899. The elimination of the unfit as illustrated by the introduced sparrow, *Passer domesticus*. Biological Lectures, Marine Biology Laboratory, Woods Hole, MA **6**:209–226.

Bunnell, F.L. and D.E. N. Tait 1981. Population dynamics of bears—implications. pp75–98, *In* C.W. Fowler and T.D. Smith (editors), "Dynamics of Large Mammal Populations," John Wiley and Sons, New York.

Bürger, R. 1988. Mutation-selection balance of continuum-of-alleles models. Mathematical Bioscience **91**:67–84.

Bürger, R. 1989. Linkage and the maintenance of heritable variation by mutation-selection balance. Genetics **121**:175–184.

Burk, T. 1982. Evolutionary significance of predation on sexually signalling males. Florida Entomologist **65**:90–104.

Burnet, B., D. Sewell and M. Bos. 1977. Genetic analysis of larval feeding behaviour in *Drosophila melanogaster*. Genetical Research **30**:149–161.

Burrough, R.J. and C.R. Kennedy. 1979. The occurrence and natural alleviation of stunting in a population of roach, *Rutilus rutilus* (L.). Journal of Fish Biology **15**:93–109.

Busack, C. and G. Gall. 1983. An initial description of the quantitative genetics of growth and reproduction in the mosquito fish. Aquaculture **32**:123–140.

Buskirk, R.E. 1975. Aggressive display and orb defense in a colonial spider, *Metabus gravidus*. Animal Behaviour **23**:560–567.

Butterworth, B.B. 1961. A comparative study of growth and development of the kangaroo rats, *Dipodomys deserti* Stephens and *Dipodomys merriami* Mearns. Growth **25**:127–139.

Cade, W. 1975. Acoustically orienting parasitoids: fly phonotaxis to cricket song. Science **190**:1312–1313.

Cade, W. 1979. The evolution of alternatiave male reproductive strategies in field crickets. pp343–379. *In* M.S. Blum and N.A. Blum (editors), "Sexual Selection and Reproductive Competition in Insects," Plenum Press, New York.

Cade, W.H. 1984. Effects of fly parasitoids on nightly calling duration in field crickets. Canadian Journal of Zoology **62**:226–228.

Calder, W.A. III. 1976. Aging in vertebrates: allometric considerations of spleen size and life span. Federation Proceedings **35**:96–97.

Calder, W.A. III. 1984. Size, Function and Life History. Harvard University Press, Cambridge.

Calder, W.A., III. 1983. Body size, mortality and longevity. Journal of Theoretical Biology **102**:135–144.

Caldwell, J.P., J.H. Thorp and T.O. Jervey. 1980. Predator-prey relationships among larval dragonflies, salamanders and frogs. Oecologia **46**:285–289.

Calef, G.W. 1973. Natural mortality of tadpoles in a population of *Rana aurora*. Ecology **54**:741–758.

Calow, P. 1977. Ecology, evolution and energetics: a study in metabolic adaptation. Advances in Ecological Research **10**:1–61.

Calow, P. 1978. The evolution of life-cycle strategies in fresh-water gastropods. Malacologia **17**:351–364.

Calow, P. and A.S. Woolhead. 1977. The relationship between ration, reproductive effort and age-specific mortality in the evolution of life-history strategies—some observations on freshwater triclads. Journal of Animal Ecology **46**:765–781.

Cals-Usciati, J. 1964. Étude comparative de caractères biométriques en fonction e l'origine géographique de diverse souches de *Drosophila melanogaster*. Annales de Genetique **7**:56–66.

Cameron, G.N. 1973. Effect of litter size on postnatal growth and survival in the desert wood rat. Journal of Mammalogy **54**:489–493.

Campbell, I.M. 1962. Reproductive capacity in the genus *Choristoneura* Led. (Lepidoptera: Tortricidae). I. Quantitative inheritance and genes as controllers of rates. Canadian Journal of Genetics and Cytology **4**:272–288.

Capinera, J.L. 1979. Quantitative variation in plants and insects: effect of propagule size on ecological plasticity. American Naturalist **114**:350–361.

Carey, K. and F.R. Ganders. 1980. Heterozygote advantage at the fruit wing locus in *Plectritis congesta* (Valerianaceae). Evolution **34**:601–607.

Carlander, K.D. 1969. Handbook of Freshwater Fish Biology. Vol. 1. Iowa State University Press, Ames.

Carlander, K.D. 1977. Handbook of Freshwater Fishery Biology. Vol. 2. Iowa State University Press, Ames.

Carson, H.L. 1961. Heterosis and fitness in experimental populations of *Drosophila melanogaster*. Evolution **15**:496–509.

Carson, H.L., D.E. Hardy, H.T. Spieth and W.S. Stone 1970. The evolutionary biology of the Hawaiian Drosophilidae. pp437–453, *In* M.K. Hecht and W.C. Steere (editors), "Essays in Evolution and Genetics in Honor of Theodosius Dobzhansky," Appleton-Century-Crofts, New York.

Carter, E.R. 1956. Investigations and management of the Dewey Lake fishery, Kentucky Department of Fish and Wildlife, Fisheries Bulletin **19**:1–20.

Case, T.J. 1978a. Endothermy and parental care in the terrestrial vertebrates. American Naturalist **112**:861–874.

Case, T.J. 1978b. On the evolution and adaptive significance of postnatal growth rates in the terrestrial vertebrates. Quarterly Review of Biology **53**:243–281.

Castilla, A.M., and D. Bauwens. 1989. Reproductive characteristics of the lacertid lizard *Lacerta lepida*. Amphibia-Reptilia **10**:445–452.

Caswell, H. 1978. A general formula for the sensitivity of population growth rate to changes in life history parameters. Theoretical Population Biology **14**:215–230.

Caswell, H. 1980. On the equivalence of maximizing reproductive value and fitness. Ecology **61**:19–24.

Caswell, H. 1982a. Optimal life histories and the maximization of reproductive value: a general theorem for complex life cycles. Ecology **63**:1218–1222.

Caswell, H. 1982b. Life history theory and the equilibrium status of populations. American Naturalist **120**:317–339.

Caswell, H. 1982c. Optimal life histories and the age-specific costs of reproduction. Journal of Theoretical Biology **98**:519–529.

Caswell, H. 1983. Phenotypic plasticity in life-history traits: demographic effects and evolutionary consequences. American Zoologist **23**:35–46.

Caswell, H. 1984. Optimal life histories and age-specific costs of reproduction: two extensions. Journal of theoretical Biology **107**:169–172.

Caswell, H. 1989. Matrix Population Models. Sinauer Associates, Inc., Sunderland, Massachusetts.

Caswell, H. and A. Hastings. 1980. Fecundity, developmental time, and population growth rate: an analytical solution. Theoretical Population Biology **17**:71–79.

Caswell, H. and L.A. Real. 1987. An approach to the pertubation analysis of optimal life histories. Ecology **68**:1045–1050.

Caswell, H. and P.A. Werner. 1978. Transient behavior and life history analysis of teasel (*Dipsacus sylvestris* Hnds.). Ecology **59**:53–66.

Caswell, H., R.J. Naiman and R. Morin. 1984. Evaluating the consequences of reproduction in complex salmonid life cycles. Aquaculture **43**:123–134.

Caughley, G. 1966. Mortality patterns in mammals, Ecology **47**:906–918.

Cavalli-Sforza, L.L., and W.F. Bodmer. 1971. The Genetics of Human Populations. W.H. Freeman and Co., San Francisco.

Cavers, P.B. and J.L. Harper. 1966. Germination polymorphism in *Rumex crispus* and *Rumex obtusifolius*. Journal of Ecology **54**:376–382.

Cavers, P.B. and M.G. Steel. 1984. Patterns of change in seed weight over time in individual plants. American Naturalist **124**:324–335.

Chapleau, F., P.H. Johansen and M. Williamson. 1988. The distinction between pattern and process in evolutionary biology: the use and abuse of the term 'strategy'. Oikos **53**:136–138.

Charlesworth, B. 1971. Selection in density-regulated populations. Ecology **52**:469–474.

Charlesworth, B. 1972. Selection in populations with overlapping generations. III Conditions for genetic equilibrium. Theoretical Population Biology **3**:377–395.

Charlesworth, B. 1973. Selection in populations with overlapping populations. V. Natural selection and life histories. American Naturalist **107**:303–311.

Charlesworth, B. 1974. Selection in populations with overlapping generations. VI. Rates of change of gene frequency and population growth rate. Theoretical Population Biology **6**:108–132.

Charlesworth, B. 1980. Evolution in Age Structured Populations. Cambridge University Press, Cambridge.

Charlesworth, B. 1984. The evolutionary genetics of life histories. pp117–133, *In* B. Shorrocks (editor). "Evolutionary Ecology," Blackwell, Oxford.

Charlesworth, B. 1990. Optimization models, quantitative genetics, and mutation. Evolution **44**:520–538.

Charlesworth, B. and J.A. Léon. 1976. The relation of reproductive effort to age. American Naturalist **110**:449–459.

Charlesworth, B. and J.A. Williamson. 1975. The probability of survival of a mutant gene in an age-structured population and implications for the evolution of life-histories. Genetical Research **26**:1–10.

Charlesworth, B. and J.T. Giesel. 1972a. Selection in populations with overlapping generations. II. Relations between gene frequency and demographic variables. American Naturalist **106**:388–401.

Charlesworth, B. and J.T. Giesel, 1972b. Selection in populations with overlapping generations. IV. Fluctuations in gene frequency with density dependent selection. American Naturalist **106**:402–411.

Charnov, E.L. 1976. Optimal foraging: the marginal value theorem. Theoretical Population Biology **9**:129–136.

Charnov, E.L. 1989a. Phenotypic evolution under Fisher's Fundamental Theorem of Natural Selection. Heredity **62**:113–116.

Charnov, E.L. 1989b. Natural selection on age at maturity in shrimp. Evolutionary Ecology **3**:236–239.

Charnov, E.L. 1991a. Pure numbers, invariants, and symmetry in the evolution of life histories. Evolutionary Ecology **5**:339–342.

Charnov, E.L. 1991b. Evolution of life history variation among female mammals. Proceedings of the National Academy of Sciences, USA **88**:1134–1137.

Charnov, E.L. and D. Berrigan. 1990. Dimensionless numbers and life history evolution: age at maturity versus the adult lifespan. Evolutionary Ecology **4**:273–275.

Charnov, E.L. and D. Berrigan. 1991a. Evolution of life history parameters in animals with indeterminate growth, particularly fish. Evolutionary Ecology **5**:63–68.

Charnov, E.L. and D. Berrigan. 1991b. Dimensionless numbers and the assembly rules for life histories. Philosophical Transactions of the Royal Society of London (B) **332**:41–48.

Charnov, E.L. and J.R. Krebs. 1973. On clutch size and fitness. Ibis **116**:217–219.

Charnov, E.L. and W.M. Schaffer. 1973. Life history consequences of natural selection: Cole's result revisited. American Naturalist **107**:791–793.

Charnov, E.L. and S.W. Skinner. 1984. Evolution of host selection and clutch size in parasitoid wasps. Florida Entomologist **67**:5–21.

Charnov, E.L. and S.W. Skinner. 1985. Complementary approaches to the understanding of parasitoid oviposition decisions. Ecological Entomology **14**:383–391.

Chen, S., S. Watanabe and K. Takagi. 1988. Growth analysis on fish population in the senescence with special reference to an estimation of age at end of reproductive span and life span. Nippon Suisan Gakkaishi **54**:1567–1572.

Chen, S.B. and S. Watanabe. 1989. Age dependence of natural mortality coefficient in fish population dynamics. Nippon Suisan Gakkaishi **55**:205–208.

Cheung, T.K. and R.J. Parker. 1974 Effect of selection on heritability and genetic correlation of two quantitative traits in mice. Canadian Journal of Genetics and Cytology **16**:599–609.

Cheverud, J.M. 1984. Quantitative genetics and developmental constraints on evolution by selection. Journal of Theoretical Biology **110**:155–171.

Cheverud, J.M. 1988. A comparison of genetic and phenotypic correlations. Evolution **42**:958–968.

Chiang, H.C. and A.C. Hodson. 1950. An analytical study of population growth in *Drosophila melanogaster*. Ecological Monographs **20**:173–206.

Childs, J.E. 1991. And the cat shall lie down with the rat. Natural History **6**:16–19.

Chimits, P. 1955. *Tilapia* and its culture. Fishery Bulletin of F. A. O. **8**:1–33.

Christiansen, F.B. 1974. Sufficient conditions for protected polymorphism in a subdivided populations. American Naturalist **108**:157–166.

Christiansen, F.B. 1985. Selection and population regulation with habitat variation. American Naturalist **126**:418–429.

Christiansen, F.B. and T.M. Fenchel. 1977. Theories of Populations in Biological Communities. Springer-Verlag, Berlin.

Christie, B.R. and R.R. Kalton. 1960. Recurrent selection for seed weight in Bromegrass, *Bromus inermis*, Leyss. Agronomy Journal **52**:575–578.

Clark, A.G. 1987. Senescence and the genetic-correlation hang-up. American Naturalist **129**:932–940.

Clark, D.R. 1970. Age specific "reproductive effort" in the worm snake *Carphophis vermis* (Kennicott). Transactions of the Kansas Academy of Science **73**:20–24.

Clark, R.J.B.S. 1984. The Life and Work of J.B.S. Haldane. Oxford University Press, Oxford.

Clarke B. 1972. Density-dependent selection. American Naturalist **106**:1–13.

Clarke, B.C. 1979. The evolution of genetic diversity. Proceedings of the Royal Society of London B **205**:453–474.

Clifford, H.F. and H. Boerger. 1974. Fecundity of mayflies (Ephemeroptera), with special reference to mayflies of a brown-water stream of Alberta, Canada. Canadian Entomologist **106**:1111–1119.

Clutton-Brock. T.H. 1984. Reproductive effort and terminal investment in iteroparous animals. American Naturalist **123**:212–229.

Clutton-Brock, T.H. 1991. The Evolution of Parental Care. Princeton University Press, Princeton, New Jersey.

Clutton-Brock, T.H., E.F. Guinnes and S.D. Albon. 1982. Red Deer—Behavior and Ecology of the Two Sexes. University of Chicago Press, Chicago.

Clutton-Brock, T.H., F.E. Guinness, and S.D. Albon. 1983. The costs of reproduction in red deer hinds. Journal of Animal Ecology **52**:367–383.

Coates, D. 1988. Length-dependent changes in egg size and fecundity in females, and brooded embryo size in males, of fork-tailed catfishes (Pisces: Ariidae) from the Sepik River, Papua New Guinea with some implications for stock assessments. Journal of Fish Biology **33**:455–464.

Cockerham, C.C., P.M. Burrows, S.S. Young and T. Prout. 1972. Frequency-dependent selection in random-mating populations. American Naturalist **106**:493–515.

Cody, M. 1974. Optimization in ecology. Science **183**:1156–1164.

Cody, M.L. 1966. A general theory of clutch size. Evolution **20**:174–184.

Cody, M.L. 1971. Ecological aspects of reproduction. pp461–512, *In* D.S. Farner and J.R. King (editors), "Avian Biology," Academic Press, New York.

Cohen, D. 1966. Optimizing reproduction in a randomly varying environment. Journal of theoretical Biology 12:119–129.

Cohen, D. 1971. Maximising final yield when growth is limited by time or by limiting resources. Journal of Theoretical Biology 33:299–307.

Cohen, D. 1976. The optimal timing of reproduction. American Naturalist 110:801–807.

Cole, K.S. 1982. Male reproductive behavior and spawning success in a temperate zone goby, *Coryphoterus nicholsi*. Canadian Journal of Zoology 60:2309–2316.

Cole, L.C. 1954. The population consequences of life history phenomena. Quarterly Review of Biology 29:103–137.

Collette, B.B. 1961. Correlations between ecology and morphology in anoline lizards from Havana, Cuba and southern Florida. Bulletin of the Museum of Comparative Zoology 125:137–162.

Congdon, J .D. 1989. Proximate and evolutionary constraints on energy relations of reptiles. Physiological Zoology 62:356–373.

Congdon, J.D. and J.W. Gibbons. 1987. Morphological constraint on egg size: a challenge of optimal egg size theory? Proceedings of the National Academy of Sciences, USA 84:4145–4147.

Congdon, J.D., J.W. Gibbons and J.L. Green. 1983. Parental investment in the chicken turtle (*Deirochelys reticularia*). Ecology 64:419–425.

Connell, J.H. 1970. A predator-prey system in the marine intertidal region. I. *Balanus glandula* and several predatory species of *Thais*. Ecological Monographs 40:49–78.

Connell, J.H. 1972. Community interactions on rocky intertidal shores. Annual Review of Ecology and Systematics 3:169:192.

Constantz, G.D. 1975. Behavioral ecology of mating in the Gila topminnow, *Poeciliopsis occidentalis* (Cyprinodontiformes: Poeciliidae). Ecology 56:966–973.

Cooke, D. 1988. Sexual selection in dung beetles. II. Female fecundity as an estimate of male reproductive success in relation to horn size, and alternative behavioral strategies in *Onthophagus binodis* Thunberg (Scarabaediae: Onthophagini). Australian Journal of Zoology 36:521–532.

Cooke, F., P.D. Taylor, C.M. Francis and R.F. Rockwell. 1990. Directional selection and clutch size in birds. American Naturalist 136:261–267.

Cooper, O.B., C.C. Wagner and G.E. Krantz. 1971. Bluegills dominate production in a mixed population of fishes. Ecology 52:280–290.

Corey, S. 1981. Comparative fecundity and reproductive strategies in seventeen species of the Cumacea (Crustacea: Peracarida). Marine Biology 62:65–72.

Corkhill, P. 1973. Food and feeding ecology of puffins. Bird Study 20:207–220.

Cote, I.M. and W. Hunte. 1989. Male and female mate choice in the redlip blenny: why bigger is better. Animal Behaviour 38:78–88.

Coulson, J.S. 1963. Egg size and shape in the Kittiwake (*Rissa tridactyla*) and their use in estimating age composition of populations. Proceedings of the Zoological Society of London B 140:211–227.

Courtney, S.P. 1984. The evolution of egg clustering by butterflies and other insects. American Naturalist **123**:276–281.

Cowley, D.E. and W.R. Atchley. 1990. Development and quantitative genetics of correlation structure among body parts of *Drosophila melanogaster*. American Naturalist **135**:242–268.

Creaser, E.P. Jr. 1973. Reproduction of the bloodworm (*Glycera dibranchiata*) in the Sheepscot Estuary, Maine. Journal of the Fisheries Research Board of Canada **30**:161–166.

Creel, S. 1990. How to measure inclusive fitness. Proceedings of the Royal Society of London (B) **241**:229–231.

Crespi, B.J. 1986. Territoriality and fighting in a colonial thrips, *Hoplothrips pedicularius*, and sexual dimorphism in Thysanoptera. Ecological Entomology **11**:119–130.

Crespi, B.J. 1989. Causes of assortative mating in arthropods. Animal Behaviour **38**:980–1000.

Cressman, R. 1988a. Frequency- and density-dependent selection: the two phenotype model. Theoretical Population Biology **34**:378–398.

Cressman, R. 1988b. Complex dynamical behavior of frequency-dependent viability selection: an example. Journal of Theoretical Biology **130**:167–173.

Cressman, R. and A.T. Dash. 1987. Density dependence and evolutionary stable strategies. Journal of Theoretical Biology **126**:393–406.

Crisp, P.J. and B. Patel. 1961. The interaction between breeding and growth rate in the barnacle *Eliminius modestus* Darwin. Limnology and Oceanography **6**:105–115.

Cronin, E.W. Jr., and P.W. Sherman. 1977. A resource-based mating system: the orange-rumped honeyguide. Living Bird **15**:5–32.

Cronmiller, J.R. and C.F. Thompson. 1980. Experimental manipulation of brood size in red-winged blackbirds. Auk **97**:559–565.

Crossner, K.A. 1977. Natural selection and clutch size in the European starling. Ecology **58**:885–892.

Crow, J.F. and M. Kimura. 1964. The theory of genetic loads. Proceedings of the XI International Congress on Genetics **3**:495–505.

Crow, J.F. and M. Kimura. 1970. An Introduction to Population Genetics Theory. Harper and Row, New York.

Crump, M.L. 1981. Variation in propagule size as a function of environmental uncertainty for tree frogs. American Naturalist **117**:724–737.

Crump, M.L. 1984. Intraclutch egg size variability in *Hyla crucifer* (Anura: Hylidae). Copeia **1984**:302–308.

Crump, M.L. and R.H. Kaplan. 1979. Clutch size partitioning of tropical tree frogs (Hylidae). Copeia **1979**:626–635.

Curio, E. and K. Regelmann. 1985. The behavioural dynamics of great tits (*Parus major*) approaching a predator. Zeitschrift fuer Tierpsychologie **69**:3–18.

Curnow, R.N. and C. Smith. 1975. Multifactorial models for familial diseases in man. Journal of the Royal Statistical Society. A 138 Part 2:131–169.

Curtsinger, J.W. 1976a. Stabilizing selection in *Drosophila melanogaster*. Journal of Heredity 67:59–60.

Curtsinger, J.W. 1976b. Stabilizing or directional selection on egg lengths? A rejoinder. Journal of Heredity 67:246–247.

Cushing, B.S. 1985. Estrous mice and vulnerability to weasel predation. Ecology 66:1976–1978.

Cushing, D.H. 1967. The grouping of herring populations. Journal of the Marine Biological Association, UK 47:193–208.

Cushing, D.H. and F.R. Harden Jones. 1968. Why do fish school? Nature 218:918–920.

Daborn, G. 1975. Life history and energy relations of the giand fairy shrimp *Branchinecta gigas* Lynch 1937 (Crustacea: Anostraca). Ecology 56:1025–1039.

Dadzie, S. and B.C.C. Wangila. 1980. Reproductive biology, length-weight relationship and relative condition of pond raised *Tilapia zilli* (Gervais). Journal of Fish Biology 17:243–253.

Dahlberg, M.D. 1979. A review of survival rates of fish eggs and larvae in relation to impact assessments. Marine Fisheries Review 41:1–12.

Daoulas, C. and A.N. Economou. 1986. Seasonal variation of egg size in the sardine, *Sardina pilchardus* Walb, of the Saronikos Gulf: causes and a probable explanation. Journal of Fish Biology 28:449–457.

David, J. and J. Legay. 1977. Relation entre la variabilité génétique de la taille des oeufs et celle de la taille des femelles: comparison de trois races géographiques de *Drosophila melanogaster*. Archives de Zoologie Experimentale et Generale 118:305–314.

David, J., P. Fouillet and J. van Herrewege. 1970. Sous-alimentation quantitative chez la Drosophile. I. Action sur le développement larvaire et sur la taille des adultes. Annales de la Society Entomologique de France 6:367–378.

Davies, N.B. and T.R. Halliday. 1978. Deep croaks and fighting assessment in toads *Bufo bufo*. Nature 274:683–685.

Davies, N.B. and T.R. Halliday. 1979. Competitive mate searching in male common toads, *Bufo, bufo*. Animal Behaviour 27:1253–1267.

Davis, D.E. 1951. The analysis of population by banding. Bird-banding 22:103–107.

Davis, J.W.F. 1975. Age, egg size, and breeding success in the herring gull *Larus argentatus*. Ibis 117:460–473.

Dawkins, R. 1976. The Selfish Gene. Oxford University Press, Oxford.

Dawkins, R. and T.R. Carlisle. 1976. Parental investment, mate desertion and a fallacy. Nature 262:131–132.

Dawson, D.G. 1972. The breeding ecology of House Sparrows. Ph.D. Thesis, Oxford University, Oxford.

Dayton, P.K. 1971. Competition, disturbance, and community organisation: the provision and subsequent utilization of space in a rocky intertidal community. Ecological Monographs **41**:351–389.

De Ciechomski, J. 1966. Development of the larvae and variations in the size of the eggs of the Argentine anchovy, *Engraulis anchoita* Hubbs and Marini. Journal du Conseil Permanent International pour l'Exploration de la Mer **30**:281–290.

Deakin, M.A.B. 1966. Sufficient conditions for genetic polymorphism. American Naturalist **100**:690–691.

Dean, J.K. 1981. The relationship between lifespan and reproduction in the grasshopper *Melanoplus*. Oecologia **48**:365–368.

DeBenedictis, P. 1978. Frequency-dependent selection: what is the problem? Evolution **32**:915–916.

Deelder, C.L. 1951. A contribution to the knowledge of the stunted growth of perch (*Perca fluviatilis* L.) in Holland. Hydrobiologia **3**:357–378.

Deevey, E.S. 1947. Life tables for natural populations of animals. Quarterly Review of Biology **22**:283–314.

Delcour, J. 1969. Influence de l'age parental sur la dimension des oeufs, la duree de developpement, et la taille thoracique des descendants, chez *Drosophila melanogaster*. Journal of Insect Physiology **15**:1999–2011.

DeLoach, C.J. and R.L. Rabb. 1972. Seasonal abundance and natural mortality of *Winthemia manducae* (Diptera: Tachinidae) and degree of parasitization of its host, the tobacco hornworm. Annals of the Entomological Society of America **65**:779–790.

DeMartini, E.E. 1988. Spawning success of the male plainfin midshipman. I. Influences of male body size and area of spawning site. Journal of Experimental Marine Biology and Ecology **121**:177–192.

DeMartini, E.E. 1991. Annual variations in fecundity, egg size, and the gonadal and somatic conditions of queenfish *Seriphus politus* (Sciaenidae). Fishery Bulletin **89**:9–18.

Demetrius, L. 1977. Adaptedness and fitness. American Naturalist **111**:1163–1168.

Dempster, E.R. 1955. Maintenance of genetic heterogeneity. Cold Spring Harbor Symposium on Quantitative Biology **20**:25–32.

Dempster, E.R., I.M. Lerner and D.C. Lowry. 1952. Continuous selection for egg production in poultry. Genetics **37**:693–708.

Den Boer, P.J. 1968. Spreading of risk and stabilization of animal numbers. Acta Biotheoretica **18**:165–194.

Denny, M.W., T.L. Daniel and M.A.R. Koehl. 1985. Mechanical limits to size in wave-swept organisms. Ecological Monographs **55**:69–102.

Derickson, W.K. 1976. Ecological and physiological aspects of reproductive strategies in two lizards. Ecology **57**:445–458.

Derr, J.A. 1980. The nature of variation in life history characters of *Dysdercus bimaculatus* (Heteroptera: Pyrrhocoridae), a colonizing species. Evolution **34**:548–557.

Desharnais, R.A. and R.F. Costantino. 1983. Natural selection and density-dependent population growth. Genetics **105**:1029–1040.

DeSteven, D. 1980. Clutch size, breeding success, and parental survival in the tree swallow (*Iridoprocne bicolor*). Evolution **34**:278–291.

Dewsbury, D.A. 1982. Ejaculate cost and male choice. American Naturalist **119**:601–610.

Diana, J.S. 1987. Simulation of mechanisms causing stunting in northern pike populations. Transactions of the American Fisheries Society **116**:612–617.

Dickerson, G.E. 1955. Genetic slippage in response to selection for multiple objectives. Cold Spring Harbour Symposium on Quantitative Biology **20**:213–223.

Dijkstra, C., A. Bijlsma, S. Daan, T. Meijer and M. Zijltra. 1990 Brood size manipulations in the kestrel (*Falco tinnunculus*): effects on offspring and parent survival. Journal of Animal Ecology **59**:269–285.

Dingle, H. 1985. Migration and life histories. Contributions to Marine Science **27**(supplement):27–44.

Dingle, H. and K.E. Evans. 1987. Responses in flight to selection on wing length in non-migratory milkweed bugs, *Oncopeltus fasciatus*. Entomologica Experimentalis et Applicata **45**:289–296.

Dingle, H., K.E. Evans and J.O. Palmer. 1988. Responses to selection among life-history traits in a nonmigratory population of milkweed bugs (*Oncopeltus fasciatus*). Evolution **42**:79–92.

Dingle, H., N.R. Blakley and E.R. Miller. 1980. Variation in body size and flight performance in milkweed bugs (*Oncopeltus*). Evolution **34**:371–385.

Dobzhansky, T. 1950. Evolution in the tropics. American Scientist **38**:209–221.

Dobzhansky, T. and H. Levene. 1955. Genetics of natural populations. XXIV. Developmental homeostasis in natural populations of *Drosophila pseudoobscura*. Genetics **40**:797–808.

Dodson, G.N. and L. Marshall. 1984. Mating patterns in an ambush bug *Phymata fasciata* (Phymatidae). American Midland Naturalist **112**:50–57.

Dominey, W. 1980. Female mimicry in male bluegill sunfish—a genetic polymorphism? Nature **284**:546–548.

Dominey, W.J. 1984. Alternative mating tactics and evolutionarily stable strategies. American Zoologist **24**:385–396.

Donald, D.B. and D.J. Alger. 1986. Dynamics of unexploited and lightly exploited populations of rainbow trout (*Salmo gairdneri*) for coastal, montane, and subalpine lakes in western Canada. Canadian Journal of Fisheries and Aquatic Sciences **43**:1773–1741.

Dorwood, D.F. 1962. Comparative biology of the white booby and brown booby *Sula* spp. at Ascension. Ibis **103b**:174–220.

Downhower, J.F., L. Brown, R. Pederson and G. Staples. 1983. Sexual selection and sexual dimorphism in mottled sculpins. Evolution **37**:96–103.

Draper, A.D. and C.P. Wilsie. 1965. Recurrent selection for seed size in Birdsfoot Trefoil, *Lotus corniculatus* L. Crop Science **5**:313–315.

Drent, R. and S. Daan. 1980. The prudent parent: energetic adjustments in avian biology. Ardea **68**:225–252.

Drickhamer, L.C. 1974. A ten-year summary of reproductive data for free-ranging *Macaca mulatta*. Folia Primatologica **21**:61–80.

Drori, D. and Y. Folman. 1969. The effect of mating on the longevity of male rats. Experimental Gerontology **4**:263–266.

Duarte, C.M. and M. Alcaraz. 1989. To produce many small or few large eggs: a size-independent reproductive tactic of fish. Oecologia **80**:401–404.

Dufresne, F., G.J. Fitzgerald and S. Lachance. 1990. Inter-male competition and reproductive costs in threespine stickleback. Behavioral Ecology **1**:140–147.

Dunbrack, R.L. and M.A. Ramsay. 1989a. Constraints on brain growth. Nature **340**:194.

Dunbrack, R.L. and M.A. Ramsay. 1989b. The evolution of viviparity in amniote vertebrates: egg retention versus egg size reduction. American Naturalist **133**:138–148.

Duncan, D.C. 1988. On the analysis of life-history traits of North American game birds. Canadian Journal of Zoology **64**:2739–2749.

Dunham, A.E. 1982. Demographic variation among populations of the iguanid lizard *Urosaurus ornatus*: implications for the study of life-history phenomena in lizards. Herpetologica **38**:208–221.

Dunham, A.E. and D.B. Miles. 1985. Patterns of covariation in life history traits of squamate reptiles: the effects of size and phylogeny reconsidered. American Naturalist **126**:231–257.

Dutil, J.D. 1986. Energetic constraints and spawning interval in the anadromous Arctic charr (*Salvelinus alpinus*). Copeia **1986**:945–955.

Dyar, M.I. 1975. The effects of red-winged blackbirds (*Agelaius phoeniceus* L.) on biomass production of corn grains (*Zea mays* L.). Journal of Applied Ecology **12**:719–729.

Dybas, H.S. 1976. The larval characteristics of featherwing and limulodid beetles and their family relationships in the Staphylinoidea (Coleoptera: Ptiliidae and Limulodidae). Fieldiana Zoology **70**:29–78.

Eberhard, W.G. 1977. Aggressive chemical mimicry by a bolas spider. Science **198**:1173–1175.

Eberhard, W.G. 1980. The natural history and behavior of the bolas spider *Mastophora dizzydeani* sp. n. (Araneidae). Psyche **87**:143–169.

Eberhard, W.G. 1982. Beetle horn dimorphism: making the best of a bad lot. American Naturalist **119**:420–426.

Echo, J.B. 1954. Some ecological relationships between yellow perch and cutthroat trout in Thompson Lakes, Montana. Transactions of the American Fisheries Society **84**:239–378.

Edley, M.T. and R. Law. 1988. Evolution of life histories and yields in experimental populations of *Daphnia magna*. Biological Journal of the Linnean Society **34**:309–326.

Eigenbrodt, H.J. and P.A. Zahl. 1939. The relationship of body size and egg size in *Drosophila*. Transactions of the Illinois State Academy of Science **1939**:204–205.

Elgar, M.A. and C.P. Catterall. 1989. Density-dependent natural selection. Trends in Ecology and Evolution **4**:95–96.

Elgar, M.A. and L.J. Heaphy. 1989. Covariation between clutch size, egg weight and egg shape: comparative evidence for chelonias. Journal of Zoology **219**:137–152.

Emlen, S.T. 1968. Territoriality in the bullfrog, *Rana catesbiana*. Copeia **1968**:240–243.

Emlen, S.T. 1976. Lek organisation and mating strategies in the bullfrog. Behavioral Ecology and Sociobiology **1**:283–313.

Endler, J.A. 1986. Natural Selection in the Wild. Princeton University Press, Princeton, New Jersey.

Endler, J.A. 1987. Predation, light intensity and courtship behaviour in *Poecilia reticulata* (Pisces: Poeciliidae). Animal Behaviour **35**:1376–1385.

Ennos, R.A. 1983. Maintenance of genetic variation in plant populations. Evolutionary Biology **16**:129–155.

Erasmus, J.E. 1962. Part-period selection for egg production. Proceedings of the 12th World Poultry Congress, Sydney **1962**:17–18.

Eschmeyer, R.W. 1936. Some characteristics of a population of stunted perch. Michigan Academy of Science, Arts and Letters **22**:613–678.

Eshel, I. 1982. Evolutionarily stable strategies and viability selection in Mendelian populations. Theoretical Population Biology **22**:204–217.

Ewens, W.J. and G. Thomson. 1970. Heterozygote selective advantage. Annals of Human Genetics **33**:365–376.

Ewing, A.W. 1961. Body size and courtship behaviour in *Drosophila melanogaster*. Animal Behaviour **9**:93–99.

Fagen, R.M. 1972. An optimal life history in which reproductive effort decreases with age. American Naturalist **106**:258–261.

Fairbairn, D.J. 1977. Why breed early? A study of reproductive tactics in *Peromyscus*. Canadian Journal of Zoology **55**:862–871.

Fairbairn, D.J. 1984. Microgeographic variation in body size and development time in the waterstrider, *Limnoporus notabilis*. Oecologia **61**:126–133.

Fairbairn, D.J. 1988. Sexual selection for homogamy in the Gerridae: an extension of Ridley's comparative approach. Evolution **42**:1212–1222.

Fairbairn, D.J. 1990. Factors influencing sexual size dimorphism in temperate waterstriders. American Naturalist **136**:61–86.

Fairchild, L. 1981. Mate selection and behavioral thermoregulation in Fowler's toads. Science **212**:950–951.

Falconer, D.S. 1981. Introduction to Quantitative Genetics, Second Edition. Longman, London.

Falconer, D.S. 1989. Introduction to Quantitative Genetics, Longmans, New York.

Farris, M.A. 1988. Quantitative genetic variation and natural selection in *Cleome serrulata* growing along a mild soil moisture gradient. Canadian Journal of Botany **66**:1870–1876.

Fehr, W.R. and C.R. Weber. 1968. Mass selection by seed size and specific gravity in soybean populations. Crop Science **8**:551–554.

Feifarek, B.P., C.A. Wyngaard and J.D.G. Allen. 1983. The cost of reproduction in a freshwater copepod. Oecologia **60**:166–168.

Feijen, H.R. and G.G.M. Schulten. 1981. Egg parasitoids (Hymen., Trichogrammatidae) of *Diopsis macrophthalma* in Malawi. Netherlands Journal of Zoology **31**:381–417.

Feldman, M.W. and R.C. Lewontin. 1975. The heritability hang-up. Science **190**:1163–1168.

Feller, W. 1940. On the logistic law of growth and its empirical verification. Acta Biotheoretica **5**:51–65.

Fellers, G.M. 1975. Behavioral interactions in North American treefrogs (Hylidae). Chesapeake Science **16**:218–219.

Fellers, G.M. 1979a. Aggression, territoriality and mating behaviour in North American treefrogs. Animal Behaviour **27**:107–119.

Fellers, G.M. 1979b. Mate selection in the gray tree frog, *Hyla versicolor*. Copeia **1979**:286–290.

Ferguson, G.W. and S.F. Fox. 1984. Annual variation of survival advantage of large juvenile side-blotched lizards *Uta stansburiana*: its causes and evolutionary significance. Evolution **38**:342–349.

Ferguson, G.W. and T. Brockman. 1980. Geographic differences of growth rate potential of *Sceloporus* lizards (Sauria: Iguanidae). Copeia **1980**:259–264.

Findlay, C.S. and F. Cooke. 1983. Genetic and environmental components of clutch size variance in a wild population of lesser snow geese (*Anser caerulescens caerulescens*). Evolution **37**:724–734.

Findley, C.S. and F. Cooke. 1987. Repeatability and heritability of clutch size in lesser snow geese. Evolution **41**:453.

Finke, M.A., D.J. Milinkovich and C.F. Thompson. 1987. Evolution of clutch size: an experimental test in the House Wren (*Troglodytes aedon*). Journal of Animal Ecology **56**:99–114.

Fisher, R.A. 1918. The correlation between relatives on the supposition of Mendelian inheritance. Transactions of the Royal Society of Edinburgh **52**:399–433.

Fisher, R.A. 1930. The Genetical Theory of Natural Selection. Clarendon Press, Oxford.

Fitch, H.S. 1957. Aspects of reproduction and development in the prairie vole (*Microtus ochrogaster*). University of Kansas, Publications of the Museum of Natural History **10**:129–161.

Fitt, G.P. 1990. Comparative fecundity, clutch size, ovariole number and egg size of *Dacus tryoni* and *D. jarvisi*, and their relationship to body size. Entomologica Experimentalis et Applicata **55**:11–21.

Flecker, A.S., J.D. Allen and N.L. McClintock. 1988. Male body size and mating success in swarms of the mayfly *Epeorus longimanus*. Holarctic Ecology **11**:280–285.

Fleischer, R.C. and R.F. Johnston. 1984. The relationship between winter climate and selection on body size of house sparrows. Canadian Journal of Zoology **62**:405–410.

Fleming, I.A. and M.R. Gross. 1990. Latitudinal clines: A trade-off between egg number and size in Pacific salmon. Ecology **71**:1–11.

Fleming, T.H. and R.J. Rauscher. 1978. On the evolution of litter size in *Peromyscus leucopus*. Evolution **32**:45–55.

Flux, J.E.C. and M.M. Flux. 1982. Artificial selection and gene flow in wild starlings, *Sturnus vulgaris*. Naturwissensch **69**:96–97.

Foote, C.J. 1988. Male mate choice dependent on male size in salmon. Behaviour **106**:63–80.

Foote, C.J. and P.A. Larkin. 1988. The role of male choice in the assortative mating of anadromous and non-anadromous sockeye salmong (*Oncorynchus nerka*). Behaviour **106**:43–62.

Ford, N.B. and R.A. Seigel. 1989. Relationships among body size, clutch size, and egg size in three species of oviparous snakes. Herpetologica **45**:75–83.

Fordham, R.H. 1971. Field populations of deermice with supplemental food. Ecology **52**:138–146.

Fowler, K. and L. Partridge. 1989. A cost of mating in female fruitflies. Nature **338**:760–761.

Fowler, L.G. 1972. Growth and mortality of fingerling chinook salmon as affected by egg size. Progressive Fish-Culturist **34**:66–69.

Freeman, G.H. 1973. Statistical methods for the analysis of genotype-environment interactions. Heredity **31**:339–354.

Fritz, R.S. and D.H. Morse. 1985. Reproductive success and foraging of the crab spider *Misumena vatia*. Oecologia **65**:194–200.

Frost, R.A. 1971. Aspects of the comparative biology of three weedy species of *Amaranthus* in southwestern Ontario. Ph.D. Thesis, University of Western Ontario, Ontario, Canada.

Frost, W.E. and C. Kipling. 1980. The growth of charr, *Salvelinus willughbii* Gunther, in Windermere. Journal of Fish Biology **16**:279–289.

Futuyama, D.J. and S.C. Peterson. 1985. Genetic variation in the use of resources by insects. Annual Review of Entomology **30**:217–238.

Gadgil, M. and W.H. Bossert. 1970. Life historical consequences of natural selection. American Naturalist **104**:1–24.

Gadgil, S., V. Nanjundiah and M. Gadgil. 1980. On evolutionarily stable compositions of populations of interacting genotypes. Journal of Theoretical Biology **84**:737–759.

Gage, T.B. and B. Dyke. 1988. Model life tables for the larger Old World Monkeys. American Journal of Primatology **16**:305–320.

Gaillard, J.-M., D. Pontier, D. Allaine, J.D. Lebreton, J. Trouvilliez and J. Clobert. 1989. An analysis of demographic tactics in birds and mammals. Oikos **56**:59–76.

Galbraith, M.G. Jr. 1967. Size-selective predation on *Daphnia* by rainbow trout and yellow perch. Transactions of the American Fisheries Society **96**:1–10.

Gall, G.A.E. 1974. Influence of size of eggs and age of female on hatchability and growth in rainbow trout. California Fish and Game **60**:26–35.

Gans, C. 1989. On phylogenetic constraints. Acta Morphologica Neerlando-Scandinavica **27**:133–138.

Garland, T.J. 1983. The relation between maximal running speed and body mass in terrestrial mammals. Journal of Zoology **199**:157–170.

Garland, T.J. 1985. Ontogenetic and individual variation in size, shape and speed in the Australian agamid lizard *Amphibolurus nuchalis*. Journal of Zoology **207**:425–439.

Garland, T.J. and P.L. Else. 1987. Seasonal, sexual, and individual variation in endurance and activity metabolism in lizards. American Journal of Physiology **252**:R439–R449.

Gäumann, E. 1935. Der stoffhaushalt der busche (*Fagus sylvatica* L.) in laufe eines jahres. Berichte de Deutschen Botanischen Gesellschaft **53**:366–377.

Geber, M.A. 1990. The cost of meristem limitation in *Polygonum arenastrum*: negative genetic correlations between fecundity and growth. Evolution **44**:799–819.

Genould, M. and P. Vogel. 1990. Energy requirements during reproduction and reproductive effort in shrews (Soricidae). Journal of Zoology **220**:41–60.

Gerhardt, H.C., R.E. Daniel, S.A. Perrill and S. Schram. 1987. Mating behaviour and male mating success in the green treefrog. Animal Behaviour **35**:1490–1503.

Ghent, A. 1960. A study of the group-feeding behavior of larvae of the jack pine sawfly, *Neodiprion pratti banksianae* Roh. Behaviour **16**:110–148.

Gibb, J.A. 1961. Bird populations. pp. 413–416. *In* A.J. Marshall (editor), "Biology and Comparative Physiology of Birds," Academic Press, New York.

Gibbons, J.W., J.L. Greene and K.K. Patterson. 1982. Variation in reproductive characteristics of aquatic turtles. Copeia **1982**:776–784.

Gibbs, H.L. 1988. Heritability and selection of clutch size in Darwin's medium ground finches (*Geospiza fortis*). Evolution **42**:750–762.

Gibson, R.L. and J.W. Bradbury. 1985. Sexual selection in lekking sage grouse: phenotypic correlates of male mating success. Behavioral Ecology and Sociobiology **18**:117–123.

Giesel, J.T. 1974. Fitness and polymorphism for net frequency distribution in iteroparous populations. American Naturalist **108**:321–331.

Giesel, J.T. 1979. Genetic co-variation of survivorship and other fitness indices in *Drosophila melanogaster*. Experimental Gerontology **14**:323–328.

Giesel, J.T. and E. Zettler. 1980. Genetic correlation of life historical parameters on certain fitness indices in *Drosophila melanogaster*: r vs diet breadth. Oecologia **47**:299–302.

Giesel, J.T., P.A. Murphy and M.N. Manlove. 1982. The influence of temperature on genetic interrelationships of life history traits in a population of *Drosophila melangaster*: what tangled data sets we weave. American Naturalist **119**:464–479.

Giles, B.E. 1990. The effects of variation in seed size on growth and reproduction in the wild barley *Hordeum vulgare* spp. *spontaneum*. Heredity **64**:239–250.

Giles, N. 1984. Implications of parental care of offspring for the anti-predator behaviour of adult male and female three-spined sticklebacks, *Gasterosteus aculeatus* L. pp. 275–289, *In* G.W. Potts and R.J. Wootton (editors), "Fish Reproduction, Strategies and Tactics," Academic Press, London.

Gill, D.E. 1972. Intrinsic rate of increase, saturation densities and competitive ability. I. An experiment with *Paramecium*. American Naturalist **106**:461–471.

Gill, D.E. 1974. Intrinsic rate of increase, saturation density, and competitive ability II. The evolution of competitive ability. American Naturalist **108**:103–116.

Gillespie, J.H. 1972. The effects of stochastic environment on allele frequencies in natural populations. Theoretical Population Biology **3**:241.

Gillespie, J.H. 1973a. Natural selection with varying selection coefficients—a haploid model. Genetical Research **21**:115–120.

Gillespie, J.H. 1973b. Polymorphism in random environments. Theoretical Population Biology **4**:193–195.

Gillespie, J.H. 1975. The role of migration in the genetic structure of populations in temporally and spatially varying environments I. Conditions for polymorphism. American Naturalist **109**:127–135.

Gillespie, J.H. 1976. A general model to account for enzyme variation in natural populations II. Characterization of the fitness function. American Naturalist **110**:809–821.

Gillespie, J.H. 1977. Natural selection for variance in offspring numbers: a new evolutionary principle. American Naturalist **111**:1010–1014.

Gillespie, J.H. 1978. A general model to account for enzyme variation in natural populations. V. The SAS-CFF model. Theoretical Population Biology **14**:1–45.

Gillespie, J.H. 1981. The role of migration in the genetic structure of populations in temporally and spatially varying environments. III Migration modification. American Naturalist **117**:223–233.

Gillespie, J.H. 1984. Pleotropic overdominance and the maintenance of genetic variation in polygenic characters. Genetics **107**:321–330.

Gillespie, J.H. and M. Turelli. 1989. Genotype-environment interactions and the maintenance of polygenic variation. Genetics **121**:129–138.

Gittleman, J.L. 1985. Carnivore body size: ecological and taxonomic correlates. Oecologia **67**:540–554.

Gittleman, J.L. and M. Kot. 1990. Adaptation: statistics and a null model for estimating phylogenetic effects. Systematic Zoology **39**:227–241.

Gittleman, J.L. and S.D. Thompson. 1988. Energy allocations in mammalian reproduction. American Zoologist **28**:863–875.

Given, M.F. 1987. Vocalizations and acoustic interactions of the carpenter frog, *Rana virgatipes*. Herpetologica **43**:467–481.

Given, M.F. 1988a. Territoriality and aggressive interactions of male carpenter frogs, *Rana virgatipes*. Copeia **1988**:411–421.

Given, M.F. 1988b. Growth rate and the cost of calling activity in male carpenter frogs, *Rana virgatipes*. Behavioral Ecology and Sociobiology **22**:153–160.

Glynn, P.W. 1970. On the ecology of the Caribbean chitons *Acanthopleura granulata* Gmelin and *Chiton tuberculatus* Linné: density, mortality, feeding, reproduction and growth. Smithsonian Contributions in Zoology **66**:1–21.

Godfray, H.C.J. 1986. Clutch size in a leaf-mining fly (*Pegomya nigritarsis*: Anthomyiidae). Ecological Entomology **11**:75–81.

Godfray, H.C.J. and A.B. Harper. 1990. The evolution of brood reduction by siblicide in birds. Journal of Theoretical Biology **145**:163–175.

Godin, J.J. and S.A. Smith. 1988. A fitness cost to foraging in the guppy. Nature **333**:69–71.

Gonor, J.J. 1972. Gonad growth in the sea urchin, *Strongylocentrotus purpuratus* (Stimpson) (Echinodermata: Echinoidea) and the assumptions of gonad index methods. Journal of Experimental Marine Biology and Ecology **10**:89–103.

Goodman, D. 1974. Natural selection and a cost ceiling on reproductive effort. American Naturalist **108**:247–268.

Goodman, D. 1981. Life history analysis of large mammals. pp. 415–436, *In* C.W. Fowler and T. Smith (editors), "Dynamics of Large Mammal Populations," John Wiley and Sons, New York.

Goodman, D. 1982. Optimal life histories, optimal notation, and the value of reproductive value. American Naturalist **119**:803–823.

Goodman, D. 1984. Risk spreading as an adaptive strategy in iteroparous life histories. Theoretical Population Biology **25**:1–20.

Goodman, L.A. 1971. On the sensitivity of intrinsic growth rate to changes in the age-specific birth and death rates. Theoretical Population Biology **2**:339–354.

Gordi, M.T. and J.A. Herrera. 1983. Estudio estadistico de la poblacion de *Larus argentatus* nidificante en las islas Medas (Estartit, Gerona). Alytes **1**:329–342.

Gould, S.J. and R.C. Lewontin. 1979. The spandrels of San Marco and the Panglossian paradigm—a critique of the adaptionist program. Proceedings of the Royal Society of London (B) **205**:581–598.

Grafen, A. 1982. How not to measure inclusive fitness. Nature **298**:425–426.

Grafen, A. 1984. Natural selection, kin selection and group selection. pp. 62–84, *In* J.R. Krebs and N.B. Davies (editors), "Behavioural Ecology," Sinauer Associates, Inc., Sunderland, Massachusetts.

Green, J. 1954. Growth, size and reproduction in *Daphnia magma* (Crustacea: Cladocera). Proceedings of the Zoological Society of London **124**:535–545.

Green, J. 1966. Seasonal variation in egg production by Cladocera. Journal of Animal Ecology **35**:77–104.

Green, J. 1967. The distribution and variation of *Daphnia lumholtzii* (Crustacea: Cladocera) in relation to fish predation in Lake Albert, East Africa. Journal of Zoology **151**:181–197.

Green, R. and P.R. Painter. 1975. Selection for fertility and development. American Naturalist **109**:1–10.

Green, R.F. 1980. A note on r- and K-selection. American Naturalist **116**:291–296.

Green, R.H. and K.D. Hobson. 1970. Spatial and temporal structure in a temperate intertidal community, with special emphasis on *Gemma gemma* (Pelecypoda: Mollusca). Ecology **51**:999–1011.

Griffiths, M.G. 1979. The Biology of the Monotremes. Academic Press, New York.

Grimaldi, J. and G. Leduc. 1973. The growth of yellow perch in various Quebec waters. Naturalist Canadien **100**:165–176.

Groeters, F.R. and H. Dingle. 1987. Genetic and mutational influences on life history plasticity in response to photoperiod by milkweed bugs (*Oncopeltus fasciatus*). American Naturalist **129**:332–341.

Gromko, M.H. 1977. What is frequency-dependent selection? Evolution **31**:438–442.

Groot, S.J. de. 1968. The Greenland halibut, a round flatfish or a flat roundfish? ICES C. M. 1968, Document No. F9.

Gross, H.L. 1972. Crown deterioration and reduced growth associated with excessive seed production by birch. Canadian Journal of Botany **50**:2431–2437.

Gross, K.L. 1984. Effects of seed size and growth form on seedling establishment of six monocarpic perennials. Journal of Ecology **72**:369–387.

Gross, M.R. 1982. Sneakers, satellites and parentals: polymorphic mating strategies in North American sunfishes. Zeitschrift fuer Tierpsychologie **60**:1–26.

Gross, M.R. 1985. Disruptive selection for alternative life histories in salmon. Nature **313**:47–48.

Gross, M.R. and E.L. Charnov. 1980. Alternative male life histories in bluegill sunfish. Proceedings of the National Academy of Sciences, USA **77**:6937–6940.

Gross, M.R., R.C. Coleman and R. McDowall. 1988. Aquatic productivity and the evolution of diadromous fish migration. Science **239**:1291–1293.

Gross, M.R. and R.G. Sargent. 1985. The evolution of male and female parental care in fishes. American Zoologist **25**:807–822.

Gunderson, D.R. 1980. Using r-K selection theory to predict natural mortality. Canadian Journal of Fisheries and Aquatic Sciences **37**:2266–2271.

Gunnarsson, B. 1988. Body size and survival: implications for an overwintering spider. Oikos **52**:274–282.

Gupta, A.P. and R.C. Lewontin. 1982. A study of reaction norms in natural populations of *Drosophila pseudobscura*. Evolution **36**:934–948.

Gustafsson, L. 1986. Lifetime reproductive success and heritabilities: empirical support for Fisher's fundamental theorem. American Naturalist **128**:761–764.

Gustafsson, L. and T. Part. 1990. Acceleration of senescence in the collared fly-catcher *Ficedula albicollis* by reproductive costs. Nature **347**:279–281.

Gustafsson, L. and W.J. Sutherland. 1988. The costs of reproduction in the Collared Flycatcher *Ficedula albicollis*. Nature **335**:813–815.

Gwynne, D.T. 1982. Mate selection by female katydids (Orthoptera: Tettigonidae, *Conocephalus nigoplurum*). Animal Behaviour **30**:734–738.

Gwynne, D.T. 1987. Sex-biased predation and the risky mate-locating behaviour of male tick-tock cicadas (Homoptera: Cicadidae). Animal Behaviour **35**:571–576.

Gwynne, D.T. and G.N. Dodson. 1983. Nonrandom provisioning by the digger wasp, *Palmedes laeviventris* (Hymenoptera, Sphecidae). Annals of the Entomological Society of America **76**:434–436.

Gwynne, D.T. and K.M. O'Neill. 1980. Territoriality in digger wasps results in sex biased predation on males (Hymenoptera: Sphecidae, *Philanthus*). Journal of the Kansas Entomological Society **53**:220–224.

Hagen, D.W. 1967. Isolating mechanisms in threespine sticklebacks (*Gasterosteus aculeatus*). Journal of the Fisheries Research Board of Canada **24**:1637–1692.

Hairston, N.G., Jr., W.E. Walton and K.T. Li. 1983. The causes and consequences of sex-specific mortality in a freshwater copepod. Limnology and Oceanography **28**:935–947.

Hairston, N.G., D.W. Tinkle and H.M. Wilbur. 1970. Natural selection and the parameters of population growth. Journal of Wildlife Management **34**:681–690.

Haldane, J.B.S. and S.D. Jayakar. 1963. Polymorphism due to selection of varying direction. Journal of Genetics **58**:237–242.

Hall, C.A.S. 1988. An assessment of several of the historically most influential theoretical models used in ecology and of the data provided in their support. Ecological Modelling **43**:5–31.

Hall, D.L., S.T. Threlkeld, C.W. Burns and P.H. Crowley. 1976. The size efficiency hypothesis and the size structure of zooplankton communities. Annual Review of Ecology and Systematics **7**:177–208.

Hall, G.O. and D.R. Marble. 1931. The relationship between the first year egg production and the egg production of later years. Poultry Science **10**:194–203.

Hamilton, W.D. 1964. The genetical evolution of social behavior. I. Journal of Theoretical Biology **7**:1–16.

Hamilton, W.D. 1966. The moulding of senescence by natural selection. Journal of Theoretical Biology **12**:12–45.

Hamilton, W.J. 1962. Reproductive adaptations in the red tree mouse. Journal of Mammalogy **43**:486–504.

Hammer, G.L. and R.L. Vanderlip. 1989. Genotype-by-environment interaction in grain sorghum. I. Effects of temperature on radiation use efficiency. Crop Science **29**:370–376.

Hammer, G.L., R.L. Vanderlip, G. Gibson, L.J. Wade, R.G. Herzell, D.R. Younger, J. Wanen and A.B. Dale. 1989. Genotype-by-environment interaction in grain sorghum. II. Effects of temperature and photoperiod on ontogeny. Crop Science **29**:376–384.

Hanson, D.A., B.J. Belonger and D.L. Schoenike. 1983. Evaluation of a mechanical reduction of black crappie and black bullheads in a small Wisconsin lake. North American Journal of Fisheries Management **3**:41–47.

Hansson, S. 1985. Local growth differences in perch (*Perca fluviatilis* L.) in a Baltic archipelago. Hydrobiologia **121**:3–8.

Hanwell, A. and M. Peaker. 1977. Physiological effects of lactation on the mother. Symposium of the Zoological Society of London **41**:297–312.

Harcourt, A.H., P.H. Harvey, S.G. Larson and R.V. Short. 1981. Testis weight, body weight and breeding system in primates. Nature **293**:55–57.

Harper, J.L. 1977. Population Biology of Plants. Academic Press, London.

Harper, J.L. and J. Ogden. 1970. The reproductive strategies of higher plants. I. The concept of strategy with special reference to *Senecio vulgaris*. Journal of Ecology **58**:681–698.

Harper, J.L. and J. White. 1974. The demography of plants. Annual Review of Ecology and Systematics **5**:419–463.

Harper, J.L., P.H. Lovell and K.G. Moore. 1970. The shapes and sizes of seeds. Annual Review of Ecology and Systematics **1**:327–356.

Harris, M.P. 1966. Breeding biology of the Manx Shearwater *Puffinus puffinus*. Ibis **108**:17–33.

Harris, M.P. 1969. The biology of stormpetrels in the Galapagos Islands. Proceedings of the California Academy of Science **37**:95–166.

Harris, M.P. 1970. The breeding ecology of the swallow-tailed Gull, *Creagrus furcatus*. Auk **87**:215–243.

Harris, M.P. and W.J. Plumb. 1965. Experiments on the ability of Herring Gulls *Larus argentatus* and Lesser Back-backed Gulls *Larus fuscus* to raise larger than normal broods. Ibis **107**:256–257.

Harris, V.E. and J.W. Todd. 1980. Male-mediated aggregation of male, female and 5th instar Southern green stink bugs and concomitant attraction of a tachinid parasite. Entomologica Experimentalis et Applicata **27**:117–126.

Harrison, R.G. 1980. Dispersal polymorphism in insects. Annual Review of Ecology and Systematics **11**:95–118.

Hart, J.L. 1973. Pacific fishes of Canada. Fisheries Research Board of Canada Bulletin **180**:1–740.

Hart, R. 1977. Why are biennials so few? American Naturalist **111**:792–799.

Hart, R.C. and I.A. McLaren. 1978. Temperature acclimation and other influences on embryonic duration in the copepod, *Pseudocalanus* sp. Marine Biology **45(23–30)**:

Hartl, D.L. 1980. Principles of Population Genetics. Sinauer Associates, Inc., Sunderland, Massachusetts.

Hartl, D.L. and R.D. Cook 1976. Stochastic selection and the maintenance of genetic variation. pp. 593–615, *In* S. Karlin and E. Nevo (editors), "Population Genetics and Ecology," Academic Press, New York.

Harvey, G.T. 1977. Mean weight and rearing performance of successive egg clusters of eastern spruce budworm (Lepidoptera: Tortricidae). Canadian Entomologist **109**:487–496.

Harvey, P.H., M.J. Stenning and B. Campbell. 1985. Individual variation in seasonal breeding success of pied flycatchers (*Ficedula hypoleuca*). Journal of Animal Ecology **54**:391–398.

Harvey, P.H. and A.E. Keymer. 1991. Comparing life histories using phylogenies. Philosophical Transactions of the Royal Society of London (B) **332**:31–39.

Harvey, P.H. and M.D. Pagel. 1991. The Comparative Method in Evolutionary Biology. Oxford University Press, Oxford.

Harvey, P.H. and T.H. Clutton-Brock. 1985. Life history variation in primates. Evolution **39**:559–581.

Hasler, J.F. and E.M. Basnks. 1975. Reproductive performance and growth in captive collared lemmings (*Dicostonyx groenlandicus*). Canadian Journal of Zoology **53**:777–787.

Hastings, A. 1978. Evolutionarily stable strategies and the evolution of life history strategies: 1. Density dependent models. Journal of Theoretical Biology **75**:527–536.

Hastings, A. and H. Caswell. 1979. Environmental variability in the evolution of life history strategies. Proceedings of the National Academy of Sciences, USA **76**:4700–4703.

Hastings, P.A. 1988. Female choice and male reproductive success in the angel blenny, *Coralliozetus angelica* (Teleostei: Chaenopsidae). Animal Behaviour **36**:115–124.

Haukioja, E. and T. Hakala. 1978. Life history evolution in *Anodonta piscinalis* (Mollusca, Pelecypoda). Oecologia **35**:253–266.

Haymes, G.T. and R.D. Morris. 1977. Brood size manipulations in Herring Gulls. Canadian Journal of Zoology **55**:1762–1766.

Hazlett, B.A. 1968. Size relationships and aggressive behavior in the hermit crab *Clibanarius vittatus*. Zeitschrift fuer Tierpsychologie **25**:608–614.

Heath, D. and D.A. Roff. 1987. A test of genetic differentiation in growth of stunted and non-stunted populations of perch and pumkinseed. Transactions of the American Fisheries Society 116:98–102.

Heath, D.D. 1986. An experimental and theoretical investigation of stunting in freshwater fishes, M.Sc. Thesis, McGill University, Montreal, Canada.

Hedrick, P.W. 1983. Genetics of Populations. Van Nostrand Reinhold Company, New York.

Hedrick, P.W. 1986. Genetic polymorphism in heterogeneous environments: a decade later. Annual Review of Ecology and Systematics 17:535–566.

Hedrick, P.W., M.E. Genevan and E.P. Ewing. 1976. Genetic polymorphism in heterogeneous environments. Annual Review of Ecology and Systematics 7:1–32.

Hegmann, J.P. and H. Dingle. 1982. Phenotypic and genetic covariance structure in Milkweed Bug life history traits. pp. 117–186, In H. Dingle and J.P. Hegmann (editors), "Evolution and Genetics of Life Histories," Springer-Verlag, New York.

Hegmann, J.P. and V.C. De Fries. 1970. Are genetic correlations and environmental correlations correlated? Nature 226:284–285.

Hegner, R.E. and J.C. Wingfield. 1987. Effects of brood-size manipulations on parental investment, breeding success, and reproductive endocrinology of house sparrows. Auk 104:470–480.

Hempel, G. and J.H.S. Blaxter. 1967. Egg weight in Atlantic herring (Clupea harengus L.). Journal du Conseil Permanent International pour l'Exploration de la Mer 31:170–195.

Henderson, P.A., R.H.A. Holmes and R.N. Bamber. 1988. Size-selective overwintering mortality in the sand smelt, Atherina boyeri Risso, and its role in population regulation. Journal of Fish Biology 33:221–233.

Hendrix, S.D. 1979. Compensatory reproduction in a biennial herb following insect defoliation. Oecologia 42:107–118.

Hendrix, S.D. 1984. Variation in seed weight and its effects on germination in Pastinaca sativa L. (Umbelliferae). American Journal of Botany 71:795–802.

Henrich, S. 1988. Variation in offspring sizes of the poeciliid fish Heterandria formosa in relation to fitness. Oikos 51:13–18.

Henrich, S. and J. Travis. 1988. Genetic variation in reproductive traits in a population of Heterandria formosa (Pisces: Poeciliidae). Journal of Evolutionary Biology 1:275–280.

Hermanutz, L.A. and R.W. Steele. 1983. Seed maturation patterns in Solidago sempervirens L. Abstract, Ontario Ecology and Ethology Colloqium, Peterborough.

Hessen, D.O. 1985. Selective zooplankton predation by pre-adult roach (Rutilus rutilus): the size-selective hypothesis versus the visibility-selective hypothesis. Hydrobiologia 124:73–79.

Hester, F.J. 1964. Effects of food supply on fecundity in the female guppy *Lebistes reticularis* (Peters). Journal of the Fisheries Research Board of Canada **21**:757–764.

Hickey, J.J. 1952. Survival studies of banded birds. USDA, Fish and Wildlife Service, Special Scientific Report, Wildlife No. 15, Washington, D.C., 117 pp.

Hickman, J.C. 1975. Environmental unpredictability and the plastic energy allocation strategies in the annual *Polygonum cascadense* (Polygonaceae). Journal of Ecology **63**:689–702.

Hikada, T. and S. Takahashi. 1987. Reproductive strategy and interspecific competition in the lake-living gobiid fish Isaza, *Chaenogobius isaza*. Journal of Ethology **5**:185–196.

Hill, J. 1975. Genotype-environment interactions—a challenge for plant breeding. Journal of Agricultural Science **85**:477–493.

Hill, R.W. 1972. The amount of maternal care in *Peromyscus leucopus* and its thermal significance for the young. Journal of Mammalogy **53**:774–790.

Hillesheim, E. 1984. Heritability of physiological characters of the Cape honeybee *Apis mellifera capensis*. Apidologie **15**:271–273.

Hindar, K. and B. Jonsson. 1982. Habitat and food segregation of the dwarf and normal arctic charr (*Salvelinus alpinus*) from Vangsvtnet Lake, Western Norway. Canadian Journal of Fisheries of Aquatic Science **93**:1030–1045.

Hines, W.G.S. 1980. An evolutionarily stable strategy for randomly mating diploid populations. Journal of Theoretical Biology **87**:379–384.

Hines, W.G.S. 1987. Can and will a sexual diploid population evolve to an ESS: the multi-locu linkage equilibrium case. Journal of Theoretical Biology **126**:1–5.

Hirschfield, M.F. 1980. An experimental analysis of reproductive effort and cost in the Japanese mendaka *Oryzias latipes*. Ecology **61**:282–293.

Hirshfield, M.F. and D.W. Tinkle. 1975. Natural selection and the evolution of reproductive effort. Proceedings of the National Academy of Sciences, USA **72**:2227–2231.

Hislop, J.R.G., A.P. Robb and J.A. Gauld. 1978. Observations on effects of feeding level on growth and reproduction in haddock *Melanogrammus aeglefinus* in captivity. Journal of Fish Biology **13**:85–89.

Hodgson, G.L. and G.E. Blackman. 1956. An analysis of the influence of plant density on the growth of *Vicia faba*. Part 1. The influence of density on the pattern of development. Journal of Experimental Botany **7**:147–165.

Hoekstra, R.F., R. Bijlsma and A.J. Dolman. 1985. Polymorphism from environmental heterogeneity: models are only robust if the heterozygote is close in fitness to the favored homozygote in each environment. Genetical Research **45**:299–314.

Hogan-Warburg, A.J. 1966. Social behavior of the ruff, *Philomachus pugnax* (L.). Ardea **54**:109–229.

Högstedt, G. 1980. Evolution of clutch size in birds: adaptive variation in relation to territory quality. Science **210**:1148–1150.

Högstedt, G. 1981. Should there be a positive or negative correlation between survival of adults in a bird population and their clutch size? American Naturalist **118**:568–571.

Holcomb, L.C. 1969. Breeding biology of the American Goldfinch in Ohio. Bird-banding **40**:26–44.

Hooper, F.F., J.E. Williams, M.H. Patriarche, F. Kent and J.C. Schneider. 1964. Report of the Michigan Department of Conservation. Michigan Department of Conservation, Institute of Fisheries Research Report No. **1688**:1–56.

Horsfall, J.A. 1984. Food supply and egg mass variation in the European coot. Ecology **65**:89–95.

Horvitz, C.C. and D.W. Schemske. 1988. Demographic cost of reproduction in a neotropical herb: An experimental field study. Ecology **69**:1741–1745.

Houle, D. 1989. The maintenance of polygenic variation in finite populations. Evolution **43**:1767–1780.

Houston, A.I. and J.M. McNamara. 1986. The influence of mortality on the behavior that maximizes fitness in a patchy environment. Oikos **47**:267–274.

How, R.A., J. Dell and B.D. Wellington. 1986. Comparative biology of eight species of *Diplodactylus* gecko in Western Australia. Herpetologica **42**:471–482.

Howard, R.D. 1978. The evolution of mating strategies in bullfrogs, *Rana catesbeiana*. Evolution **32**:850–871.

Howard, R.D. 1980. Mating behavior and mating success in wood frogs *Rana sylvatica*. Animal Behaviour **28**:705–716.

Howard, R.D. 1984. Alternative mating behaviors of young male bullfrogs. American Zologist **24**:397–406.

Howard, R.D. 1988. Sexual selection on male body size and mating behaviour in American toads, *Bufo americanus*. Animal Behaviour **36**:1796–1808.

Howe, H. 1976. Egg size, hatching asynchrony, sex and brood reduction in the common grackle. Ecology **57**:1195–1207.

Howe, H.F. 1978. Initial investment, clutch size, and brood reduction in the common grackle (*Quiscalus quiscula* L.). Ecology **59**:1109–1122.

Howell, D.J. and B.S. Roth. 1981. Sexual reproduction in agaves: the benefits of bats; the costs of semelparous advertising. Ecology **62**:1–7.

Howell, N. 1981. The effect of seed size and relative emergence time on fitness in a natural population of *Impatiens capensis* Meerb (Balsaminaceae). American Midland Naturalist **105**:312–320.

Hoyle, J.A. and A. Keast. 1987. The effect of prey morphology and size on handling time in a piscivore, the largemouth bass (*Micropterus salmoides*). Canadian Journal of Zoology **65**:1972–1977.

Hrbáček, J. and M. Hrbáčková-Esslová. 1960. Fish stock as a protective agent in the occurrence of slow-developing dwarf species and strains of the genus *Daphnia*. Internationale Revue de Gesamten Hydrobiologie **45**:355–358.

Hubbell, S.P. 1971. Of sowbugs and systems: the ecological bioenergetics of a terrestrial isopod. pp. 269–324, *In* B.C. Patten (editor), "Systems Analysis and Simulation in Ecology," Academic Press, New York.

Hubbs, C., M.M. Stevenson and A.E. Peden. 1968. Fecundity and egg size in two central Texas darter populations. Southwestern Naturalist **13**:310–323.

Hudson, J.W. 1974. The estrous cycle, reproduction, growth and development of temperature regulation in the pygmy mouse, *Baiomys taylori.* Journal of Mammalogy **55**:572–588.

Huey, R.B. 1987. Physiology, history and the comparative method. pp. 76–98, *In* A.F. Bennett, W.W. Burggren and R.B. Huey, M.E. Feder (editors), "New Directions in Ecological Physiology," Cambridge University Press, Cambridge.

Huey, R.B. and E.R. Pianka. 1981. Ecological consequences of foraging mode. Ecology **62**:991–999.

Hughes, R.N. 1971. Ecological energetics of the key-hole limpet *Fissurella barbadensis* Gmelin. Journal of Experimental Marine Biology **6**:167–178.

Hughes, R.N. and D.J. Roberts. 1980. Reproductive effort of winkles (*Littorina* spp.) with contrasted methods of reproduction. Oecologia **47**:130–136.

Hurka, H. and M. Benneweg. 1979. Patterns of seed size variation in populations of the common weed *Capsella bursa-pastoris* (Brassicaceae). Biologisches Zentralblatt **98**:699–709.

Hussell, D.J.T. 1972. Factors affecting clutch size in arctic passerines. Ecological Monographs **42**:317–364.

Hutchings, J.A. and D.W. Morris. 1985. The influence of phylogeny, size and behavior on patterns of covariation in salmonid life histories. Oikos **45**:118–124.

Hutchings, J.A. and R.A. Myers. 1987. Escalation of an asymmetric contest: mortality resulting from mate competition in Atlantic salmon, *Salmo salar.* Canadian Journal of Zoology **65**:766–768.

Hutchings, J.A. and R.A. Myers. 1988. Mating success of alternative maturation phenotypes in male Atlantic salmon, *Salmo salar.* Oecologia **75**:169–174.

Iason, G.R. 1990. The effects of size, age and cost of early breeding on reproduction in female mountain hares. Holarctic Ecology **13**:81–89.

Iles, T.D. 1971. Growth studies on the North Sea herring. 3. The growth of East Anglian herring during the adult stage of the life history for the years 1940–1967. Journal du Conseil International pour l'Exploration de la Mer **33**:386–420.

Iles, T.D. 1973. Dwarfing or stunting in the genus *Tilapia* (Cichlidae); a possibly unique recruitment mechanism. Rapports et Process-Verbeaux des Reunions de la Conseil Permanent International pour l'Exploration de la Mer **164**:247–254.

Innes, D.G. and J.S. Millar. 1979. Growth of *Clethrionomys gapperi* and *Microtus pennsylvanicus* in captivity. Growth **43**:208–217.

Itô, Y. 1980. Comparative Ecology, 2nd ed. (Trans. by J. Kikkawa). Cambridge University Press, Cambridge.

Itô, Y., Y. Tsubaki and M. Osada. 1982. Why do *Luehdorfia* butterflies lay eggs in clusters? Researches in Population Ecology **24**:375–387.

Itô, Y. and Y. Iwasa. 1981. Evolution of litter size. I. Conceptual reexamination. Researches in Population Ecology **23**:344–356.

Ives, A.R. 1989. The optimal clutch size of insects when many females oviposit per patch. American Naturalist **133**:671–687.

Ives, A.R. and R.M. May. 1985. Competition within and between species in a patchy environment: relations between microscopic and macroscopic models. Journal of Theoretical Biology **115**:65–92.

Iwasa, Y. and E. Teramoto. 1980. A criterion of life history evolution based on density dependent selection. Journal of Theoretical Biology **84**:545–566.

Iwasa, Y., Y. Suzuki and H. Matsuda. 1984. Theory of oviposition strategy of parasitoids. I. Effect of mortality and limited egg number. Theoretical Population Biology **26**:205–227.

Jackson, D.J. 1928. The inheritance of long and short wings in the weevil, *Sitona hispidula*, with a discussion of wing reduction among beetles. Transactions of the Royal Society of Edinburgh **55**:655–735.

Jago, N.D. 1973. The genesis and nature of tropical forest and savanna grasshopper faunas, with special reference to Africa. pp. 187–196, *In* B.J. Meggers, E.S. Ayensu and D. Duckworth (editors), "Tropical Forest Ecosystems in Africa and South America: A Comparative Review," Smithsonian Institution Press, Washington, D.C.

Jalaluddin, M. and S. A. Harrison. 1989. Heritability, genetic correlation, and genotype × environment interactions of soft red winter wheat yield and test weight. Cereal Research Communications **17**:43–49.

James, J.W. 1961. Selection in two environments. Heredity **16**:145–152.

Janzen, D.H. 1977. Variation in seed size within a crop of a Costa Rican *Micuna andreana* (Leguminosae). American Journal of Botany **64**:347–349.

Jarvis, M.J.F. 1974. The ecological significance of clutch size in the South African gannet (*Sula capensis*, Lichtenstein). Journal of Animal Ecology **43**: 1–17.

Jenkins, R.M. 1956. Some results of the partial fish population removal techniques in lake management. Proceedings of the Oklahoma Academy of Science **37**:164–173.

Jennings, H.S. and R.S. Lynch. 1928. Age, mortality, fertility and individual diversities in the rotifer *Proales sordida* Grosse II. Life history in relation to mortality and fecundity. Journal of Experimental Zoology **51**:339–381.

Jensen, J.P. 1958. The relation between body size and number of eggs in marine malacostrakes. Meddelelser fra Danmarks Fiskeri-og Havundersogelser **2**:1–25.

Johnson, L.K. 1982. Sexual selection in a brentid weevil. Evolution **36**:251–262.

Johnston, R.F., D.M. Niles and S.A. Rowher. 1972. Hermon Bumpus and natural selection in the house sparrow *Passer domesticus*. Evolution **26**:20–31.

Jones, J.S., B.H. Leith and P. Rawlings. 1977. Polymorphism in *Cepea*: a problem with too many solutions. Annual Review of Ecology and Systematics **8**:109–143.

Jones, P.J. 1973. Some aspects of the feeding ecology of the Great Tit *Parus major* L. Ph.D. Thesis, University of Oxford, Oxford.

Jones, R.E., J.R. Hart and G.D. Bull. 1982. Temperature, size and egg production in the cabbage butterfly, *Pieris rapae*. Australian Journal of Zoology **30**:223–232.

Jones, R.E. 1987. Behavioural evolution in the cabbage butterfly (*Pieris rapae*). Oecologia **72**:69–76.

Jones, R.E. and P.M. Ives. 1979. The adaptiveness of searching and oviposition behaviour in *Pieris rapae* L. Australian Journal of Ecology **4**:75–86.

Jones, W.T. 1985. Body size and life-history variables in Heteromyids. Journal of Mammalogy **66**:128–132.

Jong, G. de. 1979. The influence of the distribution of juveniles over patches of food on the dynamics of a population. Netherlands Journal of Zoology **29**:33–51.

Jong, G. de 1982. Fecundity selection and maximization of equilibrium number. Netherlands Journal of Zoology **32**:572–585.

Jong, G. de. 1984. Selection and numbers in models of life histories. pp. 87–101, *In* K. Wohrman and V. Loeschke (editors), "Population Biology and Evolution," Springer-Verlag, Berlin.

Jong, G. de. 1990a. Genotype-by-environment interaction and the genetic covariance between environments: multilocus genetics. Genetica **81**:171–177.

Jong, G. de. 1990b. Quantitative genetics of reaction norms. Journal of Evolutionary Biology **3**:447–468.

Jonsson, B. and K. Hinder. 1982. Reproductive strategy of dwarf and normal Arctic charr (*Salvelinus alpinus*) from Vangsvatnet Lake, western Norway. Canadian Journal of Fisheries and Aquatic Sciences **39**:1404–1413.

Jonsson, P.E. 1987. Sexual size dimorphism and disassortative mating in the Dunlin *Calidris alpina schinzii* in southern Sweden. Ornis Scandinavica **18**:257–264.

Joshi, A. and L.D. Mueller. 1988. Evolution of higher feeding rate in *Drosophila* due to density-dependent natural selection. Evolution **42**:1090–1093.

Kachi, N. and T. Hirose. 1985. Population dynamics of *Oenothera glazioviana* in a sand-dune system with special reference to the adaptive significance of size-dependent reproduction. Journal of Ecology, **73**:887–901.

Kaczmarski, F. 1966. Bioenergetics of pregnancy and lactation in the bank vole. Acta Theriologica **11**:409–417.

Kalela, O. 1957. Regulation of reproductive rate in subarctic populations of the vole *Clethrionomys rufocanus* (Sund.). Annales Academiae Scientiarum Fennicae, Series A **4**:1–60.

Kalin, M. and G. Knerer. 1977. Group and mass effects in diprionid sawflies. Nature **267**:427–429.

Kalisz, S. 1989. Fitness consequences of mating systems, seed weight, and emergence date in a winter annual, *Collinsia verna*. Evolution **43**:1263–1272.

Kallman, K.D. 1983. The sex determining mechanism of the poecilid fish, *Xiphophorus montezumae*, and the genetic control of the sexual maturation process and adult size. Copeia **1983**:755–769.

Kallman, K.D., M.P. Schreibman and V. Borkoski. 1973. Genetic control of gonadotroph differentiation in the platyfish, *Xiphophorus maculatus* (Poeciliidae). Science **181**:678–680.

Kallman, K.D. and V. Borkoski. 1978. A sex-linked gene controlling the onset of sexual maturity in female and male platyfish (*Xiphophorus maculatus*), fecundity in females and adult size in males. Genetics **89**:79–119.

Kambysellis, M.P. and W.B. Heed. 1971. Studies of oogenesis in natural populations of Drosophilidae. I. Relation of ovarian development and ecological habitats of the Hawaiian species. American Naturalist **105**:31–49.

Kaplan, R.H. 1980. The implications of ovum size variability for offspring fitness and clutch size within several populations of salamanders (*Ambystoma*). Evolution **34**:51–64.

Kaplan, R.H. 1985. Maternal influences on offspring development in the California Newt, *Taricha torosa*. Copeia **1985**:1028–1035.

Kaplan, R.H. and S.N. Salthe. 1979. The allometry of reproduction: an empirical view in salamanders. American Naturalist **113**:671–689.

Kaplan, R.H. and W.S. Cooper. 1984. The evolution of developmental plasticity in reproductive characteristics: an application of the "adaptive coin-flipping" principle. American Naturalist **123**:393–410.

Karlin, S. and R.B. Campbell. 1981. The existence of a protected polymorphism under condition of soft as opposed to hard selection in a multideme population system. American Naturalist **117**:262–275.

Karlsson, B. and C. Wiklund. 1984. Egg weight variation and lack of correlation between egg weight and offspring fitness in the wall brown butterfly *Lasiommata megera*. Oikos **43**:376–385.

Karn, M.N. and L.S. Penrose. 1951. Birth weight and gestation time in relation to maternal age, parity and infant survival. Annals of Eugenics **16**:147–164.

Kaufman, D.W. and G.A. Kaufman. 1987. Reproduction by *Peromyscus polinotus*: number, size, and survival of young. Journal of Mammalogy **68**:275–280.

Kavanagh, M.W. 1987. The efficiency of sound production in two cricket species, *Gryllotalpa australis* and *Telegryllus commodus* (Orthoptera: Grylloidea). Journal of Experimental Zoology **130**:107–119.

Kawasaki, T. 1980. Fundamental relations among the selections of life history in the marine teleosts. Bulletin of the Japanese Society of Scientific Fisheries **46**:289–293.

Kazakov, R.V. 1981. The effect of the size of the Atlantic salmon, *Salmo salar* L., eggs on embryos and alevins. Journal of Fish Biology **19**:353–360.

Keenleyside, M.H.A., R.W. Rangley and B.U. Kuppers. 1985. Female mate choice and male parental defense behaviour in the cichlid fish *Cichlasoma nigrofasciatum*. Canadian Journal of Zoology **63**:2489–2493.

Keightley, P.D. and W.G. Hill. 1988. Quantitative variability maintained by mutation-stabilizing selection balance in finite populations. Genetical Research **52**:33–43.

Keightley, P.D. and W.G. Hill. 1989. Quantitative genetic variability maintained by mutation-stabilizing selection balance: sampling variation and response to subsequent directional selection. Genetical Research **54**:45–57.

Keller, B.L. and C.J. Krebs. 1970. *Microtus* population biology; III. Reproductive changes in fluctuating populations of *M. ochrogaster* and *M. pennsylvanicus* in southern Indiana. Ecological Monographs **40**:263–294.

Kempthorne, O. 1983. Evolution of current population genetics theory. American Zoologist **23**:111–121.

Kempthorne, O. 1977. Status of quantitative genetic theory. pp. 719–767, *In* E. Pollack, O. Kempthorne, and T.B. Baily Jr. (editors), "Proceedings of the International Conference on Quantitative Genetics," Iowa State University Press, Ames.

Kenagy, G.J. 1987. Energy allocation for reproduction in the golden-mantled ground squirrel. Symposium of the Zoological Society of London **57**:259–273.

Kenagy, G.J., D. Masman, S.M. Sharbaugh and K.A. Nagy. 1990. Energy expenditure during lactation in relation to litter size in free-living golden-mantled ground squirrels. Journal of Animal Ecology **59**:73–88.

Kenagy, G.J. and S.C. Trombulak. 1986. Size and function of mammalian testes in relation to body size. Journal of Mammalogy **67**:1–22.

Kendeigh, S.C. 1972. Energy control of size limits in birds. American Naturalist **106**:79–88.

Kendeigh, S.C., T.C. Kramer and F. Hamerstrom. 1956. Variations in egg characteristics of the House Wren. Auk **73**:42–65.

Kerfoot, W.C. 1974. Egg size cycle of a cladoceran. Ecology **55**:1259–1270.

Kerfoot, W.C. 1977. Competition in cladoceran communities: the cost of evolving defenses against copepod predation. Ecology **58**:303–313.

Kessler, A. 1971. Relation between egg production and food consumption in species of the genus *Pardosa* (Lycosidae, Araneae) under experimental conditions of food abundance and food shortage. Oecologia **8**:93–109.

Khan, M.I. 1967. The genetic control of canalisation of seed size in plants. Ph.D. Thesis, University of Wales, Bangor, Wales.

Kidwell, J.F. and L. Malick. 1967. The effect of genotype, mating status, weight and egg production on longevity in *Drosophila melanogaster*. Journal of Heredity **58**:169–172.

Kilgore, D.L. 1970. The effects of northward dispersal on growth rate of young, size of young at birth, and litter size in *Sigmodon hispidus*. American Midland Naturalist **84**:510–520.

Kimura, J. and S. Masaki. 1977. Brachypterism and seasonal adaptation in *Orgyia thyellina* Butler (Lepidoptera, Lymantridae). Kontyu **45**:97–106.

King, C.E. and W.W. Anderson. 1971. Age-specific selection. II. The interaction between r and K during population growth. American Naturalist **105**:137–156.

King, D. and J. Roughgarden. 1982. Graded allocation between vegetative and reproductive growth for annual plants in growing seasons of random length. Theoretical Population Biology **22**:1–16.

King, J.R. 1973. Energetics of reproduction in birds. pp. 78–120, *In* D.S. Farner (editor), "Breeding Biology of Birds." National Academy of Sciences, Washington, D.C.

Kingsland, S.E. 1985. Modeling Nature. University of Chicago Press, Chicago.

Kinney, T.B. Jr. 1969. A summary of reported estimates of heritabilities and of genetic and phenotypic correlations for traits of chickens. Washington, U.S. Department of Agriculture Agricultural Handbook **363**:1–49.

Kirpichnikov, V.S. 1966. Goals and methods in carp selection. Selective breeding of carp and intensification of fish breeding in ponds. Bulletin of the State Scientific Research Institute for Lake River Fish **61**:41–42 (translated from Russian, Jerusalem, 1970).

Kirpichnikov, V.S. 1981. Genetic Basis of Fish Selection. Springer-Verlag, New York.

Kitahara, T., Y. Hiyama and T. Tokai. 1987. A preliminary study on quantitative relations among growth, reproduction and mortality in fishes. Researches in Population Ecology **29**:85–95.

Klein, T.W., J.C. DeFries and C.T. Finkbeiner. 1973. Heritability and genetic correlation: standard error of estimates and sample size. Behavior Genetics **3**:355–364.

Klinkhamer, P.G.L., T.J. de Jong and E. Meelis. 1990. How to test for proportionality in the reproductive effort of plants. American Naturalist **135**:291–300.

Klomp, H. 1970. The determination of clutch-size in birds: a review. Ardea **58**:1–124.

Klomp, H. and B.J. Teerink. 1962. Host selection and number of eggs per oviposition in the egg-parasite *Trichogramma embyophagum* Htg. Nature **195**:1020–1021.

Klomp, H. and B.J. Teerink. 1967. The significance of oviposition rates in the egg parasite, *Trichogramma embryophagum*. Archives Neerlandaises de Zoologie **17**:350–375.

Kluyver, H.N. 1963. The determination of reproductive rate in Paridae. Proceedings of the International Ornithologial Congress **13**:706–716.

Knoppien, P. 1985. Rare male mating advantage: A review. Biological Reviews **60**:81–117.

Knutsen, G.M. and S. Tilseth. 1985. Growth, development and feeding success of Atlantic cod larvae *Gadus morhua* related to egg size. Transactions of the American Fisheries Society **114**:507–511.

Koelink, A.F. 1972. Bioenergetics of growth in the Pigeon Guillemot. M.Sc. Thesis, University of British Columbia, Vancouver, Canada.

Kohler, W. 1977. Investigations on the phototactic behavior of *Drosophila melanogaster*. I. Selection response in the presence of multiply marked X-chromosome. Genetica **47**:93–100.

Kohn, L.A.P. and W.R. Atchley. 1988. How similar are genetic correlation structures? Data from mice and rats. Evolution **42**:467–482.

Koivunen, P., E.S. Nyholm and S. Sulkava. 1975. Occurrence and breeding of the Little Bunting *Emberiza puzilla* in Kuusamo (NE Finland). Ornis Fennoscandinavica **52**:85–96.

Kolding, S. and T.M. Fenchel. 1981. Patterns of reproduction in different populations of five species of the amphipod genus *Gammarus*. Oikos **37**:167–172.

Konig, B. and H. Mart. 1987. Maternal care in house mice. I. The weaning strategy as a means for parental manipulation of offspring quality. Behavioral Ecology and Sociobiology **20**:1–9.

Kooijman, S.A.L. 1986. Energy budgets can explain body size relations. Journal of Theoretical Biology **121**:269–282.

Koopman, B.O. 1956. The theory of search. II. Target detection. Operations Research **4**:503–531.

Korpimäki, E. 1987. Clutch size, breeding success and brood size experiments in Tenmalm's owl *Aegolius funereus*: a test of hypotheses. Ornis Scandinavica **18**:277–284.

Korpimäki, E. 1988. Costs of reproduction and success of manipulated broods under varying food conditions in Tengmalm's owl. Journal of Animal Ecology **57**:1027–1039.

Koufopanou, V. and G. Bell. 1984. Measuring the cost of reproduction. IV Predation experiments with *Daphnia pulex*. Oecologia **64**:81–86.

Kovacs, K.M. and D.M. Lavigne. 1985. Neonatal growth and organ allometry of northwest Atlantic harp seals (*Phoca grenlandica*). Canadian Journal of Zoology **63**:2793–2799.

Kovacs, K.M., D.M. Lavigne and S. Innes. 1990. Mass transfer efficiency between harp seal (*Phoca groenlandica*) mothers and their pups during lactation. Journal of Zoology **223**:213–221.

Kozlowski, J. and J. Uchmanski. 1987. Optimal individual growth and reproduction in perennial species with indeterminate growth. Evolutionary Ecology **1**:214–230.

Kozlowski, J. and R.G. Wiegert. 1986. Optimal allocation of energy to growth and reproduction. Theoretical Population Biology **29**:16–37.

Kozlowski, J. and R.G. Wiegert. 1987. Optimal age and size at maturity in annuals and perennials with determinate growth. Evolutionary Ecology **1**:231–244.

Kozlowski, J. and M. Ziolko. 1988. Gradual transition from vegetative to reproductive growth is optimal when maximum rate of reproductive growth is limited. Theoretical Population Biology **34**:118–129.

Kozlowski, T.T. and T. Keller. 1966. Food relations of woody plants. Botanical Review **32**:293–382.

Krebs, C.J. 1985. Ecology: The Experimental Analysis of Distribution and Abundance. Harper and Row, New York.

Krebs, J.R. 1971. Territory and breeding density in the Great Tit, *Parus major* L. Ecology **52**:2–22.

Krohne, D.T. 1981. Intraspecific litter size variation in *Microtus californicus*: variation within populations. Journal of Mammalogy **62**:29–40.

Kulesza, G. 1990. An analysis of clutch-size in New World passerine birds. Ibis **132**:407–422.

Kuno, E. 1983. Factors governing dynamical behaviour of insect populations: a theoretical enquiry. Researches in Population Ecology Supplement **3**:27–45.

Kuno, E. 1988. Aggregation pattern of individuals and the outcomes of competition within and between species: Differential equation models. Researches in Population Ecology **30**:69–82.

Kuroda, N. 1959. Field studies of the grey starling *Sturnus cineraeus* Temminck. 2. Breeding biology (pt. 3). Miscellaneous report of the Yamashina Institute **13**:535–552.

Kurta, A. and T.H. Kunz 1987. Size of bats at birth and maternal investment during pregnancy. pp. 79–106, *In* A.S.I. Loudon and P.A. Racey (editors), "Reproductive Energetics in Mammals," Zoological Society of London, Clarendon Press, Oxford.

Kusano, T. 1982. Postmetamorphic growth, survival, and age at first reproduction of the salamander, *Hynobius nebulosus tokyoensis* Tago in relation to a consideration on the optimal timing of first reproduction. Researches on Population Ecology **24**:329–344.

L'Abee-Lund, J.H., B. Jonsson, A.J. Jensen, L.M. Saettem, T.G. Heggberget, B.O. Johnsen and T.F. Naesje. 1989. Latitudinal variation in life-history characteristics of sea-run migrant brown trout *Salmo trutta*. Journal of Animal Ecology **58**:525–542.

Lacey, E.P., L. Real, J. Antonovics and D.G. Heckel. 1983. Variance models in the study of life histories. American Naturalist **122**:114–131.

Lack, D. 1943a. The age of the blackbird. British Birds **36**:166–172.

Lack, D. 1943b. The age of some more British birds. British Birds **36**:193–197, 214–221.

Lack, D. 1947. The significance of clutch size 1. Intraspecific variation. Ibis **89**:302–352.

Lack, D. 1948. The significance of litter size. Journal of Animal Ecology **17**:45–50.

Lack, D. 1954. The Natural Regulation of Animal Numbers. Clarendon Press, Oxford.

Lack, D. 1966. Population Studies of Birds. Clarendon Press, Oxford.

Lack, D. 1967. The Natural Regulation of Animal Numbers. Oxford University Press, Oxford.

Lack, D. 1968. Ecological Adaptations for Breeding in Birds. Methuen and Company, London.

Lackey, J.A. 1976. Reproduction, growth and development in the Yucatan deer mouse, *Peromyscus yucatanicus*. Journal of Mammalogy 57:638–655.

Lackey, J.A. 1978. Reproduction, growth, and development in high-latitude populations of *Peromyscus leucopus* (Rodentia). Journal of Mammalogy **59**:69–83.

Lamb, M. 1964. The effects of radiation on the longevity of female *Drosophila subobscura*. Journal of Insect Physiology **10**:487–497.

Landahl, J.T. and R.B. Root. 1969. Differences in the life tables of tropical and temperate milkweed bugs, genus *Oncopeltus* (Hemiptera, Lygaeidae). Ecology **50**:734–737.

Lande, R. 1975. The maintenance of genetic variation by mutation in a polygenic character with linked loci. Genetical Research **26**:221–235.

Lande, R. 1976. Natural selection and random genetic drift in phenotypic evolution. Evolution **30**:314–334.

Lande, R. 1979. Quantitative genetic analysis of multivariate evolution applied to brain:body size allometry. Evolution **33**:402–416.

Lande, R. 1980. The genetic covariance between characters maintained by pleiotropic mutations. Genetics **94**:203–215.

Lande, R. 1982. A quantitative genetic theory of life history evolution. Ecology **63**:607–615.

Lande, R. and S.H. Orzack. 1988. Extinction dynamics of age-structured populations in a fluctuating environment. Proceedings of the National Academy of Sciences, USA **85**:7418–7421.

Lande, R. and S.J. Arnold. 1983. The measurement of selection on correlated characters. Evolution **37**:1210–1226.

Langridge, J. 1962. A genetic and molecular basis for heterosis in *Arabidopsis* and *Drosophila*. American Naturalist **96**:5–27.

Larsson, F.K. 1989. Female longevity and body size as predictors of fecundity and egg length in *Graphosoma lineatum* L. Deutsche Entomologische Zeitschrift, N.F. **36**:329–334.

Larsson, F.K. 1990. Female body size relationships with fecundity and egg size in two solitary species of fossorial Hymenoptera (Colletidae and Sphecidae). Entomologia Generalis **15**:167–171.

Larsson, F.K. and V. Kustvall. 1990. Temperature reverses size-dependent male mating success of a cerambycid beetle. Functional Ecology **4**:85–90.

Latter, B.D.H. 1960. Natural selection for an intermediate optimum. Australian Journal of Biological Sciences **13**:30–35.

Law, R. 1979a. The cost of reproduction in annual meadow grass. American Naturalist **113**:3–16.

Law, R. 1979b. Optimal life histories under age-specific predation. American Naturalist **114**:399–417.

Law, R., A.D. Bradshaw and P.D. Putwain. 1977. Life history variation in *Poa annua*. Evolution **31**:233–247.

Lawlor, L.R. 1976. Parental investment and offspring fitness in the terrestrial isopod *Armadillidium vulgare* (Latr.), (Crustacea: Oniscoidea). Evolution **30**:775–785.

Laws, R.M. 1969. The Tsavo research project. Journal of Reproduction and Fertility, Supplement **6**:495–531.

Laws, R.M. 1970. Biology of African elephants. Science Progress (Oxford) **58**:251–262.

LeBoeuf, B.J. 1974. Male-male competition and reproductive success in elephant seals. American Zoologist **14**:163–176.

LeCren, E.D. 1958. Observations on the growth of perch (*Perca fluviatilis* L.) over twenty-two years with special reference to the effects of temperature and changes in population density. Journal of Animal Ecology **27**:287–334.

Lee, B.T.O. and P.A. Parsons. 1968. Selection, prediction and response. Biological Reviews **43**:139–174.

Leggett, W.C. and G. Power. 1969. Differences between two populations of landlocked Atlantic salmon (*Salmo salar*) in Newfoundland. Journal of the Fisheries Research Board of Canada **26**:1585–1596.

Leggett, W.C. and J.E. Carscadden. 1978. Latitudinal variation in reproductive characteristics of American Shad (*Alosa sapidissima*): evidence for population specific life history strategies in fish. Journal of the Fisheries Research Board of Canada **35**:1469–1478.

Leitch, I., F.E. Hytten and W.Z. Billewicz. 1959. Maternal and neonatal weights of some Mammalia. Proceedings of the Zoological Society of London **133**:11–78.

Leithold, L. 1969. The Calculus with Analytic Geometry. Harper and Row, New York.

Lemen, C.A. and H.K. Voris. 1981. A comparison of reproductive strategies among marine snakes. Journal of Animal Ecology **50**:89–101.

León, J.A. 1976. Life histories as adaptive strategies. Journal of Theoretical Biology **60**:301–335.

León, J.A. and B. Charlesworth. 1978. Ecological versions of Fisher's fundamental theorem of natural selection. Ecology **59**:457–464.

Leonard, D.E. 1970. Intrinsic factors causing quantitative changes in populations of *Porthetria dispar* (Lepidoptera: Lymantridae). Canadian Entomologist **102**:239–249.

Leopold, A.C. 1961. Senescence in plant development. Science **134**:1727–1732.

Lerner, I.M. 1951. Natural selection and egg size in poultry. American Naturalist **85**:365–372.

Lerner, I.M. 1958. The Genetic Basis of Selection. Wiley, New York.

Lerner, I.M. and C.A. Gunns. 1952. Egg size and reproductive fitness. Poultry Science **31**:537–544.

Leslie, P.H. 1945. On the use of matrices in certain population mathematics. Biometrika **33**:183–212.

Leslie, P.H. 1948. Some further notes on the use of matrices in population mathematics. Biometrika **35**:213–245.

Lessells, C.M. 1986. Brood size in Canada geese: a manipulation experiment. Journal of Animal Ecology **55**:669–689.

Lessels, C.M., F. Cooke and R.F. Rockwell. 1989. Is there a trade-off between egg weight and clutch size in wild Lesser Snow Geese (*Anser c. caerulescens*)? Journal of Evolutionary Biology **2**:457–472.

Levene, H. 1953. Genetic equilibrium when more than one ecological niche is available. American Naturalist **87**:331–333.

Levins, R. 1962. Theory of fitness in a heterogeneous environment. I. The fitness set and adaptive function. American Naturalist **96**:361–378.

Levins, R. 1963. Theory of fitness in a heterogeneous environment. II. Developmental flexibility and niche selection. American Naturalist **97**:75–90.

Levins, R. 1966. The strategy of model building in population biology. American Scientist **54**:421–431.

Levins, R. 1968. Evolution in Changing Environments. Princeton University Press, Princeton, New Jersey.

Levins, R. 1969. The effect of random variations of different types on population growth. Proceedings of the National Academy of Sciences, USA **62**:1061–1065.

Levins, R. 1970. Fitness and optimization. pp. 389–400, *In* K. Kojima (editor), "Mathematical Topics in Population Genetics," Springer-Verlag, Berlin.

Lewontin, R. 1958. A general method for investing the equilibrium of gene frequencies in a population. Genetics **43**:419–434.

Lewontin, R.C. 1964. The interaction of selection and linkage. II. Optimum models. Genetics **50**:757–782.

Lewontin, R.C. 1965. Selection for colonizing ability. pp. 77–94, *In* H.G. Baker and G.L. Stebbins (editors), "The Genetics of Colonizing Species," Academic Press, New York.

Lewontin, R.C. 1986. How important is genetics for an understanding of evolution? American Zoologist **26**:811–820.

Lewontin, R.C. and D. Cohen. 1969. On population growth in a randomly varying environment. Proceedings of the National Academy of Sciences, USA **62**:1056–1060.

Lewontin, R.C., L.R. Ginsburg and S.D. Taljaparkar. 1978. Heterosis as an explanation for large amounts of genic polymorphism. Genetics **88**:149–170.

Licht, L.E. 1974. Survival of embryos, tadpoles and adults of the frogs *Rana aurora aurora* and *Rana pretiosa pretiosa* sympatric in southwestern British Columbia. Canadian Journal of Zoology **52**:613–627.

Lima, S.L. and L.M. Dill. 1990. Behavioral decisions made under risk of predation: a review and prospectus. Canadian Journal of Zoology **68**:619–640.

Lindén, M. 1988. Reproductive trade-off between first and second clutches in the great tit *Parus major*: an experimental study. Oikos **51**:285–290.

Lindén, M. and A.P. Moller. 1989. Costs of reproduction and covariation of life history traits in birds. Trends in Ecology and Evolution **4**:367–371.

Lindroth, C.H. 1946. Inheritance of wing dimorphism in *Pterostichus anthracinus* Ill. Hereditas **32**:37–40.

Lindstedt, S. and W.A. Calder. 1976. Body size and longevity in birds. Condor **78**:91–94.

Lindstedt, S.L. and W.A. Calder III. 1981. Body size, physiological time, and longevity of homeothermic animals. Quarterly Review of Biology **56**:1–16.

Linfield, R.S. 1979. Changes in the rate of growth in a stunted roach, *Rutilus rutilus*, population. Journal of Fish Biology **15**:275–289.

Linzey, A.R. 1970. Postnatal growth and development of *Peromyscus maniculatus nubiterrae*. Journal of Mammalogy **51**:152–155.

Lloyd, C.S. 1977. The ability of the razorbill *Alco torda* to raise an additional chick to fledging. Ornis Scandinavica **8**:155–159.

Lloyd, C.S. 1979. Factors affecting breeding of Razorbills *Alca torda* on Skokholm. Ibis **121**:165–176.

Lloyd, D.C. 1987. Selection of offspring size at independence and other size-versus-number strategies. American Naturalist **129**:800–817.

Lloyd, J.E. 1965. Aggressive mimicry in *Photuris*: firefly femmes fatales. Science **149**:653–654.

Lloyd, J.E. and S.R. Wing. 1983. Nocturnal aerial predation of fireflies by light-seeking fireflies. Science **222**:634–635.

Loftsvold, D. 1986. Quantitative genetics of morphological differentiation in *Peromyscus*. I. Tests of the homogeneity of genetic covariance structure among species and subspecies. Evolution **40**:559–573.

Loman, J. 1980. Brood size optimization and adaptation among hooded crows *Corvus corone*. Ibis **122**:494–500.

Loman, J. 1982. A model of clutch size determination in birds. Oecologia **52**:253–257.

Long, D.R. 1986. Clutch formation in the turtle, *Kinosternum flavescens* (Testudines: Kinosternidae). Southwestern Naturalist **31**:1–8.

Long, D.R. and F.L. Rose. 1989. Pelvic girdle size relationships in three turtle species. Journal of Herpetology **23**:315–318.

Losos, J.B. 1990. Ecomorphology, performance capability, and scaling of West Indian, *Anolis* lizards: an evolutionary analysis. Ecological Monographs **60**:369–388.

Loudon, A. and P.A. Racey (editors). 1987. The reproductive energetics of mammals. Oxford University Press, Oxford.

Luck, R.F., H. Podoler and R. Kfir. 1982. Host selection and egg allocation behavior by *Aphytis melinus* and *A. lingnanensis:* comparison of two facultatively gregarious parasitoids. Ecological Entomology **7**:397–408.

Luckinbill, L.S. 1984. An experimental analysis of a life history theory. Ecology **65**:1170–1184.

Lundberg, C.A. and R. A. Vaiisanen. 1978. Selective correlation of egg size with chick mortality in the Black-headed Gull (*Larus ridibundus*). Condor **81**:146–156.

Lundqvist, H. and G. Fridberg. 1982. Sexual maturation versus immaturity: different tactics with adaptive values in Baltic salmon (*Salmo salar*) male smolts. Canadian Journal of Zoology **60**:1822–1827.

Lynch, M. 1977. Fitness and optimal body size in zooplankton populations. Ecology **58**:763–774.

Lynch, M. 1980a. The evolution of cladoceran life histories. Quarterly Review of Biology **55**:23–42.

Lynch, M. 1980b. Predation, enrichment and the evolution of cladoceran life histories: a theoretical approach. pp. 367–376, *In* W.C. Kerfoot (editor), "The Evolution and Ecology of Zooplankton Communities," University Press of New England, Hanover, New Hampshire.

Lynch, M. 1984. The limits to life history evolution in *Daphnia*. Evolution **38**:465–482.

MacArthur, R.H. 1982. Some generalized theorems of natural selection. Proceedings of the National Academy of Sciences. USA **48**:1893–1897.

MacArthur, R.H. and E.O. Wilson. 1967. The Theory of Island Biography. Princeton University Press, Princeton, New Jersey.

MacArthur, R.H. and E.R. Pianka. 1966. On optimal use of a patchy environment. American Naturalist **100**:603–609.

Machin, D. and S. Page. 1973. Effects of reduction of litter size on subsequent growth and reproductive performance in mice. Animal Production **16**:1–6.

Mackay, T.F.C. 1980. Genetic variance, fitness, and homeostasis in varying environments: an experimental check of the theory. Evolution **34**:1219–1222.

Mackay, T.F.C. 1981. Genetic variation in varying environments. Genetical Research **37**:79–93.

Mackinnon, J.C. 1972. Summer storage of energy and its use for winter metabolism and gonad maturation in American Plaice (*Hippoglossoides platessoides*). Journal of Fisheries Research Board of Canada **29**:1749–1759.

MacNally, R.C. 1981. On the reproductive energetics of chorusing males: energy depletion profiles, restoration and growth in two sympatric species of *Ranidella* (Anura). Oecologia **51**:181–188.

MacNally, R. and D. Young. 1981. Song energetics of the bladder cicada *Cystosoma saundersili*. Journal of Experimental Biology **90**:185–196.

Maekawa, K. and H. Onozato. 1986. Reproductive tactics and fertilization success of mature male Miyabe charr. *Salvelinus malma miyabei*. Environmental Biology of Fishes **15**:119–129.

Maekawa, K. and T. Hino. 1987. Effect of cannibalism on alternative life histories in charr. Evolution **41**:1120–1123.

Magnhagen, C. 1990. Reproduction under predation risk in the sand goby, *Pomatoschistus minutus*, and the black goby, *Gobius niger:* the effect of age and longevity. Behavioral Ecology and Sociobiology **26**:331–335.

Magnhagen, C. and L. Kuarnemo. 1989. Big is better: the importance of size for reproductive success in male *Pomatoschistus minutus* (Pallas) (Pisces, Gobiidae). Journal of Fish Biology **35**:755–763.

Magnuson, J.J. 1962. An analysis of aggressive behavior, growth and competition for food in mendaka (*Oryzias latipes* [Pisces: Cyprinodontidae]). Canadian Journal of Zoology **40**:313–363.

Magnusson, W.E.L., L.J. de Paiva, R.M. da Rocha, C.R. Franke, L.A. Kasper and A.P. Lima. 1985. The correlates of foraging mode in a community of Brazilian lizards. Herpetologica **41**:324–322.

Mangel, M. 1987. Oviposition site selection and clutch size in insects. Journal of Mathematical Biology **25**:1–22.

Mangel, M. and C.W. Clark. 1988. Dynamic Modeling in Behavioral Ecology. Princeton University Press, Princeton, New Jersey.

Mangold, J.R. 1978. Attraction of *Euphasiopteryx ochracea, Corethrella* sp. and gryllids to broadcast songs of the southern mole cricket. Florida Entomologist **61**:57–61.

Mann, G.J. and P.J. McCart. 1981. Comparison of sympatric dwarf and normal populations of least ciscoe (*Coregonus sardinella*) inhabiting Trout Lake, Yukon Territory. Canadian Journal of Fisheries and Aquatic Sciences **38**:240–244.

Mann, R.H.K. and C.A. Mills. 1979. Demographic aspects of fish fecundity. Symposium of the Zoological Society of London **44**:161–177.

Mannes, J.C. and D.B. Jester. 1980. Age and growth, abundance, and biomass production of green sunfish, *Lepomis cyanellus* (Centrarchidae), in a eutrophic desert pond. Southwestern Naturalist **25**:297–311.

Marconato, A., A. Bisazza and G. Marin. 1989. Correlates of male reproductive success in *Padogobius martensi* (Gobiidae). Journal of Fish Biology **34**:889–899.

Marden, J.H. 1989. Bodybuilding dragonflies: Costs and benefits of maximizing flight muscles. Physiological Zoology **62**:505–521.

Markow, T.A. and A.G. Clark. 1984. Correlated response to phototactic selection. Behavior Genetics **14**:279–293.

Marks, R.W. 1982. Genetic variability for density sensitivity of three components of fitness in *Drosophila melanogaster*. Genetics **101**:301–316.

Markus, H.C. 1934. Life history of the black-head minnow, *Pimephales promelas*. Copeia **1934**:116–122.

Marsh, E. 1986. Effects of egg size on offspring fitness and maternal fecundity in the orangethroat darter, *Etheostoma spectabile* (Pisces: Percidae). Copeia **1986**:18–30.

Marshall, D. 1986. Effects of seed size on seedling success in three species of *Sesbania* (Fabaceae). American Journal of Botany **73**:457–464.

Marshall, J.A. 1953. Egg size in arctic, antarctic and deep sea fishes. Evolution 7:328–341.

Marshall, L.D. 1988. Small male advantage in mating in *Parapediasia teterrella* and *Agriphila plumbifimbriella* (Lepidoptera: Pyralidae). American Midland Naturalist 119:412–419.

Marshall, L.D. 1990. Intra-specific variation in reproductive effort by female *Parapediasia teterrella* (Lepidoptera: Pyralidae) and its relations to body size. Canadian Journal of Zoology 68:44–48.

Martin, R.D. 1984. Scaling effects and adaptive strategies in mammalian lactation. Symposium of the Zoological Society of London 51:87–117.

Martin, T.E. 1988. Nest placement: implications for selected life-history traits, with special reference to clutch size. American Naturalist 132:900–910.

Masaki, S. 1967. Geographic variation and climatic adaptation in a field cricket (Orthoptera: Gryllidae). Evolution 21:725–741.

Masaki, S. 1973. Climatic adaptation and photoperiodic response in the band-legged ground cricket. Evolution 26:587–600.

Masaki, S. 1978a. Climatic adaptation and species status in the Lawn Ground Cricket. II. Body Size. Oecologia 35:343–356.

Masaki, S. 1978b. Seasonal and latitudinal adaptations in the life cycles of crickets. pp72–100, *In* H. Dingle (editor) "Evolution of Insect Migration and Diapause," Springer-Verlag, New York.

Mason, L.G. 1964. Stabilizing selection for mating fitness in natural populations of *Tetropes*. Evolution 18:492–497.

Mason, L.G. 1969. Mating selection in the California Oak moth (Lepidoptera, Dioptidae). Evolution 23:55–58.

Mather, K. and B.J. Harrison. 1949. The manifold effects of selection. Part 1. Heredity 3:1–52.

Mattingly, D.K. and P.A. McClure. 1982. Energetics of reproduction in large-littered cotton rats (*Sigmodon hispidus*). Ecology 63:183–195.

May, R.M. 1971. Stability in model ecosystems. Proceedings of the Ecological Society of Australia 6:18–56.

May, R.M. 1973. Stability in randomly fluctuating versus deterministic environments. American Naturalist 107:621–650.

Maynard Smith, J. 1958. The effects of temperature and of egg-laying on the longevity of *Drosophila subobscura*. Journal of Experimental Biology 35:832–842.

Maynard Smith, J. 1966. Sympatric speciation. American Naturalist 100:637–650.

Maynard Smith, J. 1970. Genetic polymorphism in a varied environment. American Naturalist 104:487–490.

Maynard Smith, J. 1977. Parental investment: a prospective analysis. Animal Behaviour 25:1–9.

Maynard Smith, J. 1978. Optimization in evolution, Annual Review of Ecology and Systematics **9**:31–56.

Maynard Smith, J. 1981. Will a sexual population evolve to an ESS? American Naturalist **117**:1015–1018.

Maynard Smith, J. 1982. Evolution and the Theory of Games. Cambridge University Press, Cambridge.

Maynard Smith, J. and G. R. Price. 1973. The logic of animal conflict. Nature **246**:15–18.

Maynard Smith, J., R. Burian, S. Kauffman, P. Alberch, J. Campbell, B. Goodwin, R. Lande, D. Raup and L. Wolpert. 1985. Developmental constraints and evolution. Quarterly Review of Biology **60**:265–287.

Maynard Smith, J. and R. Hoekstra. 1980. Polymorphism in a varied environment: how robust are the models? Genetical Research **35**:45–57.

Maynard Smith, J. and R.J.G. Savage. 1956. Some locomotory adaptations in mammals. Zoological Journal of the Linnean Society **42**:603–622.

Maynard Smith, J. and R.L. Brown. 1986. Competition and body size. Theoretical Population Biology **30**:166–179.

Mayo, O. 1983. Natural Selection and its Constraints. Academic Press, London.

Mayr, E. 1983. How to carry out the adaptationist program? American Naturalist **121**:324–334.

Mazer, S.J. 1989. Ecological, taxonomic, and life history correlates of seed mass among Indiana dune antiosperms. Ecological Monographs **59**:153–175.

McCauley, D.E. 1982. The behavioural components of sexual selection in the milkweed beetle *Tetraopes tetraophthalmus*. Animal Behaviour **30**:23–28.

McCauley, D.E. 1983. An estimate of the relative opportunities for natural and sexual selection in a population of milkweed beetles. Evolution **37**:701–707.

McCauley, D.E. and M.J. Wade. 1978. Female choice and the mating structure of a natural population of the soldier beetle, *Chauliognathus pennsylvanicus*. Evolution **32**:771–775.

McClenaghan, L.R. Jr. and M.S. Gaines. 1978. Reproduction in marginal populations of the hispid cotton rat (*Sigmodon hispidus*) in northeastern Kansas. Occasional Papers of the Museum of Natural History, University of Kansas **74**:1–16.

McClure, P.A. 1981. Sex-biased litter reduction in food-restricted wood rats (*Neotoma floridana*). Science **211**:1058–1060.

McClure, P.A. 1987. The energetics of reproduction and life histories of cricetine rodents *Neotoma floridana* and *Sigmodon hispidus*). pp. 241–258, *In* A.S.I. Loudon and P.A. Racey (editors). "Reproductive Energetics in Mammals," Zoological Society of London, Clarendon Press, Oxford.

McGillivray, W.B. 1983. Intraseasonal reproductive costs for the house sparrow (*Passer domesticus*) Auk **100**:25–32.

McGinley, M.A. 1989. The influence of a positive correlation between clutch size and offspring fitness on the optimal offspring size. Evolutionary Ecology **3**:150–156.

McGinley, M.A., D.H. Temme and M.A. Geber. 1987. Parental investment in offspring in variable environments: theoretical and empirical considerations. American Naturalist **130**:370–398.

McGinley, M.A. and E.L. Charnov. 1988. Multiple resources and the optimal balance between size and number of offspring. Evolutionary Ecology **2**:77–84.

McGurk, M.D. 1986. Natural mortality of marine pelagic fish eggs and larvae: role of spatial patchiness. Marine Ecology—Progress Series **34**:227–242.

McKenzie, R.A. 1964. Smelt life history and fishery in the Miramichi River, New Brunswick. Fisheries Research Board of Canada Bulletin **144**:1–77.

McKeown, T., T. Marshall and R.G. Record. 1976. Influences on fetal growth. Journal of Reproduction and Fertility **47**:167–181.

McLachlan, A. 1983. Life-history tactics of rain-pool dwellers. Journal of Animal Ecology **52**:545–561.

McLaren, I.A. 1965. Some relationships between temperature and egg size, body size, development rate, and fecundity, of the copepod *Pseudocalanus*. Limnology and Oceanography **10**:528–538.

McLaren, I.A. 1966. Adaptive significance of large size and long life of the chaetognath *Sagitta elegans* in the Arctic. Ecology **47**:852–856.

McLaughlin, R.L. 1989. Search modes of birds and lizards: evidence for alternative movement patterns. American Naturalist **133**:654–670.

McLaughlin, R.L. and R.D. Montgomerie. 1989 Early nest departure does not improve the survival of Lapland Longspur chicks. Auk **106**:738–741.

McMahon, T. 1973. Size and shape in biology. Science **179**:1201–1204.

McMillan, I., M. Fitz-Earle and D.S. Robson. 1970a. Quantitative genetics of fertility. I. Lifetime egg production of *Drosophila melanogaster*—Theoretical. Genetics **65**:349–353.

McMillan, I., M. Fitz-Earle and D.S. Robson. 1970b. Quantitative genetics of fertility. II Lifetime egg production of *Drosophila melanogaster*—Experimental. Genetics **65**:355–369.

McNab, B.K. 1980. Food habits, energetics, and the population biology of mammals. American Naturalist **116**:106–124.

McNab, B.K. 1984. Energetics: the behavioral and ecological consequences of body size. Florida Entomologist **67**:68–73.

McNaughton, S.J. 1975. r- and K-selection in *Typha*. American Naturalist **109**:251–261.

Meats, A. 1971. The relative importance to population increase of fluctuations in mortality, fecundity and time variables of the reproductive schedule. Oecologia **6**:223–237.

Medway, L. 1972. Reproductive cycles of the flat-headed bats *Tylonycteris pachypus* and *T. robustula* (Chiroptera: Vespertilioninae) in a humid equatorial environment. Zoological Journal of the Linnean Society **5**:33–61.

Mellors, W.K. 1975. Selective predation on ephippial *Daphnia* and the resistance of ephippial eggs to digestion. Ecology **56**:974–980.

Menge, B.A. 1973. Effect of predation and environmental patchiness on the body size of a tropical pulmonate limpet. Velinger **16**:87–92.

Menge, B.A. 1974. Effect of wave action and competition on brooding and reproductive effort in the sea star *Leptasterias hexactis*. Ecology **55**:84–93.

Merritt, E.S. 1962. Selection for egg production in geese. Proceedings of the 12th World Poultry Congress **1962**:85–87.

Mertz, D.B. 1971. The mathematical demography of the California condor population. American Naturalist **105**:437–453.

Mertz, D.R. 1975. Senescent decline in flour beetle strains selected for early adult fitness. Physiological Zoology **48**:1–23.

Messina, F.J. 1989. Genetic basis of variable oviposition behavior in *Callosobruchus maculatus* (Coleoptera: Bruchidae). Annals of the Entomological Society of America **82**:792–796.

Michod, R.E. 1979. Evolution of life histories in response to age-specific mortality factors. American Naturalist **113**:531–550.

Migula, P. 1969. Bioenergetics of pregnancy and lactation in European common vole. Acta Theriologica **14**:167–179.

Millar, J.S. 1973. Evolution of litter-size in the pika *Ochotoma princeps* (Richardson). Evolution **27**:134–143.

Millar, J.S. 1975. Tactics of energy partitioning in breeding *Peromyscus*. Canadian Journal of Zoology **53**:967–976.

Millar, J.S. 1978. Energetics of reproduction in *Peromyscus leucopus*: the cost of lactation. Ecology **59**:1055–1061.

Millar, J.S. 1981. Pre-partum reproductive characteristics of eutherian mammals. Evolution **35**:1149–1163.

Millar, J.S. and R.M. Zammuto. 1983. Life histories of mammals: an analysis of life tables. Ecology **64**:631–635.

Miller, K.M. and T.H. Carefoot. 1989. The role of spatial and size refuges in the interaction between juvenile barnacles and grazing limpets. Journal of Experimental Marine Biology and Ecology **134**:157–174.

Milonoff, M. 1989. Can nest predation limit clutch size in precocial birds? Oikos **55**:424–427.

Minchella, D.J. and P.T. Loverde. 1981. A cost of increased early reproductive effort in the snail *Biomphalaria glabrata*. American Naturalist **118**:876–881.

Miranda, J.R. de and P. Eggleston. 1989. Analysis of dominance for competitive ability in *Drosophila melanogaster*. Heredity **63**:221–229.

Mitchell, R. 1975. The evolution of oviposition tactics in the bean weevil, *Callosobruchus maculatus*. Ecology **56**:696–702.

Mitchell, W.A. and T.J. Valone. 1990. The optimization research program: studying adaptations by their function. Quarterly Review of Biology **65**:43–52.

Mitchell-Olds, T. and J.J. Rutledge. 1986. Quantitative genetics in natural plant populations: a review of the theory. American Naturalist **127**:379–402.

Mittelbach. G.G. 1981. Patterns of invertebrate size and abundance in aquatic habitats. Canadian Journal of Fisheries and Aquatic Sciences **38**:896–904.

Mitton, J.B. 1978. Relationship between hetcrozygosity for enzyme loci and variation of morphological characters in natural populations. Nature **273**:661–662.

Moav, R. and G. Wohlfarth. 1976. Two-way selection for growth rate in the common carp (*Cyprinus carpio* L.) Genetics **82**:83–101.

Mock, D.W. and G.A. Parker. 1986. Advantages and disadvantages of egret and heron brood reduction. Evolution **40**:459–470.

Mock, D.W., H. Drummond and C.H. Stinson. 1990. Avian siblicide. American Scientist **78**:438–449.

Moller, H., R.H. Smith and R.M. Sibly. 1989. Evolutionary demography of a bruchid beetle. I. Quantitative genetical analysis of the female life history. Functional Ecology **3**:673–681.

Monro, J. 1967. The exploitation and conservation of resources by populations of insects. Journal of Animal Ecology **36**:531–547.

Montalvo, A.M. and J.D. Ackerman. 1987. Limitations to fruit production in *Ionopsis utricularioides* (Orchidaceae). Biotropica **19**:24–31.

Monteleone, D.M. and E.D. Houde. 1990. Influence of maternal size on survival and growth of striped bass *Morone saxatilis* Walbaum eggs and larvae. Journal of Experimental Marine Biology and Ecology **140**:1–11.

Moore, R.A. and M.C. Singer. 1987. Effects of maternal age and adult diet on egg weight in the butterfly *Euphydras editha*. Ecological Entomology **12**:401–408.

Moreau, R.E. 1944. Clutch size: a comparative study, with special reference to African birds. Ibis **1944**:286–347.

Morris, D.W. 1986. Proximate and ultimate controls of life-history variation: the evolution of litter size in white-footed mice (*Peromyscus leucopus*). Evolution **40**:169–181.

Morris, J.A. 1963. Continuous selection for egg production using short term records. Australian Journal of Agricultural Research **14**:909–925.

Morris, M.R. 1989. Female mate choice of large males in the treefrog *Hyla chrysoscelis*: the importance of identifying the scale of choice. Behavioral Ecology and Sociobiology **25**:275–281.

Moss, R. and A. Watson. 1982. Heritability of egg size, hatch weight, body weight and viability in red grouse (*Lagopus lagopus scoticus*). Auk **99**:683–686.

Moss, R., A. Watson, P. Rothery and W.W. Glennie. 1981. Clutch size, egg size, hatch weight and laying date in relation to early mortality in Red Grouse *Lagopus lagopus scoticus* chicks. Ibis **123**:450–462.

Mountford, M.D. 1968. The significance of litter size. Journal of Animal Ecology **37**:363–367.

Mousseau, T.A. and D.A. Roff. 1987. Natural selection and the heritability of fitness components. Heredity **59**:181–198.

Mousseau, T.A. and D.A. Roff. 1989. Adaptation to seasonality in a cricket: patterns of phenotypic and genotypic variation in body size and diapause expression along a cline in season length. Evolution **43**:1483–1496.

Mousseau, T.A. and H. Dingle. 1991. Maternal effects in insect life histories. Annual Review of Entomology **36**:511–534.

Mousseau, T.A., N.C. Collins and G. Cabana. 1987. A comparative study of sexual selection and reproductive investment in the slimy sculpin, *Cottus cognatus*, Oikos **51**:156–162.

Mueller, L.D. 1988a Density-dependent population growth and natural selection in food-limited environments: the *Drosophila* model. American Naturalist **132**:786–809.

Mueller, L.D. 1988b. Evolution of competitve ability in *Drosophila* by density-dependent natural selection. Proceedings of the National Academy of Sciences, USA **85**:4383–4386.

Mueller, L.D. 1990. Density-dependent natural selection does not increase efficiency. Evolutionary Ecology **4**:290–297.

Mueller, L.D. 1991. Ecological determinants of life-history evolution. Philosophical Transactions of the Royal Society of London (B) **332**:25–30.

Mueller, L.D. and F.J. Ayala. 1981. Trade-off between r-selection and K-selection in *Drosophila* populations. Proceedings of the National Academy of Sciences, USA **78**:1303–1305.

Mueller, L.D., P. Guo and F.J. Ayala. 1991. Density-dependent natural selection and trade-offs in life history traits. Science **253**:433–435.

Mueller, L.D. and V.F. Sweet. 1986. Density-dependent natural selection in *Drosophila*: evolution of pupation height. Evolution **40**:1354–1356.

Muhlbrock. O. 1959. Factors influencing the life-span of inbred mice. Gerontologica **3**:177–183.

Mundy, P.J. and A.W. Cook. 1975. Hatching and rearing of two chicks by the hooded vulture. Ostrich **46**:45–50.

Murdoch, W.W. and A.A. Sih. 1978. Age-dependent interference in a predatory insect. Journal of Animal Ecology **47**:581–592.

Murphy, D.D., A.E. Launer and P.R. Ehrlich. 1983. The role of adult feeding in egg production and population dynamics of the checkerspot butterfly *Euphydryas editha*. Oecologia **56**:257–263.

Murphy, E.C. 1985. Bergmann's rule, seasonality, and geographic variation in body size of house sparrows. Evolution **39**:1327–1334.

Murphy, G.I. 1968. Pattern in life history and the environment. American Naturalist **102**:391–403.

Murphy, P.A., J.T. Giesel and M.N. Manlove. 1983. Temperature effects on life history variation in *Drosophila simulans*. Evolution **37**:1182–1192.

Murton, R.K., N.J. Westwood and A.J. Isaacson. 1974. Factors affecting egg-weight, body-weight and moult of the woodpigeon *Columba palumbus*. Ibis **116**:52–73.

Myers, J.H. 1976. Distribution and dispersal in populations capable of resource depletion. Oecologia **23**:255–269.

Myers, P. and L.L. Master. 1983. Reproduction by *Peromyscus maniculatus:* size and compromise. Journal of Mammalogy **64**:1–18.

Myers, R.A. 1986. Game theory and the evolution of Atlantic salmon (*Salmo salar*) age at maturation. Special Publication of Canadian Fisheries and Aquatic Sciences **89**:53–61.

Myers, R.A. and J.A. Hutchings. 1987. Mating of anadromous Atlantic salmon, *Salmo salar*. L., with mature male parr. Journal of Fish Biology **31**:143–146.

Myers, R.A., J.A. Hutchings and R.J. Gibson. 1986. Variation in male parr maturation within and among populations of Atlantic salmon, *Salmo salar*. Journal of Fisheries and Aquatic Sciences **43**:1242–1248.

Myers, R.A. and R.W. Doyle. 1983. Predicting natural mortality rates and reproduction–mortality trade-offs from fish life history data. Canadian Journal of Fisheries and Aquatic Sciences **40**:612–620.

Myrcha, A., L. Ryszkowski and W. Walkowa. 1969. Bioenergetics of pregnancy and lactation in white mouse. Acta Theriologica **14**:161–166.

Nagy, K.A. 1983. Ecological energetics. pp. 24–54, *In* E.R. Pianka and T.W. Schoener R.B. Huey (editors), "Lizard Ecology: Studies of a Model Organism," Harvard University Press, Cambridge, Massachusetts.

Nakashima, B.S. and W.C. Leggett. 1975. Yellow perch (*Perca flavescens*) biomass responses to different levels of phytoplankton and benthic biomass in Lake Memphremagog, Quebec–Vermont. Journal of the Fisheries Research Board of Canada **32**:1785–1797.

Nakasuji, F. 1982. Seasonal changes in native host plants of a migrant skipper, *Parnara guttata* Bremer et Grey (Lepidoptera: Hesperiidae). Applied Entomology and Zoology **17**:146–148.

Nakasuji, F. 1987. Egg size of skippers (Lepidoptera: Hesperiidae) in relation to their host specificity and to leaf toughness of host plants. Ecological Research **2**:175–183.

Nakasuji, F. and M. Kimura. 1984. Seasonal polymorphism of egg size in a migrant skipper. *Parnara guttata guttata* (Lepidoptera: Hesperiidae). Kontyu **52**:253–259.

Namkoong, G. and H.R. Gregorius. 1985. Conditions for protected polymorphisms in subdivided plant populations. 2. Seed versus pollen migration. American Naturalist **125**:521–534.

Nelson, J.B. 1964. Factors influencing clutch size and chick growth in the North Atlantic gannet, *Sula bassana*. Ibis **106**:63–77.

Nelson, J.B. 1966a. Clutch size in the Sulidae. Nature **210**:435–436.

Nelson, J.B. 1966b. The breeding biology of the Gannet *Sula bassana* on the Bass Rock, Scotland. Ibis **108**:584–626.

Nesse, R.M. 1988. Life table tests of evolutionary theories of senescence. Experimental Gerontology **23**:445–453.

Nestor, K.E., K.I. Brown and C.R. Weaver. 1972. Egg quality and poult production in turkeys. 2. Inheritance and relationship among traits. Poultry Science **51**:147–158.

Nettleship, D.N. 1972. Breeding success of the common puffin (*Fratercula arctica* L.) on different habitats at Great Island Newfoundland. Ecological Monographs **42**:239–268.

Neville, A.C. 1963. Daily growth layers for determining the age of grasshopper populations. Oikos **14**:1–8.

Nevo, E. 1988a. Genetic diversity in nature: patterns and theory. Evolutionary Biology **23**:217–246.

Nevo, E. 1988b. Genetic differentiation in evolution. ISI Atlas of Science: Aminal and Plant Sciences **1988**:195–202.

Nevo, E. and A. Beiles. 1988. Genetic parallelism of protein polymorphism in nature: ecological test of the neutral theory of molecular evolution. Biological Journal of the Linnean Society **35**:229–245.

Nevo, E., A. Beiles and R. Ben-Shlomo 1984. The evolutionary significance of genetic diversity: ecological, demographic and life history correlates. pp. 13–213, *In* G.S. Mani (editor), "Evolutionary Dynamics of Genetic Diversity. Lecture Notes in Biomathematics 53," Springer-Verlag, New York.

Nevo, E. and A. Beiles. 1989. Genetic diversity of the desert: patterns and testable hypotheses. Journal of Arid Environments **17**:241–244.

Newlin, M.E. 1976. Reproduction in the bunch grass lizard *Sceloporus scalaris*. Herpetologica **32**:171–184.

Newman, R.A. 1988. Genetic variation for larval anuran (*Scaphiopus couchii*) development time in an uncertain environment. Evolution 42:763–773.

Newton, I. 1985. Lifetime reproductive output of female sparrowhawks. Journal of Animal Ecology **54**:241–253.

Ni, I. 1978. Comparative fish population studies. Ph.D. Thesis, University of British Columbia, Vancouver, Canada.

Nice, M.M. 1937. Studies in the life history of the song sparrow. Part 1. Transactions of the Linnaean Society of New York **4**:1–247.

Nickell, C.D. and J.E. Grafius. 1969. Analysis of a vegetative response to selection for high yield in winter barley, *Hordeum vulgare* L. Crop Science **9**:447–451.

Nijhout, H.F. 1975. A threshold size for metamorphosis in the tobacco hornworm, *Manduca sexta* (L.) Biological Bulletin **149**:214–225.

Nijhout, H.F. 1979. Stretch-induced moulting in *Oncopeltus fasciatus*. Journal of Insect Physiology **25**:277–281.

Nijhout, H.F. and C.M. Wiliams. 1974a. Control of moulting and metamorphosis in the tobacco hornworm, *Manduca sexta* (L): cessation of juvenile hormone secretion as a trigger for pupation. Journal of Experimental Biology **61**:493–501.

Nijhout, H.F. and C.M. Williams. 1974b. Control of moulting and metamorphosis in the tobacco hornworm, *Manduca sexta* (L): growth of the last-instar larva and the decision to pupate. Journal of Experimental Biology **61**:481–491.

Nisbet, I.C.T. 1973. Courtship-feeding, egg size and breeding success in Common Terns. Nature **241**:141–142.

Nisbet, I.C.T. 1978. Dependence of fledging success on egg size, parental performance and egg composition among Common and Roseate Terns, *Sterna hirundo* and *S. dougallii*. Ibis **120**:207–215.

Noonan, K.C. 1983. Female mate choice in the cichlid fish *Cichlasoma nigrofasciatum*. Animal Behaviour **31**:1005–1010.

Noordwijk, A.J. van. 1989. Reaction norms in genetical ecology. BioScience **39**:453–458.

Noordwijk, A.J. van and G. de Jong. 1986. Acquisition and allocation of resources: their influence on variation in life history tactics. American Naturalist **128**:127–142.

Noordwijk, A.J. van, J.H. van Balen and W. Sharloo. 1980. Heritability of ecologically important traits in the great tit. Ardea **68**:193–203.

Noordwijk, A.J. van, J.H. van Balen and W. Scharloo. 1981. Genetic and environmental variation in clutch size of the great tit (*Parus major*). Netherlands Journal of Zoology :342–372.

Nordeng, H. 1983. Solution to the "Char Problem" based on Arctic Char (*Salvelinus alpinus*) in Norway. Canadian Journal of Fisheries and Aquatic Sciences **40**:1372–1387.

Nordskog, A.W. 1977. Success and Failure of quantitative genetic theory in poultry. pp. 569–586, *In* E. Pollack, O. Kempthorne, and T.B. Bailey Jr. (editors), "Proceedings of the International Conference on Quantitative Genetics," Iowa State University Press.

Nordskog, A.W. and M. Festing. 1962. Selection and correlated responses in the fowl. Proceedings of the 12th World Poultry Congress, Sydney **1962**:25–29.

Norman, F.I. and M.D. Gottisch. 1969. Artificial twinning in the short-tailed shearwater *Puffinus tenuirostris*. Ibis **111**:391–393.

Nur, N. 1984a. The consequences of brood size for breeding blue tits. I. Adult survival, weight change and cost of reproduction. Journal of Animal Ecology **53**:479–496.

Nur, N. 1984b. The consequences of brood size for breeding blue tits. II. Nestling weight, offspring survival and optimal brood size. Journal of Animal Ecology **53**:497–517.

Nur, N. 1986. Is clutch size variation in the blue tit (*Parus caeruleus*) adaptive? Journal of Animal Ecology **55**:983–999.

Nur, N. 1988a. The consequences of brood size for breeding blue tits. III. Measuring the cost of reproduction: survival, fecundity, and differential dispersal. Evolution **42**:351–362.

Nur, N. 1988b. The cost of reproduction in birds: an examination of the evidence. Ardea **76**:155–168.

Nussbaum, R.A. 1981. Seasonal shifts in clutch-size and egg-size in the side-blotched lizard. *Uta stansburiana,* Baird and Girard. Oecologia **49**:8–13.

Nussbaum, R.A. 1987. Parental care and egg size in salamanders: an examination of the safe harbor hypothesis. Researches in Population Ecology **29**:27–44.

Nussbaum, R.A. and L.V. Diller. 1976. The life history of the side-blotched lizard, *Uta stansburiana* Baird and Girard, in north-central Oregon. Northwest Science **50**:243–260.

Nussbaum, R.A. 1985. The evolution of parental care in salamanders. Miscellaneous Publications of the Museum of Zoology, University of Michigan **169**:1–50.

Nussbaum, R.A. and D.L. Schultz. 1989. Coevolution of parental care and egg size. American Naturalist **133**:591–603.

O'Brien, W.J., N.A. Slade and G.L. Vinyard. 1976. Apparent size as the determinant of prey selection by bluegill sunfish (*Lepomis macrochirus*). Ecology **57**:1304–1310.

O'Connor, R.J. 1978. Brood reduction in birds: selection for fratricide, infanticide or suicide? Animal Behaviour **26**:79–96.

O'Connor, R.J. 1979. Egg weights and brood reduction in the European swift (*Apus apus*). Condor **81**:133–145.

O'Neill, K.M. and S.W. Skinner. 1990. Ovarian egg size in five species of parasitoid wasps. Journal of Zoology, London **220**:115–122.

O'Toole, J.J. 1982. Seed banks of *Panicum miliaceum* L. in three crops. M.Sc. Thesis, University of Western Ontario, Ontario.

Oftedal, O.T. 1984. Body size and reproductive strategy as correlates of milk energy output in lactating mammals. Acta Zoologica Fennica **171**:183–186.

Oftedal, O.T., D.J. Boness and R.A. Tedman 1987. The behavior, physiology and anatomy of lactation in the pinnipedia. pp. 175–245, *In* H.H. Genoways (editor), "Current Mammalogy, Vol. 1," Plenum Press, New York.

Ohsumi, S. 1979. Interspecies relationships among some biological parameters in cetaceans and estimation of the natural mortality coefficient of the southern hemisphere minke whale. Report of the International Whaling Commission **29**:297–406.

Ojanen, M., M. Orell and R.A. Vaisanen. 1978. Egg and clutch sizes in four passerine species in northern Finland. Ornis Fennica **55**:60–68.

Ojanen, M., M. Orell and R.A. Vaisanen. 1979. Role of heredity in egg size variation in the Great Tit *Parus major* and the Pied Flycatcher *Ficedula hypoleuca*. Ornis Scandinavica **10**:22–28.

Oksengorn-Proust, J. 1954. Etude bimetrique de la taille des oeufs de *Drosophila melanogaster*. Comptes Rendus des Seances de l'Academie des Sciences **238**:1356–1358.

Olson, F.C.W. 1964. The survival value of fish schooling. Journal du Conseil Permanent International pour l'Exploration de la Mer **XXIX**:115–116.

Olsson, G. 1960. Some relations between number of seeds per pod, seed size and oil content and the effects of selection for these characters in *Brassica* and *Sinapis*. Hereditas **46**:29–70.

Oosthuizen, E. and N. Daan. 1974. Egg fecundity and maturity of North Sea Cod, *Gadus morhua*. Netherlands Journal of Sea Research **8**:378–397.

Orell, M. and K. Koivula. 1988. Cost of reproduction: parental survival and production of recruits in the willow tit *Parus montanus*. Oecologia **77**:423–432.

Orians, G.H. 1961. The ecology of blackbird (*Agelaius*) social systems. Ecological Monographs **31**:285–312.

Orozco, F. 1976. A dynamic study of genotype environment interaction with egg laying of *Tribolium casteneum*. Heredity **37**:157–171.

Orton, R.A. and R.M. Sibly. 1990. Egg size and growth rate in *Theodoxus fluviatilis* (L). Sunctional Ecology **4**:91–94.

Orzack, S.H. and S. Tuljapurkar. 1989. Population dynamics in variable environments. VII. The demography and evolution of iteroparity. American Naturalist **133**:901–923.

Oster, G.F. and S.M. Rocklin. 1979. Optimization models in evolutionary biology. Lectures on Mathematics in the Life Sciences **11**:21–88.

Otronen, M. 1984. The effect of differences in body size on the male territorial system of the fly *Dryomyza anilis*. Animal Behaviour **32**:882–890.

Otte, D. 1972. Simple *versus* elaborate behavior in grasshoppers: an analysis of communication in the genus *Syrbula*. Behaviour **42**:291–322.

Pagel, M.D. and P.H. Harvey. 1988. Recent developments in the analysis of comparative data. Quarterly Review of Biology **63**:413–440.

Paine, M.D. 1990. Life history tactics of darter (Percidae: Etheostomatiini) and their relationship with body size, reproductive behaviour, latitude and rarity. Journal of Fish Biology **37**:437–488.

Paine, R.T. 1965. Natural history, limiting factors and energetics of the opisthobranch *Navanax inermis*. Ecology **46**:603–619.

Paine, R.T. 1976. Size-limited predation: an observational and experimental approach with the *Mytilus–Pisaster* interaction. Ecology **57**:858–873.

Pak, G.A. and E.R. Oatman. 1982. Biology of *Trichogramma brevicapillum*. Entomologica Experimentalis Applicata **32**:61–67.

Palmer, A.R. 1990. Predation size, prey size, and the scaling of vulnerability: hatching gastropods vs barnacles. Ecology **71**:759–775.

Palmer, A.R. and C. Strobeck. 1986. Fluctuating asymmetry: measurement, analysis, patterns, Annual Review of Ecology and Systematics 17:391–421.

Palmer, J.O. and H. Dingle. 1986. Direct and correlated responses to selection among life-history traits in milkweed bugs (*Oncopeltus fasciatus*). Evolution 40:767–777.

Pamilo. 1988. Genetic variation in heterogeneous environments. Annales Zoologici Fennici 25:99–106.

Parker, G.A., D.W. Mock and T.C. Lamey. 1989. How selfish should stronger sibs be? American Naturalist 133:846–868.

Parker, G.A. and J. Maynard Smith. 1990. Optimality theory in evolutionary biology. Nature 348:27–33.

Parker, G.A. and M. Begon. 1986. Optimal egg size and clutch size: effects of environmental and maternal phenotype. American Naturalist 128:573–592.

Parker, G.A. and S.P. Courtney. 1984. Models of clutch size in insect oviposition. Theoretical Population Biology 26:27–48.

Parker, R.R. and P.A. Larkin, 1959. A concept of growth in fishes. Journal of the Fishes Research Board of Canada 16:721–745.

Parkes, A.S. 1926. The growth of young mice according to size of litter. Annals of Applied Biology 13:374–394.

Parry, G.D. 1981. The meanings of r- and K-selection. Oecologia 48:260–264.

Parsons, P.A. 1964. Egg lengths in *Drosophila melanogaster* and correlated responses to selection. Genetica 35:175–181.

Partridge, B.L. 1982. The structure and function of fish schools. Scientific American 246:114–123.

Partridge, L. and M. Farquhar. 1981. Sexual activity reduces life span of male fruit flies. Nature 294:580–582.

Partridge, L. and M. Farquar. 1983. Lifetime mating success of male fruitflies (*Drosophila melanogaster*) is related to their size. Animal Behaviour 31:871–877.

Partridge, L. and P.H. Harvey. 1985. Costs of reproduction. Nature 316:20.

Partridge, L. and R. Andrews. 1985. The effect of reproductive activity on the longevity of male *Drosophila melanogaster* is not caused by acceleration of ageing. Journal of Insect Physiology 31:393–395.

Patterson, C.B., W.J. Erckmann and G.H. Orians. 1980. An experimental study of parental investment and polygyny in male blackbirds. American Naturalist 116:757–769.

Pauly, D. 1978. A preliminary compilation of fish length growth parameters. Berichte aus dem Institute fur Meereskunde an der Christian-Albrechts-Universitat Kiel 55:1–200.

Pauly, D. 1980. On the interrelationships between natural mortality, growth parameters, and mean environmental temperature in 175 fish stocks. Journal du Conseil Permanent International pour l'Exploration de la Mer 39:175–192.

Payne, R.B. 1965. Clutch size and numbers of eggs laid by brown-headed cowbirds. Condor **67**:44–60.

Payne, R.B. 1974. The evolution of clutch size and reproductive rates in parasitic cuckoos. Evolution **28**:169–181.

Pearl, R. 1940. Introduction to Medical Biometry and Statistics. Saunders, Philadelphia.

Pearl, R. and C.R. Doering. 1923. A comparison of the mortality of certain organisms with that of man. Science **57**:209–212.

Pearl, R. and J.R. Miner. 1935. Experimental studies on the duration of life. XIV. The comparative mortality of certain lower organisms. Quarterly Review of Biology **10**:60–79.

Pearson, O.P. 1948. Metabolism of small mammals, with remarks on the lower limit of mammalian size. Science **108**:44.

Pease, C.M. and J.J. Bull. 1988. A critique of methods for measuring life history trade-offs. Journal of Evolutionary Biology **1**:293–303.

Perril, S.A., H.C. Gerhardt and R. Daniel. 1982. Mating strategy shifts in male green treefrogs (*Hyla cinerea*): an experimental study. Animal Behaviour **30**:43–48.

Perrin, N. 1988. Why are offspring born larger when it is colder? Phenotypic plasticity for offspring size in the clandoceran *Simocephalus vetulus* (Muller). Functional Ecology **2**:283–288.

Perrin, N. 1989. Population density and offspring size in the cladoceran *Simocephalus vetulus* (Müller). Functional Ecology **3**:29–36.

Perrin, N. and J.F. Rubin. 1990. On dome-shaped norms of reaction for size-at-age at maturity in fishes. Functional Ecology **4**:53–57.

Perrins, C. 1964. Survival of young swifts in relation to brood size. Nature **203**:1147–1148.

Perrins, C.M. 1970. The timing of birds' breeding seasons. Ibis **112**:242–255.

Perrins, C.M. 1977. The role of predation in the evolution of clutch size. pp. 181–191, *In* B. Stonehouse and C. Perrins (editors), "Evolutionary Ecology," University Park Press, Baltimore.

Perrins, C.M. and M.P. Harris and C.K. Britton. 1973. Survival of Manx Shearwaters *Puffinus puffinus*. Ibis **115**:535–548.

Perrins, C.M. and P.J. Jones. 1974. The inheritance of clutch size in the great tit (*Parus major* L.) Condor **76**:225–229.

Perrone, M.J. 1978. Male size and breeding success in a monogamous cichlid fish. Environmental Biology of Fishes **3**:193–201.

Peterman, R.M. 1990. Statistical power analysis can improve fisheries research and management. Canadian Journal of Fisheries and Aquatic Sciences **47**:2–15.

Peters, R.H. 1983. The Ecological Implications of Body Size. Cambridge University Press, Cambridge.

Petersen, C.W. 1988. Male mating success, sexual size dimorphism, and site fidelity in two species of *Malacoctenus* (Labrisomidae). Environmental Biology of Fishes **21**:173–183.

Petersen, C.W. 1990. The relationships among population density, individual size, mating tactics, and reproductive success in a hermaphrodite fish, *Serranus fasciatus*. Behaviour **113**:57–80.

Peterson, B. 1950. The relation between size of mother and number of eggs and young in some spiders. Experientia **6**:96–98.

Peterson, I. and J.S. Wroblewski. 1984. Mortality rate of fishes in the pelagic ecosystem. Canadian Journal of Fisheries and Aquatic Sciences **41**:1117–1120.

Pettifor, R.A., C.M. Perrins and R.H. McCleery. 1988. Individual optimization of clutch size in great tits. Nature **336**:160–162.

Philippi, T. and J. Seger. 1989. Hedging one's evolutionary bets, revisited. Trends in Ecology and Evolution **4**:41–44.

Pianka, E.R. 1970. On r- and K-selection. American Naturalist **104**:592–597.

Pianka, E.R. and W.S. Parker. 1975. Age specific reproductive tactics. American Naturalist **109**:453–464.

Pierce, J.R. and D.B. Ralin. 1972. Vocalization behavior of the males of three species in the *Hyla versicolor* complex. Herpetologica **28**:329–337.

Pimm, S.L. and A. Redfearn. 1988. The variability of population densities. Nature **334**:613–614.

Piñero, D., J. Sarukhan and P. Alberdi. 1982. The costs of reproduction in a tropical palm *Astrocaryum mexicanum*. Journal of Ecology **70**:473–481.

Pinkowski, B.C. 1975. Growth and development of Eastern Bluebirds. Bird-Banding **46**:273–289.

Pinkowski, B.C. 1977. Breeding adaptations in the Eastern Bluebird. Condor **79**:289–302.

Pitafi, K.D., R. Simpson, J.J. Stephen and T.H. Day. 1990. Adult size and mate choice in seaweed flies (*Coleopa frigida*) Heredity **65**:91–97.

Pitcher, T.J. 1986. Functions of shoaling behaviour in teleosts. pp. 297–337, *In* T.J. Pitcher (editor). "The Behaviour of Teleost Fishes," Croom Helm, Kent.

Pitelka, L.F. 1977. Energy allocation in annual and perennial lupines (*Lupinus:* Leguminosae). Ecology **58**:1055–1065.

Pitelka, L.F., M.E. Thayer and S.B. Hansen. 1983. Variation in achene weight in *Aster acuminatus*. Caddadian Journal of Botany **61**:1415–1420.

Pitman, R.W. 1979. Effects of female age and egg size on growth and mortality in rainbow trout. Progressive Fish Culturist **41**:202–204.

Pitt, T.K. 1966. Sexual maturity and spawning of the American plaice, *Hippoglossoides platessoides* (Fabr.), from Grand Bank and Newfoundland areas. Journal of the Fisheries Research Board of Canada **23**:651–672.

Plumb, W.J. 1965. Observations on the breeding biology of the razorbill. British Birds **48**:449–456.

Pollack, E. and O. Kempthorne. 1970. Malthusian parameters in genetic populations. Part I. Haploid and selfing models. Theoretical Population Biology 1:315–345.

Poole, W.E. 1960. Breeding of the wild rabbit, *Oryctolagus caniculatus* (L.) in relation to the environment. C.S.I.R.O. Wildlife Research 5:21–43.

Pope, J.A., D.H. Mills and W.M. Shearer. 1961. The fecundity of Atlantic salmon (*Salmo salar* Linn). Freshwater Salmon Fisheries Research 26:1–12.

Post, J.R. and D.O. Evans. 1989. Size-dependent overwinter mortality of young-of-the-year yellow perch (*Perca flavescens*): laboratory, in situ enclosure, and field experiments. Canadian Journal of Fisheries and Aquatic Sciences 46:1958–1968.

Potvin, C. and D.A. Roff. 1992. Distribution-free and robust statistical methods: viable alternatives to parametric statistics? Ecology, in press.

Powell. J.R. 1971. Genetic polymorphism in varied environments. Science 174:1035–1036.

Powles, P.M. 1958. Studies of reproduction and feeding of Atlantic cod (*Gadus callarias* L.) in the Southwestern Gulf of St. Lawrence. Journal of the Fisheries Research Board of Canada 15:1383–1402.

Pressley, P.H. 1981. Parental effort and the evolution of nest-guarding tactics in the threespine stickleback, *Gasterosteus aculeatus* L. Evolution 35:282–295.

Prestwich, K.N. and T.J. Walker. 1981. Energetics of singing in crickets: effects of temperature in three trilling species (Orthoptera: Gryllidae). Journal of Comparative Physiology 143:199–212.

Price, T. and L. Liou. 1989. Selection on clutch size in birds. American Naturalist 134:950–959.

Price, T., M. Kirkpatrick and S.J. Arnold. 1988. Directional selection and the evolution of breeding date in birds. Science 240:798–799.

Price, T.D. 1984. Sexual selection on body size, territory and plumage variables in a population of Darwin's finches. Evolution 38:327–341.

Price, T.D. and P.R. Grant: 1984. Life history traits and natural selection for small body size in a population of Darwin's finches. Evolution 38:483–494.

Primack, R.B. 1979. Reproductive effort in annual and perennial species of *Plantago* (Plantaginaceae). American Naturalist 114:51–62.

Primack, R.B. and J. Antonovics. 1979. Experimental ecological genetics in *Plantago*. VII. Reproductive effort in populations of *P. lanceolata*. Evolution :742–752.

Prince, H.H., P.B. Siegel and G.W. Cornwell. 1970. Inheritance of egg production and juvenile growth in mallards. Auk 87:342–352.

Pritchard, G. 1965. Prey capture by dragonfly larvae (Odonata: Anisoptera). Canadian Journal of Zoology 43:271–289.

Promislow, D.E.L. and P.H. Harvey. 1990. Living fast and dying young: a comparative analysis of life-history variation among mammals. Journal of Zoology 220:417–437.

Prout, T. 1968. Sufficient conditions for multiple niche polymorphism. American Naturalist **102**:493–496.

Prout, T. 1971. The relation between fitness components and population prediction. I. The estimation of fitness components. Genetics **68**:127–149.

Prout, T. 1980. Some relationships between density-dependent selection and density dependent population growth. Evolutionary Biology **13**:1–68.

Provine, W.B. 1971. Origins of Theoretical Population Genetics. University of Chicago Press, Chicago.

Provine, W.B. 1986. Sewall Wright and Evolutionary Biology. University of Chicago Press, Chicago.

Pugesek, B.H. 1981. Increased reproductive effort with age in the California gull (*Larus californicus*). Science **212**:822–823.

Pugesek, B.H. and K.L. Diem. 1990. The relationship between reproduction and survival in known-aged California gulls. Ecology **71**:811–817.

Purrington, F.F. and J.S. Uleman. 1972. Brood size of the parasitic wasp, *Hyssopus thymus:* functional correlation with the mass of a cryptic host. Annals of the Entomological Society of America **65**:280–281.

Pyle, D.W. 1976. Effects of artificial selection on reproductive fitness in *Drosophila*. Nature **263**:317–319.

Pyle, D.W. 1978. Correlated responses to selection for a behavioral trait in *Drysophila melanogaster*. Behavior Genetics **8**:333–340.

Quinn, J.S. and R.D. Morris. 1986. Intra-clutch egg-weight apportionment and chick survival in Caspian terns. Canadian Journal of Zoology **64**:2116–2122.

Quiring, D.T. and J.N. McNeil. 1984. Influence of intraspecific larval competition and mating on the longevity and reproductive performance of females of the leaf miner *Agromyza frontella* (Rodani) (Diptera: Agromyzidae). Canadian Journal of Zoology **62**:2197–2200.

Rabe, F.W. III. 1957. Brook trout populations in Colorado beaver ponds. M.Sc. Thesis, Colorado State University, Colorado.

Rahn, H., P.R. Sotherland and C.V. Paganelli. 1985. Interrelationships between egg mass and adult body mass and metabolism among passerine birds. Journal of Ornithology **126**:263–271.

Ramer, J.D., T.A. Jenssen and C.J. Hurst. 1983. Size-related variation in the advertisement call of *Rana clamitans* (Anura: Ranidae), and its effect on conspecific males. Copeia **1983**141–155.

Ramsay, M.A. and R.L. Dunbrack. 1987. Is the giant panda a bear? Oikos **50**:267.

Ramsay, M.A. and R.L. Dunbrack. 1986. Physiological constraints on life history phenomena: the example of small bear cubs at birth. American Naturalist **127**:735–743.

Rana, K.J. 1985. Influences of egg size on the growth, onset of feeding, point-of-no-return, and survival of unfed *Oreochromis mossambicus* Fry. Aquaculture **46**:119–131.

Randolph, P.A., J.C. Randolph, K. Mattingly and M.M. Foster. 1977. Energy costs of reproduction in the cotton rat, *Sigmodon hispidus*. Ecology **58**:31–45.

Random, A.B. 1967. Reproductive biology of white-tailed deer in Manitoba. Journal of Wildlife Management **31**:114–123.

Rapoport, A. 1985. Applications of game-theoretic concepts in biology. Bulletin of Mathematical Biology **47**:161–192.

Rask, M. 1983. Differences in growth of perch (*Perca fluviatilis*) in two small forest lakes. Hydrobiologia **101**:139–144.

Read, A.F. and P.H. Harvey. 1989. Life history differences among the eutherian radiations. Journal of Zoology **219**:329–353.

Real, L.A. 1980. Fitness, uncertainty, and the role of diversification in evolution and behavior. American Naturalist **115**:623–638.

Reed, J. and N.C. Stenseth. 1984. On evolutionarily stable strategies. Journal of Theoretical Biology **108**:491–508.

Reeve, E.C.R. and F.W. Robertson. 1953. Analysis of environmental variability in quantitative inheritance. Nature **171**:874–875.

Reid, W.V. 1987. The cost of reproduction in the glaucous-winged gull. Oecologia **74**:458–467.

Reid, W.V. 1988. Age-specific patterns of reproduction in the glaucous-winged gull: increased effort with age? Ecology **69**:1454–1465.

Reimchen, T.E. 1988. Inefficient predators and prey injuries in a population of giant stickleback. Canadian Journal of Zoology **66**:2036–2044.

Reimers, N. 1979. A history of a stunted brook trout population in an alpine lake: a lifespan of 24 years. California Fish and Game **65**:196–215.

Reiss, M.J. 1985. The allometry of reproduction: why larger species invest relatively less in their offspring. Journal of Theoretical Biology **113**:529–544.

Reznick, D. 1981. "Grandfather effects": the genetics of offspring size in mosquito fish *Gambusia affinis*. Evolution **35**:941–953.

Reznick, D. 1982. Genetic determination of offspring size in the guppy (*Poecilia reticulata*). American Naturalist **120**:181–188.

Reznick, D. 1983. The structure of guppy life histories: the tradeoff between growth and reproduction. Ecology **64**:862–873.

Reznick, D. 1985. Costs of reproduction: an evaluation of the empirical evidence. Oikos **44**:257–267.

Reznick, D., E. Perry and J. Travis. 1986. Measuring the cost of reproduction: a comment on papers by Bell. Evolution **40**:1338–1344.

Reznick, D., H. Bryga and J.A. Endler. 1990. Experimentally induced life-history evolution in a natural population. Nature **346**:357–359.

Reznick, D. and J.A. Endler. 1982. The impact of predation on life history evolution in 'Trinidadian guppies (*Poecilia reticulata*). Evolution **36**:160–177.

Rhodes, C.P. and D.M. Holdich. 1982. Observations on the fecundity of the freshwater crayfish, *Austropotamobius pallipes* (Lereboullet), in the British Isles. Hydrobiologia **89**:231–236.

Rice, D.W. and K. Kenyon. 1962. Breeding cycles and behaviour of Laysan and Black-footed albatrosses. Auk **79**:517–567.

Richards, F.J. 1959. A flexible growth function for empirical use. Journal of Experimental Botany **10**:290–300.

Richards, L.J. and J.H. Myers. 1980. Maternal influences on size and emergence time in the cinnabar moth. Canadian Journal of Zoology **58**:1452–1457.

Richart, E.A. 1977. Reproduction, growth and development in two species of cloud forest *Peromyscus* from southern Mexico. Occasional Papers of the Museum of Natural History, University of Kansas **67**:1–22.

Ricker, W.E. 1975. Computation and Interpretation of Biological Statistics of Fish Populations. Fisheries Research Board of Canada, Bulletin 191.

Ricklefs, R.E. 1968. Patterns of growth in birds. Ibis **110**:419–451.

Ricklefs, R.E. 1969. An analysis of nestling mortality in birds. Smithsonian Contributions in Zoology **9**:1–48.

Ricklefs, R.E. 1973a. Patterns of growth in birds II. Growth rate and mode of development. Ibis **115**:177–210.

Ricklefs, R.E. 1973b. Fecundity, mortality and avian demography. pp. 366–435, *In* D.S. Farmer (editor), "Breeding Biology of Birds," Proceedings of the National Academy of Sciences USA, Washington, D.C.

Ricklefs, R.E. 1974. Energetics of reproduction in birds. pp. 152–292. *In* J.R.A. Paynter (editor). "Avian Energetics," Nuttall Onithological Club, Cambridge, Massachusetts.

Ricklefs, R.E. 1977a. A note on the evolution of clutch size. pp. 193–214, *In* B. Stonehouse and C. Perrins (editors), "Evolutionary Ecology," University Park Press, London.

Ricklefs, R.E. 1977b. On the evolution of reproductive strategies in birds: reproductive effort. American Naturalist **111**:453–478.

Ricklefs, R. 1982. A comment on the optimization of body size in *Drosophila* to Roff's life history model. American Naturalist **120**:686–688.

Ricklefs, R.E. 1983. A comment on the regulation of reproductive success towards e^{-1}. Oikos **41**:284–285.

Ricklefs, R.E., D.C. Hahn and W.A. Montevecchi. 1978. The relationship between egg size and clutch size in the Laughing gull and Japanese Quail. Auk **95**:135–144.

Riddle, R.A., P.S. Dawson and D.F. Zirkle. 1986. An experimental test of the relationship between genetic variation and environmental variation in *Tribolium* flour beetles. Genetics **113**:391–404.

Ridley, M. 1983. The Explanation of Organic Diversity. Clarendon Press, Oxford.

Ridley, M. and D.J. Thompson. 1979. Size and mating in *Asellus aquaticus* (Crustacea: Isopoda). Zeitschrift fuer Tierpsychologie **51**:380–397.

Riechert, S.E. 1978. Games spiders play: behavioral variability in territorial disputes. Behavioral Ecology and Sociobiology **3**:135–162.

Riechert, S.E. and P. Hammerstein. 1983. Game theory in the ecological context. Annual Review of Ecology and Systematics **14**:377–409.

Rijnsdorp, A.D., F. van Lent and K. Groeneveld. 1983. Fecundity and the energetics of reproduction and growth of North Sea Plaice (*Pleuronectes platessa* L.) International Council for the Exploration of the Sea C.M. 1983/g:**31**:1–12.

Riley, J.G. 1979. Evolutionary equilibrium strategies. Journal of Theoretical Biology **76**:109–123.

Riska, B. 1986. Some models for development, growth and morphometric correlation. Evolution **40**:1303–1311.

Riska, B. 1989. Composite traits, selection response and evolution. Evolution **43**:1172–1191.

Rismiller, P.D. and R.S. Seymour. 1991. The echidna. Scientific American **264**:96–103.

Roach, D.A. and R.D. Wulff. 1987. Maternal effects in plants. Annual Review of Ecology and Systematics **18**:209–235.

Robertson, A. 1955. Selection in animals: synthesis. Cold Spring Harbor Symposium on Quantitative Biology **20**:225–229.

Robertson, A. 1977. The effect of selection on the estimation of genetic parameters. Zeitschrift fuer Tierzuechtung Zuechtungsbiologie **94**:131–135.

Robertson, F.W. 1957. Studies in quantitative inheritance. XI. Genetic and environmental correlation between body size and egg production in *Drosophila melanogaster*. Journal of Genetics **55**:428–443.

Robertson, F.W., M. Shook, G. Takei and H. Gaines. 1968. Observations on the biology and nutrition of *Drosophila disticha*, Hardy, and indigenous Hawaiian species. Studies in Genetics, IV. Research Reports **4**:279–299.

Robertson, I. 1971. The influence of brood-size on reproductive success in two species of cormorant, *Phalocrocorax auritus* and *P. pelagicus*, and its relation to the problem of clutch-size. M.Sc. thesis, University of British Columbia, Vancouver, Canada.

Robertson, J.G.M. 1986a. Male territoriality, fighting and assessment of fighting ability in the Australian frog *Uperoleia rugosa*. Animal Behaviour **34**:763–772.

Robertson, J.G.M. 1986b. Female choice, male strategies and the role of vocalizations in the Australian frog *Uperoleia rugosa*. Animal Behaviour **34**:773–784.

Roby, D.D. and R.E. Ricklefs. 1986. Energy expenditure in adult least auklets and diving petrels during the chick-rearing period. Physiological Zoology **59**:661–678.

Rockwell, R.F., C.S. Findlay and F. Cooke. 1987. Is there an optimal clutch size in snow geese? American Naturalist **130**:839–863.

Roff, D.A. 1976. Stabilizing selection in *Drosophila melanogaster*: a comment. Journal of Heredity **67**:245–246.

Roff, D.A. 1977. Disperal in dipterans: its costs and consequences. Journal of Animal Ecology **46**:443–456.

Roff, D.A. 1978. Size and survival in a stochastic environment. Oecologia **36**:163–172.

Roff, D.A. 1980. Optimizing development time in a seasonal environment: the "ups and downs" of clinal variation. Oecologia **45**:202–208.

Roff, D.A. 1981a. On being the right size. American Naturalist **118**:405–422.

Roff, D.A. 1981b. Reproductive uncertainty and the evolution of iteroparity: why don't flatfish put all their eggs in one basket? Canadian Journal of Fisheries and Aquatic Sciences **38**:968–977.

Roff, D.A. 1982. Reproductive strategies in flatfish: a first synthesis. Canadian Journal of Fisheries and Aquatic Sciences **39**:1686–1698.

Roff, D.A. 1983a. Phenological adaptation in a seasonal environment: a theoretical perspective. pp. 253–270. In V.K. Brown and I. Hodek (editors), "Diapause and Life Cycle Strategies in Insects,: Dr. W. Junk, The Hague.

Roff, D.A. 1983b. An allocation model of growth and reproduction in fish. Canadian Journal of Fisheries and Aquatic Sciences **40**:1395–1404.

Roff, D.A. 1983c. Development rates and the optimal body size in *Drosophila*: a reply to Ricklefs. American Naturalist **122**:570–575.

Roff, D.A. 1984a. The evolution of life history parameters in teleosts. Canadian Journal of Fisheries and Aquatice Sciences **41**:984–1000.

Roff, D.A. 1984b. The cost of being able to fly: a study of wing polymorphism in two species of crickets. Oecologia **63**:30–37.

Roff, D.A. 1986a. The evolution of wing dimorphism in insects. Evolution **40**:1009–1020.

Roff, D.A. 1986b. The evolution of wing polymorphisms and its impact on life cycle adaptation in insects. pp. 209–221, In F. Taylor and R. Karban (editors), "The Evolution of Insect Life Cycles," Springer-Verlag, New York.

Roff, D.A. 1986c. Predicting body size with life history models. BioScience **36**:316–323.

Roff, D.A. 1988. The evolution of migration and some life history parameters in marine fishes. Environmental Biology of Fishes **22**:133–146.

Roff, D.A. 1990a. The evolution of flightlessness in insects. Ecological Monographs **60**:389–421.

Roff, D.A. 1990b. Understanding the evolution of insect life cycles: the role of genetical analysis. pp. 5–27, In F. Gilbert (editor), "Genetics, Evolution and Coordination of Insect Life Cycles," Springer-Verlag, New York.

Roff, D.A. 1991a. The evolution of life history variation in fishes, with particular reference to flatfishes. Netherlands Journal of Sea Research in press.

Roff, D.A. 1991b. Life history consequences of bioenergetic and biomechanical constraints on migration. American Zoologist **31**:205–215.

Roff, D.A. and T.A. Mousseau. 1987. Quantitative genetics and fitness: lessons from *Drosophila*. Heredity **58**:103–118.

Rohwer, F.C. 1988. Inter- and intraspecific relationships between egg size and clutch size in waterfowl. Auk **105**:161–176.

Rohwer, F.C. and D.I. Eisenhauer. 1989. Egg mass and clutch size relationships in geese, eiders, and swans. Ornis Scandinavica **20**:43–48.

Roitberg, B.D. 1989. The cost of reproduction in rosehip flies, *Rhagoletis basiola*: eggs are time. Evolutionary Ecology **3**:183–188.

Rose, M.R. 1982. Antagonistic pleiotropy, dominance, and genetic variation. Heredity **48**:63–78.

Rose, M.R. 1984a. Artificial selection on a fitness-component in *Drosophila melanogaster*. Evolution **38**:516–526.

Rose, M.R. 1984b. Genetic covariation in *Drosophila*: untangling the data. American Naturalist **123**:N565–569.

Rose, M.R. 1984c. Laboratory evolution of postponed senescence in *Drosophila melanogaster*. Evolution **38**:1004–1010.

Rose, M.R. 1985. Life history evolution with antagonistic pleiotropy and overlapping generations. Theoretical Population Biology **28**:342–358.

Rose, M.R. 1991. Evolutionary Biology of Aging. Oxford University Press, New York.

Rose, M.R. and B. Charlesworth. 1981a. Genetics of life history in *Drosophila melanogaster*. I. Sib analysis of adult females. Genetics **97**:173–186.

Rose, M.R. and B. Charlesworth. 1981b. Genetics of life history in *Drosophila melanogaster*. II. Exploratory selection experiments. Genetics **97**:187–196.

Rose, M.R., P.M. Service and E.W. Hutchinson 1987. Three approaches to trade-offs in life-history evolution. pp. 91–105, *In* V. Loeschke (editor), "Genetic Constraints on Adaptive Evolution," Springer-Verlag, Berlin.

Rosen, R. 1967. Optimality Principles in Biology. Butterworths, London.

Røskaft, E. 1985. The effect of enlarged brood size on the future reproductive potential of the rook. Journal of Animal Ecology **54**:255–260.

Rothlisberg, P.C. 1985. Life history strategies. Bulletin of Marine Science **37**:761–762.

Roughgarden, J. 1971. Density-dependent natural selection. Ecology **52**:453–468.

Rovner, J.S. 1968. Territoriality in the sheet-web spider *Linyphia triangularis* (Clerck) (Araneae. Linyphidae). Zeitschrift fuer Tierpsychologie **25**:232–242.

Rutherford, J.C. 1973. Reproduction, growth and mortality of the holothurian *Cucumaria pseudocurata*. Marine Biology **22**:167–176.

Rutledge, W.P. and J.C. Barron. 1972. The effects of the removal of stunted white crappie on the remaining crappie population of Meridian State Park Lake, Bosque, Texas. Texas Parks and Wildlife Department Technical Series **12**:1–41.

Ryan, M.J 1980. Female mate choice in a neotropical frog. Science **209**:523–525.

Ryan, M.J. 1983. Sexual selection and communication in a Neotropical frog, *Physalaemus pustulosus*. Evolution **37**:261–272.

Ryan, M.J., G.A. Bartholomew and A.S. Rand, 1983. Energetics of reproduction in a neotropical frog, *Physalaemus pustulosus*. Ecology **64**:1456–1462.

Ryan, M.J., M.D. Tuttle and A.S. Rand. 1982. Bat predation and sexual advertisement in a Neotropical anuran. American Naturalist **119**:136–139.

Ryan, M.J., M.D. Tuttle and L.K. Taft. 1981. The costs and benefits of frog chorusing behavior. Behavioral Ecology and Sociobiology **8**:273–278.

Ryder, J.P. 1975. Egg laying, egg size, and success in relation to immature-mature plumage of Ring-billed Gulls. Wilson Bulletin **87**:534–542.

Ryser, J. 1988. Determination of growth and maturation in the common frog, *Rana temporaria*, by skeletochrondology. Journal of Zoology **216**:673–685.

Ryser, J. 1989. Weight loss, reproductive output, and the cost of reproduction in the common frog, *Rana temporaria*. Oecologia **78**:264–268.

Sacher, G.A. 1959. Relation of lifespan to brain weight. pp. 115–141, *In* G.E.W. Wolstenholme and M. O'Connor (editors), "The Lifespan of Animals," Little Brown, Boston.

Sadleir, R.M.F.S. 1974. The ecology of the deer mouse *Peromyscus maniculatus* in a coastal coniferous forest. II. Reproduction. Canadian Journal of Zoology **52**:119–131.

Saether, B.E. 1987. The influence of body weight on the covariation between reproductive traits in European birds. Oikos **48**:79–88.

Saether, B.E. 1988. Pattern of covariation between life-history traits of European birds. Nature **331**:616–617.

Saether, B.E. 1989. Survival rates in relation to body weight in European birds Ornis Scandinavica **20**:13–21.

Sakaluk, S.K. and J.J. Belwood. 1984. Gecko phonotaxis to cricket calling song: a case of satellite predation. Animal Behaviour (659–662):

Sakaluk, S.K. and W.A. Snedden. 1990. Nightly calling durations of male sagebrush crickets. *Cyphoderris strepitans*: size, mating and seasonal effects. Oikos **57**:153–160.

Salt, G. 1961. Competition among insect parasitoids. Symposium of the Society of Experimental Biology **14**:96–119

Salthe, S.N. 1969. Reproductive modes and the number and size of ova in the urodeles. American Midland Naturalist **81**:467–490.

Samollow, P.B. 1980. Selective mortality and reproduction in a natural population of *Bufo boreas*. Evolution **34**:18–39.

Samollow, P.E. and M. Soulé. 1983. A case of stress related heterozygote superiority in nature. Evolution **37**:646–649.

Sandell, M. 1989. Ecological energetics, optimal body size and sexual size dimorphism: A model applied to the stoat, *Mustela erminea* L. Functional Ecology **3**:315–324.

Sandoz, O. 1956. Changes in the fish population of Lake Murray following the reduction of gizzard shad numbers. Proceedings of the Oklahoma Academy of Science **37**:174–181.

Sanfriel, U.N. 1975. On the significance of clutch size in nidifugous birds. Ecology **56**:703–708.

Sang, J.H. 1949. The ecological determinants of population growth in a *Drosophila* culture. Physiological Zoology **XXII**:183–201.

Sang, J.H. 1950. Population growth in *Drosophila* cultures. Biological Review **25**:188–217.

Santos, M., A. Ruiz, A. Barbadilla, J.E. Quezada-Diaz, E. Hasson and A. Fondevila. 1988. The evolutionary history of *Drosophila buzzatii*. XIV. Larger flies mate more often in nature. Heredity **61**:255–262.

Sergeant, A.B., S.H. Allen and R.T. Eberhard. 1984. Red fox predation on breeding ducks in midcontinental North America. Wildlife Monographs **89**:1–41.

Sargent, P.C. and M.R. Gross. 1985. Parental investment decision rules and the Concorde fallacy. Behavioral Ecology and Sociobiology **17**:43–45.

Sargent, R.C., P.D. Taylor and M.R. Gross. 1987. Parental care and the evolution of egg size in fishes. American Naturalist **129**:32–46.

Sarukhán, J. and J.L. Harper. 1973. Studies on plant demography: *Ranunculus repens* L., *R. Bulbosus* L., and *R. acris* L. I. Population flux and survivorship. Journal of Ecology **61**:675–716.

Saudray, Y. 1954. Utilisation des reserves lipidiques au cours de la ponte et due developpement embryonnaire chez deux Crustaces: *Ligia oceanica* Fab. et *Homarus vulgaris* Edw. Comptes Rendus des Seances de la Societe de Biologie et de ses Filiales **148**:814–816.

Sauer, J.R. and N.A. Slade. 1987. Size-based demography of vertebrates. Annual Review of Ecology and Systematics **18**:71–90.

Schaal, B. 1980. Reproductive capacity and seed size in *Lupinus texensis*. American Journal of Botany **67**:703–709.

Schaal, B.A. and W.J. Leverich 1984. Age-specific fitness components in plants: genotype and phenotype. pp. 173–182. *In* K. Wöhrmann and V. Loeschke (editors), "Population Biology and Evolution," Springer-Verlag, Berlin.

Schaffer, W.M. 1983. The application of optimal control theory to the general life history problem. American Naturalist **121**:418–431.

Schaffer, W.M. 1974a. Selection for optimal life histories: the effects of age structure. Ecology **55**:291–303.

Schaffer, W.M. 1974b. Optimal reproductive effort in fluctuating environments. American Naturalist **108**:783–790.

Schaffer, W.M. 1979a. Equivalence of maximizing reproductive value and fitness in the case of reproductive strategies. Proceedings of the National Academy of Sciences, USA **76**:3567–3569.

Schaffer, W.M. 1979b. The theory of life-history evolution and its application to Atlantic Salmon. Symposium of the Zoological Society of London **44**:307–326.

Schaffer, W.M. 1981. On reproductive value and fitness. Ecology 62:1683–1685.

Schaffer, W.M. and M.L. Rosenzweig. 1977. Selection for optimal life histories II. Multiple equilibria and the evolution of alternative reproductive strategies. Ecology 58:60–72.

Schaffer, W.M. and P.F. Elson. 1975. The adaptive significance of variations in life history among local populations of Atlantic Salmon in North America. Ecology 56:577–590.

Schall, J.J. 1978. Reproductive strategies in sympatric whiptail lizards (Cnemidorphorus): two parthenogenetic and three bisexual species. Copeia 1978:108–116.

Schaller, G.B. 1985. The Giant Pandas of Wolong. University of Chicago Press, Chicago.

Scheiner, S.M. and R.F. Lyman. 1989. The genetics of phenotypic plasticity 1. Heritability. Journal of Evolutionary Biology 2::95–107.

Scheiner, S.M., R.L. Caplan and R.F. Lyman. 1989. A search for trade-offs among life history traits in Drosophila melanogaster. Evolutionary Ecology 3:51–63.

Schiering, J. E. 1977. Stabilizing selection for size as related to mating fitness in Tetraopes. Evolution 31:447–449.

Schifferli. L. 1973. The effect of egg weight on the subsequent growth of Great Tits Parus major. Ibis 115:549–558.

Schifferli, L. 1978. Experimental modification of brood size among House Sparrows Passer domesticus. Ibis 120:365–369.

Schlichting, C.D. 1986. The evolution of phenotypic plasticity in plants. Annual Review of Ecology and Systematics 17:667–693.

Schmale, M.C. 1981. Sexual selection and reproductive success in males of the bicolour damselfish, Eupomacentrus partitus. Animal Behaviour 29:1172–1184.

Schmidt-Nielsen, K. 1984. Scaling: Why is Animal Size so Important? Cambridge University Press, Cambridge.

Schoener, T. 1971. Theory of feeding strategies. Annual Review of Ecology and Systematics 2:369–403.

Schoener, T.W. and A. Schoener. 1978. Estimating and interpreting body-size growth in some Anolis lizards. Copeia 1978:390–405.

Schwaegerle, K.E. 1991. Quantitative genetics of fitness traits in a wild population of Phlox. Evolution 45:169–177.

Schwaegerle, K.E. and D.A. Levin. 1990. Quantitative genetics of seed size variation in Phlox. Evolutionary Ecology 4:143–148.

Scott, D.P. 1962. Effect of food quantity on fecundity of rainbow trout Salmo gairdneri. Journal of the Fisheries Research Board of Canada 19:715–731.

Scott, W.B. and E.J. Crossman. 1973. Freshwater fishes of Canada. Fisheries Research Board of Canada Bulletin 184:1–966.

Scribner, J.M. and F. Slansky Jr. 1981. The nutritional ecology of immature insects. Annual Review of Entomology 26:183–211.

Seed, R.A. and R.A. Brown. 1978. Growth as a strategy of survival in two marine bivalves, *Cerastoderma* and *Modiolus modiolus*. Journal of Animal Ecology **47**:283–292.

Seger, J. and H.J. Brockman. 1987. What is bet-hedging? Oxford Surveys in Evolutionary Biology **4**:182–211.

Seigel, R.A. and H.S. Fitch. 1984. Ecological patterns of relative clutch mass in snakes. Oecologia **61**:293–301.

Seigel, R.A., H.S. Fitch and N.B. Ford. 1986. Variation in relative clutch mass in snakes among and within species. Herpetologica **42**:179–185.

Seigel, R.A., M.M. Huggins and N.B. Ford. 1987. Reduction in locomotor ability as a cost of reproduction in gravid snakes. Oecologia **73**:481–485.

Selcer, K.W. 1990. Egg-size relationships in a lizard with fixed clutch size: variation in a population of the Mediterranean gecko. Herpetologica **46**:15–21.

Semlitsch, R.D. 1985. Reproductive strategy of a facultatively paedomorphic salamander *Ambystoma talpoideum*. Oecologia **65**:305–313.

Semlitsch, R.D. and J.W. Gibbons. 1990. Effects of egg size on success of larval salamanders in complex aquatic environments. Ecology **71**:1789–1795.

Service, P. 1984. Genotypic interaction in an aphid-host plant relationship, *Uroleucon rudbeckiae* and *Rudbeckia laciniata*. Oecologia **61**:271–276.

Service, P.M. and M.R. Rose. 1985. Genetic covariation among life-history components: the effects of novel environments. Evolution **39**:943–945.

Sewell, D., B. Burnet and K. Connolly. 1975. Genetic analysis of larval feeding behaviour in *Drosophila melanogaster*. Genetical Research **24**:163–173.

Seymour, R. 1974. Convective and evaporative cooling in sawfly larvae. Journal of Insect Physiology **20**:2447–2457.

Shafi, M. and P.S. Maitland. 1971. The age and growth of perch (*Perca fluviatilis* L.) in two Scottish lochs. Journal of Fish Biology **3**:39–57.

Shaw, R.G. 1986. Response to density in a wild population of the perennial herb *Salvia lyrata*: variation among families. Evolution **40**:492–505.

Sheridan, A.K. and J.S.F. Barker. 1974. Two-trait selection and the genetic correlation. II. Changes in the genetic correlation during two-trait selection. Australian Journal of Biological Sciences **27**:89–101.

Shine, R. 1978. Propagule size and parental care: the "safe harbor" hypothesis. Journal of Theoretical Biology **75**:417–424.

Shine, R. 1980. Costs' of reproduction in reptiles. Oecologia **46**:92–100.

Shine, R. 1988. Constraints on reproductive investment: a comparison between aquatic and terrestrial snakes. Evolution **42**:17–27.

Shine, R. 1989. Alternative models for the evolution of offspring size. American Naturalist **134**:311–317.

Shine, R. and J.J. Bull. 1979. The evolution of live-bearing in lizards and snakes. American Naturalist **113**:905–923.

Shorrocks, B. 1970. Population fluctuations in the fruitfly (*Drosophila melanogaster*) maintained in the laboratory. Journal of Animal Ecology **34**:229–253.

Sibly, R. and K. Monk. 1987. A theory of grasshopper life cycles. Oikos **48**:186–194.

Sibly, R. and P. Calow. 1983. An integrated approach to life-cycle evolution using selective landscapes. Journal of Theoretical Biology **102**:527–547.

Sibly, R. and P. Calow. 1985. Classification of habitats by selection pressures: a synthesis of life cycle and r/K theory. pp. 75–90, *In* R.M. Sibly and R.H. Smith (editors), "Behavioural Ecology," Blackwell, Oxford.

Sibly, R., P. Calow and R.H. Smith. 1988. Optimal size of seasonal breeders. Journal of Theoretical Biology **133**:13–21.

Sih, A. 1980. Optimal behavior: can foragers balance two conflicting demands? Science **210**:1041–1043.

Sih, A. 1982. Foraging strategies and the avoidance of predation by an aquatic insect, *Notonecta hoffmanni*. Ecology **63**:786–796.

Sih, A., J. Krupa and S. Travers. 1990. An experimental study on the effects of predation risk and feeding regime on the mating behavior of the water strider. American Naturalist **135**:284–290.

Siler, W. 1979. A competing-risk model for animal mortality. Ecology **60**:750–757.

Silvertown, J.W. 1984. Phenotypic variety in seed germination behavior: the ontogeny and evolution of somatic polymorphism in seeds. American Naturalist **124**:1–16.

Simmons, L.W. 1986. Intermale competition and mating success in the field cricket, *Gryllus bimaculatus* (De Geer). Animal Behaviour **34**:567–579.

Simmons, L.W. 1988. The calling song of the field cricket, *Gryllus bimaculatus* (De Geer): constraints on transmission and its role in intermale competition and female choice. Animal Behaviour **36**:380–394.

Simmons, M.J., C.R. Preston and W.R. Engels. 1980. Pleiotropic effects on fitness of mutations affecting viability in *Drosophila melanogaster*. Genetics **94**:467–475.

Simmons, R. 1986. Food provisioning, nestling growth and experimental manipulation of brood size in the African redbreasted sparrowhawk *Accipiter rufiventris*. Ornis Scandinavica **17**:31–40.

Sinervo, B. 1990. The evolution of maternal investment in lizards: an experimental and comparative analysis of egg size and its effects on offspring performance. Evolution **44**:279–294.

Sinervo, B. and L.R. McEdward. 1988. Developmental consequences of an evolutionary change in egg size: an experimental test. Evolution **42**:885–899.

Sinervo, B. and R.B. Huey. 1990. Allometric engineering: an experimental test of the causes of interpopulational differences in performance. Science **248**:1106–1109.

Singh, S.M. and E. Zouros. 1978. Genetic variation associated with growth rate in the American oyster (*Crassostrea virginica*). Evolution **32**:342–353.

Skinner, S.W. 1985. Clutch size as an optimal foraging problem for insects. Behavioral Ecology and Sociobiology **17**:231–238.

Skogland, T. 1989. Natural selection of wild reindeer life history traits by food limitation and predation. Oikos **55**:101–110.

Skoglund, W.C., K.C. Seeger and A.T. Ringrose. 1952. Growth of broiler chicks hatched from various sized eggs when reared in competition with each other. Poultry Science **31**:796–799.

Slagsvold, T. 1982a. Clutch Size, nest size, and hatching asynchrony in birds: experiments with the fieldfare (*Turdus pilaris*). Ecology **63**:1389–1399.

Slagsvold, T. 1982b. Clutch size variation in passerine birds: the nest predation hypothesis. Oecologia **54**:159–169.

Slagsvold, T. 1984. Clutch size variation in birds in relation to nest predation: on the costs of reproduction. Journal of Animal Ecology **53**:945–953.

Slagsvold, T. 1989a. Experiments on clutch size and nest size in passerine birds. Oecologia **80**:297–302.

Slagsvold, T. 1989b. On the evolution of clutch and nest size in passerine birds. Oecologia **179**:300–305.

Slatkin, M. 1974. Hedging one's evolutionary bets. Nature **250**:704–705.

Slobodkin, L.B. 1966. Growth and Regulation of Animal Populations. Holt, Reinhart, and Winston, New York.

Slonaker, J.R. 1924. The effect of copulation, pregnancy, pseudo-pregnancy and lactation on the voluntary activity and food consumption of the albino rat. American Journal of Physiology **71**:362.

Smidt, E.L.B. 1969. The greenland halibut, *Reinhardtius hippoglossoides* (Walb.), biology and exploitation in Greenland waters. Meddelelser fra Danmarks Fiskeri- og Havundersógelser **6**:79–148.

Smith, A.P. and T.P. Young. 1982. The cost of reproduction in *Senecio keniodendron*, a giant rosette species of Mt Kenya. Oecologia **55**:243–247.

Smith, C.C. and S.D. Fretwell. 1974. The optimal balance between size and number of offspring. American Naturalist **108**:499–506.

Smith, F.E. 1963. Population dynamics in *Daphnia magna* and a new model for population growth. Ecology **44**:651–663.

Smith, H.G., H. Källander and G.A. Nilsson. 1987. Effects of experimentally altered brood size on frequency and timing of second clutches in the great tit. Auk **104**:700–706.

Smith, H.G., H. Källander and J.A. Nilsson. 1989. The trade-off between offspring number and quality in the great tit *Parus major*. Journal of Animal Ecology **58**:383–401.

Smith, J.N.M. 1981. Does high fecundity reduce survival in song sparrows. Evolution **35**:1142–1148.

Smith, J.N.M. and D.A. Roff. 1980. Temporal spacing of broods, brood size, and parental care in song sparrows (*Melospiza melodia*). Canadian Journal of Zoology **58**:1007–1015.

Smith, R.H. and C.M. Lessells 1985. Oviposition, ovicide and larval competition in granivorous insects. pp. 423–448, *In* R.M. and R.H. Smith (editors), "Behavioural Ecology: Ecological Consequences of Adaptive Behaviour," Balckwell, Oxford.

Smith, R.H. and M.H. Bass. 1972. Relation of artificial pod removal to soybean yields. Journal of Economic Entomology **65**:606–608.

Smith, W. and J.J. McManus. 1975. The effects of litter size on the bioenergetics and water requirements of lactating *Mus musculus*. Comparative Biochemistry and Physiology **51**:111–115.

Snell, T.W. 1977. Clonal selection: competition among clones. Archiv fuer Hydrobiologie **8**:202–204.

Snell, T.W. 1978. Fecundity, development time, and population growth rate. Oecologia **32**:119–125.

Snell, T.W. and C.E. King. 1977. Lifespan and fecundity patterns in rotifers: the cost of reproduction. Evolution **31**:882–890.

Snell, T.W. and D.G. Burch. 1975. The effects of density on resource partitioning in *Chamaesyce hirta* (Euphorbiaceae). Ecology **56**:742–746.

Snyder, R.J. 1990. Clutch size of anadromous and freshwater threespine sticklebacks: a reassessment. Canadian Journal of Zoology **68**:2027–2030.

Snyder, R.J. 1991a. Migration and life histories of the threespine stickleback: evidence for adaptive variation in growth rate between populations. Environmental Biology of Fishes **31**:381–388.

Snyder, R.J. 1991b. Quantitative genetic analysis of life histories in two freshwater populations of the threespine stickleback. Copeia **1991**:526–529.

Snyder, R.J. and H. Dingle. 1989. Adaptive, genetically-based differences in life history between estuary and freshwater threespine sticklebacks (*Gasterosteus aculeatus*). Canadian Journal of Zoology **67**:2448–2454.

Snyder, R.J. and H. Dingle. 1990. Effects of freshwater and marine overwintering environments on life histories of threespine sticklebacks: evidence for adaptive variation between anadromous and resident freshwater populations. Oecologia **84**:386–390.

Sohn, J.J. and D. Policansky. 1977. The costs of reproduction in the mayapple *Podophyllum peltatus* (Berberidaceae). Ecology **58**:742–746.

Sokal, R.R. 1970. Senescence and genetic load: evidence from *Tribolium*. Science **167**:1733–1734.

Sokal, R.R. and I. Karten. 1964. Competition among genotypes in *Tribolium casteneum* at varying densities and gene frequencies. Genetics **49**:195–211.

Solbreck, C., R. Olsson, D.B. Anderson and J. Förare. 1989. Size, life history and responses to food shortage in the two geographical strains of a seed bug *Lygaeus equestris*. Oikos **55**:387–396.

Solbrig, O.T. and B.B. Simpson. 1974. Components of regulation of dandelions in Michigan. Journal of Ecology **62**:473–486.

Soler, M. 1988. Egg size variation in the Jackdaw *Corvus monedula* in Granada, Spain. Bird Study **35**:69–76.

Soliman, M.H. 1972. Correlated response to natural selection in laboratory populations of *Tribolium casteneum*. Canadian Journal of Genetics and Cytology **14**:971–978.

Soliman, M.H. 1982. Directional and stabilizing selection for developmental time and correlated response in reproductive fitness in *Tribolium casteneum*. Theoretical and Applied Genetics **63**:111–116.

Soller, M., T. Brody, Y. Eitan, T. Agursky and C. Wexler. 1984. Effect of diet and early quantitative feed restriction on the minimum weight requirement for onset of sexual maturity in White Rock broiler breeds. Poultry Science **63**:2103–2113.

Soper, R.S., G.E. Shewell and D. Tyrrell. 1976. *Colidonamyia auditrix* Nov. sp. (Diptera: Sarcophagidae), a parasite which is attracted by the mating song of its host, *Okanagana rimosa* (Homoptera: Cicadae). Canadian Entomologist **108**:61–68.

Soulé, M.E. 1979. Heterozygosity and developmental stability: another look. Evolution **33**:396–401.

Soulé, M.E. 1982. Allometric variation. I. The theory and some consequences. American Naturalist **120**:751–764.

Southwood, T.R.E. 1962. Migration of terrestrial arthropods in relation to habitat. Biological Reviews **37**:171–214.

Sparholt, H. 1985. The population, survival, growth, reproduction and food of arctic char. *Salvelinus alpinus* (L.), in four unexploited lakes in Greenland. Journal of Fish Biology **26**313–330.

Spencer, H.G. and R.W. Marks. 1988. The maintenance of single-locus polymorphism. 1. Numerical studies of a viability selection model. Genetics **120**:605–613.

Spight, T.M. and J. Emlen. 1976. Clutch size of two marine snails with a changing food supply. Ecology **57**:1162–1178.

Spinage, C.A. 1972. African ungulate life tables. Ecology **53**:645–652.

Stamp, N.E. 1980. Egg deposition patterns in butterflies: why do some species cluster their eggs rather than deposit them singly? American Naturalist **115**:367–380.

Stamp, N.E. and J.R. Lucas. 1983. Ecological correlates of explosive seed dispersal. Oecologia **59**:272–278.

Stanton, M.L. 1984. Seed variation in wild radish: effect of seed size on components of seedling and adult fitness. Ecology **65**:1105–1112.

Stearns, S.C. 1976. Life-history tactics: a review of the ideas. Quarterly Review of Biology **51**:3–46.

Stearns, S.C. 1977. The evolution of life history traits: a critique of the theory and a review of the data. Annual Review of Ecology and Systematics **8**:145–171.

Stearns, S.C. 1980. A new view of life-history evolution. Oikos 35:266–281.

Stearns, S.C. 1983. The influence of size and phylogeny on patterns of covariation among life-history traits in the mammals. Oikos 41:173–187.

Stearns, S.C. 1984. The effects of size and phylogeny on patterns of covariation in the life history traits of lizards and snakes. American Naturalist 123:56–72.

Stearns, S.C. 1989. The evolutionary significance of phenotypic plasticity. BioScience 39:436–445.

Stearns, S.C., G. de Jong and B. Newman. 1991. The effects of phenotypic plasticity on genetic correlations. Trends in Ecology and Evolution 6:122–126.

Stearns, S.C. and J.C. Koella. 1986. The evolution of phenotypic plasticity in life-history traits: predictions of reaction norms for age and size at maturity. Evolution 40:893–913.

Stearns, S.C. and P. Schmid-Hempel. 1987. Evolutionary insights should not be wasted. Oikos 49:118–125.

Stearns, S.C. and R.E. Crandall. 1981. Quantitative predictions of delayed maturity. Evolution 35:455–463.

Steele, D.H. 1977. Correlation between egg size and development period. American Naturalist 111:371–372.

Steinwascher, K. 1984. Egg size variation in *Aedes aegypti*: relationship to body size and other variables. American Midland Naturalist 112:76–84.

Stenhouse, S.L., N.G. Hairston and A.E. Cobey. 1983. Predation and competition in *Ambystoma* larvae: field and laboratory experiments. Journal of Herpetology 17:210–220.

Stenseth, N.C. 1984. Optimal reproductive success in animals with parental care. Oikos 43:251–253.

Steveninck, R.F.M. van. 1957. Factors affecting the abscission of reproductive organs in yellow lupins (*Lupinus luteus* L.) Part 1. The effect of different patterns of flower removal. Journal of Experimental Botany 8:373–381.

Stevens, E.D. and R.K. Josephson. 1977. Metabolic rate and body temperature in singing katydids. Physiological Zoology 50:31–42.

Stewart, R.E.A. 1986. Energetics of age-specific reproductive effort in female harp seals, *Phoca groenlandica*. Journal of Zoology A208:503–517.

Stinson, C.H. 1979. On the selective advantage of fratricide in raptors. Evolution 33:1219–1225.

Strathmann, R.R. 1977. Egg size, larval development, and juvenile size in benthic marine invertebrates. American Naturalist 111:373–376.

Streams, F.A. and T.P. Shubeck. 1982. Spatial structure and intraspecific interactions in *Notonecta* populations. Environmental Entomology 11:652–659.

Strong, D. 1972. Life history variation among the populations of an amphipod. Ecology 53:1103–1111.

Sue, K., D.N. Ferro and R.M. Emberson. 1980. Life history and seasonal ovarian development of *Sitona humeralis* (Coleoptera: Curculionidae) in New Zealand. New Zealand Entomologist 7:165–169.

Sullivan. T.P. 1976. Demography and dispersal in island and mainland populations of the deer mouse. *Peromyscus maniculatus*. M.Sc. Thesis, University of British Columbia, Vancouver, Canada.

Suntzeff, F., E. Cowdry and P.B. Hixon. 1962. Influence of maternal age on offspring in mice. Journal of Gerontology **17**:2–7.

Sutherland, W.J., A. Grafen and P.H. Harvey. 1986. Life history correlations and demography. Nature **320**:88.

Svendsen, G. 1964. Comparative reproduction and development in two species of mice in the genus *Peromyscus*. Transactions of the Kansas Academy of Science **67**:527–538.

Svensson, I. 1988. Reproductive costs in two sex-role reversed pipefish species (Syngnathidae). Journal of Animal Ecology **57**:929–942.

Swihart. R.K. 1984. Body size, breeding season length, and life history tactics of lagomorphs. Oikos **43**:282–290.

Swingland, I.R. 1977. Reproductive effort and life history strategy of the Aldabran giant tortoise. Nature **269**:402–404.

Swingland, I.R. and M. Coe. 1978. The natural regulation of Giant tortoise populations on Aldabra Atoll. Reproduction. Journal of Zoology **186**:285–309.

Taborsky, M., B. Hudde and P. Wirtz. 1987. Reproductive behaviour and ecology of *Symphodus* (*Crenilabrus*) *ocellatus*, a European wrasse with four types of male behaviour. Behaviour **102**:82–118.

Tachida, H. and T. Mukai. 1985. The genetic structure of natural populations of *Drosophila melanogaster*. XIX. Genotype-environment interaction in viability. Genetics **111**:43–55.

Taigen, T.L. and K.D. Wells. 1985. Energetics of vocalization by an anuran amphibian (*Hyla versicolor*). Journal of Comparative Physiology **155**:163–170.

Tait, D.E.N. 1980. Abandonment as a reproductive tactic—the example of grizzly bears. American Naturalist **115**:800–808.

Takahashi, H. and H. Iwasawa. 1988a. Interpopulation variations in clutch size and egg size in the Japanese salamander, *Hynobius nigrescens*. Zoological Science (Tokyo) **5**:1073–1081.

Takahashi, H. and H. Iwasawa. 1988b. Intraclutch egg size variability in *Hynobius nigrescens* and *Hynobius lichenatus* (Urodela: Hynobiidae). Science reports of Niigata University, Series D (Biology) **25**:19–29.

Takahashi, H. and H. Iwasawa. 1989. Clutch size and egg size variations in salamanders: traits of the variations in some levels of biological organization. Current Herpetology in East Asia 282–291.

Takano, T., S. Kusakabe and T. Mukai. 1987. The genetic structure of natural populations of *Drosophila melanogaster*. XX. Comparison of genotype-environment interaction in viability between a northern and southern population. Genetics **117**:245–254.

Tallamy, D.W. 1982. Age specific maternal defense in *Gargaphia solani* (Hemiptera: Tinigidae). Behavioral Ecology and Sociobiology **11**:7–11.

Tallamy, D.W. and R.F. Denno. 1982. Life history trade-offs in *Gargaphia solani* (Hemiptera: Tingidae): the cost of reproduction. Ecology **63**:616–620.

Tallis, G.M. 1959. Sampling errors of genetic correlation coefficients calculated from the analyses of variance and covariance. Australian Journal of Statistics **1**:35–43.

Tanaka, T., K. Tokuda and S. Kotero. 1970. Effects of infant loss on the interbirth interval of Japanese monkeys. Primates **11**:113–117.

Tantawy, A.O. and F.A. Rakha. 1964. Studies on natural populations of *Drosophila*. IV. Genetic variances of and correlations between four characters in *D. melanogaster* and *D. simulans*. Genetics **50**:1349–1355.

Tantawy, A.O. and M.R. El-Helw. 1966. Studies on natural populations of *Drosophila*. V. Correlated response to selection in *Drosophila melanogaster*. Genetics **53**:97–110.

Tarburton, M.K. 1987. An experimental manipulation of clutch and brood size of white-rumped swiftlets: *Aerodramus spodiopygius* of Fiji. Ibis **129**:107–114.

Taylor, C.E. 1976. Genetic variation in heterogeneous environments. Genetics **83**:887–894.

Taylor, H.M., R.S. Gourby, C.E. Lawrence and R.S. Kaplan. 1974. Natural selection of life history attributes: an analytical approach. Theoretical Population Biology **5**:104–122.

Taylor, T.H.C. 1937. The Biological Control of an Insect in Fiji. An Account of the Coconut Leaf-Mining Beetle and Its Parasite Complex. Richard Clay and Sons, Ltd., Suffolk.

Tedin, O. 1925. Verebung, variation, und systematik in der gatung *Camelina*. Hereditas **6**:275–386.

Telfer, W.H. and L.D. Rutberg. 1960. The effects of blood protein depletion on the growth of the oocytes in the cecropia moth. Biological Bulletin **118**:352–366.

Temin, R.G. 1966. Homozygous viability and fertility loads in *Drosophila melanogaster*. Genetics **53**:27–56.

Tessier, A., L.L. Henry, C.E. Goulden and M.W. Durand. 1983. Starvation in *Daphnia*: energy reserves and reproductive allocation. Limnology and Oceanography **28**:667–676.

Tessier, A.J. and N.L. Consolatti. 1989. Variation in offspring size in *Daphnia* and consequences for individual fitness. Oikos **56**:269–276.

Thoday, J.M. 1953. Components of fitness. Symposium of the Society for Experimental Biology **7**:96–113.

Thompson, J.N. 1984. Variation among individual seed masses in *Lomatium grayi* (Umbelliferae) under controlled conditions: magnitude and partitioning of the variance. Ecology **65**:626–631.

Thompson, K. and A.J.A. Stewart. 1981. The measurement and meaning of reproductive effort in plants. American Naturalist **117**:205–211.

Thompson, P.A. 1981. Variation in seed size within populations of *Silene dioica* (L.) Claiv. in relation to habitat. Annals of Botany **47**:623–634.

Thompson, S. 1986. Male spawning success and female choice in the mottled triplefin *Forsterygion varium* (Pisces: Triptergiidae). Animal Behaviour **34**:580–589.

Thornhill, R. 1975. Scorpionflies as kleptoparasites of web-building spiders. Nature **253**:709–711.

Thornhill, R. 1978. Some arthropod predators and parasites of adult mecoptera. Environmental Entomology **7**:714–716.

Thornhill, R. 1979a. Male and female sexual selection and the evolution of mating strategies in insects. pp. 81–121, *In* M.S. Blum and N.A. Blum (editors), "Sexual Selection and Reproductive Competition in Insects," Academic Press, New York.

Thornhill, R. 1979b. Male pair-formation pheromones in *Panorpa* scorpionflies (Mecoptera: Panorpidae). Environmental Entomology **8**:886–889.

Thornhill, R. 1979c. Adaptive female-mimicking behavior in a scorpionfly. Science **205**:412–414.

Thornhill, R. 1980a. Sexual selection within mating swarms of the lovebug, *Plecia nearctica* (Diptera: Bibionidae). Animal Behaviour **28**:405–412.

Thornhill, R. 1980b. Mate choice in *Hylobittacus apicalis* (Insecta: Mecoptera) and its relation to some models of female choice. Evolution **34**:519–538.

Thornhill, R. 1980c. Competition and coexistence among *Panorpa* scorpionflies (Mecoptera: Panorpidae). Ecological Monographs **50**:179–197.

Thornhill, R. 1980d. Rape in *Panorpa* scorpionflies and a general rape hypothesis. ANimal Behaviour **28**:52–59.

Thornhill, R. 1981a. Alternative hypotheses for traits presumed to have evolved by sperm competition. pp. 151–178, *In* R.L. Smith (editor), "Sperm Competition and the Evolution of Animal Mating Systems," Academic Press, New York.

Thornhill, R. 1981b. *Panorpa* (Mecoptera: Panorpidae) scorpionflies: systems for understanding resource-defense polygyny and alternative male reproductive efforts. Annual Review of Ecology and Systematics **12**:355–386.

Thornhill, R. and Alcock, J. 1983. The Evolution of Insect Mating Systems. Harvard University Press, Cambridge.

Thorpe, J.E. and K.A. Mitchell. 1981. Stocks of Atlantic salmon (*Salmo salar*) in Britain and Ireland: discreteness and current management. Canadian Journal of Fisheries and Aquatic Sciences **38**:1576–1580.

Tiainen, J., I.K. Hanski, T. Pakkala, J. Piiroinen and R. Yrjola. 1989. Clutch size, nestling growth and nestling mortality of the starling *Sturnus vulgaris* in south Finnish agroenvironments. Ornis Fennica **66**:41–48.

Tilley, S.G. 1968. Size-fecundity relationships in five desmognathine salamanders. Evolution **22**:806–816.

Tilley, S.G. 1972. Aspects of parental care and embryonic development in *Desmognathus ochrophaeus*. Copeia **1972**:532–540.

Tinbergen, J.M. 1987. Costs of reproduction in the great tit: intraseasonal costs associated with brood size. Ardea **75**:111–122.

Tinbergen, J.M., J.H. van Balen and H.M. van Eck. 1985. Density-dependent survival in an isolated great tit population: Kluyvers data reanalyzed. Ardea **73**:38–48.

Tinbergen, J.M. and M.C. Boerlijst. 1990. Nestling weight and survival in individual great tits (*Parus major*). Journal of Animal Ecology **59**:1113–1127.

Tinkle, D.W. and R.E. Ballinger. 1972. *Sceloporus undulatus*: a study of the intraspecific comparative demography of a lizard. Ecology **53**:570–584.

Tinkle, D.W. 1967. The life and demography of the side-blotched lizard. Miscellaneous Publications of the Museum of Zoology, University of Michigan **132**:1–182.

Tinkle, D.W. 1972. The dynamics of a Utah population of *Sceloporus undulatus*. Herpetologica **28**:351–359.

Tinkle, D.W. 1973. A population analysis of the sagebush lizard, *Sceloporus graciosus*, in southern Utah. Copeia **1973**:284–295.

Tinkle, D.W. and J.W. Gibbons. 1977. The distribution and evolution of viviparity in reptiles. Miscellaneous Publications of the Museum of Zoology, University of Michigan **154**:1–55.

Tinkle, D.W. and N.F. Hadley. 1973. Reproductive effort and winter activity in the viviparous montane lizard *Sceloporus jarrovi*. Copeia **1973**:272–276.

Tinkle, D.W. and N.F. Hadley. 1975. Lizard reproductive effort: calorie estimates and comments on its evolution. Ecology **56**:427–434.

Tomovic, R. 1963. Sensitivity Analysis of Dynamic Systems. McGraw-Hill, New York.

Townsend, D.S. 1986. The costs of male parental care and its evolution in a neotropical frog. Behavioral Ecology and Sociobiology **19**:187–195.

Townsend, D.S., M.M. Stewart and F.H. Pough. 1984. Male parental care and its adaptive significance in a neotropical frog. Animal Behaviour **32**:421–431.

Tracy, C.R. 1977. Minimum size of mammalian homeotherms: role of the thermal environment. Science **198**:1034–1035.

Trail, P.W. 1987. Predation and antipredator behaviour at Guianan Cock-of-the-Rock leks. Auk **104**:496–507.

Travis, J. 1983a. Variation in development patterns of larval anurans in temporary ponds 1. Persistent variations within a *Hyla gratiosa* population. Evolution **37**:496–512.

Travis, J. 1983b. Variation in growth and survival of *Hyla gratiosa* larvae in experimental enclosures. Copeia **1983**:232–237.

Travis, J., S.B. Emerson and M. Blouin. 1987. A quantitative genetic analysis of larval life-history traits in *Hyla crucifer*. Evolution **41**:145–156.

Trehan, P.K., D.S. Dhir and B. Singh. 1983. Inheritance of sex ratio and its relationship with some reproductive traits in egg-type chickens. Indian Journal of Animal Science **53**:220–221.

Trivers, R.L. 1972. Parental investment and sexual selection. pp. 136–179, *In* B.G. Campbell (editor), "Sexual Selection and the Descent of Man 1871–1971," Aldine, Chicago.

Tucker, J.K., R.S. Funk and G.L. Paukstis. 1978. The adaptive significance of egg morphology in two turtles (*Chrysemys picta* and *Terrapene carolina*). Bulletin of the Maryland Herpetological Society **14**:10–22.

Tuomi, J., T. Hakala and E. Haukioja. 1983. Alternative concepts of reproductive effort, costs of reproduction, and selection in life-history evolution. American Zoologist **23**:25–34.

Turelli, M. 1977. Random environments and stochastic calculus. Theoretical Population Biology **12**:140–178.

Turelli, M. 1984. Heritable genetic variation via mutation-selection balance: Lerch's zeta meets the abdominal bristle. Theoretical Population Biology **25**:138–193.

Turelli, M. 1985. Effects of pleiotropy on preditions concerning mutation-selection balance for polygenic traits. Genetics **111**:163–195.

Turelli, M. 1988. Phenotypic evolution, constant covariance, and the maintenance of additive variance. Evolution **42**:1342–1347.

Turelli, M. and D. Petry. 1980. Density-dependent selection in a random environment: an evolutionary process that can maintain stable population dynamics. Proceedings of the National Academy of Science (US) **77**:7501–7505.

Turnbull, A.L. 1962. Quantitative studies of the food of *Linyphia triangularis* Clerck (Araneae: Linyphiidae). Canadian Entomologist **94**:1233–1249.

Turner, J.H. Jr., J.S. Worley, J.H.H. Ramey, P.E. Hoskinson and J.M. Stewart. 1979. Relationship of week of flowering and parameters of boll yield in cotton. Agronomy Journal **71**:248–251.

Tuttle, M.D. and M.J. Ryan. 1981. Bat predation and the evolution of frog vocalizations in the neotropics. Science **214**:677–678.

Tyndale-Biscoe, M. and R.D. Hughes. 1968. Change in the female reproductive system as age indicators in the bushfly *Musca retustissima* Wlk. Bulletin of Entomological Research **59**:129–141.

Underwood, A.J. 1974. On models for reproductive strategy in marine benthic invertebrates. American Naturalist **108**:874–878.

Vadas, R.L. 1977. Preferential feeding: an optimization strategy in sea urchins. Ecological Monographs **47**:337–371.

Vahl, O. 1981. Age specific residual reproductive value and reproductive effort in the Iceland scallop, *Chlamys islandica* (D.F. Muller). Oecologia **51**:53–56.

Väisänen, R.A., O. Hildén, M. Soikkeli and S. Vuolanto. 1972. Egg dimension variation in five wader species: the role of heredity. Ornis Fennica **49**:25–44.

Van Damme, R., D. Bauwens and R.F. Verheyen. 1989. Effect of relative clutch mass on sprint speed in the lizard *Lacerta vivipara*. Journal of Herpetology **23**:459–461.

Van Delden, W. 1982. The alcohol dehydrogenase polymorphism in *Drosophila melanogaster*. Selection at an enzyme locus. Evolutionary Biology **15**:187–222.

Van Dijk, T.S. 1979. On the relationship between reproduction, age and survival in two carabid beetles: *Calathus melanocephalus* L. and *Pterostichus caerulescens* L. (Coleoptera, Carabidae). Oecologia **40**:63–80.

Van Rhijn, J.G. 1973. Behavioural dimorphism in male ruffs, *Philomachus pugnax* (L.). Behaviour **47**:153–229.

Van Tijen, W.F. and A.R. Kuit. 1970. The heritability of characteristics of egg quality, their mutual correlation and the relationship with productivity. Archiv fuer Gefluegelkunde **34**:201–210.

Van Valen, L. and G.W. Mellin. 1967. Selection in natural populations. 7. New York babies (Fetal life study). Annals of Human Genetics **31**:109–127.

Vance, R.R. 1973a. On reproductive strategies in marine benthic invertebrates. American Naturalist **107**:339–352.

Vance, R.R. 1973b. More on reproductive strategies in marine benthic invertebrates. American Naturalist **107**:353–361.

Vance, R.R. 1973. Reply to Underwood. American Naturalist **108**:879–880.

VanVleck L.D. and C.R. Henderson. 1961. Empirical sampling estimates of genetic correlations. Biometrics **17**:359–371.

Vehrencamp, S.L., J.W. Bradbury and R.M. Gibson. 1989. The energetic cost of display in male sage grouse. Animal Behaviour **38**:885–896.

Vermeer, K. 1963. The breeding ecology of the glaucous-winged gull *Larus glaucescens* on Mandarte Island. Occasional Papers British Columbia Provincial Museum **13**:1–104.

Vermeij, G.J. 1974. Marine faunal dominance and molluscan shell form. Evolution **28**:656–664.

Verner, J. 1963. Song rates and polygamy in the Long-billed Marsh wren. Proceedings of the XIIIth International Ornithological Congress 299–307.

Vetter, E.F. 1988. Estimation of natural mortality in fish stocks: a review. Fishery Bulletin **86**:25–43.

Via, S. 1984. The quantitative genetics of polyphagy in an insect herbivore. 1. Genotype-environment interaction in larval performance on different host plant species. Evolution **38**:881–895.

Via, S. and R. Lande. 1985. Genotype-environment interaction and the evolution of phenotypic plasticity. Evolution **39**:505–522.

Vickers, G.T. and C. Cannings. 1987. On the definition of an evolutionarily stable strategy. Journal of Theoretical Biology **129**:349–353.

Vincint, T.L. and H.R. Pulliam. 1980. Evolution of life history strategies for an asexual annual plant model. Theoretical Population Biology **17**:215–231.

Vincent, T.L. and J.S. Brown. 1988. The evolution of ESS theory. Annual Review of Ecology and Systematics **19**:423–443.

Vinegar, M.B. 1975. Demography of the striped plateau lizard, *Sceloporus virgatus*. Ecology **56**:172–182.

Vitt, L.J. 1981. Lizard reproduction: habitat specificity and constraints on relative clutch mass. American Naturalist **117**:506–514.

Vitt, L.J. 1986. Reproductive tactics of sympatric gekkonid lizards with a comment on the evolutionary and ecological consequences of invariant clutch size. Copeia **1986**:773–786.

Vitt, L.J. and H.J. Price. 1982. Ecological and evolutionary determinants of relative clutch mass in lizards. Herpetologia **38**:237–255.

Vitt, L.J. and J.D. Congdon. 1978. Body shape, reproductive effort, and relative clutch mass in lizards: resolution of a paradox. American Naturalist **112**:595–608.

Vitt, L.J. and R.A. Seigel. 1985. Life history traits of lizards and snakes. American Naturalist **125**:480–484.

Voigt, R.L., C.O. Gardner and O.J. Webster. 1966. Inheritance of seed size in Sorghum, *Sorghum vulgare* Pers. Crop Science **6**:582–586.

Von Foerster, H., M.M. Mora and L.W. Amiot. 1960. Doomsday: Friday, 13 November, A.D. 2026. Science **132**:1291–1295.

Von Foerster, H., P.M. Mora and LW. Amiot. 1961. Doomsday. Science **133**:943–946.

Von Haartman, L. 1954. Der Trauerfliegenschnapper. III. Die Nahrungsbiologie. Acta Zoologica Fennica **83**:1–196.

Von Haartman, L. 1967. Clutch-size in the pied flycatcher. Proceedings of the International Ornithological Congress **14**:155–164.

Vuouren, I., M. Rajasilta and J. Salo. 1983. Selective predation and habitat shift in a copepod species—support for the predation hypothesis. Oecologia **59**:62–64.

Waage, J.K. and H.C.J. Godfray 1985. Reproductive strategies and population ecology of insect parasitoids. pp. 449–470. *In* R.M. Smith and R.H. Smith (editors), "Behavioural Ecology: Ecological Consequences of Adaptive Behaviour," Blackwell, Oxford.

Waage, J.K. and S.M. Ng. 1984. The reproductive strategy of a parasitic wasp. 1. Optimal progeny and sex allocation in *Trichogramma evanescens*. Journal of Animal Ecology **53**:401–416.

Wagner, W.E. Jr. 1989. Fighting, assessment and frequency alteration in Blanchard's cricket frog. Behavioural Ecology and Sociobiology **25**:429–436.

Walker, T.J. 1964. Experimental demonstration of a cat locating orthopteran prey by the prey's call. Florida Entomologist **47**:163–165.

Waller, D.M. 1988. Plant morphology and reproduction. pp. 203–227, *In* J. Lovett Doust and L. Lovett Doust (editors), "Plant Reproductive Ecology: Patterns and Strategies," Oxford University Press, Oxford.

Wallinga, J.H. and H. Bakker. 1978. Effect of long-term selection for litter size in mice on lifetime reproduction rate. Journal of Animal Science **46**:1563–1571.

Walsh, J.B. 1984. Hard lessons from soft selection. American Naturalist **124**:518–526.

Walton, J. 1983. Growth parameters for typical anadromous and dwarf stocks of alewives, *Alosa pseudoharengus* (Pisces, Clupeidae). Environmental Biology of Fishes **9**:277–287.

Wanless, S. 1984. The growth and food of young gannets *Sula bassana* on Aila Craig. Seabird **7**:62–70.

Ward, J.G. 1973. Reproductive success, food supply, and the evolution of clutch size in the glaucous-winged gull. Ph.D. thesis, University of British Columbia, Vancouver.

Ward, P. 1965. The breeding biology of the black-faced dioch *Quelea quelea* in Nigeria. Ibis **107**:326–349.

Ward, P.I. 1989. Mate choice in *Gammarus* (Amphipoda). Journal of Zoology **218**:633–635.

Ware, D.M. 1975a. Growth, metabolism, and optimal swimming speed of a pelagic fish. Journal of the Fisheries Research Board of Canada **32**:33–41.

Ware, D.M. 1975b. Relation between egg size, growth and natural mortality of larval fish. Journal of the Fisheries Research Board of Canada **32**:2503–2512.

Ware, D.M. 1977. Spawning time and egg size of Atlantic mackerel, *Scomber scombrus*, in relation to the plankton. Journal of the Fisheries Research Baord of Canada **34**:2308–2315.

Ware, D.M. 1978. Bioenergetics of pelagic fish: theoretical change in swimming speed and ration with body size. Journal of the Fisheries Research Board of Canada **35**:220–228.

Ware, D.M. 1980. Bioenergetics of stock and recruitment. Canadian Journal of Fisheries and Aquatic Sciences **37**:1012–1024.

Warner, R.R. 1984a. Deferred reproduction as a response to sexual selection in a coral reef fish: a test of the life historical consequences. Evolution **38**:148–162.

Warner, R.R. 1984b. Mating behavior and hermaphroditism in coral reef fish. American Scientist **72**:128–136.

Warner, R.R. and D.R. Robertson. 1978. Sexual patterns in the labroid fishes of the Western Caribbeasn, I: The wrasses (Labridae). Smithsonian Contributions to Zoology **254**:1–27.

Warner, R.R. and I.F. Downs. 1977. Comparative life histories: growth vs reproduction in normal males and sex-changing hermaphrodites of the striped parrotfish, *Scarus croicensis*. Proceedings, Third International Coral Reef Symposium 275–281.

Warner, R.R. and S.G. Hoffman. 1980a. Population density and the economics of territorial defence in a coral reef fish. Ecology **61**:772–780.

Warner, R.R. and S.G. Hoffman. 1980b. Local population size as a determinant of mating system and sexual composition in two tropical reef fishes (*Thalassoma* sp.). Evolution **34**:508–518.

Warren, D.C. 1924. Inheritance of egg size in *Drosophila melanogaster*. Genetics **9**:41–69.

Warshaw, S.J. 1972. Effects of alewives (*Alosa pseudoharengus*) on the zooplankton of Lake Wononskopomac, Connecticut. Limnology and Oceanography **17**:816–825.

Watanabe, T.K. 1969. Persistence of a visible mutant in natural populations of *Drosophila melanogaster*. Japanese Journal of Genetics **44**:15–22.

Watkinson, A. and J. White. 1985. Some life history consequences of modular construction in plants. Philosophical Transactions of the Royal Society of London (B) **313**:31–51.

Wehle, D.H.S. 1983. The food, feeding, and development of young tufted and horned puffins in Alaska. Condor **85**:427–442.

Weiner, J. 1987. Limits to energy budget and tactics in energy investments during reproduction in the Djungarian hamster (*Phodopus sungorus sungorus* Pallas 1770). pp. 79–106. *In* A.S.I. Loudon and P.A. Racey (editors), "Reproductive Energetics in Mammals," Zoological Society of London, Clarendon Press, Oxford.

Weis, I.M. 1982. The effects of propagule size on germination and seedling growth in *Mirabilis hirsuta*. Canadian Journal of Botany **60**:1868–1874.

Weis, A.F., P.W. Price and M. Lynch. 1983. Selective pressures on clutch size in the gall maker *Asteromyia carbonifera*. Ecology **64**:688–695.

Weiser, W., H. Forstner, N. Medgyesy and S. Hinterleitner. 1988. To switch or not to switch: partitioning of energy between growth and activity in larval cyprinids (Cyprinidae: Teleostei). Functional Ecology **2**:499–507.

Weissburg, M. 1986. Risky business: on the ecological relevance of risk-sensitive foraging. Oikos **46**:261–262.

Wells, K.D. 1977a. Territoriality and male mating success in the green frog (*Rana clamitans*). Ecology **58**:750–762.

Wells, K.D. 1977b. The social behaviour of anuran amphibians. Animal Behaviour **25**:666–693.

Wells, K.D. 1978. Territoriality in the green frog (*Rana clamitans*). Vocalizations and agonistic behavior. Animal Behaviour **26**:1051–1063.

Wells, L. 1970. Effects of alewife predation on zooplankton populations in Lake Michigan. Limnology and Oceanography **15**:556–565.

Werner, E.E. 1974. The fish size, prey size and handling time relation in several sunfishes and some implications. Journal of the Fisheries Research Board of Canada **31**:1531–1536.

Werner, E.E. 1986. Amphian metamorphosis: growth rate, predation risk, and the optimal size at transformation. American Naturalist **128**:319–341.

Werner, E.E., J.F. Gilliam, D.J. Hall and G.G. Mittelbach. 1983. An experimental test of the effects of predation on habitat use in fish. Ecology **64**:1540–1548.

Werner, P.A. 1975. Predictions of fate from rosette size in teasel, *Dipsacus fullonum* L. Oecologia **20**:197–201.

Werner, P.A. 1976. Ecology of plant populations in successional environments. Systematic Botanist **1**:246–268.

Werner, P.A. 1979. Competition and coexistence of similar species. pp. 287–310, *In* S.J. and P.H. Raven O.T. Solbrig (editors), "Topics in Plant Population Biology," Columbia University Press, New York.

Werner, P.A. and H. Caswell. 1977. Population rates and age versus stage distribution models for teasel (*Dipsacus sylvestris* Huds.). Ecology **58**:1103–1111.

West-Eberhard, M.J. 1989. Phenotypic plasticity and the origins of diversity. Annual Review of Biology **20**:249–278.

Westcott, B. 1986. Some methods of analyzing genotype-environment interaction. Heredity **56**:243–253.

Western, D. 1979. Size, life history and ecology in mammals. African Journal of Ecology **17**:185–204.

Westoby, M. 1974. An analysis of diet selection by large generalist herbivores. American Naturalist **108**:290–304.

Westoby, M. 1981. How diversified seed germination behaviour is selected. American Naturalist **118**:882–885.

Wheeler, A. 1969. The Fish of the British Isles and North-West Europe. Macmillan, London.

Whitehead, P.J., W.J. Freeland and K. Tschirner. 19909. Early growth of Magpie geese, *Anseranus semipalmata*: sex differences and influence of egg size. Australian Journal of Zoology **38**:249–262.

Whittaker, R.H. and D. Goodman. 1979. Classifying species according to their demographic strategy 1. Population fluctuations and environmental heterogeneity. American Naturalist **113**:185–200.

Wigglesworth, V.B. 1934. The physiology of ecdysis in *Rhodnius prolixus* (Hemiptera). II. Factors controlling moulting and 'metamorphosis'. Quarterly Journal of the Microscopic Society **77**:191–222.

Wiklund, C. and A. Persson. 1983. Fecundity, and the relation of egg weight variation to offspring fitness in the speckled butterfly *Pararge aegeria*, or why don't butterfly females lay more eggs? Oikos **40**:53–63.

Wilbur, H.M. 1977. Propagule size, number and dispersion pattern in *Ambystoma* and *Asclepias*. American Naturalist **111**:43–68.

Wilbur, H.M., D.W. Tinkle and J.P. Collins. 1974. Environmental certainty, trophic level, and resource availability in life history evolution. American Naturalist **108**:805–817.

Wilbur, H.M. and J.P. Collins. 1973. Ecological aspects of amphibian metamorphosis. Science **182**:1305–1314.

Wilburg, H.M., D.I. Rubenstein and L. Fairchild. 1978. Sexual selection in toads: the roles of female choice and male body size. Evolution **32**:264–270.

Williams, G.C. 1966. Natural selection, the costs of reproduction and a refinement of Lack's principle. American Naturalist **100**:687–690.

Williams, J. 1988. Field metabolism of tree swallows during the breeding season. Auk **105**:706–714.

Williams, W.P. 1967. The growth and mortality of four species of fish in the River Thames at Reading. Journal of Animal Ecology **36**:695–720.

Willis, J.H., J.A. Coyne and M. Kirkpatrick. 1991. Can one predict the evolution of quantitative characters without genetics? Evolution **45**:441–444.

Wilson, M.F. 1972. Evolutionary ecology of plants: a review. II. Ecological life histories. The Biologist **54**:148–162.

Wilson, A.M. and K. Thompson. 1989. A comparative study of reproductive allocation in 40 British grasses. Functional Ecology **3**:297–302.

Wilson, D.C. and R.E. Milleman. 1969. Relationships of female age and size to embryo number and size in the shiner perch. *Cymatogaster aggregata*. Journal of the Fisheries Research Board of Canada **26**:2339–2344.

Wilson, D.S. 1988. Holism and reductionism in evolutionary ecology. Oikos **53**:269–273.

Wilson, D.S., M. Leighton and D.R. Leighton. 1978. Interference competition in a tropical ripple bug (Hemiptera: Veliidae). Biotropica **10**:302–306.

Wilson, M.E., T.P. Gordon and I.S. Bernstein. 1978. Timing of births and reproductive success in rhesus monkey social groups. Journal of Medical Primatology **7**:202–212.

Wilson, T.A. 1965. Natural mortality and reproduction for a food supply at minimum metabolism. American Naturalist **99**:373–376.

Winfield, I.J. and C.R. Townsend. 1983. The cost of copepod reproduction: increased susceptibility to fish predation. Oecologia **60**:406–411.

Wing, S.R. 1988. Cost of mating for female insects: risk of predation in *Photinus collustrans* (Coleoptera: Lampyridae). American Naturalist **131**:139–142.

Winkler, D.W. and K. Wallin. 1987. Offspring size and number: a life history model linking effort per offspring and total effort. American Naturalist **129**:708–720.

Winn, A.A. 1988. Ecological and evolutionary consequences of seed size in *Prunella vulgaris*. Ecology **69**:1537–1544.

Winters, G.H. and J.S. Campbell. 1974. Some biological aspects and population parameters of Grand Bank capelin. ICNAF Research Document **74/76**:1–23.

Wissinger, S.A. 1988. Spatial distribution, life history and estimates of survivorship in a fourteen-species assemblage of larval dragonflies (Odonata: Anisoptera). Freshwater Biology **20**:329–340.

Wittenberger, J. 1979. A model for delayed reproduction in iteroparous animals. American Naturalist **114**:439–446.

Wodinsky, J. 1977. Hormonal inhibition of feeding and death in *Octopus*: control by optic gland secretion. Science **198**:948–951.

Wolff, K. and J. Haeck. 1990. Genetic analysis of ecologically relevant morphological variability in *Plantago lanceolata* L. VI. The relation between allozyme heterozygosity and some fitness components. Journal of Evolutionary Biology **3**:243–255.

Woodring, J.P. 1983. Control of moulting in the house cricket, *Acheta domesticus*. Journal of Insect Physiology **29**:461–464.

Woombs, M. and J. Laybourn-Parry. 1984. Growth, reproduction and longevity in nematodes from sewage treatment plants. Oecologia **64**:168–172.

Wootton, R.J. 1973. The effect of size of food ration on egg production in the female three-spined stickleback, *Gasterosteus aculeatus* L. Journal of Fish Biology **5**:89–96.

Wootton, R.J. 1979. Energy costs of egg production and environmental determinants of fecundity in teleost fishes. Symposium of the Zoological Society of London **44**:133–159.

Wootton, R.J. 1984. Introduction: strategies and tactics in fish reproduction. *In* G.W. Potts and R.J. Wootton (editors), "Fish Reproduction: Strategies and Tactics", pp. 1–12, Academic Press, London.

Wootton, R.J. 1990. Ecology of Teleost Fishes. Chapman and Hall, London.

Wright, S. and O.N. Eaton. 1929. The persistence of differentiation among inbred families of guinea pigs. Technical Bulletin of the U.S. Department of Agriculture **103**:1–45.

Wyatt, T. 1973. The biology of *Oikopleura dioica* and *Fritillaria borealis* in the Southern Bight. Marine Biology **22**:137–158.

Wydoski, R.S. and E.L. Cooper. 1966. Maturation and fecundity of brook trout from infertile streams. Journal of the Fisheries Research Board of Canada **23**:623–649.

Yamada, Y. 1962. Genotype by environment interaction and genetic correlation of the same trait under different environments. Japanese Journal of Genetics **37**:498–509.

Yamamura, K. 1989. Effect of aggregation on the reproductive rate of populations. Researches on Population Ecology **31**:161–168.

Ydenberg, R.C. 1989. Growth-mortality trade-offs, parent-offspring conflict, and the evolution of juvenile life histories in the avian family, Alcidae. Ecology **70**:1494–1506.

Ydenberg, R.C. and D.F. Bertram. 1989. Lack's clutch size hypothesis and brood enlargement studies of colonial seabirds. Colonial Waterbirds **12**:134–137.

Yodzis, P. 1981. Concerning the sense in which maximizing fitness is equivalent to maximizing reproductive value. Ecology **62**:1681–1682.

Yokoi, Y. 1989. An analysis of age- and size-dependent flowering: a critical-production model. Ecological Research **4**:387–397.

Young, T.P. 1981. A general model of comparative fecundity for semelparous and iteroparous life histories. American Naturalist **118**:27–36.

Yule, G.U. 1902. Mendel's laws and their probable relations to intra-racial heredity. New Phytologist **1**:193–207, 222–238.

Yule, G.U. 1906. On the theory of inheritance of quantitative compound characters on the basis of Mendel's laws—a preliminary note. Report of the 3rd International Conference of Genetics, 140–142.

Zaret, T.M. 1980. Predation in Freshwater Communities. Yale University Press, New Haven, Connecticut.

Zaret, T.M. and W.C. Kerfoot. 1975. Fish predation on *Bosmina longirostris*: body-size selection versus visibility selection. Ecology **56**:232–237.

Zastrow, C.E., E.D. Houde and E.H. Saunders. 1990. Quality of striped bass (*Morone saxatilis*) eggs in relation to river source and female weight. Rapports et Proces-Verbeaux des Re(/)unions de la Conseil International de L'Exploration de la Mer **191**:34–42.

Ziehe, M. and H.-R. Gregorius. 1985. The significance of over- and underdominance for the maintenance of genetic polymorphisms. 1. Underdominance and stability. Journal of Theoretical Biology **117**:483–504.

Zimmerer, E.J. and K.D. Kallman. 1989. Genetic basis for alternative reproductive tactics in the pygmy swordtall, *Xiphophorus nigrensis*. Evolution **43**:1298–1307.

Zimmerman, J.K. and I.M. Weis. 1983. Fruit size variation and its effects on germination and seedling growth in *Xanthium strumarium*. Canadian Journal of Botany **61**:2309–2315.

Zimmerman, J.L. 1971. The territory and its density dependent effect in *Spiza americana*. Auk **88**:591–612.

Ziolko, M. and J. Kozlowski. 1983. Evolution of body size: an optimization model. Mathematical Bioscience **64**:127–143.

Zouros, E.S., M. Singh and H.E. Miles. 1980. Growth rate in oysters: an overdominant phenotype and its possible explanations. Evolution **34**:856–867.

Zuberi, M.I. and T.S. Gale. 1976. Variation in wild populations of *Papaver dubium* X. Genotype × environment interaction associated with differences in soil. Heredity **36**:359–368.

Zullini, A. and M. Pagani. 1989. The ecological meaning of relative egg size in soil and freshwater nematodes. Nematologica **35**:90–96.

Taxonomic Index

Subject Index